67630 OB.-
B 03

MEN OF MATHEMATICS

by

E. T. BELL

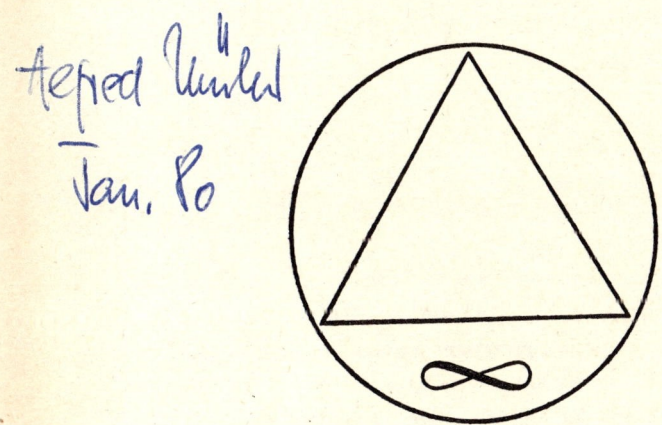

SIMON AND SCHUSTER · NEW YORK

TO TOBY

COPYRIGHT © 1937 BY E. T. BELL
COPYRIGHT RENEWED © 1965 BY TAINE T. BELL
ALL RIGHTS RESERVED
PUBLISHED BY SIMON AND SCHUSTER
ROCKEFELLER CENTER, 630 FIFTH AVENUE
NEW YORK, N.Y. 10020
MANUFACTURED IN THE UNITED STATES OF AMERICA

TWENTY-SECOND PRINTING

Acknowledgments

WITHOUT A MASS OF FOOTNOTES it would be impossible to cite authority for every statement of historical fact in the following pages. But little of the material consulted is available outside of large university libraries, and most of it is in foreign languages. For the principal dates and leading facts in the life of a particular man I have consulted the obituary notices (of the moderns); these are found in the proceedings of the learned societies of which the man in question was a member. Other details of interest are given in the correspondence between mathematicians and in their collected works. In addition to the few specific sources cited presently, bibliographies and references in the following have been especially helpful.

(1) The numerous historical notes and papers abstracted in the *Jahrbuch über die Fortschritte der Mathematik* (section on history of mathematics).

(2) The same in *Bibliotheca Mathematica*.

Only three of the sources are sufficiently "private" to need explicit citation. The life of Galois is based on the classic account by P. Dupuy in the *Annales scientifiques de l' École normale supérieure* (3me série, tome 13, 1896), and the edited notes by Jules Tannery. The correspondence between Weierstrass and Sonja Kowalewski was published by Mittag-Leffler in the *Acta Mathematica* (also partly in the *Comptes rendus du 2me Congrès international des Mathématiciens*, Paris, 1902). Many of the details concerning Gauss are taken from the book by W. Sartorius von Waltershausen, *Gauss zum Gedächtniss*, Leipzig, 1856.

It would be rash to claim that every date or spelling of proper names in the book is correct. Dates are used chiefly with the purpose of orienting the reader as to a man's age when he made his most original inventions. As to spellings, I confess my helplessness in the face of such variants as Basle, Bâle, Basel for one Swiss town, or Utzendorff, Uitzisdorf for another, each preferred by some admittedly reputable authority. When it comes to choosing between James and Johann, or between Wolfgang and Farkas, I take the easier way and identify the man otherwise.

Most of the portraits are reproduced from those in the David Eugene Smith Collection, Columbia University. The portrait of Newton is from an original mezzotint loaned by Professor E. C. Watson. The drawings have been constructed accurately by Mr. Eugene Edwards.

As on a previous occasion (*The Search for Truth*), it gives me great pleasure to thank Doctor Edwin Hubble and his wife, Grace, for their invaluable assistance. While I alone am responsible for all statements in the book, nevertheless it was a great help to have scholarly criticism (even if I did not always profit by it) from two experts in fields in which I cannot claim to be expert, and I trust that their constructive criticisms have lightened my own deficiencies. Doctor Morgan Ward also has criticized certain of the chapters and has made many helpful suggestions on matters in which he is expert. Toby, as before, has contributed much; in acknowledgment for what she has given, I have dedicated the book to her (if she will have it)—it is as much hers as mine.

Last, I wish to thank the staffs of the various libraries which have generously helped with the loan of rare books and bibliographical material. In particular I should like to thank the librarians at Stanford University, the University of California, the University of Chicago, Harvard University, Brown University, Princeton University, Yale University, The John Crerar Library (Chicago), and the California Institute of Technology.

E. T. BELL

Contents

1. INTRODUCTION – – – – – – – – – – – – – 3
    For the reader's comfort. The beginning of modern mathematics. Are mathematicians human? Witless parodies. Illimitable scope of mathematical evolution. Pioneers and scouts. A clue through the maze. Continuity and discreteness. Remarkable rarity of common sense. Vivid mathematics or vague mysticism? Four great ages of mathematics. Our own the Golden Age.

2. MODERN MINDS IN ANCIENT BODIES – – – – – 19
    ZENO (fifth century B.C.), EUDOXUS (408–355 B.C.), ARCHIMEDES (287?–212 B.C.)
    Modern ancients and ancient moderns. Pythagoras, great mystic, greater mathematician. Proof or intuition? The taproot of modern analysis. A bumpkin upsets the philosophers. Zeno's unresolved riddles. Plato's needy young friend. Inexhaustible exhaustion. The useful conics. Archimedes, aristocrat, greatest scientist of antiquity. Legends of his life and personality. His discoveries and claim to modernity. A sturdy Roman. Defeat of Archimedes and triumph of Rome.

3. GENTLEMAN, SOLDIER, AND MATHEMATICIAN – – 35
    DESCARTES (1596–1650)
    The good old days. A child philosopher but no prig. Inestimable advantages of lying in bed. Invigorating doubts. Peace in war. Converted by a nightmare. Revelation of analytic geometry. More butchering. Circuses, professional jealousy, swashbuckling, accommodating lady friends. Distaste for hell-fire and respect for the Church. Saved by a brace of cardinals. A Pope brains himself. Twenty years a recluse. The Method. Betrayed by fame. Doting Elisabeth. What Descartes really thought of her. Conceited Christine. What she did to Descartes. Creative simplicity of his geometry.

4. THE PRINCE OF AMATEURS — — — — — — — — 56
FERMAT (1601–1665)

Greatest mathematician of the seventeenth century. Fermat's busy, practical life. Mathematics his hobby. His flick to the calculus. His profound physical principle. Analytic geometry again. Arithmetica and logistica. Fermat's supremacy in arithmetic. An unsolved problem on primes. Why are some theorems "important"? An intelligence test. "Infinite descent." Fermat's unanswered challenge to posterity.

5. "GREATNESS AND MISERY OF MAN" — — — — — 73
PASCAL (1623–1662)

An infant prodigy buries his talent. At seventeen a great geometer. Pascal's wonderful theorem. Vile health and religious inebriety. The first calculating Frankenstein. Pascal's brilliance in physics. Holy sister Jacqueline, soul-saver. Wine and women? "Get thee to a nunnery!" Converted on a spree. Literature prostituted to bigotry. The Helen of Geometry. A celestial toothache. What the post-mortem revealed. A gambler makes mathematical history. Scope of the theory of probability. Pascal creates the theory with Fermat. Folly of betting against God or the Devil.

6. ON THE SEASHORE — — — — — — — — — — — 90
NEWTON (1642–1727)

Newton's estimate of himself. An uncertified youthful genius. Chaos of his times. On the shoulders of giants. His one attachment. Cambridge days. Young Newton masters futility of suffering fools gladly. The Great Plague a greater blessing. Immortal at twenty four (or less). The calculus. Newton unsurpassed in pure mathematics, supreme in natural philosophy. Gnats, hornets, and exasperation. The Principia. Samuel Pepys and other fussers. The flattest anticlimax in history. Controversy, theology, chronology, alchemy, public office, death.

7. MASTER OF ALL TRADES — — — — — — — — — 117
LEIBNIZ (1646–1716)

Two superb contributions. A politician's offspring. Genius at fifteen. Seduced by the law. The "universal characteristic." Symbolic reasoning. Sold out to ambition. A master diplomat. Diplomacy being what it is, the diplomatic exploits of the master are left to the historians. Fox into historian, statesman into mathematician. Applied ethics. Existence of God. Optimism. Forty years of futility. Discarded like a dirty rag.

Contents

8. **NATURE OR NURTURE?** — — — — — — — — — 131
   THE BERNOULLIS (seventeenth-and eighteenth centuries)
   Eight mathematicians in three generations. Clinical evidence for heredity. The calculus of variations.

9. **ANALYSIS INCARNATE** — — — — — — — — — 139
   EULER (1707–1783)
   The most prolific mathematician in history. Snatched from theology. Rulers foot the bills. Practicality of the unpractical. Celestial mechanics and naval warfare. A mathematician by chance and foreordination. Trapped in St. Petersburg. The virtues of silence. Half blind in his morning. Flight to liberal Prussia. Generosity and boorishness of Frederick the Great. Return to hospitable Russia. Generosity and graciousness of Catherine the Great. Total blindness at noon. Master and inspirer of masters for a century.

10. **A LOFTY PYRAMID** — — — — — — — — — — 153
    LAGRANGE (1736–1813)
    Greatest and most modest mathematician of the eighteenth century. Financial ruin his opportunity. Conceives his masterpiece at nineteen. Magnanimity of Euler. Turin, to Paris, to Berlin: a grateful bastard aids a genius. Conquests in celestial mechanics. Frederick the Great condescends. Absent-minded marriage. Work as a vice. A classic in arithmetic. The Mécanique analytique *a living masterpiece. A landmark in the theory of equations. Welcomed in Paris by Marie Antoinette. Nervous exhaustion, melancholia, and universal disgust in middle life. Reawakened by the French Revolution and a young girl. What Lagrange thought of the Revolution. The metric system. What the revolutionists thought of Lagrange. How a philosopher dies.*

11. **FROM PEASANT TO SNOB** — — — — — — — — 172
    LAPLACE (1749–1827)
    Humble as Lincoln, proud as Lucifer. A chilly reception and a warm welcome. Laplace grandiosely attacks the solar system. The Mécanique céleste *His estimate of himself. What others have thought of him. The "potential" fundamental in physics. Laplace in the French Revolution. Intimacy with Napoleon. Laplace's political realism superior to Napoleon's.*

12. **FRIENDS OF AN EMPEROR** — — — — — — — 183
    MONGE (1746–1818), FOURIER (1768–1830)
    A knife grinder's son and a tailor's boy help Napoleon to upset the aristocrats' applecart. Comic opera in Egypt. Monge's descriptive geometry and the Machine Age. Fourier's analysis and modern

physics. Imbecility of trusting in princes or proletarians. Boring to death and bored to death.

13. THE DAY OF GLORY — — — — — — — — — — 206
PONCELET (1788–1867)
Resurrected from a Napoleonic shambles. The path of glory leads to jail. Wintering in Russia in 1812. What genius does in prison. Two years of geometry in hell. The rewards of genius: stupidities of routine. Poncelet's projective geometry. Principles of continuity and duality.

14. THE PRINCE OF MATHEMATICIANS — — — — — 218
GAUSS (1777–1855)
Gauss the mathematical peer of Archimedes and Newton. Humble origin. Paternal brutality. Unequalled intellectual precocity. His chance, at ten. By twelve he dreams revolutionary discoveries, by eighteen achieves them. The Disquisitiones Arithmeticae. Other epochal works summarized. The Ceres disaster. Napoleon, indirectly robbing Gauss, takes second best. Fundamental advances in all branches of mathematics due to Gauss too numerous for citation: see the account given. A sage of sages. Unwelcome death.

15. MATHEMATICS AND WINDMILLS — — — — — — 270
CAUCHY (1789–1857)
Change in nature of mathematics with nineteenth century. Childhood in the French Revolution. Cauchy's early miseducation. Lagrange's prophecy. The young Christian engineer. Prophetic acuteness of Malus. The theory of groups. In the front rank at twenty seven. One of Fermat's enigmas solved. The pious hippopotamus. Butted by Charles the Goat. Memoirs on astronomy and mathematical physics. Sweetness and obstinacy invincible. The French Government makes a fool of itself. Cauchy's place in mathematics. Drawbacks of an irreproachable character.

16. THE COPERNICUS OF GEOMETRY — — — — — — 294
LOBATCHEWSKY (1793–1856)
The widow's mite. Kazan. Appointed professor and spy. Universal ability. Lobatchewsky as an administrator. Reason and incense combat the cholera. Russian gratitude. Humiliated in his prime. Blind as Milton, Lobatchewsky dictates his masterpiece. His advance beyond Euclid. Non-Euclidean geometry. A Copernicus of the intellect.

17. GENIUS AND POVERTY — — — — — — — — — — 307
ABEL (1802–1829)
Norway in 1802. Smothered by clerical fecundity. Abel's awakening. Generosity of a teacher. A pupil of the masters. His lucky

blunder. Abel and the quintic. The Government to the rescue. Abel's grand tour of mathematical Europe not so grand. French civility and German cordiality. Crelle and his Journal. *Cauchy's unpardonable sin. "Abel's Theorem." Something to keep mathematicians busy 500 years. Crowning a corpse.*

18. THE GREAT ALGORIST — — — — — — — — — — 327
    JACOBI (1804–1851)
    Galvanoplastics versus mathematics. Born rich. Jacobi's philological ability. Dedicates himself to mathematics. Early work. Cleaned out. A goose among foxes. Hard times. Elliptic functions. Their place in the general development. Inversion. Work in arithmetic, dynamics, algebra, and Abelian functions. Fourier's pontification. Jacobi's retort.

19. AN IRISH TRAGEDY — — — — — — — — — — — 340
    HAMILTON (1805–1865)
    Ireland's greatest. Elaborate miseducation. Discoveries at seventeen. A unique university career. Disappointed in love. Hamilton and the poets. Appointed at Dunsink. Systems of rays. The Principia of optics. Prediction of conical refraction. Marriage and alcohol. Fields. Complex numbers. The commutative law repealed. Quaternions. Mountains of paper.

20. GENIUS AND STUPIDITY — — — — — — — — — 362
    GALOIS (1811–1832)
    An all-time world record in stupidity. Galois' childhood. The pedagogues surpass themselves. At sixteen Galois repeats Abel's mistake. Politics and education. Examinations as arbiters of genius. Hounded to death by a priest. More academic ineptitude. Absent-minded Cauchy again. Driven to rebellion. A master mathematician at nineteen. "A carcase to stir up the people." The foulest sewer in Paris. Patriots rush to the field of honor. Galois' last night. The riddle of equations solved. Buried like a dog.

21. INVARIANT TWINS — — — — — — — — — — — 378
    SYLVESTER (1814–1897); CAYLEY (1821–1895)
    Cayley's contributions. Early life. Cambridge. Recreations. Called to the Bar. Fourteen years in the law. Cayley meets his collaborator. Sylvester's stormier life. Hamstrung by religion. Cayley and Sylvester contrasted. Sylvester's mission to the Virginians. Further false steps. The theory of invariants. Called to Johns Hopkins University. Inextinguishable vitality. "Rosalind." Cayley's unification of geometry. Space of n *dimensions. Matrices. Oxford endorses Sylvester. Respectable at last.*

Contents

22. MASTER AND PUPIL — 406
WEIERSTRASS (1815–1897); SONJA KOWALEWSKI (1850–1891)
  The father of modern analysis. Relations of Weierstrass to his contemporaries. The penalties of brilliance. Forced into law, forces himself out. Beer and broadswords. A fresh start. Debt to Gudermann. Fifteen years in the mud. Miraculous extrication. Weierstrass' life problem. Too much success. Sonja storms the master. His favorite pupil. Their friendship. A woman's gratitude. Repenting, Sonja wins Paris prize. Weierstrass universally honored. Power series. Arithmetization of analysis. Doubts.

23. COMPLETE INDEPENDENCE — 433
BOOLE (1815–1864)
  British mathematics. Damned at birth by snobbery. Boole's struggle for education. False diagnoses. Providence intervenes. Discovery of invariants. What is algebra? A philosopher attacks a mathematician. Frightful carnage. Boole's chance. "The Laws of Thought." Symbolic logic. Its mathematical significance. Boolean algebra. Dead in his prime.

24. THE MAN, NOT THE METHOD — 448
HERMITE (1822–1901)
  Old problems and new methods. Hermite's masterful mother. His detestation of examinations. Instructs himself. Higher mathematics sometimes easier than elementary. Educational disasters. Letters to Jacobi. A master at twenty one. Revenge on his examiners. Abelian functions. Pestered by Cauchy. Hermite's mysticism. Solution of the general quintic. Transcendental numbers. A hint to circle-squarers. Hermite's internationalism.

25. THE DOUBTER — 466
KRONECKER (1823–1891)
  Legend of an American saint. Lucky Kronecker. School triumphs. Great gifts. Algebraic numbers. Battles with Weierstrass. Kronecker's business career. Returns rich to mathematics. The Galois theory. Kronecker's lectures. His skepticism his most original contribution.

26. ANIMA CANDIDA — 484
RIEMANN (1826–1866)
  Poor but happy. Riemann's chronic shyness. Destined for the church. Saved. A famous hypothesis. Career at Göttingen. "A new mathematic." Physical researches. Application of topology to analysis. Epoch-making essay on foundations of geometry. Gauss enthusiastic. The blessings of poverty. A root of tensor analysis. Quest for health.

Under a fig tree. Riemann's landmark in geometry. Curvature of space. Pathbreaking for relativity.

27. ARITHMETIC THE SECOND — — — — — — — — — 510
    KUMMER (1810–1893), DEDEKIND (1831–1916)
    Aged in the wood. Napoleonic warp to Kummer's geniality. Equally gifted in the abstract and the concrete. What Fermat's Last Theorem started. Theory of ideal numbers. Kummer's invention comparable to Lobatchewsky's. Wave surface in four dimensions. Big of body, mind, and heart. Dedekind, last pupil of Gauss. First expositor of Galois. Early interest in science. Turns to mathematics. Dedekind's work on continuity. His creation of the theory of ideals.

28. THE LAST UNIVERSALIST — — — — — — — — — 526
    POINCARÉ (1854–1912)
    Poincaré's universality and methods. Childhood setbacks. Seized by mathematics. Keeps his sanity in Franco-Prussian war. Starts as mining engineer. First great work. Automorphic functions. "The keys of the algebraic cosmos." The problem of n bodies. Is Finland civilized? Poincaré's new methods in celestial mechanics. Cosmogony. How mathematical discoveries are made. Poincaré's account. Forebodings and premature death.

29. PARADISE LOST? — — — — — — — — — — — — 555
    CANTOR (1845–1918)
    Old foes with new faces. Rotting creeds. Cantor's artistic inheritance and father-fixation. Escape, but too late. His revolutionary work gets him nowhere. Academic pettiness. Disastrous consequences of "safety first." An epochal result. Paradox or truth? Infinite existence of transcendentals. Aggressiveness advances, timidity retires. Further spectacular claims. Two types of mathematicians. Insane? Counter-revolution. The battle grows fiercer. Cursing the enemy. Universal loss of temper. Where stands mathematics today? And where will it stand tomorrow? Invictus.

INDEX — — — — — — — — — — — — — — — — — — 581

THEY SAY, WHAT THEY SAY, LET THEM SAY

(Motto of Marischal College, Aberdeen)

The science of Pure Mathematics, in its modern developments, may claim to be the most original creation of the human spirit.—A. N. WHITEHEAD (*Science and the Modern World*, 1925)

A mathematical truth is neither simple nor complicated in itself, it is.—ÉMILE LEMOINE

A mathematician who is not also something of a poet will never be a complete mathematician.—KARL WEIERSTRASS

I have heard myself accused of being an opponent, an enemy of mathematics, which no one can value more highly than I, for it accomplishes the very thing whose achievement has been denied me.—GOETHE

Mathematicians are like lovers. . . . Grant a mathematician the least principle, and he will draw from it a consequence which you must also grant him, and from this consequence another.—FONTENELLE

It is easier to square the circle than to get round a mathematician.—AUGUSTUS DE MORGAN

I regret that it has been necessary for me in this lecture to administer such a large dose of four-dimensional geometry. I do not apologise, because I am really not responsible for the fact that nature in its most fundamental aspect is four-dimensional. Things are what they are. . . .—A. N. WHITEHEAD (*The Concept of Nature*, 1920)

* * *

Number rules the universe.—THE PYTHAGOREANS

Mathematics is the Queen of the Sciences, and Arithmetic the Queen of Mathematics.—C. F. GAUSS

Thus number may be said to rule the whole world of quantity, and the four rules of arithmetic may be regarded as the complete equipment of the mathematician. JAMES CLERK MAXWELL

The different branches of Arithmetic—Ambition, Distraction, Uglification, and Derision.—THE MOCK TURTLE (*Alice in Wonderland*)

God made the integers, all the rest is the work of man.—LEOPOLD KRONECKER

[Arithmetic] is one of the oldest branches, perhaps the very oldest branch, of human knowledge; and yet some of its most abstruse secrets lie close to its tritest truths.—H. J. S. SMITH

* * *

Plato's writings do not convince any mathematician that their author was strongly addicted to geometry. . . . We know that he encouraged mathematics. . . . But if—which nobody believes—the μηδείς ἀγεωμέτρητος εἰσίτω [Let no man ignorant of geometry enter] of Tzetzes had been written over his gate, it would no more have indicated the geometry within than a warning not to forget to bring a packet of sandwiches would now give promise of a good dinner.—AUGUSTUS DE MORGAN

There is no royal road to geometry.—MENAECHMUS (*to* ALEXANDER THE GREAT)

* * *

He studied and nearly mastered the six books of Euclid since he was a member of Congress.

He began a course of rigid mental discipline with the intent to improve his faculties, especially his powers of logic and language. Hence his fondness for Euclid, which he carried with him on the circuit till he could demonstrate with ease all the propositions in the six books; often studying far into the night, with a candle near his pillow, while his fellow-lawyers, half a dozen in a room, filled the air with interminable snoring.—ABRAHAM LINCOLN (*Short Autobiography*, 1860)

* * *

Strange as it may sound, the power of mathematics rests on its evasion of all unnecessary thought and on its wonderful saving of mental operations. —ERNST MACH

A single curve, drawn in the manner of the curve of prices of cotton, describes all that the ear can possibly hear as the result of the most complicated musical performance. . . . That to my mind is a wonderful proof of the potency of mathematics.—LORD KELVIN

* * *

The mathematician, carried along on his flood of symbols, dealing apparently with purely formal truths, may still reach results of endless importance for our description of the physical universe.—KARL PEARSON

Examples . . . which might be multiplied *ad libitum*, show how difficult it often is for an experimenter to interpret his results without the aid of mathematics.—LORD RAYLEIGH

But there is another reason for the high repute of mathematics: it is mathematics that offers the exact natural sciences a certain measure of security which, without mathematics, they could not attain.—ALBERT EINSTEIN

Mathematics is the tool specially suited for dealing with abstract concepts of any kind and there is no limit to its power in this field. For this reason a book on the new physics, if not purely descriptive of experimental work, must be essentially mathematical.—P. A. M. DIRAC (*Quantum Mechanics*, 1930)

As I proceeded with the study of Faraday, I perceived that his method of conceiving the phenomena [of electromagnetism] was also a mathematical one, though not exhibited in the conventional form of mathematical symbols. I also found that these methods were capable of being expressed in the ordinary mathematical forms, and thus compared with those of the professed mathematicians.—JAMES CLERK MAXWELL (*A Treatise on Electricity and Magnetism*, 1873)

* * *

Query 64. . . . Whether mathematicians . . . have not their mysteries, and, what is more, their repugnances and contradictions?—BISHOP BERKELEY

To create a healthy philosophy you should renounce metaphysics but be a good mathematician.—BERTRAND RUSSELL (*in a lecture*, 1935)

Mathematics is the only good metaphysics.—LORD KELVIN

How can it be that mathematics, being after all a product of human thought independent of experience, is so admirably adapted to the objects of reality?—ALBERT EINSTEIN (1920)

Every *new* body of discovery is mathematical in form, because there is no other guidance we can have.—C. G. DARWIN (1931)

The infinite! No other question has ever moved so profoundly the spirit of man.—DAVID HILBERT (1921)

The notion of infinity is our greatest friend; it is also the greatest enemy of our peace of mind. . . . Weierstrass taught us to believe that we had at last thoroughly tamed and domesticated this unruly element. Such however is not the case; it has broken loose again. Hilbert and Brouwer have set out to tame it once more. For how long? We wonder.—JAMES PIERPONT (*Bulletin of the American Mathematical Society*, 1928)

In my opinion a mathematician, in so far as he is a mathematician, need not preoccupy himself with philosophy—an opinion, moreover, which has been expressed by many philosophers.—HENRI LEBESGUE (1936)

God ever geometrizes.—PLATO

God ever arithmetizes.—C. G. J. JACOBI

The Great Architect of the Universe now begins to appear as a pure mathematician.—J. H. JEANS (*The Mysterious Universe*, 1930)

Mathematics is the most exact science, and its conclusions are capable of absolute proof. But this is so only because mathematics does not *attempt* to draw absolute conclusions. All mathematical truths are relative, conditional.—CHARLES PROTEUS STEINMETZ (1923)

It is a safe rule to apply that, when a mathematical or philosophical author writes with a misty profundity, he is talking nonsense.—A. N. WHITEHEAD (1911)

CHAPTER ONE

Introduction

THIS SECTION IS HEADED *Introduction* rather than *Preface* (which it really is) in the hope of decoying habitual preface-skippers into reading—for their own comfort—at least the following paragraphs down to the first row of stars before going on to meet some of the great mathematicians. I should like to emphasize first that this book is not intended, in any sense, to be a history of mathematics, or any section of such a history.

The lives of mathematicians presented here are addressed to the general reader and to others who may wish to see what sort of human beings the men were who created *modern* mathematics. Our object is to lead up to some of the dominating ideas governing vast tracts of mathematics as it exists today and to do this through the lives of the men responsible for those ideas.

Two criteria have been applied in selecting names for inclusion: the importance for modern mathematics of a man's work; the human appeal of the man's life and character. Some qualify under both heads, for example Pascal, Abel, and Galois; others, like Gauss and Cayley, chiefly under the first, although both had interesting lives. When these criteria clash or overlap in the case of several claimants to remembrance for a particular advance, the second has been given precedence as we are primarily interested here in mathematicians as human beings.

Of recent years there has been a tremendous surge of general interest in science, particularly physical science, and its bearing on our rapidly changing philosophical outlook on the universe. Numerous excellent accounts of current advances in science, written in as untechnical language as possible, have served to lessen the gap between the professional scientist and those who must make their livings at something other than science. In many of these expositions, especially those concerned with relativity and the modern quantum theory, names occur with which the general reader cannot be expected to be familiar—Gauss, Cayley, Riemann, and Hermite, for instance. With a knowledge of who these men were, their part in preparing for the explosive growth of physical science since 1900, and an appreciation of their rich personalities, the magnificent achievements of science fall into a truer perspective and take on a new significance.

The great mathematicians have played a part in the evolution of scientific and philosophic thought comparable to that of the philosophers and scientists themselves. To portray the leading features of that part through the lives of master mathematicians, presented against a background of some of the dominant problems of their times, is the purpose of the following chapters. The emphasis is wholly on modern mathematics, that is, on those great and simple guiding ideas of mathematical thought that are still of vital importance in living, creative science and mathematics.

It must not be imagined that the sole function of mathematics—"the handmaiden of the sciences"—is to serve science. Mathematics has also been called "the Queen of the Sciences." If occasionally the Queen has seemed to beg from the sciences she has been a very proud sort of beggar, neither asking nor accepting favors from any of her more affluent sister sciences. What she gets she pays for. Mathematics has a light and wisdom of its own, above any possible application to science, and it will richly reward any intelligent human being to catch a glimpse of what mathematics means to itself. This is not the old doctrine of art for art's sake; it is art for humanity's sake. After all, the whole purpose of science is not technology—God knows we have gadgets enough already; science also explores depths of a universe that will never, by any stretch of the imagination, be visited by human beings or affect our material existence. So we shall attend also to some of the things which the great mathematicians have considered worthy of loving understanding for their intrinsic beauty.

Plato is said to have inscribed "Let no man ignorant of geometry enter here" above the entrance to his Academy. No similar warning need be posted here, but a word of advice may save some overconscientious reader unnecessary anguish. The gist of the story is in the lives and personalities of the creators of modern mathematics, not in the handful of formulas and diagrams scattered through the text. The basic ideas of modern mathematics, from which the whole vast and intricate complexity has been woven by thousands of workers,

Introduction

are simple, of boundless scope, and well within the understanding of any human being with normal intelligence. Lagrange (whom we shall meet later) believed that a mathematician has not thoroughly understood his own work till he has made it so clear that he can go out and explain it effectively to the first man he meets on the street.

This of course is an ideal and not always attainable. But it may be recalled that only a few years before Lagrange said this the Newtonian "law" of gravitation was an incomprehensible mystery to even highly educated persons. Yesterday the Newtonian "law" was a commonplace which every educated person accepted as simple and true; today Einstein's relativistic theory of gravitation is where Newton's "law" was in the early decades of the eighteenth century; tomorrow or the day after Einstein's theory will seem as "natural" as Newton's "law" seemed yesterday. With the help of time Lagrange's ideal is not unattainable.

Another great French mathematician, conscious of his own difficulties no less than his readers', counselled the conscientious not to linger too long over anything hard but to "Go on, and faith will come to you." In brief, if occasionally a formula, a diagram, or a paragraph seems too technical, skip it. There is ample in what remains.

Students of mathematics are familiar with the phenomenon of "slow development," or subconscious assimilation: the first time something new is studied the details seem too numerous and hopelessly confused, and no coherent impression of the whole is left on the mind. Then, on returning after a rest, it is found that everything has fallen into place with its proper emphasis—like the development of a photographic film. The majority of those who attack analytic geometry seriously for the first time experience something of the sort. The calculus on the other hand, with its aims clearly stated from the beginning, is usually grasped quickly. Even professional mathematicians often skim the work of others to gain a broad, comprehensive view of the whole before concentrating on the details of interest to them. Skipping is not a vice, as some of us were told by our puritan teachers, but a virtue of common sense.

As to the amount of mathematical knowledge necessary to understand *everything* that some will wisely skip, I believe it may be said honestly that a high school course in mathematics is sufficient. Matters far beyond such a course are frequently mentioned, but wherever they are, enough description has been given to enable anyone with high school mathematics to follow. For some of the most important ideas discussed in connection with their originators—groups, space of many dimensions, non-Euclidean geometry, and symbolic logic, for example —*less* than a high school course is ample for an understanding of the basic concepts. All that is needed is interest and an undistracted head. Assimilation of some of these invigorating ideas of modern mathematical thought will be found as refreshing as a drink of cold water on a hot day and as inspiring as any art.

To facilitate the reading, important definitions have been repeated where necessary, and frequent references to earlier chapters have been included from time to time.

The chapters need not be read consecutively. In fact, those with a speculative or philosophical turn of mind may prefer to read the last chapter first. With a few trivial displacements to fit the social background the chapters follow the chronological order.

It would be impossible to describe *all* the work of even the least prolific of the men considered, nor would it be profitable in an account for the general reader to attempt to do so. Moreover, much of the work of even the greater mathematicians of the past is now of only historical interest, having been included in more general points of view. Accordingly only some of the conspicuously new things each man did are described, and these have been selected for their originality and importance in modern thought.

Of the topics selected for description we may mention the following (among others) as likely to interest the general reader: the modern doctrine of the infinite (chapters 2, 29); the origin of mathematical probability (chapter 5); the concept and importance of a group (chapter 15); the meanings of invariance (chapter 21); non-Euclidean geometry (chapter 16 and part of 14); the origin of the mathematics of general relativity (last part of chapter 26); properties of the common whole numbers (chapter 4), and their modern generalization (chapter 25); the meaning and usefulness of so-called imaginary numbers—like $\sqrt{-1}$ (chapters 14, 19); symbolic reasoning (chapter 23). But anyone who wishes to get a glimpse of the power of the mathematical method, especially as applied to science, will be repaid by seeing what the calculus is about (chapters 2, 6).

Modern mathematics began with two great advances, analytic geometry and the calculus. The former took definite shape in 1637, the latter about 1666, although it did not become public property till a decade later. Though the idea behind it all is childishly simple, yet the method of analytic geometry is so powerful that very ordinary boys of seventeen can use it to prove results which would have baffled the greatest of the Greek geometers—Euclid, Archimedes, and Apollonius. The man, Descartes, who finally crystallized this great method had a particularly full and interesting life.

In saying that Descartes was responsible for the creation of analytic geometry we do not mean to imply that the new method sprang full-armed from his mind alone. Many before him had made significant advances toward the new method, but it remained for Descartes to take the final step and actually to put out the method as a definitely workable engine of geometrical proof, discovery, and invention. But even Descartes must share the honor with Fermat.

Similar remarks apply to most of the other advances of modern mathematics. A new concept may be "in the air" for generations until some one man—occasionally two or three together—sees clearly the essential detail that his predecessors missed, and the new thing comes into being. Relativity, for example, is sometimes said to have been the great invention reserved by time for the genius of Minkowski. The fact is, however, that Minkowski did not create the theory of relativity and that Einstein did. It seems rather meaningless to say that So-and-so might have done this or that if circumstances had been other than they were. Any one of us no doubt could jump over the moon if we and the physical universe were different from what we and it are, but the truth is that we do not make the jump.

In other instances, however, the credit for some great advance is not always justly placed, and the man who first used a new method more powerfully than its inventor sometimes gets more than his due. This seems to be the case, for instance, in the highly important matter of the calculus. Archimedes had the fundamental notion of limiting sums from which the integral calculus springs, and he not only had the notion but showed that he could apply it. Archimedes also used the method of the differential calculus in one of his problems. As we approach Newton and Leibniz in the seventeenth century the history of the calculus becomes extremely involved. The new method was more than merely "in the air" before Newton and Leibniz brought it down to earth; Fermat actually had it. He also invented the method of Cartesian geometry independently of Descartes. In spite of indubitable facts such as these we shall follow tradition and ascribe to each great leader what a majority vote says he should have, even at the risk of giving him a little more than his just due. Priority after all gradually loses its irritating importance as we recede in time from the men to whom it was a hotly contested cause of verbal battles while they and their partisans lived.

* * *

Those who have never known a professional mathematician may be rather surprised on meeting some, for mathematicians as a class are probably less familiar to the general reader than any other group of brain workers. The mathematician is a much rarer character in fiction than his cousin the scientist, and when he does appear in the pages of a novel or on the screen he is only too apt to be a slovenly dreamer totally devoid of common sense—comic relief. What sort of mortal is he in real life? Only by seeing in detail what manner of men some of the *great* mathematicians were and what kind of lives they lived, can we recognize the ludicrous untruth of the traditional portrait of a mathematician.

Strange as it may seem, not all of the great mathematicians have been professors in colleges or universities. Quite a few were soldiers by profession; others went into mathematics from theology, the law, and medicine, and one of the greatest was as crooked a diplomat as ever lied for the good of his country. A few have had no profession at all. Stranger yet, not all professors of mathematics have been mathematicians. But this should not surprise us when we think of the gulf between the average professor of poetry drawing a comfortable salary and the poet starving to death in his garret.

The lives that follow will at least suggest that a mathematician can be as human as anybody else—sometimes distressingly more so. In ordinary social contacts the majority have been normal. There have been eccentrics in mathematics, of course; but the percentage is no higher than in commerce or the professions. As a group the great mathematicians have been men of all-round ability, vigorous, alert, keenly interested in many things outside of mathematics and, in a fight, men with their full share of backbone. As a rule mathematicians have been bad customers to persecute; they have usually been capable of returning what they received with compound interest. For the rest they were geniuses of tremendous accomplishment marked off from the majority of their gifted fellowmen only by an irresistible impulse to do mathematics. On occasion mathematicians have been (and some still are in France) extremely able administrators.

In their politics the great mathematicians have ranged over the whole spectrum from reactionary conservatism to radical liberalism. It is probably correct to say that as a class they have tended slightly to the left in their political opinions. Their religious beliefs have included everything from the narrowest orthodoxy—sometimes shading into the blackest bigotry—to complete skepticism. A few were dogmatic and positive in their assertions concerning things about which they knew nothing, but most have tended to echo the great Lagrange's "I do not know."

Another characteristic calls for mention here, as several writers and artists (some from Hollywood) have asked that it be treated—the sex life of great mathematicians. In particular these inquirers wish to know how many of the great mathematicians have been perverts—a somewhat indelicate question, possibly, but legitimate enough to merit a serious answer in these times of preoccupation with such topics. None. Some lived celibate lives, usually on account of economic disabilities, but the majority were happily married and brought up their children in a civilized, intelligent manner. The children, it may be noted in passing, were often gifted far above the average. A few of the great mathematicians of bygone centuries kept mistresses when such was the fashionable custom of their times. The only mathematician discussed here whose life might offer something of interest to a Freudian is Pascal.

Returning for a moment to the movie ideal of a mathematician, we note that sloppy clothes have not been the invariable attire of great mathematicians. All through the long history of mathematics about which we have fairly detailed knowledge, mathematicians have paid the same amount of attention to their personal appearance as any other equally numerous group of men. Some have been fops, others slovens; the majority, decently inconspicuous. If today some earnest individual affecting spectacular clothes, long hair, a black sombrero, or any other mark of exhibitionism, assures you that he is a mathematician, you may safely wager that he is a psychologist turned numerologist.

The psychological peculiarities of great mathematicians is another topic in which there is considerable interest. Poincaré will tell us something about the psychology of mathematical creation in a later chapter. But on the general question not much can be said till psychologists call a truce and agree among themselves as to what is what. On the whole the great mathematicians have lived richer, more virile lives than those that fall to the lot of the ordinary hard-working mortal. Nor has this richness been wholly on the side of intellectual adventuresomeness. Several of the greater mathematicians have had more than their share of physical danger and excitement, and some of them have been implacable haters—or, what is ultimately the same, expert controversialists. Many have known the lust of battle in their prime, reprehensibly enough, no doubt, but still humanly enough, and in knowing it they have experienced something no jellyfish has ever felt: "Damn braces, Bless relaxes," as that devout Christian William Blake put it in his *Proverbs of Hell.*

This brings us to what at first sight (from the conduct of several of the men considered here) may seem like a significant trait of mathematicians—their hair-trigger quarrelsomeness. Following the lives of several of these men we get the impression that a great mathematician is more likely than not to think others are stealing his work, or disparaging it, or not doing him sufficient honor, and to start a row to recover imaginary rights. Men who should have been above such brawls seem to have gone out of their way to court battles over priority in discovery and to accuse their competitors of plagiarism. We shall see enough dishonesty to discount the superstition that the pursuit of truth necessarily makes a man truthful, but we shall not find indubitable evidence that mathematics makes a man bad-tempered and quarrelsome.

Another "psychological" detail of a similar sort is more disturbing. Envy is carried up to a higher level. Narrow nationalism and international jealousies, even in impersonal pure mathematics, have marred the history of discovery and invention to such an extent that it is almost impossible in some important instances to get at the facts or to form a just estimate of the significance of a particular man's work for modern thought. Racial fanaticism—especially in recent times—has also complicated the task of anyone who may attempt to give an unbiased account of the lives and work of scientific men outside his own race or nation.

An impartial account of western mathematics, including the award to each man and to each nation of its just share in the intricate development, could be written only by a Chinese historian. He alone would have the patience and the detached cynicism necessary for disentangling the curiously perverted pattern to discover whatever truth may be concealed in our variegated occidental boasting.

* * *

Even in restricting our attention to the modern phase of mathematics we are faced with a problem of selection that must be solved somehow. Before the solution adopted here is indicated it will be of interest to estimate the amount of labor that would be required for a detailed history of mathematics on a scale similar to that of a political history for any important epoch, say that of the French Revolution or the American Civil War.

When we begin unravelling a particular thread in the history of mathematics we soon get a discouraged feeling that mathematics itself is like a vast necropolis to which constant additions are being made for the eternal preservation of the newly dead. The recent arrivals, like some of the few who were shelved for perpetual remembrance 5000 years ago, must be so displayed that they shall seem to retain the full vigor of the manhood in which they died; in fact the illusion must be created that they have not yet ceased living. And the deception must be so natural that even the most skeptical archaeologist prowling through the mausoleums shall be moved to exclaim with living mathematicians themselves that mathematical truths are immortal, imperishable; the same yesterday, today, and forever; the very stuff of which eternal verities are fashioned and the one glimpse of changelessness behind all the recurrent cycles of birth, death, and decay our race has ever caught. Such may indeed be the fact; many, especially those of the older generation of mathematicians, hold it to be no less.

But the mere spectator of mathematical history is soon overwhelmed by the appalling mass of mathematical inventions that still maintain their vitality and importance for modern work, as discoveries of the past in any other field of scientific endeavor do not, after centuries and tens of centuries.

A span of less than a hundred years covers everything of significance in the French Revolution or the American Civil War, and less than five hundred leaders in either played parts sufficiently memorable to merit recording. But the army of those who have made at least one definite contribution to mathematics as we know it soon becomes a mob as we look back over history; 6000 or 8000 names press forward for some word from us to preserve them from oblivion, and once the bolder leaders have been recognized it becomes largely a matter of arbitrary, illogical legislation to judge who of the clamoring multitude shall be permitted to survive and who be condemned to be forgotten.

This problem scarcely presents itself in describing the development of the physical sciences. They also reach far back into antiquity; yet for the most of them 350 years is a sufficient span to cover everything of importance to modern thought. But whoever attempts to do full, human justice to mathematics and mathematicians will have a wilderness of 6000 years in which to exercise such talents as he may have, with that mob of 6000 to 8000 claimants before him for discrimination and attempted justice.

The problem becomes more desperate as we approach our own times. This is by no means due to our closer proximity to the men of the two centuries immediately preceding our own, but to the universally acknowledged fact (among professional mathematicians) that the nineteenth century, prolonged into the twentieth, was, and is, the greatest age of mathematics the world has ever known. Compared to what glorious Greece did in mathematics the nineteenth century is a bonfire beside a penny candle.

What threads shall we follow to guide us through this labyrinth of mathematical inventions? The main thread has already been indicated: that which leads from the half-forgotten past to some of those dominating concepts which now govern boundless empires of mathematics—but which may themselves be dethroned tomorrow to make room for yet vaster generalizations. Following this main thread we shall pass by the *developers* in favor of the *originators*.

Both inventors and perfectors are necessary to the progress of any science. Every explorer must have, in addition to his scouts, his followers to inform the world as to what he has discovered. But to the majority of human beings, whether justly or not is beside the point, the explorer who first shows the new way is the more arresting personality, even if he himself stumbles forward but half a step. We shall follow the originators in preference to the developers. Fortunately for historical justice the majority of the great originators in mathematics have also been peerless developers.

Introduction

Even with this restriction the path from the past to the present may not always be clear to those who have not already followed it. So we may state here briefly what the main guiding clue through the whole history of mathematics is.

From the earliest times two opposing tendencies, sometimes helping one another, have governed the whole involved development of mathematics. Roughly these are the *discrete* and the *continuous*.

The discrete struggles to describe all nature and all mathematics atomistically, in terms of distinct, recognizable individual elements, like the bricks in a wall, or the numbers 1,2,3, . . . The continuous seeks to apprehend natural phenomena—the course of a planet in its orbit, the flow of a current of electricity, the rise and fall of the tides, and a multitude of other appearances which delude us into believing that we know nature—in the mystical formula of Heraclitus: "All things flow." Today (as will be seen in the concluding chapter), "flow," or its equivalent, "continuity," is so unclear as to be almost devoid of meaning. However, let this pass for the moment.

Intuitively we *feel* that we *know* what is meant by "continuous motion"—as of a bird or a bullet through the air, or the fall of a raindrop. The motion is *smooth*; it *does not proceed by jerks*; it is *unbroken*. In *continuous* motion or, more generally, in the concept of continuity itself, the *individualized* numbers 1,2,3, . . . , are *not* the appropriate mathematical image. *All* the points on a segment of a straight line, for instance, have no such clear-cut individualities as have the numbers of the sequence 1,2,3, . . . , where *the step from one member of the sequence to the next is the same* (namely 1: $1 + 2 = 3$, $1 + 3 = 4$, and so on); for *between* any two points on the line segment, no matter how close together the points may be, we can always *find*, or at least *imagine*, another point: *there is no "shortest" step from one point to the "next."* In fact there is no *next* point at all.

The last—the conception of *continuity*, "no nextness"—when developed in the manner of Newton, Leibniz, and their successors leads out into the boundless domain of *the calculus* and its innumerable applications to science and technology, and to all that is today called *mathematical analysis*. The other, the *discrete* pattern based on 1,2,3, . . . , is the domain of algebra, the theory of numbers, and symbolic logic. Geometry partakes of both the continuous and the discrete.

A major task of mathematics today is to harmonize the continuous and the discrete, to include them in one comprehensive mathematics, and to eliminate obscurity from both.

* * *

It may be doing our predecessors an injustice to emphasize modern mathematical thought with but little reference to the pioneers who took the first and possibly the most difficult steps. But nearly everything useful that was done in mathematics before the seventeenth century has suffered one of two fates: either it has been so greatly simplified that it is now part of every regular school course, or it was long since absorbed as a detail in work of greater generality.

Things that now seem as simple as common sense—our way of writing numbers, for instance, with its "place system" of value and the introduction of a symbol for zero, which put the essential finishing touch to the place system—cost incredible labor to invent. Even simpler things, containing the very essence of mathematical thought —*abstractness and generality*, must have cost centuries of struggle to devise; yet their originators have vanished leaving not a trace of their lives and personalities. For example, as Bertrand Russell observed, "It must have taken many ages to discover that a brace of pheasants and a couple of days were both instances of the number two." And it took some twenty five centuries of *civilization* to evolve Russell's own logical definition of "two" or of any cardinal number (reported in the concluding chapter).

Again, the conception of a point, which we (erroneously) think we fully understand when we begin school geometry must have come very late in man's career as an artistic, cave-painting animal. Horace Lamb, an English mathematical physicist, would "erect a monument to the unknown mathematical inventor of the mathematical point as the supreme type of that abstraction which has been a necessary condition of scientific work from the beginning."

Who, by the way, *did* invent the mathematical point? In one sense Lamb's forgotten man; in another, Euclid with his definition "a point is that which has no parts and which has no magnitude"; in yet a third sense Descartes with his invention of the "coordinates of a point"; until finally in geometry as experts practise it today the mysterious "point" has joined the forgotten man and all his gods in everlasting oblivion, to be replaced by something more usable—*a set of numbers written in a definite order.*

The last is a modern instance of the abstractness and precision toward which mathematics strives constantly, only to realize when abstractness and precision are attained that a higher degree of abstractness and a sharper precision are demanded for clear understanding. Our own conception of a "point" will no doubt evolve into something yet more abstract. Indeed the "numbers" in terms of which points are described today dissolved about the beginning of this century into the shimmering blue of pure logic, which in its turn seems about to vanish in something rarer and even less substantial.

It is not necessarily true then that a step-by-step following of our predecessors is the sure way to understand either their conception of mathematics or our own. Such a retracing of the path that has led up to our present outlook would undoubtedly be of great interest in itself. But it is quicker to glance back over the terrain from the hilltop on which we now stand. The false steps, the crooked trails, and the roads that led nowhere fade out in the distance, and only the broad highways are seen leading straight back to the past, where we lose them in the mists of uncertainty and conjecture. Neither space nor number, nor even time, have the same significance for us that they had for the men whose great figures appear dimly through the mist.

A Pythagorean of the sixth century before Christ could intone "Bless us, divine Number, thou who generatest gods and men"; a Kantian of the nineteenth century could refer confidently to "space" as a form of "pure intuition"; a mathematical astronomer could announce a decade ago that the Great Architect of the Universe is a pure mathematician. The most remarkable thing about all of these profound utterances is that human beings no stupider than ourselves once thought they made sense.

To a modern mathematician such all-embracing generalities mean less than nothing. Yet in parting with its claim to be the universal generator of gods and men mathematics has gained something more substantial, a faith in itself and in its ability to create human values.

Our point of view has changed—and is still changing. To Descartes' "Give me space and motion and I will give you a world," Einstein today might retort that altogether too much is being asked, and that the demand is in fact meaningless: without a "world"—matter—there is neither "space" nor "motion." And to quell the turbulent, muddled mysticism of Leibniz in the seventeenth century, over the mysterious $\sqrt{-1}$: "The Divine Spirit found a sublime outlet in that wonder of analysis, the portent of the ideal, that mean between being and not-being, which we call the imaginary [square] root of negative unity," Hamilton in the 1840's constructed a number-couple which any intelligent child can understand and manipulate, and which does for mathematics and science all that the misnamed "imaginary" ever did. The mystical "not-being" of the seventeenth century Leibniz is seen to have a "being" as simple as ABC.

Is this a loss? Or does a modern mathematician lose anything of value when he seeks through the postulational method to track down that elusive "feeling" described by Heinrich Hertz, the discoverer of wireless waves: "One cannot escape the feeling that these mathematical formulas have an independent existence and an intelligence of their own, that they are wiser than we are, wiser even than their discoverers, that we get more out of them than was originally put into them"?

Any competent mathematician will understand Hertz' feeling, but he will also incline to the belief that whereas continents and wireless waves are discovered, dynamos and mathematics are invented and do what we make them do. We can still dream but we need not deliberately court nightmares. If it is true, as Charles Darwin asserted, that "Mathematics seems to endow one with something like a new sense," that sense is the sublimated common sense which the physicist and engineer Lord Kelvin declared mathematics to be.

Is it not closer to our own habits of thought to agree temporarily with Galileo that "Nature's great book is written in mathematical symbols" and let it go at that, than to assert with Plato that "God ever geometrizes," or with Jacobi that "God ever arithmetizes"? If we care to inspect the symbols in nature's great book through the critical eyes of modern science we soon perceive that we ourselves did the writing, and that we used the particular script we did because we invented it to fit our own understanding. Some day we may find a more expressive shorthand than mathematics for correlating our experiences of the physical universe—unless we accept the creed of the scientific mystics that everything *is* mathematics and is not merely *described* for our convenience in mathematical language. *If* "Number rules the universe" as Pythagoras asserted, Number is merely our delegate to the throne, for we rule Number.

When a modern mathematician turns aside for a moment from his symbols to communicate to others the feeling that mathematics inspires in him, he does not echo Pythagoras and Jeans, but he may quote what Bertrand Russell said about a quarter of a century ago: "Mathematics, rightly viewed, possesses not only truth but supreme beauty—a beauty cold and austere, like that of sculpture, without appeal to any part of our weaker nature, without the gorgeous trappings of painting or music, yet sublimely pure, and capable of a stern perfection such as only the greatest art can show."

Another, familiar with what has happened to our conception of mathematical "truth" in the years since Russell praised the beauty of mathematics, might refer to the "iron endurance" which some acquire from their attempt to understand what mathematics means, and quote James Thomson's lines (which close this book) in description of Dürer's *Melencolia* (the frontispiece). And if some devotee is reproached for spending his life on what to many may seem the selfish pursuit of a beauty having no immediate reflection in the lives of his fellowmen, he may repeat Poincaré's "Mathematics for mathematics' sake. People have been shocked by this formula and yet it is as good as life for life's sake, if life is but misery."

* * *

To form an estimate of what modern mathematics compared to ancient has accomplished, we may first look at the mere bulk of the work in the period after 1800 compared to that before 1800. The most extensive history of mathematics is that of Moritz Cantor, *Geschichte der Mathematik*, in three large closely printed volumes (a fourth, by collaborators, supplements the three). The four volumes total about 3600 pages. Only the outline of the development is given by Cantor; there is no attempt to go into details concerning the contributions described, nor are technical terms explained so that an outsider could understand what the whole story is about, and biography is cut to the bone; the history is addressed to those who have some technical training. This history *ends with the year 1799*—just before modern mathematics began to feel its freedom. What if the *outline* history of mathematics in the nineteenth century alone were attempted on a similar scale? It has been estimated that nineteen or twenty volumes the size of Cantor's would be required to tell the story, say about 17,000 pages. The nineteenth century, on this scale, contributed to mathematical knowledge about *five times as much* as was done in the whole of preceding history.

The beginningless period before 1800 breaks quite sharply into two. The break occurs about the year 1700, and is due mainly to Isaac Newton (1642–1727). Newton's greatest rival in mathematics was Leibniz (1646–1716). According to Leibniz, of all mathematics up to the time of Newton, the more important half is due to Newton. This estimate refers to the power of Newton's general methods rather than to the bulk of his work; the *Principia* is still rated as the most massive addition to scientific thought ever made by one man.

Continuing back into time beyond 1700 we find nothing comparable till we reach the Golden Age of Greece—a step of nearly 2000 years. Farther back than 600 B.C. we quickly pass into the shadows, coming out into the light again for a moment in ancient Egypt. Finally we arrive at the first great age of mathematics, about 2000 B.C., in the Euphrates Valley.

The descendants of the Sumerians in Babylon appear to have been the first "moderns" in mathematics; certainly their attack on algebraic equations is more in the spirit of the algebra we know than anything done by the Greeks in their Golden Age. More important than the technical algebra of these ancient Babylonians is their recognition— as shown by their work—of the necessity for *proof* in mathematics. Until recently it had been supposed that the Greeks were the first to recognize that proof is demanded for mathematical propositions. This was one of the most important steps ever taken by human beings. Unfortunately it was taken so long ago that it led nowhere in particular so far as our own civilization is concerned—unless the Greeks followed consciously, which they may well have done. They were not particularly generous to their predecessors.

Mathematics then has had four great ages: the Babylonian, the Greek, the Newtonian (to give the period around 1700 a name), and the recent, beginning about 1800 and continuing to the present day. Competent judges have called the last the Golden Age of Mathematics.

Today mathematical invention (discovery, if you prefer) is going forward more vigorously than ever. The only thing, apparently, that can stop its progress is a general collapse of what we have been pleased to call civilization. If that comes, mathematics may go underground for centuries, as it did after the decline of Babylon; but if history repeats itself, as it is said to do, we may count on the spring bursting forth again, fresher and clearer than ever, long after we and all our stupidities shall have been forgotten.

CHAPTER TWO

Modern Minds in Ancient Bodies

ZENO, EUDOXUS, ARCHIMEDES

> ... *the glory that was Greece*
> *And the grandeur that was Rome.*
> —E. A. POE

TO APPRECIATE our own Golden Age of mathematics we shall do well to have in mind a few of the great, simple guiding ideas of those whose genius prepared the way for us long ago, and we shall glance at the lives and works of three Greeks: Zeno (495–435 B.C.), Eudoxus (408–355 B.C.), and Archimedes (287–212 B.C.). Euclid will be noticed much later, where his best work comes into its own.

Zeno and Eudoxus are representative of two vigorous opposing schools of mathematical thought which flourish today, the critical-destructive and the critical-constructive. Both had minds as penetratingly critical as their successors in the nineteenth and twentieth centuries. This statement can of course be inverted: Kronecker (1823–1891) and Brouwer (1881–   ), the modern critics of mathematical analysis—the theories of the infinite and the continuous—are as ancient as Zeno; the creators of the modern theories of continuity and the infinite, Weierstrass (1815–1897), Dedekind (1831–1916), and Cantor (1845–1918) are intellectual contemporaries of Eudoxus.

Archimedes, the greatest intellect of antiquity, is modern to the core. He and Newton would have understood one another perfectly, and it is just possible that Archimedes, could he come to life long enough to take a post-graduate course in mathematics and physics, would understand Einstein, Bohr, Heisenberg, and Dirac better than they understand themselves. Of all the ancients Archimedes is the only one who habitually thought with the unfettered freedom that the greater mathematicians permit themselves today with all the hard-won gains of twenty five centuries to smooth their way, for he alone of all the Greeks had sufficient stature and strength to stride clear over the obstacles thrown in the path of mathematical progress by frightened geometers who had listened to the philosophers.

Any list of the three "greatest" mathematicians of all history would include the name of Archimedes. The other two usually associated with him are Newton (1642–1727) and Gauss (1777–1855). Some, considering the relative wealth—or poverty—of mathematics and physical science in the respective ages in which these giants lived, and estimating their achievements against the background of their times, would put Archimedes first. Had the Greek mathematicians and scientists followed Archimedes rather than Euclid, Plato, and Aristotle, they might easily have anticipated the age of modern mathematics, which began with Descartes (1596–1650) and Newton in the seventeenth century, and the age of modern physical science inaugurated by Galileo (1564–1642) in the same century, by two thousand years.

Behind all three of these precursors of the modern age looms the half-mythical figure of Pythagoras (569?–500? B.C.), mystic, mathematician, investigator of nature to the best of his self-hobbled ability, "one tenth of him genius, nine-tenths sheer fudge." His life has become a fable, rich with the incredible accretions of his prodigies; but only this much is of importance for the development of mathematics as distinguished from the bizarre number-mysticism in which he clothed his cosmic speculations: he travelled extensively in Egypt, learned much from the priests and believed more; visited Babylon and repeated his Egyptian experiences; founded a secret Brotherhood for high mathematical thinking and nonsensical physical, mental, moral, and ethical speculation at Croton in southern Italy; and, out of all this, made two of the greatest contributions to mathematics in its entire history. He died, according to one legend, in the flames of his own school fired by political and religious bigots who stirred up the masses to protest against the enlightenment which Pythagoras sought to bring them. *Sic transit gloria mundi.*

Before Pythagoras it had not been clearly realized that *proof* must proceed from *assumptions*. Pythagoras, according to persistent tradition, was the first European to insist that the *axioms*, the *postulates*, be set down first in developing geometry and that the entire development thereafter shall proceed by applications of close deductive reasoning to the axioms. Following current practice we shall use "postulate," instead of "axiom" hereafter, as "axiom" has a pernicious historical association of "self-evident, necessary truth" which "postulate" does not have; a postulate is an arbitrary assumption laid down by the mathematician himself and not by God Almighty.

Pythagoras then imported *proof* into mathematics. This is his greatest achievement. Before him geometry had been largely a collection of rules of thumb empirically arrived at without any clear indication of the mutual connections of the rules, and without the slightest suspicion that all were deducible from a comparatively small number of postulates. Proof is now so commonly taken for granted as the very spirit of mathematics that we find it difficult to imagine the primitive thing which must have preceded mathematical reasoning.

Pythagoras' second outstanding mathematical contribution brings us abreast of living problems. This was the discovery, which humiliated and devastated him, that the common whole numbers 1,2,3, . . . are insufficient for the construction of mathematics even in the rudimentary form in which he knew it. Before this capital discovery he had preached like an inspired prophet that all nature, the entire universe in fact, physical, metaphysical, mental, moral, mathematical—*everything*—is built on the *discrete* pattern of the integers 1,2,3, . . . and is interpretable in terms of these God-given bricks alone; God, he declared indeed, *is* "number," and by that he meant common whole number. A sublime conception, no doubt, and beautifully simple, but as unworkable as its echo in Plato—"God ever geometrizes," or in Jacobi—"God ever arithmetizes," or in Jeans—"The Great Architect of the Universe now begins to appear as a mathematician." One obstinate mathematical discrepancy demolished Pythagoras' discrete philosophy, mathematics, and metaphysics. But, unlike some of his successors, he finally accepted defeat—after struggling unsuccessfully to suppress the discovery which abolished his creed.

This was what knocked his theory flat: it is impossible to find two whole numbers such that the square of one of them is equal to twice the square of the other. This can be proved by a simple argument* within the reach of anyone who has had a few weeks of algebra,

---

*Let $a^2 = 2b^2$, where, without loss of generality, a, b are whole numbers without any common factor greater than 1 (such a factor could be cancelled from the assumed equation). If a is *odd*, we have an immediate contradiction, since $2b^2$ is *even*; if a is *even*, say $2c$, then $4c^2 = 2b^2$, or $2c^2 = b^2$, so b is *even*, and hence a, b have the common factor 2, again a contradiction.

or even by anyone who thoroughly understands elementary arithmetic. Actually Pythagoras found his stumbling-block in geometry: the ratio of the side of a square to one of its diagonals cannot be expressed as the ratio of any two whole numbers. This is equivalent to the statement above about squares of whole numbers. In another form we would say that the square root of 2 is *irrational*, that is, is not equal to any whole number or decimal fraction, or sum of the two, got by dividing one whole number by another. Thus even so simple a geometrical concept as that of the diagonal of a square defies the integers 1,2,3, ... and negates the earlier Pythagorean philosophy. We can easily construct the diagonal *geometrically, but we cannot measure it in any finite number of steps*. This impossibility sharply and clearly brought irrational numbers and the infinite (non-terminating) processes which they seem to imply to the attention of mathematicians. Thus the square root of two can be calculated to any required *finite* number of decimal places by the process taught in school or by more powerful methods, but the decimal never "repeats" (as that for 1/7 does, for instance), nor does it ever terminate. In this discovery Pythagoras found the taproot of modern mathematical analysis.

Issues were raised by this simple problem which are not yet disposed of in a manner satisfactory to all mathematicians. These concern the mathematical concepts of the infinite (the unending, the uncountable), limits, and continuity, concepts which are at the root of modern analysis. Time after time the paradoxes and sophisms which crept into mathematics with these apparently indispensable concepts have been regarded as finally eliminated, only to reappear a generation or two later, changed but yet the same. We shall come across them, livelier than ever, in the mathematics of our time. The following is an extremely simple, intuitively obvious picture of the situation.

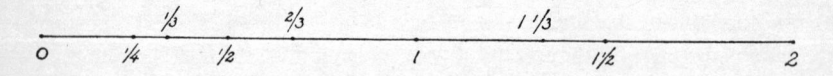

```
            ⅓        ⅔            1 ⅓
────┬───┬────┬────────┬──────────┬──────────┬────
 0    ¼   ½             1          1½           2
```

Consider a straight line two inches long, and imagine it to have been traced by the "continuous" "motion" of a "point." The words in quotes are those which conceal the difficulties. Without analysing them we easily persuade ourselves that we picture what they signify. Now label the left-hand end of the line 0 and the right-hand end 2.

Half-way between 0 and 2 we naturally put 1; half-way between 0 and 1 we put ½; half-way between 0 and ½ we put ¼, and so on. Similarly, between 1 and 2 we mark the place 1½, between 1½ and 2, the place 1¾, and so on. Having done this we may proceed in the same way to mark ⅓, ⅔, 1⅓, 1⅔, and then split each of the resulting segments into smaller equal segments. Finally, "in imagination," we can conceive of this process having been carried out for *all* the common fractions and common mixed numbers which are greater than 0 and less than 2; the conceptual division-points give us *all the rational numbers between 0 and 2*. There are an infinity of them. Do they completely "cover" the line? No. To what point does the square root of 2 correspond? No point, because this square root is not obtainable by dividing *any* whole number by another. But the square root of 2 is obviously a "number" of some sort;* its representative point lies somewhere between 1.41 and 1.42, and we can cage it down as closely as we please. To cover the line completely we are forced to imagine or to invent infinitely more "numbers" than the rationals. That is, if we accept the line as being *continuous*, and *postulate* that to each point of it corresponds one, and only one, "real number." The same kind of imagining can be carried on to the entire plane, and farther, but this is sufficient for the moment.

Simple problems such as these soon lead to very serious difficulties. With regard to these difficulties the Greeks were divided, just as we are, into two irreconcilable factions; one stopped dead in its mathematical tracks and refused to go on to analysis—the integral calculus, at which we shall glance when we come to it; the other attempted to overcome the difficulties and succeeded in convincing itself that it had done so. Those who stopped committed but few mistakes and were comparatively sterile of truth no less than of error; those who went on discovered much of the highest interest to mathematics and rational thought in general, some of which may be open to destructive criticism, however, precisely as has happened in our own generation. From the earliest times we meet these two distinct and antagonistic types of mind: the justifiably cautious who hang back because the ground quakes under their feet, and the bolder pioneers who leap the chasm to find treasure and comparative safety on the other side. We shall look first at one of those who refused to leap. For penetrating

* The inherent viciousness of such an assumption is obvious.

subtlety of thought we shall not meet his equal till we reach the twentieth century and encounter Brouwer.

Zeno of Elea (495–435 B.C.) was a friend of the philosopher Parmenides, who, when he visited Athens with his patron, shocked the philosophers out of their complacency by inventing four innocent paradoxes which they could not dissipate in words. Zeno is said to have been a self-taught country boy. Without attempting to decide what was his purpose in inventing his paradoxes—authorities hold widely divergent opinions—we shall merely state them. With these before us it will be fairly obvious that Zeno would have objected to our "infinitely continued" division of that two-inch line a moment ago. This will appear from the first two of his paradoxes, the *Dichotomy* and the *Achilles*. The last two, however, show that he would have objected with equal vehemence to the *opposite* hypothesis, namely that the line is *not* "infinitely divisible" but is composed of a *discrete* set of points that can be counted off 1,2,3, ... All four together constitute an iron wall beyond which progress appears to be impossible.

First, the *Dichotomy*. Motion is impossible, because whatever moves must reach the middle of its course *before* it reaches the end; but *before* it has reached the middle it must have reached the quarter-mark, and so on, *indefinitely*. Hence the motion can never even start.

Second, the *Achilles*. Achilles running to overtake a crawling tortoise ahead of him can never overtake it, because he must first reach the place from which the tortoise started; when Achilles reaches that place, the tortoise has departed and so is still ahead. Repeating the argument we easily see that the tortoise will always be ahead.

Now for the other side.

The *Arrow*. A moving arrow at any instant is either at rest or not at rest, that is, moving. If the instant is indivisible, the arrow cannot move, for if it did the instant would immediately be divided. But time is made up of instants. As the arrow cannot move in any one instant, it cannot move in any time. Hence it always remains at rest.

The *Stadium*. "To prove that half the time may be equal to double the time. Consider three rows of bodies

	First Position					Second Position				
(A)	o	o	o	o	(A)		o	o	o	o
(B)	o	o	o	o	(B)	o	o	o	o	
(C)	o	o	o	o	(C)			o	o	o	o

one of which (A) is at rest while the other two (B), (C) are moving with equal velocities in opposite directions. By the time they are all in the same part of the course (B) will have passed twice as many of the bodies in (C) as in (A). Therefore the time which it takes to pass (A) is twice as long as the time it takes to pass (C). But the time which (B) and (C) take to reach the position of (A) is the same. Therefore double the time is equal to half the time." (Burnet's translation.) It is helpful to imagine (A) as a circular picket fence.

These, in non-mathematical language, are the sort of difficulties the early grapplers with continuity and infinity encountered. In books written twenty years or so ago it was said that "the positive theory of infinity" created by Cantor, and the like for "irrational" numbers, such as the square root of 2, invented by Eudoxus, Weierstrass, and Dedekind, had disposed of all these difficulties once and forever. Such a statement would not be accepted today by all schools of mathematical thought. So in dwelling upon Zeno we have in fact been discussing ourselves. Those who wish to see any more of him may consult Plato's *Parmenides*. We need remark only that Zeno finally lost his head for treason or something of the sort, and pass on to those who did not lose their heads over his arguments. Those who stayed behind with Zeno did comparatively little for the advancement of mathematics, although their successors have done much to shake its foundations.

Eudoxus (408–355 B.C.) of Cnidus inherited the mess which Zeno bequeathed the world and not much more. Like more than one man who has left his mark on mathematics, Eudoxus suffered from extreme poverty in his youth. Plato was in his prime while Eudoxus lived and Aristotle was about thirty when Eudoxus died. Both Plato and Aristotle, the leading philosophers of antiquity, were much concerned over the doubts which Zeno had injected into mathematical reasoning and which Eudoxus, in his theory of proportion—"the crown of Greek mathematics"—was to allay till the last quarter of the nineteenth century.

As a young man Eudoxus moved to Athens from Tarentum, where he had studied with Archytas (428–347 B.C.), a first-rate mathematician, administrator, and soldier. Arriving in Athens, Eudoxus soon fell in with Plato. Being too poor to live near the Academy, Eudoxus trudged back and forth every day from the Piraeus where fish and olive oil were cheap and lodging was to be had for a smile in the right place.

Although he himself was not a mathematician in the technical sense, Plato has been called "the maker of mathematicians," and it cannot be denied that he did irritate many infinitely better mathematicians than himself into creating some real mathematics. As we shall see, his total influence on the development of mathematics was probably baneful. But he did recognize what Eudoxus was and became his devoted friend until he began to exhibit something like jealousy toward his brilliant protégé. It is said that Plato and Eudoxus made a journey to Egypt together. If so, Eudoxus seems to have been less credulous than his predecessor Pythagoras; Plato however shows the effects of having swallowed vast quantities of the number-mysticism of the East. Finding himself unpopular in Athens, Eudoxus finally settled and taught at Cyzicus, where he spent his last years. He studied medicine and is said to have been a practising physician and legislator on top of his mathematics. As if all this were not enough to keep one man busy he undertook a serious study of astronomy, to which he made outstanding contributions. In his scientific outlook he was centuries ahead of his verbalizing, philosophizing contemporaries. Like Galileo and Newton he had a contempt for speculations about the physical universe which could not be checked by observation and experience. If by getting to the sun, he said, he could ascertain its shape, size, and nature, he would gladly share the fate of Phaëthon, but in the meantime he would not guess.

Some idea of what Eudoxus did can be seen from a very simple problem. To find the area of a rectangle we multiply the length by the breadth. Although this sounds intelligible it presents serious difficulties unless both sides are measurable by *rational* numbers. Passing these particular difficulties we see them in a more evident form in the next simplest type of problem, that of finding the length of a *curved* line, or the area of a *curved* surface, or the volume enclosed by *curved* surfaces.

Any young genius wishing to test his mathematical powers may try to devise a method for doing these things. Provided he has never seen it done in school, how would he proceed to give a rigorous proof of the formula for the circumference of a circle of any given radius? Whoever does that entirely on his own initiative may justly claim to be a mathematician of the first rank. The moment we pass from figures bounded by *straight* lines or *flat* surfaces we run slap into all the problems of continuity, the riddles of the infinite and the mazes of irrational numbers. Eudoxus devised the first logically satisfactory method, which Euclid reproduced in Book V of his *Elements*, for handling such problems. In his *method of exhaustion*, applied to the computation of areas and volumes, Eudoxus showed that we need not assume the "existence" of "infinitely small quantities." It is sufficient for the purposes of mathematics to be able to reach a magnitude *as small as we please* by the continued division of a given magnitude.

To finish with Eudoxus we shall state his epochal definition of equal ratios which enabled mathematicians to treat irrational numbers as rigorously as the rationals. This was, essentially, the starting-point of one modern theory of irrationals.

"The first of four magnitudes is said to have the *same ratio* to the second that the third has to the fourth when, any whatever equimultiples [the same multiples] of the *first* and *third* being taken, and any other equimultiples of the *second* and *fourth*, the multiple of the *first* is greater than, equal to, or less than the multiple of the *second*, according as the multiple of the *third* is greater than, equal to, or less than the multiple of the *fourth*."

Of the Greeks not yet named whose work influenced mathematics after the year 1600 only Apollonius need be mentioned here. Apollonius (260?–200? B.C.) carried geometry in the manner of Euclid—the way it is still taught to hapless beginners—far beyond the state in which Euclid (330?–275? B.C.) left it. As a geometer of this type—a *synthetic*, "pure" geometer—Apollonius is without a peer till Steiner in the nineteenth century.

If a cone standing on a circular base and extending indefinitely in both directions through its vertex is cut by a plane, the curve in which the plane intersects the surface of the cone is called a conic section. There are five possible kinds of conic sections: the ellipse; the hyperbola, consisting of two branches; the parabola, the path of a projectile in a vacuum; the circle; and a pair of intersecting straight lines. The ellipse, parabola and hyperbola are "mechanical curves" according to the Platonic formula; that is, these curves cannot be constructed by the use of straightedge and compass alone, although it is easy, with these implements, to construct any desired number of points lying on any one of these curves. The geometry of the conic sections, worked out to a high degree of perfection by Apollonius and his successors, proved to be of the highest importance in the celestial mechanics of the seventeenth and succeeding centuries. Indeed, had not the Greek geometers run ahead of Kepler it is unlikely that Newton could ever have come upon his law of universal gravitation, for which Kepler had

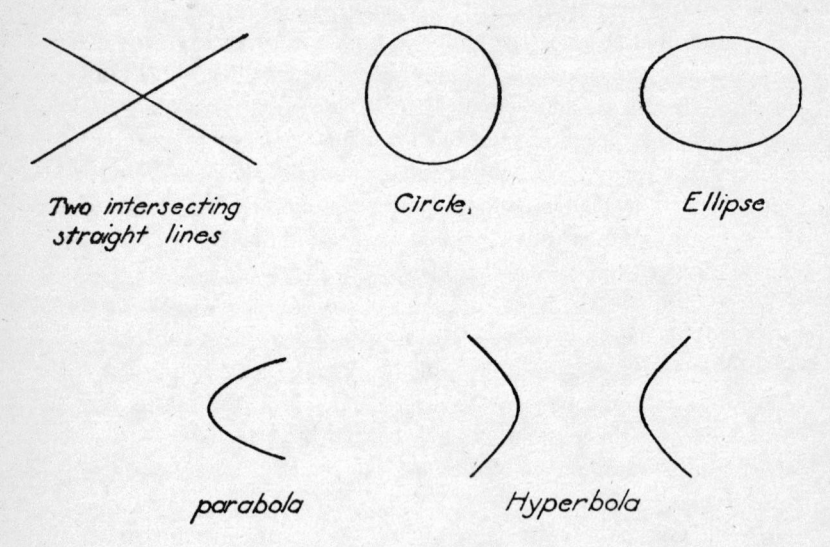

Two intersecting straight lines   *Circle*   *Ellipse*

parabola   *Hyperbola*

prepared the way with his laboriously ingenious calculations on the orbits of the planets.

Among the later Greeks and the Arabs of the Middle Ages Archimedes seems to have inspired the same awe and reverence that Gauss did among his contemporaries and followers in the nineteenth century, and that Newton did in the seventeenth and eighteenth. Archimedes was the undisputed chieftain of them all, "the old man," "the wise one," "the master," "the *great* geometer." To recall his dates, he lived in 287–212 B.C. Thanks to Plutarch more is known about his death than his life, and it is perhaps not unfair to suggest that the typical historical biographer Plutarch evidently thought the King of Mathematicians a less important personage historically than the Roman soldier Marcellus, into whose *Life* the account of Archimedes is slipped like a tissue-thin shaving of ham in a bull-choking sandwich. Yet Archimedes is today Marcellus' chief title to remembrance—and execration. In the death of Archimedes we shall see the first impact of a crassly practical civilization upon the greater thing which it destroyed—Rome, having half-demolished Carthage, swollen with victory and imperially purple with valor, falling upon Greece to shatter its fine fragility.

In body and mind Archimedes was an aristocrat. The son of the astronomer Pheidias, he was born at Syracuse, Sicily, and is said to have been related to Hieron II, tyrant (or king) of Syracuse. At any rate he was on intimate terms with Hieron and his son Gelon, both of whom had a high admiration for the king of mathematicians. His essentially aristocratic temperament expressed itself in his attitude to what would today be called applied science. Although he was one of the greatest mechanical geniuses of all time, if not the greatest when we consider how little he had to go on, the aristocratic Archimedes had a sincere contempt for his own practical inventions. From one point of view he was justified. Books could be written on what Archimedes did for applied mechanics; but great as this work was from our own mechanically biased point of view, it is completely overshadowed by his contributions to pure mathematics. We look first at the few known facts about him and the legend of his personality.

According to tradition Archimedes is a perfect museum specimen of the popular conception of what a great mathematician should be. Like Newton and Hamilton he left his meals untouched when he was deep in his mathematics. In the matter of inattention to dress he even surpasses Newton, for on making his famous discovery that a floating body loses in weight an amount equal to that of the liquid displaced, he leaped from the bath in which he had made the discovery by observing his own floating body, and dashed through the streets of Syracuse stark naked, shouting *"Eureka, eureka!"* (I have found it, I have found it!) What he had found was the first law of hydrostatics. According to the story a dishonest goldsmith had adulterated the gold of a crown for Hieron with silver and the tyrant, suspecting fraud, had asked Archimedes to put his mind on the problem. Any high school boy knows how it is solved by a simple experiment and some easy arithmetic on specific gravity; "the principle of Archimedes" and its numerous practical applications are meat for youngsters and naval engineers today, but the man who first saw through them had more than common insight. It is not definitely known whether the goldsmith was guilty; for the sake of the story it is usually assumed that he was.

Another exclamation of Archimedes which has come down through the centuries is "Give me a place to stand on and I will move the earth" (πᾶ βῶ καὶ κινῶ τὰν γᾶν, as he said it in Doric). He himself was strongly moved by his discovery of the laws of levers when he made his boast. The phrase would make a perfect motto for a modern scientific institute; it seems strange that it has not been appropriated. There is another version in better Greek but the meaning is the same.

In one of his eccentricities Archimedes resembled another great mathematician, Weierstrass. According to a sister of Weierstrass, he could not be trusted with a pencil when he was a young school teacher if there was a square foot of clear wallpaper or a clean cuff anywhere in sight. Archimedes beats this record. A sanded floor or dusted hard smooth earth was a common sort of "blackboard" in his day. Archimedes made his own occasions. Sitting before the fire he would rake out the ashes and draw in them. After stepping from the bath he would anoint himself with olive oil, according to the custom of the time, and then, instead of putting on his clothes, proceed to lose himself in the diagrams which he traced with a fingernail on his own oily skin.

Archimedes was a lonely sort of eagle. As a young man he had studied for a short time at Alexandria, Egypt, where he made two life-long friends, Conon, a gifted mathematician for whom Archimedes had a high regard both personal and intellectual, and Eratosthenes, also a good mathematician but quite a fop. These two, particularly Conon, seem to have been the only men of his contemporaries with whom Archimedes felt he could share his thoughts and be assured of understanding. Some of his finest work was communicated by letters to Conon. Later, when Conon died, Archimedes corresponded with Dositheus, a pupil of Conon.

Leaving aside his great contributions to astronomy and mechanical invention we shall give a bare and inadequate summary of the principal additions which Archimedes made to pure and applied mathematics.

He invented general methods for finding the areas of curvilinear plane figures and volumes bounded by curved surfaces, and applied these methods to many special instances, including the circle, sphere, any segment of a parabola, the area enclosed between two radii and two successive whorls of a spiral, segments of spheres, and segments of surfaces generated by the revolution of rectangles (cylinders), triangles (cones), parabolas (paraboloids), hyperbolas (hyperboloids), and ellipses (spheroids) about their principal axes. He gave a method for calculating π (the ratio of the circumference of a circle to its diameter), and fixed π as lying between 3 1/7 and 3 10/71; he also gave methods for approximating to square roots which show that he anticipated the invention by the Hindus of what amount to periodic continued fractions. In arithmetic, far surpassing the incapacity of the unscientific Greek method of symbolizing numbers to write, or even to describe, large numbers, he invented a system of numeration capable of handling numbers as large as desired. In mechanics he laid down some of the fundamental postulates, discovered the laws of levers, and applied his mechanical principles (of levers) to calculate the areas and centers of gravity of several flat surfaces and solids of various shapes. He created the whole science of hydrostatics and applied it to find the positions of rest and of equilibrium of floating bodies of several kinds.

Archimedes composed not one masterpiece but many. How did he do it all? His severely economical, logical exposition gives no hint of the *method* by which he arrived at his wonderful results. But in 1906, J. L. Heiberg, the historian and scholar of Greek mathematics, made the dramatic discovery in Constantinople of a hitherto "lost" treatise of Archimedes addressed to his friend Eratosthenes: *On Mechanical Theorems, Method*. In it Archimedes explains how by weighing, in imagination, a figure or solid whose area or volume was unknown against a known one, he was led to the knowledge of the fact he sought; the fact being known it was then comparatively easy (for him) to prove it mathematically. In short he used his mechanics to advance his mathematics. This is one of his titles to a modern mind: *he used anything and everything that suggested itself as a weapon to attack his problems*.

To a modern all is fair in war, love, and mathematics; to many of the ancients, mathematics was a stultified game to be played according to the prim rules imposed by the philosophically-minded Plato. According to Plato only a straightedge and a pair of compasses were to be permitted as the implements of construction in geometry. No wonder the classical geometers hammered their heads for centuries against "the three problems of antiquity": to trisect an angle; to construct a cube having double the volume of a given cube; to construct a square equal to a circle. *None of these problems is possible with only straightedge and compass*, although it is hard to prove that the third is not, and the impossibility was finally proved only in 1882. All constructions effected with other implements were dubbed "mechanical" and, as such, for some mystical reason known only to Plato and his geometrizing God, were considered shockingly vulgar and were rigidly taboo in respectable geometry. Not till Descartes, 1985 years after the death of Plato, published his analytic geometry, did geometry escape from its Platonic straightjacket. Plato of course had been dead for sixty years or more before Archimedes was born, so he cannot be censured for not appreciating the lithe power and freedom of the methods of Archimedes. On the other hand, only praise is due Archimedes for not appreciating the old-maidishness of Plato's rigidly corseted conception of what the muse of geometry should be.

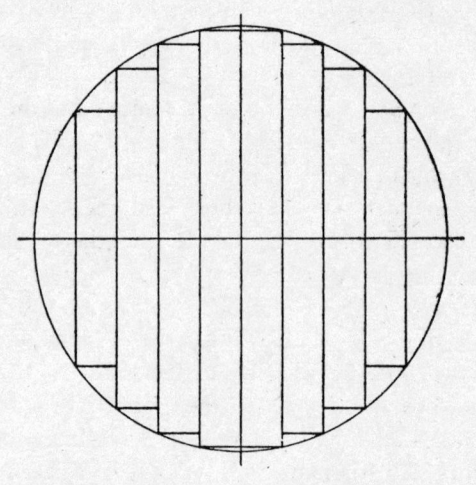

The second claim of Archimedes to modernity is also based upon his methods. Anticipating Newton and Leibniz by more than 2000 years he invented the integral calculus and in one of his problems anticipated their invention of the differential calculus. These two calculuses together constitute what is known as *the* calculus, which has been described as the most powerful instrument ever invented for the mathematical exploration of the physical universe. To take a simple example, suppose we wish to find the area of a circle. Among other ways of doing this we may slice the circle into any number of parallel strips of equal breadth, cut off the curved ends of the strips, so that the discarded bits shall total the least possible, by cuts perpendicular to the strips, and then add up the areas of all the resulting rectangles. This gives an approximation to the area sought. By increasing the number of strips indefinitely and taking the limit of the sum, we get the area of the circle. This (crudely described) process of taking the limit of the sum is called *integration*; the method of performing such summations is called the *integral calculus*. It was this calculus which Archimedes used in finding the area of a segment of a parabola and in other problems.

The problem in which he used the differential calculus was that of constructing a tangent at any given point of his spiral. If the angle which the tangent makes with any given line is known, the tangent can easily be drawn, for there is a simple construction for drawing a straight line through a given point parallel to a given straight line. The problem of finding the angle mentioned (for *any* curve, not merely for the spiral) is, in geometrical language, the main problem of the *differential* calculus. Archimedes solved this problem for his spiral. His spiral is the curve traced by a point moving with uniform speed along a straight line which revolves with uniform angular speed about a fixed point on the line. If anyone who has not studied the calculus imagines Archimedes' problem an easy one he may time himself doing it.

The life of Archimedes was as tranquil as a mathematician's should be if he is to accomplish all that is in him. All the action and tragedy of his life were crowded into its end. In 212 B.C. the second Punic war was roaring full blast. Rome and Carthage were going at one another hammer and tongs, and Syracuse, the city of Archimedes, lay temptingly near the path of the Roman fleet. Why not lay siege to it? They did.

Puffed up with conceit of himself ("relying on his own great fame," as Plutarch puts it), and trusting in the splendor of his "preparedness" rather than in brains, the Roman leader, Marcellus, anticipated a speedy conquest. The pride of his confident heart was a primitive piece of artillery on a lofty harp-shaped platform supported by eight galleys lashed together. Beholding all this fame and miscellaneous shipping descending upon them the timider citizens would have handed Marcellus the keys of the city. Not so Hieron. He too was prepared for war, and in a fashion that the practical Marcellus would never have dreamed of.

It seems that Archimedes, despising applied mathematics himself, had nevertheless yielded in peace time to the importunities of Hieron, and had demonstrated to the tyrant's satisfaction that mathematics can, on occasion, become devastatingly practical. To convince his friend that mathematics is capable of more than abstract deductions.

Archimedes had applied his laws of levers and pulleys to the manipulation of a fully loaded ship, which he himself launched singlehanded. Remembering this feat when the war clouds began to gather ominously near, Hieron begged Archimedes to prepare a suitable welcome for Marcellus. Once more desisting from his researches to oblige his friend, Archimedes constituted himself a reception committee of one to trip the precipitate Romans. When they arrived his ingenious deviltries stood grimly waiting to greet them.

The harp-shaped turtle affair on the eight quinqueremes lasted no longer than the fame of the conceited Marcellus. A succession of stone shots, each weighing over a quarter of a ton, hurled from the supercatapults of Archimedes, demolished the unwieldy contraption. Cranelike beaks and iron claws reached over the walls for the approaching ships, seized them, spun them round, and sank or shattered them against the jutting cliffs. The land forces, mowed down by the Archimedean artillery, fared no better. Camouflaging his rout in the official bulletins as a withdrawal to a previously prepared position in the rear, Marcellus backed off to confer with his staff. Unable to rally his mutinous troops for an assault on the terrible walls, the famous Roman leader retired.

At last evincing some slight signs of military common sense, Marcellus issued no further "backs against the wall" orders of the day, abandoned all thoughts of a frontal attack, captured Megara in the rear, and finally sneaked up on Syracuse from behind. This time his luck was with him. The foolish Syracusans were in the middle of a bibulous religious celebration in honor of Artemis. War and religion have always made a bilious sort of cocktail; the celebrating Syracusans were very sick indeed. They woke up to find the massacre in full swing. Archimedes participated in the blood-letting.

His first intimation that the city had been taken by theft was the shadow of a Roman soldier falling across his diagram in the dust. According to one account the soldier had stepped on the diagram, angering Archimedes to exclaim sharply, "Don't disturb my circles!" Another states that Archimedes refused to obey the soldier's order that he accompany him to Marcellus until he had worked out his problem. In any event the soldier flew into a passion, unsheathed his glorious sword, and dispatched the unarmed veteran geometer of seventy five. Thus died Archimedes.

As Whitehead has observed, "No Roman lost his life because he was absorbed in the contemplation of a mathematical diagram."

CHAPTER THREE

Gentleman, Soldier, and Mathematician

DESCARTES

[Analytic geometry], far more than any of his metaphysical speculations, immortalized the name of Descartes, and constitutes the greatest single step ever made in the progress of the exact sciences.—JOHN STUART MILL

"I DESIRE ONLY TRANQUILLITY AND REPOSE." These are the words of the man who was to deflect mathematics into new channels and change the course of scientific history. Too often in his active life René Descartes was driven to find the tranquillity he sought in military camps and to seek the repose he craved for meditation in solitary retreat from curious and exacting friends. Desiring only tranquillity and repose, he was born on March 31, 1596 at La Haye, near Tours, France, into a Europe given over to war in the throes of religious and political reconstruction.

His times were not unlike our own. An old order was rapidly passing; the new was not yet established. The predatory barons, kings, and princelings of the Middle Ages had bred a swarm of rulers with the political ethics of highway robbers and, for the most part, the intellects of stable boys. What by common justice should have been thine was mine provided my arm was strong enough to take it away from thee. This may be an unflattering picture of that glorious period of European history known as the late Renaissance, but it accords fairly well with our own changing estimate, born of intimate experience, of what should be what in a civilized society.

On top of the wars for plunder in Descartes' day there was superimposed a rich deposit of religious bigotry and intolerance which incubated further wars and made the dispassionate pursuit of science a highly hazardous enterprise. To all this was added a comprehensive ignorance of the elementary rules of common cleanliness. From the point of view of sanitation the rich man's mansion was likely to be as filthy as the slums where the poor festered in dirt and ignorance, and the recurrent plagues which aided the epidemic wars in keeping the prolific population below the famine limit paid no attention to bank accounts. So much for the good old days.

On the immaterial, enduring side of the ledger the account is brighter. The age in which Descartes lived was indeed one of the great intellectual periods in the spotted history of civilization. To mention only a few of the outstanding men whose lives partly overlapped that of Descartes, we recall that Fermat and Pascal were his contemporaries in mathematics; Shakespeare died when Descartes was twenty; Descartes outlived Galileo by eight years, and Newton was eight when Descartes died; Descartes was twelve when Milton was born, and Harvey, the discoverer of the circulation of the blood, outlived Descartes by seven years, while Gilbert, who founded the science of electromagnetism, died when Descartes was seven.

René Descartes came from an old noble family. Although René's father was not wealthy his circumstances were a little better than easy, and his sons were destined for the careers of gentlemen—*noblesse oblige* —in the service of France. René was the third and last child of his father's first wife, Jeanne Brochard, who died a few days after René's birth. The father appears to have been a man of rare sense who did everything in his power to make up to his children for the loss of their mother. An excellent nurse took the mother's place, and the father, who married again, kept a constant, watchful, intelligent eye on his "young philosopher" who always wanted to know the cause of everything under the sun and the reason for whatever his nurse told him about heaven. Descartes was not exactly a precocious child, but his frail health forced him to expend what vitality he had in intellectual curiosity.

Owing to René's delicate health his father let lessons slide. The boy however went ahead on his own initiative and his father wisely let him do as he liked. When Descartes was eight his father decided that formal education could not be put off longer. After much intelligent inquiry he chose the Jesuit college at La Flèche as the ideal school for his son. The rector, Father Charlet, took an instant liking to the pale, confiding little boy and made a special study of his case. Seeing that he must build up the boy's body if he was to educate his mind, and noticing that Descartes seemed to require much more rest than normal boys of his age, the rector told him to lie in bed as late as he pleased in the mornings and not to leave his room till he felt like joining his companions in the classroom. Thereafter, all through his life except for one unfortunate episode near its close, Descartes spent his mornings in bed when he wished to think. Looking back in middle age on his schooldays at La Flèche, he averred that those long, quiet mornings of silent meditation were the real source of his philosophy and mathematics.

His work went well and he became a proficient classicist. In line with the educational tradition of the time much attention was put on Latin, Greek, and rhetoric. But this was only a part of what Descartes got. His teachers were men of the world themselves and it was their job to train the boys under their charge to be "gentlemen"—in the best sense of that degraded word—for their rôle in the world. When he left the school in August, 1612, in his seventeenth year, Descartes had made a life-long friend in Father Charlet and was almost ready to hold his own in society. Charlet was only one of the many friends Descartes made at La Flèche; another, Mersenne (later Father), the famous amateur of science and mathematics, had been his older chum and was to become his scientific agent and protector-in-chief from bores.

Descartes' distinctive talent had made itself evident long before he left school. As early as the age of fourteen, lying meditating in bed, he had begun to suspect that the "humanities" he was mastering were comparatively barren of human significance and certainly not the sort of learning to enable human beings to control their environment and direct their own destiny. The authoritative dogmas of philosophy, ethics, and morals offered for his blind acceptance began to take on the aspect of baseless superstitions. Persisting in his childhood habit of accepting nothing on mere authority, Descartes began unostentatiously questioning the alleged demonstrations and the casuistical logic by which the good Jesuits sought to gain the assent of his reasoning faculties. From this he rapidly passed to the fundamental doubt which was to inspire his life-work: how do we *know* anything? And further, perhaps more importantly, if we cannot say definitely that we know anything, how are we ever to find out those things which we may be capable of knowing?

On leaving school Descartes thought longer, harder, and more desperately than ever. As a first fruit of his meditations he apprehended the heretical truth that logic of itself—the great method of the schoolmen of the Middle Ages which still hung on tenaciously in humanistic education—is as barren as a mule for any creative human purpose. His second conclusion was closely allied to his first: compared to the demonstrations of mathematics—to which he took like a bird to the air as soon as he found his wings—those of philosophy, ethics, and morals are tawdry shams and frauds. How then, he asked, shall we ever find out anything? By the scientific method, although Descartes did not call it that: by *controlled experiment* and the application of rigid mathematical reasoning to the results of such experiment.

It may be asked what he got out of his rational skepticism. One fact, and only one: "I exist." As he put it, *"Cogito ergo sum"* (I think, therefore I am).

By the age of eighteen Descartes was thoroughly disgusted with the aridity of the studies on which he had put so much hard labor. He resolved to see the world and learn something of life as it is lived in flesh and blood and not in paper and printers' ink. Thanking God that he was well enough off to do as he pleased he proceeded to do it. By an understandable overcorrection of his physically inhibited childhood and youth he now fell upon the pleasures appropriate to normal young men of his age and station and despoiled them with both hands. With several other young blades hungering for life in the raw he quit the depressing sobriety of the paternal estate and settled in Paris. Gambling being one of the accomplishments of a gentleman in that day, Descartes gambled with enthusiasm—and some success. Whatever he undertook he did with his whole soul.

This phase did not last long. Tiring of his bawdy companions, Descartes gave them the slip and took up his quarters in plain, comfortable lodgings in what is now the suburb of Saint-Germain where, for two years, he buried himself in incessant mathematical investigation. His gay deeds at last found him out, however, and his hare-brained friends descended whooping upon him. The studious young man looked up, recognized his friends, and saw that they were one and all intolerable bores. To get a little peace Descartes decided to go to war.

Thus began his first spell of soldiering. He went first to Breda, Holland, to learn his trade under the brilliant Prince Maurice of Orange. Being disappointed in his hopes for action under the Prince's colors, Descartes turned a disgusted back on the peaceful life of the camp, which threatened to become as exacting as the hurly-burly of Paris, and hastened to Germany. At this point of his career he first showed symptoms of an amiable weakness which he never outgrew. Like a small boy trailing a circus from village to village Descartes seized every favorable opportunity to view a gaudy spectacle. One was now about to come off at Frankfurt, where Ferdinand II was to be crowned. Descartes arrived in time to take in the whole rococo show. Considerably cheered up he again sought his profession and enlisted under the Elector of Bavaria, then waging war against Bohemia.

The army was lying inactive in its winter quarters near the little village of Neuburg on the banks of the Danube. There Descartes found in plenty what he had been seeking, tranquillity and repose. He was left to himself and he found himself.

The story of Descartes' "conversion"—if it may be called that—is extremely curious. On St. Martin's Eve, November 10, 1619, Descartes experienced three vivid dreams which, he says, changed the whole current of his life. His biographer (Baillet) records the fact that there had been considerable drinking in celebration of the saint's feast and suggests that Descartes had not fully recovered from the fumes of the wine when he retired. Descartes himself attributes his dreams to quite another source and states emphatically that he had touched no wine for three months before his elevating experience. There is no reason to doubt his word. The dreams are singularly coherent and quite unlike those (according to experts) inspired by a debauch, especially of stomach-filling wine. On the surface they are easily explicable as the subconscious resolution of a conflict between the dreamer's desire to lead an intellectual life and his realization of the futility of the life he was actually living. No doubt the Freudians have analyzed these dreams, but it seems unlikely that any analysis in the classical Viennese manner could throw further light on the invention of analytic geometry, in which we are chiefly interested here. Nor do the several mystic or religious interpretations seem likely to be of much assistance in this respect.

In the first dream Descartes was blown by evil winds from the security of his church or college toward a third party which the wind was powerless to shake or budge; in the second he found himself observing a terrific storm with the unsuperstitious eyes of science, and he noted that the storm, once seen for what it was, could do him no harm; in the third he dreamed that he was reciting the poem of Ausonius which begins, *"Quod vitae secatabor iter?"* (What way of life shall I follow?)

There was much more. Out of it all Descartes says he was filled with "enthusiasm" (probably intended in a mystic sense) and that there had been revealed to him, as in the second dream, the magic key which would unlock the treasure house of nature and put him in possession of the true foundation, at least, of all the sciences.

What was this marvelous key? Descartes himself does not seem to have told anyone explicitly, but it is usually believed to have been nothing less than the application of algebra to geometry, analytic geometry in short and, more generally, the exploration of natural phenomena by mathematics, of which mathematical physics today is the most highly developed example.

November 10, 1619, then, is the official birthday of analytic geometry and therefore also of modern mathematics. Eighteen years were to pass before the method was published. In the meantime Descartes went on with his soldiering. On his behalf mathematics may thank Mars that no half-spent shot knocked his head off at the battle of Prague. A score or so of promising young mathematicians a few years short of three centuries later were less lucky, owing to the advance of that science which Descartes' dream inspired.

As never before the young soldier of twenty two now realized that if he was ever to find truth he must first reject absolutely all ideas acquired from others and rely upon the patient questioning of his own mortal mind to show him the way. All the knowledge he had received from authority must be cast aside; the whole fabric of his inherited moral and intellectual ideas must be destroyed, to be refashioned more enduringly by the primitive, earthy strength of human reason alone. To placate his conscience he prayed the Holy Virgin to help him in his heretical project. Anticipating her assistance he vowed a pilgrimage to the shrine of Notre-Dame de Lorette and proceeded forthwith to subject the accepted truths of religion to a scorching, devastating criticism. However, he duly discharged his part of the contract when he found the opportunity.

In the meantime he continued his soldiering, and in the spring of 1620 enjoyed some very real fighting at the battle of Prague. With the rest of the victors Descartes entered the city chanting praises to

God. Among the terrified refugees was the four-year-old Princess Elisabeth,* who was later to become Descartes' favorite disciple.

At last, in the spring of 1621, Descartes got his bellyful of war. With several other gay gentlemen soldiers he had accompanied the Austrians into Transylvania, seeking glory and finding it—on the other side. But if he was through with war for the moment he was not yet ripe for philosophy. The plague in Paris and the war against the Huguenots made France even less attractive than Austria. Northern Europe was both peaceful and clean; Descartes decided to pay it a visit. Things went well enough till Descartes dismissed all but one of his bodyguard before taking boat for east Frisia. Here was a Heaven-sent opportunity for the cut-throat crew. They decided to knock their prosperous passenger on the head, loot him, and pitch his carcase to the fish. Unfortunately for their plans Descartes understood their language. Whipping out his sword he compelled them to row him back to the shore, and once again analytic geometry escaped the accidents of battle, murder, and sudden death.

The following year passed quietly enough in visits to Holland and Rennes, where Descartes' father lived. At the end of the year he returned to Paris, where his reserved manner and somewhat mysterious appearance immediately got him accused of being a Rosicrucian. Ignoring the gossip, Descartes philosophized and played politics to get himself a commission in the army. He was not really disappointed when he failed, as he was left free to visit Rome where he enjoyed the most gorgeous spectacle he had yet witnessed, the ceremony celebrated every quarter of a century by the Catholic Church. This Italian interlude is of importance in Descartes' intellectual development for two reasons. His philosophy, so far as it fails to touch the common man, was permanently biased against that lowly individual by the fill which the bewildered philosopher got of unwashed humanity gathered from all corners of Europe to receive the papal benediction. Equally important was Descartes' failure to meet Galileo. Had the mathematician been philosopher enough to sit for a week or two at the feet of the father of modern science, his own speculations on the physical universe might have been less fantastic. All that Descartes got out of his Italian journey was a grudging jealousy of his incomparable contemporary.

---

* Daughter of Frederick, Elector Palatine of the Rhine, and King of Bohemia, and a granddaughter of James I of England.

Immediately after his holiday in Rome, Descartes enjoyed another bloody spree of soldiering with the Duke of Savoy, in which he so distinguished himself that he was offered a lieutenant generalship. He had sense enough to decline. Returning to the Paris of Cardinal Richelieu and the swashing D'Artagnan—the latter near-fiction, the former less credible than a melodrama—Descartes settled down to three years of meditation. In spite of his lofty thoughts he was no gray-bearded savant in a dirty smock, but a dapper, well dressed man of the world, clad in fashionable taffeta and sporting a sword as befitted his gentlemanly rank. To put the finishing touch to his elegance he crowned himself with a sweeping, broad-brimmed, ostrich-plumed hat. Thus equipped he was ready for the cut-throats infesting church, state, and street. Once when a drunken lout insulted Descartes' lady of the evening, the irate philosopher went after the rash fool quite in the stump-stirring fashion of D'Artagnan, and having flicked the sot's sword out of his hand, spared his life, not because he was a rotten swordsman, but because he was too filthy to be butchered before a beautiful lady.

Having mentioned one of Descartes' lady friends we may dispose of all but two of the rest here. Descartes liked women well enough to have a daughter by one. The child's early death affected him deeply. Possibly his reason for never marrying may have been, as he informed one expectant lady, that he preferred truth to beauty; but it seems more probable that he was too shrewd to mortgage his tranquillity and repose to some fat, rich, Dutch widow. Descartes was only moderately well off, but he knew when he had enough. For this he has been called cold and selfish. It seems juster to say that he knew where he was going and that he realized the importance of his goal. Temperate and abstemious in his habits he was not mean, never inflicting on his household the Spartan regimen he occasionally prescribed for himself. His servants adored him, and he interested himself in their welfare long after they had left his service. The boy who was with him at his death was inconsolable for days at the loss of his master. All this does not sound like selfishness.

Descartes also has been accused of atheism. Nothing could be farther from the truth. His religious beliefs were unaffectedly simple in spite of his rational skepticism. He compared his religion, indeed, to the nurse from whom he had received it, and declared that he found it as comforting to lean upon one as on the other. A rational mind is sometimes the queerest mixture of rationality and irrationality on earth.

Another trait affected all Descartes' actions till he gradually outgrew it under the rugged discipline of soldiering. The necessary coddling of his delicate childhood infected him with a deep tinge of hypochondria, and for years he was chilled by an oppressive dread of death. This, no doubt, is the origin of his biological researches. By middle age he could say sincerely that nature is the best physician and that the secret of keeping well is to lose the fear of death. He no longer fretted to discover means of prolonging existence.

His three years of peaceful meditation in Paris were the happiest of Descartes' life. Galileo's brilliant discoveries with his crudely constructed telescope had set half the natural philosophers of Europe to pottering with lenses. Descartes amused himself in this way, but did nothing of striking novelty. His genius was essentially mathematical and abstract. One discovery which he made at this time, that of the principle of virtual velocities in mechanics, is still of scientific importance. This really was first-rate work. Finding that few understood or appreciated it, he abandoned abstract matters and turned to what he considered the highest of all studies, that of man. But, as he dryly remarks, he soon discovered that the number of those who understand man is negligible in comparison with the number of those who think they understand geometry.

Up till now Descartes had published nothing. His rapidly mounting reputation again attracted a horde of fashionable dilettantes, and once more Descartes sought tranquillity and repose on the battlefield, this time with the King of France at the siege of La Rochelle. There he met that engaging old rascal Cardinal Richelieu, who was later to do him a good turn, and was impressed, not by the Cardinal's wiliness, but by his holiness. On the victorious conclusion of the war Descartes returned with a whole skin to Paris, this time to suffer his second conversion and abandon futilities forever.

He was now (1628) thirty two, and only his miraculous luck had preserved his body from destruction and his mind from oblivion. A stray bullet at La Rochelle might easily have deprived Descartes of all claim to remembrance, and he realized at last that if he was ever to arrive it was high time that he be on his way. He was aroused from his sterile state of passive indifference by two Cardinals, De Bérulle and De Bagné, to the first of whom in particular the scientific world owes an everlasting debt of gratitude for having induced Descartes to publish.

The Catholic clergy of the time cultivated and passionately loved the sciences, in grateful contrast to the fanatical Protestants whose bigotry had extinguished the sciences in Germany. On becoming acquainted with De Bérulle and De Bagné, Descartes blossomed out like a rose under their genial encouragement. In particular, during soirées at De Bagné's, Descartes spoke freely of his new philosophy to a M. de Chandoux (who was later hanged for counterfeiting, not a result of Descartes' lessons in casuistry, let us hope). To illustrate the difficulty of distinguishing the true from the false Descartes undertook to produce twelve irrefutable arguments showing the falsity of any incontestable truth and, conversely, to do the like for the truth of any admitted falsehood. How then, the bewildered listeners asked, shall mere human beings distinguish truth from falsehood? Descartes confided that he had (what he considered) an infallible method, drawn from mathematics, for making the required distinction. He hoped and planned, he said, to show how his method could be applied to science and human welfare through the medium of mechanical invention.

De Bérulle was profoundly stirred by the vision of all the kingdoms of the earth with which Descartes had tempted him from the pinnacle of philosophic speculation. In no uncertain terms he told Descartes that it was his duty to God to share his discoveries with the world, and threatened him with hell-fire—or at least the loss of his chance of heaven—if he did not. Being a devout practising Catholic Descartes could not possibly resist such an appeal. He decided to publish. This was his second conversion, at the age of thirty two. He straightway retired to Holland, where the colder climate suited him, to bring his decision to realization.

For the next twenty years he wandered about all over Holland, never settling for long in any one place, a silent recluse in obscure villages, country hotels and out-of-the-way corners of great cities, methodically carrying on a voluminous scientific and philosophical correspondence with the leading intellects of Europe, using as intermediary the trusted friend of his school days at La Flèche, Father Mersenne, who alone knew the secret at any time of Descartes' address. The parlor of the cloister of the Minims, not far from Paris, became the exchange (through Mersenne) for questions, mathematical problems, scientific and philosophical theories, objections, and replies.

During his long vagabondage in Holland Descartes occupied himself with a number of studies in addition to his philosophy and mathematics. Optics, chemistry, physics, anatomy, embryology, medicine, astronomical observations, and meteorology, including a study of the rainbow, all claimed their share of his restless activity. Any man today spreading his effort over so diversified a miscellany would write himself down a fiddling dilettante. But it was not so in Descartes' age; a man of talent might still hope to find something of interest in almost any science that took his fancy. Everything that came Descartes' way was grist to his mill. A brief visit to England acquainted him with the mystifying behavior of the magnetic needle; forthwith magnetism had to be included in his comprehensive philosophy. The speculations of theology also called for his attention. All through his theorizing his mind was shadowed by the incubus of his early training. He would not have shaken it off if he could.

All of what Descartes had gathered and excogitated was to be incorporated into an imposing treatise, *Le Monde*. In 1634, Descartes being then thirty eight, the treatise was undergoing its final revision. It was to have been a New Year's gift to Father Mersenne. All learned Paris was agog to see the masterpiece. Mersenne had been granted many previews of selected portions but as yet he had not seen the completed, dovetailed work. Without irreverence *Le Monde* may be described as what the author of the Book of Genesis might have written had he known as much science and philosophy as Descartes did. Descartes intended his account of God's creation of the universe to supply the lack which some readers had felt in the Bible story of the six days' creation, namely, an element of rationality. From the distance of three hundred years there seems but little to choose between Genesis and Descartes, and it is somewhat difficult for us to realize that such a book as *Le Monde* could ever have caused a bishop or a pope to fly into a cold, murderous rage. As a matter of fact none did; Descartes saw to that.

Descartes was aware of the judgments of ecclesiastical justice. He also knew of the astronomical researches of Galileo and of that fearless man's championship of the Copernican system. In fact he was impatiently waiting to see Galileo's latest book before putting the final touches to his own. Instead of receiving the copy a friend had promised to send him, he got the stunning news that Galileo, in the seventieth year of his age, and in spite of the sincere friendship that the powerful Duke of Tuscany had for him, had been given up to the Inquisition and had been forced (June 22, 1633) on his knees to abjure as a heresy the Copernican doctrine that the Earth moves round the Sun. What would have happened to Galileo had he refused to forswear his scientific knowledge Descartes could only conjecture, but the names of Bruno, Vanini, and Campanella recurred to his mind.

Descartes was crushed. In his own book he had expounded the Copernican system as a matter of course. On his own account he had been far more daring than Copernicus or Galileo had ever had occasion to be, because he was interested in the theology of science whereas they were not. He had proved to his own satisfaction the *necessity* of the cosmos as it exists, and he thought he had shown that if God had created any number of distinct universes they must all, under the action of "natural law," sooner or later have fallen into line with *necessity* and have evolved into the universe as it actually is. Descartes, in short, professed with his scientific knowledge to know a great deal more about the nature and ways of God than either the author of Genesis or the theologians had ever dreamed of. If Galileo had been forced to get down on his knees for his mild and conservative heresy, what could Descartes expect?

To say that fear alone stopped Descartes from publishing *Le Monde* is to miss the more important part of the truth. He was not only afraid —as any sane man might well have been; he was deeply hurt. He was as convinced of the truth of the Copernican system as he was of his own existence. But he was also convinced of the infallibility of the Pope. Here now was the Pope making a silly ass of himself by contradicting Copernicus. This was his first thought. His casuistical schooling came to his aid. In some way, through the mystical incomprehensibilities of some superhuman synthesis, the Pope and Copernicus would yet both be proved right. From this as yet unrevealed Pisgah height Descartes confidently hoped and expected some day to look down in philosophic serenity on the apparent contradiction and see it vanish in a glory of reconciliation. It was simply impossible for him to give up either the Pope or Copernicus. So he suppressed his book and kept both his belief in the infallibility of the Pope and the truth of the Copernican system. As a sop to his subconscious self-respect he decided that *Le Monde* should be published after his death.

By that time perhaps the Pope too would be dead and the contradiction would have resolved itself.

Descartes' determination not to publish extended to all his work. But in 1637, when Descartes was forty one, his friends overcame his reluctance and induced him to permit the printing of his masterpiece, of which the title is translated as *A Discourse on the Method of rightly conducting the Reason and seeking Truth in the Sciences. Further, the Dioptric, Meteors, and Geometry, essays in this Method*. This work is known shortly as the *Method*. It was published on June 8, 1637. This is the day, then, on which analytic geometry was given to the world. Before describing wherein that geometry is superior to the synthetic geometry of the Greeks we shall finish with the life of its author.

After having given the reasons for Descartes' delay in publication it is only fair to tell now the other and brighter side of the story.

The Church which Descartes had feared but which had never actually opposed him now came most generously to his aid. Cardinal Richelieu gave Descartes the privilege of publishing either in France or abroad anything he cared to write. (In passing we may ask, however, by what right, divine, or other, did Cardinal Richelieu, or any other human being, dictate to a philosopher and man of science what he should or should not publish?) But in Utrecht, Holland, the Protestant theologians savagely condemned Descartes' work as atheistic and dangerous to that mystic entity known as "The State." The liberal Prince of Orange threw his great weight on Descartes' side and backed him to the limit.

Since the autumn of 1641 Descartes had been living at a quiet little village near the Hague in Holland, where the exiled Princess Elisabeth, now a young woman with a penchant for learning, rusticated with her mother. The Princess does indeed seem to have been a prodigy of learning. After mastering six languages and digesting much literature she had turned to mathematics and science, hoping to find more nourishing fare. One theory to account for this remarkable young woman's unusual appetite ascribes her hunger for knowledge to a disappointment in love. Neither mathematics nor science satisfied her. Then Descartes' book came her way and she knew that she had found what she needed to fill her aching void—Descartes. An interview was arranged with the somewhat reluctant philosopher.

It is very difficult to understand exactly what happened thereafter.

Descartes was a gentleman with all the awe and reverence of a gentleman of those gallant, royalty-ridden times for even the least potent prince or princess. His letters are models of courtly discretion, but somehow they do not always ring quite true. One spiteful little remark, quoted in a moment, probably tells more of what he really thought of the Princess Elisabeth's intellectual capacity than do all the reams of subtle flattery he wrote to or about his eager pupil with one eye on his style and the other on publication after his death.

Elisabeth insisted upon Descartes giving her lessons. Officially he declared that "of all my disciples she alone has understood my works completely." There is no doubt that he was genuinely fond of her in a fatherly, cat-looking-at-a-king's-female-relative sort of way, but to believe that he meant what he said as a scientific statement of fact is to stretch credulity to the limit, unless, of course, he meant it as a wry comment on his own philosophy. Elisabeth may have understood too much, for it seems to be a fact that only a philosopher thoroughly understands his own philosophy, although any fool can think he does. Anyhow, he did not propose to her nor, so far as is known, did she propose to him.

Among other parts of his philosophy which he expounded to her was the method of analytic geometry. Now there is a certain problem in elementary geometry which can be quite simply solved by pure geometry, and which looks easy enough, but which is a perfect devil for analytic geometry to handle in the strict Cartesian form. This is to construct a circle which shall touch (be tangent to) any three circles given at random whose centers do not all lie on one straight line. There are eight solutions possible. The problem is a fine specimen of the sort that are *not* adapted to the crude brute force of elementary Cartesian geometry. *Elisabeth solved it by Descartes' methods.* It was rather cruel of him to let her do it. His comment on seeing her solution gives the whole show away to any mathematician. She was quite proud of her exploit, poor girl. Descartes said he would not undertake to carry out her solution and actually construct the required tangent circle in a month. If this does not convey his estimate of her mathematical aptitude it is impossible to put the matter plainer. It was an unkind thing to say, especially as she missed the point and he knew that she would.

When Elisabeth left Holland she corresponded with Descartes to almost the day of his death. His letters contain much that is fine and sincere, but we could wish that he had not been so dazzled by the aura of royalty.

In 1646 Descartes was living in happy seclusion at Egmond, Holland, meditating, gardening in a tiny plot, and carrying on a correspondence of incredible magnitude with the intellectuals of Europe. His greatest mathematical work lay behind him, but he still continued to think about mathematics, always with penetration and originality. One problem to which he gave some attention was Zeno's of Achilles and the tortoise. His solution of the paradox would not be universally accepted today but it was ingenious for its era. He was now fifty and world-famous, far more famous in fact than he would ever have cared to be. The repose and tranquillity he had longed for all his life still eluded him. He continued to do great work, but he was not to be left in peace to do all that was in him. Queen Christine of Sweden had heard of him.

This somewhat masculine young woman was then nineteen, already a capable ruler, reputedly a good classicist (of this, more later), a wiry athlete with the physical endurance of Satan himself, a ruthless huntress, an expert horsewoman who thought nothing of ten hours in the saddle without once getting off, and finally a tough morsel of femininity who was as hardened to cold as a Swedish lumberjack. With all this she combined a certain thick obtuseness toward the frailties of less thick-skinned beings. Her own meals were sparing; so were those of her courtiers. Like a hibernating frog she could sit for hours in an unheated library in the middle of a Swedish winter; her hangers-on begged her through their chattering teeth to throw all the windows wide open and let the merry snow in. Her cabinet, she noted without a qualm, always agreed with her. She knew everything there was to be known; her ministers and tutors told her so. As she got along on only five hours' sleep she kept her toadies hopping through the hoop nineteen hours a day. The very hour this holy terror saw Descartes' philosophy she decided she must annex the poor sleepy devil as her private instructor. All her studies so far had left her empty and hungering for more. Like the erudite Elisabeth she knew that only copious douches of philosophy from the philosopher himself could assuage her raging thirst for knowledge and wisdom.

But for that unfortunate streak of snobbery in his make-up Descartes might have resisted Queen Christine's blandishments till he was ninety and sans teeth, sans hair, sans philosophy, sans everything.

Descartes held out till she sent Admiral Fleming in the spring of 1649 with a ship to fetch him. The whole outfit was generously placed at the reluctant philosopher's disposal. Descartes temporized till October. Then, with a last regretful look round his little garden, he locked up and left Egmond forever.

His reception in Stockholm was boisterous, not to say royal. Descartes did not live at the Palace; that much was spared him. Importunately kind friends, however, the Chanutes, shattered his last remaining hope of reserving a little privacy. They insisted that he live with them. Chanute was a fellow countryman, in fact the French ambassador. All might have gone well, for the Chanutes were really most considerate, had not the obtuse Christine got it into her immovable head that five o'clock in the morning was the proper hour for a busy, hardboiled young woman like herself to study philosophy. Descartes would gladly have swapped all the headstrong queens in Christendom for a month's dreaming abed at La Flèche with the enlightened Charlet unobtrusively near to see that he did not get up too soon. However, he dutifully crawled out of bed at some ungodly hour in the dark, climbed into the carriage sent to collect him, and made his way across the bleakest, windiest square in Stockholm to the palace where Christine sat in the icy library impatiently waiting for her lesson in philosophy to begin promptly at five A.M.

The oldest inhabitants said Stockholm had never in their memory suffered so severe a winter. Christine appears to have lacked a normal human skin as well as nerves. She noticed nothing, but kept Descartes unflinchingly to his ghastly rendezvous. He tried to make up his rest by lying down in the afternoons. She soon broke him of that. A Royal Swedish Academy of Sciences was gestating in her prolific activity; Descartes was hauled out of bed to deliver her.

It soon became plain to the courtiers that Descartes and their Queen were discussing much more than philosophy in these interminable conferences. The weary philosopher presently realized that he had stepped with both feet into a populous and busy hornets' nest. They stung him whenever and wherever they could. Either the Queen was too thick to notice what was happening to her new favorite or she was clever enough to sting her courtiers through her philosopher. In any event, to silence the malicious whisperings of "foreign influence," she resolved to make a Swede of Descartes. An estate was set aside for him by royal decree. Every desperate move he made to get out of the mess only bogged him deeper. By the first of January, 1650, he was up to his neck with only a miracle of rudeness as his one dim hope of ever freeing himself. But with his inbred respect for royalty he could not bring himself to speak the magic words which would send him flying back to Holland, although he said plenty, with courtly politeness, in a letter to his devoted Elisabeth. He had chanced to interrupt one of the lessons in Greek. To his amazement Descartes learned that the vaunted classicist Christine was struggling over grammatical puerilities which, he says, he had mastered by himself when he was a little boy. His opinion of her mentality thereafter appears to have been respectful but low. It was not raised by her insistence that he produce a ballet for the delectation of her guests at a court function when he resolutely refused to make a mountebank of himself by attempting at his age to master the stately capers of the Swedish lancers.

Presently Chanute fell desperately ill of inflammation of the lungs. Descartes nursed him. Chanute recovered; Descartes fell ill of the same disease. The Queen, alarmed, sent doctors. Descartes ordered them out of the room. He grew steadily worse. Unable in his debility to distinguish friend from pest he consented at last to being bled by the most persistent of the doctors, a personal friend, who all the time had been hovering about awaiting his chance. This almost finished him, but not quite.

His good friends the Chanutes, seeing that he was a very sick man, suggested that he might enjoy the last sacrament. He had expressed a desire to see his spiritual counsellor. Commending his soul to the mercy of God, Descartes faced his death calmly, saying the willing sacrifice of his life which he was making might possibly atone for his sins. La Flèche gripped him to the last. The counsellor asked him to signify whether he wished the final benediction. Descartes opened his eyes and closed them. He was given the benediction. Thus he died on February 11, 1650, aged 54, a sacrifice to the overweening vanity of a headstrong girl.

Christine lamented. Seventeen years later when she had long since given up her crown and her faith, the bones of Descartes were returned to France (all except those of the right hand, which were retained by the French Treasurer-General as a souvenir for his skill in engineering the transaction) and were re-entombed in Paris in what is now the Pantheon. There was to have been a public oration, but this was hastily forbidden by order of the crown, as the doctrines of

Descartes were deemed to be still too hot for handling before the people. Commenting on the return of Descartes' remains to his native France, Jacobi remarks that "It is often more convenient to possess the ashes of great men than to possess the men themselves during their lifetime."

Shortly after his death Descartes' books were listed in the *Index* of that Church which, accepting Cardinal Richelieu's enlightened suggestion during the author's lifetime, had permitted their publication. "Consistency, thou art a jewel!" But the faithful were not troubled by consistency, "the bugbear of little minds"—and the ratbane of inconsistent bigots.

We are not concerned here with the monumental additions which Descartes made to philosophy. Nor can his brilliant part in the dawn of the experimental method detain us. These things fall far outside the field of pure mathematics in which, perhaps, his greatest work lies. It is given to but few men to renovate a whole department of human thought. Descartes was one of those few. Not to obscure the shining simplicity of his greatest contribution, we shall briefly describe it alone and leave aside the many beautiful things he did in algebra and particularly in algebraic notation and the theory of equations. This one thing is of the highest order of excellence, marked by the sensuous simplicity of the half dozen or so greatest contributions of all time to mathematics. Descartes remade geometry and made modern geometry possible.

The basic idea, like all the really great things in mathematics, is simple to the point of obviousness. Lay down any two intersecting lines on a plane. Without loss of generality we may assume that the lines are at right angles to one another. Imagine now a city laid out on the American plan, with avenues running north and south, streets east and west. The whole plan will be laid out with respect to *one* avenue and *one* street, called the *axes*, which intersect in what is called the *origin*, from which street-avenue numbers are read consecutively. Thus it is clear without a diagram where 1002 West 126 Street is, if we note that the *ten avenues* summarized in the number 1002 are stepped off to the *west*, that is, on the map, to the *left* of the origin. This is so familiar that we visualize the position of any particular address instantly. The avenue-number and street-number, with the necessary supplements of smaller numbers (as in the "2" in "1002" above) enable us to fix definitely and uniquely the position of any *point* whatever with respect to the *axes*, by giving the *pair* of numbers which measure its *east or west* and its *north or south* from the *axes*; this pair of numbers is called the *coordinates* of the point (with respect to the axes).

Now suppose a point to wander over the map. The *coordinates* (x,y) of *all* the points on the curve over which it wanders will be connected by an *equation*, (this must be taken for granted by the reader who has never plotted a graph to fit data), which is called *the equation of the*

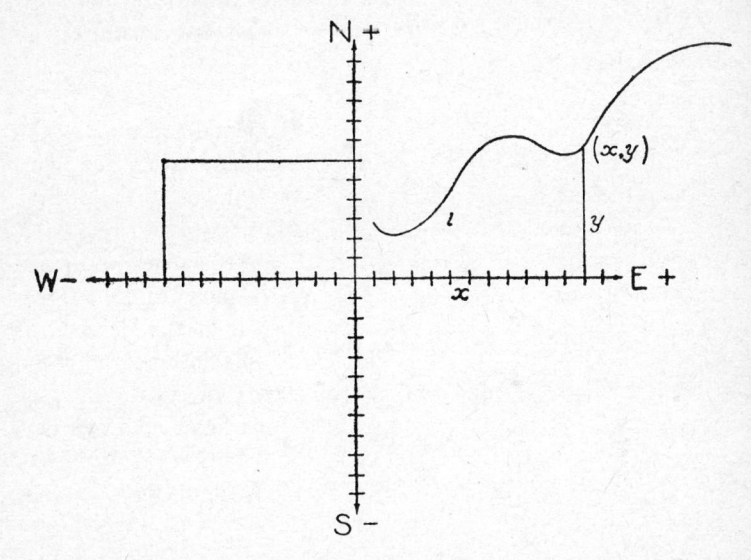

curve. Suppose now for simplicity that our curve is a circle. We have its equation. What can be done with it? Instead of this particular equation, we can write down the most general one of the same kind (for example, here, of the *second degree*, with no cross-product term, and with the coefficients of the highest powers of the coordinates equal), and then proceed to manipulate this equation algebraically. Finally we put back the results of all our algebraic manipulations into their equivalents in terms of coordinates of points on the diagram which, all this time, we have been deliberately forgetting. Algebra is easier to see through than a cobweb of lines in the Greek manner of elementary geometry. What we have done has been to *use our algebra for the discovery and investigation of geometrical theorems concerning circles*.

For straight lines and circles this may not seem very exciting; we knew how to do it all before in another, a Greek, way. Now comes the real power of the method. *We start with equations of any desired or suggested degree of complexity and interpret their algebraic and analytic properties geometrically.* Thus we have not only dropped geometry as our pilot; we have tied a sackful of bricks to his neck before pitching him overboard. *Henceforth algebra and analysis are to be our pilots to the unchartered seas of "space" and its "geometry."* All that we have done can be extended, at one stride, to space of any number of dimensions; for the plane we need *two* coordinates, for ordinary "solid" space *three*, for the geometry of mechanics and relativity, *four* coordinates,

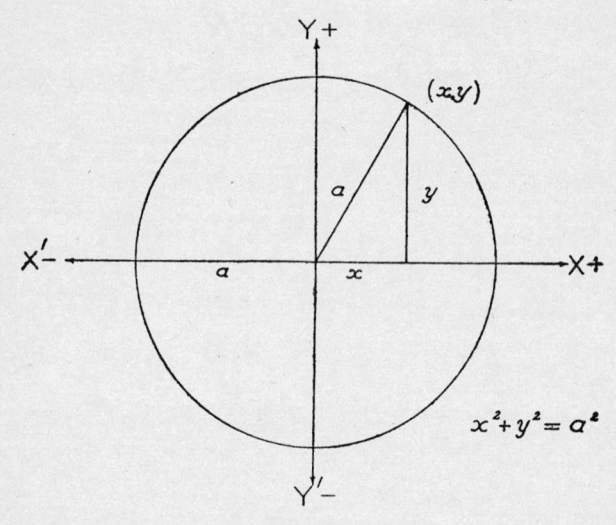

$$x^2 + y^2 = a^2$$

and finally, for "space" as mathematicians like it, either n coordinates, or as many coordinates as there are of *all* the numbers 1, 2, 3, . . . , or as many as there are of *all* the points on a line. This is beating Achilles and the tortoise in their own race.

Descartes did not revise geometry; he created it.

It seems fitting that an eminent living mathematical fellow-countryman of Descartes should have the last word, so we shall quote Jacques Hadamard. He remarks first that the mere invention of coordinates was not Descartes' greatest merit, because that had already been done "by the ancients"—a statement which is exact only if we read the unexpressed intention into the unaccomplished deed. Hell is paved with the half-baked ideas of "the ancients" which they could never quite cook through with their own steam.

"It is quite another thing to recognize [as in the use of coordinates] a general method and to follow to the end the idea which it represents. It is exactly this merit, whose importance every real mathematician knows, that was preëminently Descartes' in geometry; it was thus that he was led to what . . . is his truly great discovery in the matter; namely, the application of the method of coordinates not only to translate into equations curves already defined geometrically, but, looking at the question from an exactly opposite point of view, to the *a priori* definition of more and more complicated curves and, hence, more and more general. . . .

"Directly, with Descartes himself, later, indirectly, in the return which the following century made in the opposite direction, it is the entire conception of the object of mathematical science that was revolutionized. Descartes indeed understood thoroughly the significance of what he had done, and he was right when he boasted that he had so far surpassed all geometry before him as Cicero's rhetoric surpasses the ABC."

CHAPTER FOUR

The Prince of Amateurs

FERMAT

I have found a very great number of exceedingly beautiful theorems.
—P. FERMAT

NOT ALL OF OUR DUCKS can be swans; so after having exhibited Descartes as one of the leading mathematicians of all time, we shall have to justify the assertion, frequently made and seldom contradicted, that the greatest mathematician of the seventeenth century was Descartes' contemporary Fermat (1601?–1665). This of course leaves Newton (1642–1727) out of consideration. But it can be argued that Fermat was *at least* Newton's equal *as a pure mathematician*, and anyhow nearly a third of Newton's life fell into the eighteenth century, whereas the whole of Fermat's was lived out in the seventeenth.

Newton appears to have regarded his mathematics principally as an instrument for scientific exploration and put his main effort on the latter. Fermat on the other hand was more strongly attracted to pure mathematics although he also did notable work in the applications of mathematics to science, particularly optics.

Mathematics had just entered its modern phase with Descartes' publication of analytic geometry in 1637, and was still for many years to be of such modest extent that a gifted man could reasonably hope to do good work in both the pure and applied divisions.

As a pure mathematician Newton reached his climax in the invention of the calculus, an invention also made independently by Leibniz. More will be said on this later; for the present it may be remarked that Fermat conceived and applied the leading idea of the differential calculus thirteen years before Newton was born and seventeen before Leibniz was born, although he did not, like Leibniz, reduce his method to a set of rules of thumb that even a dolt can apply to easy problems.

As for Descartes and Fermat, each of them, entirely independently of the other, invented analytic geometry. They corresponded on the subject but this does not affect the preceding assertion. The major part of Descartes' effort went to miscellaneous scientific investigations, the elaboration of his philosophy, and his preposterous "vortex theory" of the solar system—for long a serious rival, even in England, to the beautifully simple, unmetaphysical Newtonian theory of universal gravitation. Fermat seems never to have been tempted, as both Descartes and Pascal were, by the insidious seductiveness of philosophizing about God, man, and the universe as a whole; so, after having disposed of his part in the calculus and analytic geometry, and having lived a serene life of hard work all the while to earn his living, he still was free to devote his remaining energy to his favorite amusement—pure mathematics, and to accomplish his greatest work, the foundation of the theory of numbers, on which his undisputed and undivided claim to immortality rests.

It will be seen presently that Fermat shared with Pascal the creation of the mathematical theory of probability. If all these first-rank achievements are not enough to put him at the head of his contemporaries in pure mathematics we may ask who did more? Fermat was a born originator. He was also, in the strictest sense of the word, so far as his science and mathematics were concerned, an amateur. Without doubt he is one of the foremost amateurs in the history of science, if not the very first.

Fermat's life was quiet, laborious, and uneventful, but he got a tremendous lot out of it. The essential facts of his peaceful career are quickly told. The son of the leather-merchant Dominique Fermat, second consul of Beaumont, and Claire de Long, daughter of a family of parliamentary jurists, the mathematician Pierre Fermat was born at Beaumont-de-Lomagne, France, in August, 1601 (the exact date is unknown; the baptismal day was August 20th). His earliest education was received at home in his native town; his later studies, in preparation for the magistracy, were continued at Toulouse. As Fermat lived temperately and quietly all his life, avoiding profitless disputes, and as he lacked a doting sister like Pascal's Gilberte to record his boyhood prodigies for posterity, singularly little appears to have survived of his career as a student. That it must have been brilliant will be evident from the achievements and accomplishments of his maturity; no man without a solid foundation of exact scholarship could have been the classicist and littérateur that Fermat became. His marvelous work in the theory of numbers and in mathematics generally cannot be traced to his schooling; for the fields in which he did his greatest work, not having been opened up while he was a student, could scarcely have been suggested by his studies.

The only events worth noting in his material career are his installation at Toulouse, at the age of thirty (May 14, 1631), as commissioner of requests; his marriage on June 1st of the same year to Louise de Long, his mother's cousin, who presented him with three sons, one of whom, Clément-Samuel, became his father's scientific executor, and two daughters, both of whom took the veil; his promotion in 1648 to a King's councillorship in the local parliament of Toulouse, a position which he filled with dignity, integrity, and great ability for seventeen years—his entire working life of thirty four years was spent in the exacting service of the state; and finally, his death at Castres on January 12, 1665, in his sixty fifth year, two days after he had finished conducting a case in the town of his death. "Story?" he might have said; "Bless you, sir! I have none." And yet this tranquilly living, honest, even-tempered, scrupulously just man has one of the finest stories in the history of mathematics.

His story is his work—his recreation, rather—done for the sheer love of it, and the best of it is so simple (to state, but not to carry through or imitate) that any schoolboy of normal intelligence can understand its nature and appreciate its beauty. The work of this prince of mathematical amateurs has had an irresistible appeal to amateurs of mathematics in all civilized countries during the past three centuries. This, the theory of numbers as it is called, is probably the one field of mathematics in which a talented amateur today may hope to turn up something of interest. We shall glance at his other contributions first after a passing mention of his "singular erudition" in what many call the humanities. His knowledge of the chief European languages and literatures of Continental Europe was wide and accurate, and Greek and Latin philology are indebted to him for several important corrections. In the composition of Latin, French, and Spanish verses, one of the gentlemanly accomplishments of his day, he showed great skill and a fine taste. We shall understand his even, scholarly life if we picture him as an affable man, not touchy or huffy under criticism (as Newton in his later years was), without pride, but having a certain vanity which Descartes, his opposite in all respects, characterized by saying, "Mr. de Fermat is a Gascon; I am not." The allusion to the Gascons may possibly refer to an amiable sort of braggadocio which some French writers (for example Rostand in *Cyrano de Bergerac*, Act II, Scene VII) ascribe to their men of Gascony. There may be some of this in Fermat's letters, but it is always rather naïve and inoffensive, and nothing to what he might have justly thought of his work even if his head had been as big as a balloon. And as for Descartes it must be remembered that he was not exactly an impartial judge. We shall note in a moment how his own soldierly obstinacy caused him to come off a bad second-best in his protracted row with the "Gascon" over the extremely important matter of tangents.

Considering the exacting nature of Fermat's official duties and the large amount of first-rate mathematics he did, some have been puzzled as to how he found time for it all. A French critic suggests a probable solution: Fermat's work as a King's councillor was an aid rather than a detriment to his intellectual activities. Unlike other public servants —in the army for instance—parliamentary councillors were expected to hold themselves aloof from their fellow townsmen and to abstain from unnecessary social activities lest they be corrupted by bribery or otherwise in the discharge of their office. Thus Fermat found plenty of leisure.

We now briefly state Fermat's part in the evolution of the calculus. As was remarked in the chapter on Archimedes, a geometrical equivalent of the fundamental problem of the *differential* calculus is to draw the straight line tangent to a given, unlooped, continuous arc of a curve at any given point. A sufficiently close description of what "continuous" means here is "smooth, without breaks or sudden jumps"; to give an exact, mathematical definition would require pages of definitions and subtle distinctions which, it is safe to say, would have puzzled and astonished the inventors of the calculus, including Newton and Leibniz. And it is also a fair guess that if all these subtleties which modern students demand had presented themselves to the originators, the calculus would never have got itself invented.

The creators of the calculus, including Fermat, relied on geometric and physical (mostly kinematical and dynamical) intuition to get them ahead: they *looked at* what passed in their imaginations for the *graph* of a "continuous curve," pictured the process of drawing a straight line tangent to the curve at any point P on the curve by taking another point Q, also on the curve, drawing the straight line PQ joining P and Q, and then, in imagination, letting the point Q slip along the arc of the curve from Q to P, till Q coincided with P, when the *chord* PQ, in the *limiting position* just described, became the *tangent* PP to the curve at the point P—the very thing they were looking for.

The next step was to translate all this into algebraical or analytical language. Knowing the coordinates x, y of the point P on the graph, and those, say $x + a$, $y + b$, of Q, before Q started to slip along to coincidence with P, they inspected the graph and saw that the *slope*

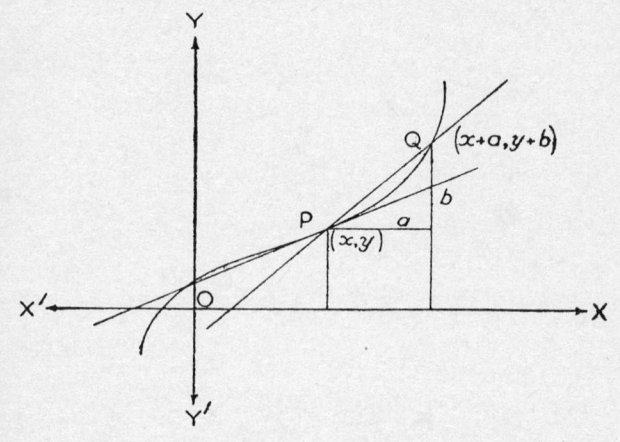

of the *chord* PQ was equal to b/a—obviously a measure of the "steepness" of the chord with relation to the x-axis (the line along which x-distances are measured); this "steepness" is precisely what is meant by slope. From this it was evident that the *required slope of the tangent at P* (after Q had slipped into coincidence with P) would be the *limiting value* of b/a as both b and a approached the value *zero* simultaneously; for $x + a$, $y + b$, the coordinates of Q, ultimately become x, y, the coordinates of P. This limiting value is the required slope. Having the slope and the point P they could now draw the tangent.

This is not exactly Fermat's process for drawing tangents but his own process was, broadly, equivalent to what has been described.

Why should all this be worth the serious attention of any rational or practical man? It is a long story, only a hint of which need be given here; more will be said when we discuss Newton. One of the fundamental ideas in dynamics is that of the *velocity* (speed) of a moving particle. If we graph the number of units of distance passed over by the particle in a unit of time against the number of units of time, we get a line, straight or curved, which pictures at a glance the *motion* of the particle, and the *steepness* of this line at any given point of it will obviously give us the *velocity* of the particle at the instant corresponding to the point; the faster the particle is moving, the steeper the *slope* of the *tangent line*. This slope does in fact measure the velocity of the particle at any point of its path. The problem in *motion*, when translated into *geometry*, is exactly that of finding the slope of the tangent line at a given point of a curve. There are similar

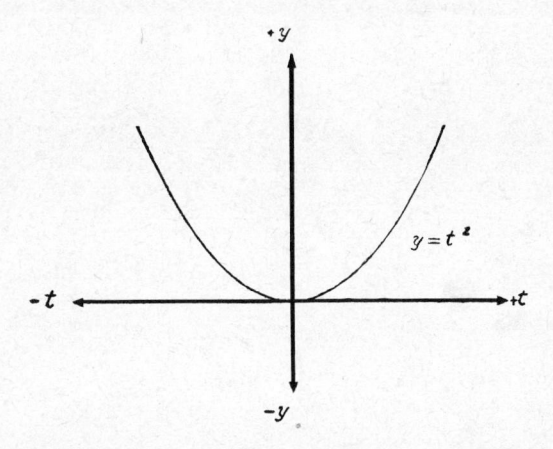

$$y = t^2$$

problems in connection with *tangent planes* to surfaces (which also have important interpretations in mechanics and mathematical physics), and all are attacked by the differential calculus—whose fundamental problem we have attempted to describe as it presented itself to Fermat and his successors.

Another use of this calculus can be indicated from what has already been said. Suppose some quantity y is a "function" of another, t, written $y = f(t)$, which means that when any definite number, say 10, is substituted for t, so that we get $f(10)$—"function f of 10"— we can calculate, from the *algebraical expression* of f, supposed given, the *corresponding* value of y, here $y = f(10)$. To be explicit, suppose $f(t)$ is that particular "function" of t which is denoted in algebra by t^2, or $t \times t$. Then, when $t = 10$, we get $y = f(10)$, and hence *here* $y = 10^2$, $= 100$, for *this* value of t; when $t = \frac{1}{2}$, $y = \frac{1}{4}$, and so on, for *any* value of t.

All this is familiar to anyone whose grammar-school education ended not more than thirty or forty years ago, but some may have forgotten what they did in arithmetic as children, just as others could not decline the Latin *mensa* to save their souls. But even the most forgetful will see that we could plot the graph of $y = f(t)$ for any particular form of f (when $f(t)$ is t^2 the graph is a parabola like an inverted arch). Imagine the graph drawn. If it has on it *maxima* (highest) or *minima* (lowest) points—points higher or lower than those *in their immediate neighborhoods*—we observe that the tangent at each of these *maxima* or *minima* is *parallel* to the t-axis. That is, the *slope* of the tangent at such an *extremum* (maximum *or* minimum) of the

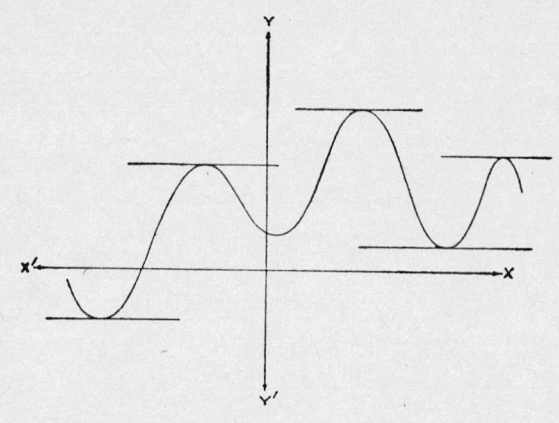

$f(t)$ we are plotting is *zero*. Thus if we were seeking the *extrema* of a given function $f(t)$ we should again have to solve our slope-problem for the particular curve $y = f(t)$ and, having found the slope for the *general* point t, y, equate to zero the algebraical expression of this slope in order to find the values of t corresponding to the extrema. This is substantially what Fermat did in his method of maxima and minima invented in 1628–29, but not made semipublic till ten years later when Fermat sent an account of it through Mersenne to Descartes.

The scientific applications of this simple device—duly elaborated, of course, to take account of far more complicated problems than that just described—are numerous and far reaching. In mechanics, for instance, as Lagrange discovered, there is a certain "function" of the positions (coordinates) and velocities of the bodies concerned in a problem which, when made an extremum, furnishes us with the "equations of motion" of the system considered, and these in turn enable us to determine the motion—to describe it completely—at any given instant. In physics there are many similar functions, each of which sums up most of an extensive branch of mathematical physics in the simple requirement that the function in question must be an extremum;* Hilbert in 1916 found one for general relativity. So Fermat was not fooling away his time when he amused himself in the leisure left from a laborious legal job by attacking the problem of maxima and minima. He himself made one beautiful and astonishing application of his principles to optics. In passing it may be noted that this particular discovery has proved to be the germ of the newer quantum theory—in its mathematical aspect, that of "wave mechanics"—elaborated since 1926. Fermat discovered what is usually called "the principle of least time." It would be more accurate to say "extreme" (least *or* greatest) instead of "least."*

According to this principle, if a ray of light passes from a point A to another point B, being reflected and refracted ("refracted," that is, bent, as in passing from air to water, or through a jelly of variable density) in any manner during the passage, the path which it must take can be calculated—all its twistings and turnings due to refraction, and all its dodgings back and forth due to reflections—from the *single* requirement that the *time* taken to pass from A to B shall be an extremum (but see the preceding footnote).

From this principle Fermat deduced the familiar laws of reflection and refraction: the angle of incidence (in reflection) is equal to the angle of reflection; the sine of the angle of incidence (in refraction) is a *constant* number times the sine of the angle of refraction in passing from one medium to another.

The matter of analytic geometry has already been mentioned; Fermat was the first to apply it to space of three dimensions. Descartes contented himself with two dimensions. The extension, familiar to all students today, would not be self-evident to even a gifted man from Descartes' developments. It may be said that there is usually greater difficulty in finding a significant extension of a particular

---

*This statement is sufficiently accurate for the present account. Actually, the values of the variables (coordinates and velocities) which make the function in question *stationary* (neither increasing nor decreasing, roughly) are those required. An *extremum* is stationary; but a *stationary* is not necessarily an extremum.

kind of geometry from space of two dimensions to three than there is in passing from three to four or five . . . , or n. Fermat corrected Descartes in an essential point (that of the classification of curves by their degrees). It seems but natural that the somewhat touchy Descartes should have rowed with the imperturbable "Gascon" Fermat. The soldier was frequently irritable and acid in his controversy over Fermat's method of tangents; the equable jurist was always unaffectedly courteous. As usually happens the man who kept his temper got the better of the argument. But Fermat deserved to win, not because he was a more skilful debater, but because he was right.

In passing, we should suppose that Newton would have heard of Fermat's use of the calculus and would have acknowledged the information. Until 1934 no evidence to this effect had been published, but in that year Professor L. T. More recorded in his biography of Newton a hitherto unnoticed letter in which Newton says explicitly that he got the hint of the method of the differential calculus from Fermat's method of drawing tangents.

We now turn to Fermat's greatest work, that which is intelligible to all, mathematicians and amateurs alike. This is the so-called "theory of numbers," or "the higher arithmetic," or finally, to use the unpedantic name which was good enough for Gauss, *arithmetic*.

The Greeks separated the miscellany which we lump together under the name "arithmetic" in elementary textbooks into two distinct compartments, *logistica*, and *arithmetica*, the first of which concerned the practical applications of reckoning to trade and daily life in general, and the second, arithmetic in the sense of Fermat and Gauss, who sought to discover the properties of numbers as such.

Arithmetic in its ultimate and probably most difficult problems investigates the mutual relationships of those common whole numbers 1, 2, 3, 4, 5, . . . which we utter almost as soon as we learn to talk. In striving to elucidate these relationships, mathematicians have been driven to the invention of subtle and abstruse theories in algebra and analysis, whose forests of technicalities obscure the initial problems— those concerning 1, 2, 3, . . . but whose real justification will be the solution of those problems. In the meantime the by-products of these apparently useless investigations amply repay those who undertake them by suggesting numerous powerful methods applicable to other fields of mathematics having direct contact with the physical universe.

To give but one instance, the latest phase of algebra, that which is cultivated today by professional algebraists and which is throwing an entirely new light on the theory of algebraic equations, traces its origin directly to attempts to settle Fermat's simple Last Theorem (which will be stated when the way has been prepared for it).

We begin with a famous statement Fermat made about prime numbers. A positive prime number, or briefly a *prime*, is any number greater than 1 which has as its divisors (without remainder) only 1 and the number itself; for example 2, 3, 5, 7, 13, 17 are primes, and so are 257, 65537. But 4294967297 is not a prime, because it has 641 as a divisor, nor is the number 18446744073709551617, because it is exactly divisible by 274177; both 641 and 274177 are primes. When we say in arithmetic that one number has as divisor another number, or is divisible by another, we mean *exactly divisible, without remainder*. Thus 14 is divisible by 7, 15 is not. The two large numbers were displayed above with malice aforethought for a reason that will be apparent in a moment. To recall another definition, the *n*th *power* of a given number, say N, is the result of multiplying together n N's, and is written N^n; thus $5^2 = 5 \times 5 = 25$; $8^4 = 8 \times 8 \times 8 \times 8 = 4096$. For uniformity N itself may be written as N^1. Again, such a pagoda as 2^{3^5} means that we are first to calculate 3^5 ($= 243$), and then "raise" 2 to this power, 2^{243}; the resulting number has seventy four digits.

The next point is of great importance in the life of Fermat, also in the history of mathematics. Consider the numbers 3, 5, 17, 257, 65537. They all belong to one "sequence" of a specific kind, because they are all generated (from 1 and 2) by the same simple process, which will be seen from

$3 = 2 + 1, 5 = 2^2 + 1, 17 = 2^4 + 1, 257 = 2^8 + 1, 65537 = 2^{16} + 1;$

and if we care to verify the calculation we easily see that the two large numbers displayed above are $2^{32} + 1$ and $2^{64} + 1$, also numbers of the sequence. We thus have seven numbers belonging to this sequence and *the first five of these numbers are primes, but the last two are not primes.*

Observing how the sequence is composed, we note the "exponents" (the upper numbers indicating what powers of 2 are taken), namely 1, 2, 4, 8, 16, 32, 64, and we observe that these are 1 (which can be written 2^0, as in algebra, if we like, for uniformity), $2^1, 2^2, 2^3$,

2^4, 2^5, 2^6. Namely, our sequence is $2^{2^n} + 1$, where n ranges over 0, 1, 2, 3, 4, 5, 6. We need not stop with $n = 6$; taking $n = 7$, 8, 9, ..., we may continue the sequence indefinitely, getting more and more enormous numbers.

Suppose we wish now to find out if a particular number of this sequence is a prime. Although there are many shortcuts, and whole classes of trial divisors can be rejected by inspection, and although modern arithmetic limits the kinds of trial divisors that need be tested, our problem is of the same order of laboriousness as would be the dividing of the given number in succession by the primes 2, 3, 5, 7, ... which are less than the square root of the number. If none of these divides the number, the number is prime. Needless to say the labor involved in such a test, even using the known shortcuts, would be prohibitive for even so small a value of n as 100. (The reader may assure himself of this by trying to settle the case $n = 8$.)

Fermat asserted that he was convinced that *all the numbers of the sequence are primes*. The displayed numbers (corresponding to $n = 5$, 6) contradict him, as we have seen. This is the point of historical interest which we wished to make: Fermat *guessed wrong, but he did not claim to have proved his guess*. Some years later he *did* make an obscure statement regarding what he had done, from which some critics infer that he had deceived himself. The importance of this fact will appear as we proceed.

As a psychological curiosity it may be mentioned that Zerah Colburn, the American lightning-calculating boy, when asked whether this sixth number of Fermat's (4294967297) was prime or not, replied after a short mental calculation that it was not, as it had the divisor 641. He was unable to explain the process by which he reached his correct conclusion. Colburn will occur again (in connection with Hamilton).

Before leaving "Fermat's numbers" $2^{2^n} + 1$ we shall glance ahead to the last decade of the eighteenth century where these mysterious numbers were partly responsible for one of the two or three most important events in all the long history of mathematics. For some time a young man in his eighteenth year had been hesitating—according to the tradition—whether to devote his superb talents to mathematics or to philology. He was equally gifted in both. What decided him was a beautiful discovery in connection with a simple problem in elementary geometry familiar to every schoolboy.

A *regular* polygon of n sides has all its n sides equal and all its n angles equal. The ancient Greeks early found out how to construct regular polygons of 3, 4, 5, 6, 8, 10 and 15 sides by the use of straightedge and compass alone, and it is an easy matter, with the same implements, to construct from a regular polygon having a given number of sides another regular polygon having twice that number of sides. The next step then would be to seek straightedge and compass constructions for regular polygons of 7, 9, 11, 13, . . . sides. Many sought, but failed to find, because such constructions are impossible, only they did not know it. After an interval of over 2200 years the young man hesitating between mathematics and philology took the next step—a long one—forward.

As has been indicated it is sufficient to consider only polygons having an *odd* number of sides. The young man proved that a straightedge and compass construction of a regular polygon having an odd number of sides is possible when, and only when, that number is either a *prime* Fermat number (that is a prime of the form $2^{2^n} + 1$), or is made up by multiplying together *different* Fermat primes. Thus the construction is possible for 3, 5, or 15 sides as the Greeks knew, but not for 7, 9, 11 or 13 sides, and is also possible for 17 or 257 or 65537 or—for what the next prime in the Fermat sequence 3, 5, 17, 257, 65537, . . . may be, *if there is one*—nobody yet (1936) knows—and the construction is also possible for 3×17, or $5 \times 257 \times 65537$ sides, and so on. It was this discovery, announced on June 1, 1796, but made on March 30th, which induced the young man to choose mathematics instead of philology as his life work. His name was Gauss.

As a discovery of another kind which Fermat made concerning numbers we state what is known as "Fermat's Theorem" (*not* his "Last Theorem"). If n is any whole number and p any prime, then $n^p - n$ is divisible by p. For example, taking $p = 3$, $n = 5$, we get $5^3 - 5$, or $125 - 5$, which is 120 and is 3×40; for $n = 2$, $p = 11$, we get $2^{11} - 2$, or $2048 - 2$, which is $2046 = 11 \times 186$.

It is difficult if not impossible to state why some theorems in arithmetic are considered "important" while others, equally difficult to prove, are dubbed trivial. One criterion, although not necessarily conclusive, is that the theorem shall be of use in other fields of mathematics. Another is that it shall suggest researches in arithmetic or in mathematics generally, and a third that it shall be in some respect universal. Fermat's theorem just stated satisfies all of these somewhat arbitrary demands: it is of indispensable use in many departments of mathematics, including the theory of groups (see Chapter 15), which in turn is at the root of the theory of algebraic equations; it has suggested many investigations, of which the entire subject of primitive roots may be recalled to mathematical readers as an important instance; and finally it is universal in the sense that it states a property of *all* prime numbers—such general statements are extremely difficult to find and very few are known.

As usual, Fermat stated his theorem about $n^p - n$ without proof. The first proof was given by Leibniz in an undated manuscript, but he appears to have known a proof before 1683. The reader may like to test his own powers on trying to devise a proof. All that is necessary are the following facts, which can be proved but may be assumed for the purpose in hand: a given whole number can be built up in one way only—apart from rearrangements of factors—by multiplying together primes; if a prime divides the product (result of multiplying) of two whole numbers, it divides at least one of them. To illustrate: $24 = 2 \times 2 \times 2 \times 3$, and 24 cannot be built up by multiplication of primes in any essentially different way—we consider $2 \times 2 \times 2 \times 3$, $2 \times 2 \times 3 \times 2$, $2 \times 3 \times 2 \times 2$ and $3 \times 2 \times 2 \times 2$ as the same; 7 divides 42, and $42 = 2 \times 21 = 3 \times 14 = 6 \times 7$, in each of which 7 divides at least one of the numbers multiplied together to give 42; again, 98 is divisible by 7, and $98 = 7 \times 14$, in which case 7 divides both 7 and 14, and hence at least one of them. From these two facts the proof can be given in less than half a page. It is within the understanding of any normal fourteen-year-old, but it is safe to wager that out of a million human beings of normal intelligence of any or all ages, less than ten of those who had had no more mathematics than grammar-grade arithmetic would succeed in finding a proof within a reasonable time—say a year.

This seems to be an appropriate place to quote some famous remarks of Gauss concerning the favorite field of Fermat's interests and his own. The translation is that of the Irish arithmetician H. J. S. Smith (1826–1883), from Gauss' introduction to the collected mathematical papers of Eisenstein published in 1847.

"The higher arithmetic presents us with an inexhaustible store of interesting truths—of truths too, which are not isolated, but stand in a close internal connection, and between which, as our knowledge increases, we are continually discovering new and sometimes wholly unexpected ties. A great part of its theories derives an additional charm from the peculiarity that important propositions, with the impress of simplicity upon them, are often easily discoverable by induction, and yet are of so profound a character that we cannot find their demonstration till after many vain attempts; and even then, when we do succeed, it is often by some tedious and artificial process, while the simpler methods may long remain concealed."

One of these interesting truths which Gauss mentions is sometimes considered the most beautiful (but not the most important) thing about numbers that Fermat discovered: every prime number of the form $4n + 1$ is a sum of two squares, and is such a sum in only one way. It is easily proved that no number of the form $4n - 1$ is a sum of two squares. As all primes greater than 2 are readily seen to be of one or other of these forms, there is nothing to add. For an example, 37 when divided by 4 yields the remainder 1, so 37 must be the sum of two squares of whole numbers. By trial (there are better ways) we find indeed that $37 = 1 + 36, = 1^2 + 6^2$, and that there are no other squares x^2 and y^2 such that $37 = x^2 + y^2$. For the prime 101 we have $1^2 + 10^2$; for 41 we find $4^2 + 5^2$. On the other hand $19, = 4 \times 5 - 1$, is not a sum of two squares.

As in nearly all of his arithmetical work, Fermat left no proof of this theorem. It was first proved by the great Euler in 1749 after he had struggled, off and on, for *seven years* to find a proof. But Fermat does describe the ingenious method, which he invented, whereby he proved this and some others of his wonderful results. This is called "infinite descent," and is infinitely more difficult to accomplish than Elijah's ascent to Heaven. His own account is both concise and clear, so we shall give a free translation from his letter of August, 1659, to Carcavi.

"For a long time I was unable to apply my method to affirmative propositions, because the twist and the trick for getting at them is much more troublesome than that which I use for negative propositions. Thus, when I had to prove that *every prime number which exceeds a multiple of 4 by 1 is composed of two squares*, I found myself in a fine torment. But at last a meditation many times repeated gave me the light I lacked, and now affirmative propositions submit to my method, with the aid of certain new principles which necessarily must be adjoined to it. The course of my reasoning in affirmative propositions is such: if an arbitrarily chosen prime of the form $4n + 1$ is not a sum of two squares, [I prove that] there will be another of the same nature, less than the one chosen, and [therefore] next a third still less, and so on. Making an infinite descent in this way we finally arrive at the number 5, the least of all the numbers of this kind [$4n + 1$]. [By the proof mentioned and the preceding argument from it], it follows that 5 is not a sum of two squares. But it is. Therefore we must infer by a *reductio ad absurdum* that all numbers of the form $4n + 1$ are sums of two squares."

All the difficulty in applying descent to a new problem lies in the first step, that of proving that *if* the assumed or conjectured proposition is *true* of any number of the kind concerned chosen at random, *then* it will be *true* of a *smaller* number of the *same kind*. There is no general method, applicable to all problems, for taking this step. Something rarer than grubby patience or the greatly overrated "infinite capacity for taking pains" is needed to find a way through the wilderness. Those who imagine genius is nothing more than the ability to be a good bookkeeper may be recommended to exert their infinite patience on Fermat's Last Theorem. Before stating the theorem we give one more example of the deceptively simple problems Fermat attacked and solved. This will introduce the topic of *Diophantine analysis*, in which Fermat excelled.

Anyone playing with numbers might well pause over the curious fact that $27 = 25 + 2$. The point of interest here is that both 27 and 25 are exact powers, namely $27 = 3^3$ and $25 = 5^2$. Thus we observe that $y^3 = x^2 + 2$ has a solution in *whole numbers* x, y; the solution is $y = 3$, $x = 5$. As a sort of superintelligence test the reader may now prove that $y = 3$, $x = 5$ are the *only* whole numbers which satisfy the equation. It is not easy. In fact it requires more innate intellectual capacity to dispose of this apparently childish thing than it does to grasp the theory of relativity.

The equation $y^3 = x^2 + 2$, *with the restriction that the solution y, x is to be in whole numbers*, is *indeterminate* (because there are more unknowns, namely two, x and y, than there are equations, namely one, connecting them) and *Diophantine*, after the Greek who was one of the first to insist upon *whole number* solutions of equations or, less stringently, on *rational* (fractional) solutions. There is no difficulty whatever in describing an infinity of solutions *without* the restriction to whole numbers: thus we may give x *any* value we please and then determine y by adding 2 to this x^2 and extracting the cube root of the result. But the *Diophantine* problem of finding *all* the *whole number* solutions is quite another matter. The solution $y = 3$, $x = 5$ is seen "by inspection"; the difficulty of the problem is to prove that there are *no other* whole numbers y, x which will satisfy the equation. Fermat proved that there are none but, as usual, suppressed his proof, and it was not until many years after his death that a proof was found.

This time he was not guessing; the problem is hard; he asserted that he had a proof; a proof was later found. And so for all of his positive assertions with the one exception of the seemingly simple one which he made in his Last Theorem and which mathematicians, struggling for nearly 300 years, have been unable to prove: whenever Fermat asserted that he had *proved* anything, the statement, with the one exception noted, has subsequently been proved. Both his scrupulously honest character and his unrivalled penetration as an arithmetician substantiate the claim made for him by some, but not by all, that he knew what he was talking about when he asserted that he possessed a proof of his theorem.

It was Fermat's custom in reading Bachet's *Diophantus* to record the results of his meditations in brief marginal notes in his copy. The margin was not suited for the writing out of proofs. Thus, in commenting on the eighth problem of the Second Book of Diophantus' Arithmetic, which asks for the solution in rational numbers (fractions or whole numbers) of the equation $x^2 + y^2 = a^2$, Fermat comments as follows:

"On the contrary, it is impossible to separate a cube into two cubes, a fourth power into two fourth powers, or, generally, any power above the second into two powers of the same degree: I have discovered a truly marvellous demonstration [of this general theorem] which this margin is too narrow to contain" (Fermat, *Oeuvres*, III, p. 241). This is his famous Last Theorem, which he discovered about the year 1637.

To restate this in modern language: Diophantus' problem is to find whole numbers or fractions x, y, a such that $x^2 + y^2 = a^2$; Fermat asserts that *no* whole numbers or fractions exist such that $x^3 + y^3 = a^3$, or $x^4 + y^4 = a^4$, or, generally, such that $x^n + y^n = a^n$ if n is a whole number greater than 2.

Diophantus' problem has an infinity of solutions; specimens are $x = 3$, $y = 4$, $a = 5$; $x = 5$, $y = 12$, $a = 13$. Fermat himself gave a proof by his method of infinite descent for the impossibility of $x^4 +$

$y^4 = a^4$. Since his day $x^n + y^n = a^n$ has been proved impossible in whole numbers (or fractions) for a great many numbers n (up to all primes* less than $n = 14000$ if none of the numbers x, y, a is divisible by n), but this is not what is required. A proof disposing of *all* n's greater than 2 is demanded. Fermat said he possessed a "marvellous" proof.

After all that has been said, is it likely that he had deceived himself? It may be left up to the reader. One great arithmetician, Gauss, voted against Fermat. However, the fox who could not get at the grapes declared they were sour. Others have voted for him. Fermat was a mathematician of the first rank, a man of unimpeachable honesty, and an arithmetician without a superior in history.†

---

*The reader can easily see that it suffices to dispose of the case where n is an odd prime, since, in algebra, $u^{ab} = (u^a)^b$, where u, a, b are any numbers.

†In 1908 the late Professor Paul Wolfskehl (German) left 100,000 marks to be awarded to the first person giving a *complete* proof of Fermat's Last Theorem. The inflation after the World War reduced this prize to a fraction of a cent, which is what the mercenary will now get for a proof.

CHAPTER FIVE

"Greatness and Misery of Man"

PASCAL

We see ... that the theory of probabilities is at bottom only common sense reduced to calculation; it makes us appreciate with exactitude what reasonable minds feel by a sort of instinct, often without being able to account for it. ... It is remarkable that [this] science, which originated in the consideration of games of chance, should have become the most important object of human knowledge.—P. S. LAPLACE

YOUNGER BY TWENTY SEVEN YEARS than his great contemporary Descartes, Blaise Pascal was born at Clermont, Auvergne, France, on June 19, 1623, and outlived Descartes by twelve years. His father Étienne Pascal, president of the court of aids at Clermont, was a man of culture and had some claim to intellectual distinction in his own times; his mother, Antoinette Bégone, died when her son was four. Pascal had two beautiful and talented sisters, Gilberte, who became Madame Périer, and Jacqueline, both of whom, the latter especially, played important parts in his life.

Blaise Pascal is best known to the general reader for his two literary classics, the *Pensées* and the *Lettres écrites par Louis de Montalte à un provincial de ses amis* commonly referred to as the "Provincial Letters," and it is customary to condense his mathematical career to a few paragraphs in the display of his religious prodigies. Here our point of view must necessarily be somewhat oblique, and we shall consider Pascal primarily as a highly gifted mathematician who let his masochistic proclivities for self-torturing and profitless speculations on the sectarian controversies of his day degrade him to what would now be called a religious neurotic.

On the mathematical side Pascal is perhaps the greatest might-have-been in history. He had the misfortune to precede Newton by only a few years and to be a contemporary of Descartes and Fermat, both more stable men than himself. His most novel work, the creation of the mathematical theory of probability, was shared with

Fermat, who could easily have done it alone. In geometry, for which he is famous as a sort of infant prodigy, the creative idea was supplied by a man—Desargues—of much lesser celebrity.

In his outlook on experimental science Pascal had a far clearer vision than Descartes—from a modern point of view—of the scientific method. But he lacked Descartes' singleness of aim, and although he did some first-rate work, allowed himself to be deflected from what he might have done by his morbid passion for religious subtleties.

It is useless to speculate on what Pascal might have done. Let his life tell what he actually did. Then, if we choose, we can sum him up as a mathematician by saying that he did what was in him and that no man can do more. His life is a running commentary on two of the stories or similes in that New Testament which was his constant companion and unfailing comfort: the parable of the talents, and the remark about new wine bursting old bottles (or skins). If ever a wonderfully gifted man buried his talent, Pascal did; and if ever a medieval mind was cracked and burst asunder by its attempt to hold the new wine of seventeenth-century science, Pascal's was. His great gifts were bestowed upon the wrong person.

At the age of seven Pascal moved from Clermont with his father and sisters to Paris. About this time the father began teaching his son. Pascal was an extremely precocious child. Both he and his sisters appear to have had more than their share of nature's gifts. But poor Blaise inherited (or acquired) a wretched physique along with his brilliant mind, and Jacqueline, the more gifted of his sisters, seems to have been of the same stripe as her brother, for she too fell a victim to morbid religiosity.

At first everything went well enough. Pascal senior, astonished at the ease with which his son absorbed the stock classical education of the day, tried to hold the boy down to a reasonable pace to avoid injuring his health. Mathematics was taboo, on the theory that the young genius might overstrain himself by using his head. His father was an excellent drillmaster but a poor psychologist. His ban on mathematics naturally excited the boy's curiosity. One day when he was about twelve Pascal demanded to know what geometry was about. His father gave him a clear description. This set Pascal off like a hare after his true vocation. Contrary to his own opinion in later life he had been called by God, not to torment the Jesuits, but to be a great mathematician. But his hearing was defective at the time and he got his orders confused.

What happened when Pascal began the study of geometry has become one of the legends of mathematical precocity. In passing it may be remarked that infant prodigies in mathematics do not invariably blow up as they are sometimes said to do. Precocity in mathematics has often been the first flush of a glorious maturity, in spite of the persistent superstition to the contrary. In Pascal's case early mathematical genius was not extinguished as he grew up but stifled under other interests. The ability to do first-class mathematics persisted, as will be seen from the episode of the cycloid, late into his all too brief life, and if anything is to be blamed for his comparatively early mathematical demise it is probably his stomach. His first spectacular feat was to prove, entirely on his own initiative, and without a hint from any book, that the sum of the angles of a triangle is equal to two right angles. This encouraged him to go ahead at a terrific pace.

Realizing that he had begotten a mathematician, Pascal senior wept with joy and gave his son a copy of Euclid's *Elements*. This was quickly devoured, not as a task, but as play. The boy gave up his games to geometrize. In connection with Pascal's rapid mastery of Euclid, sister Gilberte permits herself an overappreciative fib. It is true that Pascal had found out and proved several of Euclid's propositions for himself before he ever saw the book. But what Gilberte romances about her brilliant young brother is less probable than a throw of a billion aces in succession with one die, for the reason that it is infinitely improbable. Gilberte declared that her brother had rediscovered for himself the first thirty two propositions of Euclid, and that he had found them *in the same order* as that in which Euclid sets them forth. The thirty second proposition is indeed the famous one about the sum of the angles of a triangle which Pascal rediscovered. Now, there may be only one way of doing a thing right, but it seems more likely that there are an infinity of ways of doing it wrong. We know today that Euclid's allegedly rigorous demonstrations, even in the first four of his propositions, are no proofs at all. That Pascal faithfully duplicated all of Euclid's oversights on his own account is an easy story to tell but a hard one to believe. However, we can forgive Gilberte for bragging. Her brother was worth it. At the age of fourteen he was admitted to the weekly scientific discussions, conducted by Mersenne, out of which the French Academy of Sciences developed.

While young Pascal was fast making a geometer of himself, old Pascal was making a thorough nuisance of *himself* with the authorities on account of his honesty and general uprightness. In particular he disagreed with Cardinal Richelieu over a little matter of imposing taxes. The Cardinal was incensed; the Pascal family went into hiding till the storm blew over. It is said that the beautiful and talented Jacqueline rescued the family and restored her father to the light of the Cardinal's countenance by her brilliant acting, incognito, in a play presented for Richelieu's entertainment. On inquiring the name of the charming young artiste who had captivated his clerical fancy, and being told that she was the daughter of his minor enemy, Richelieu very handsomely forgave the whole family and planted the father in a political job at Rouen. From what is known of that wily old serpent, Cardinal Richelieu, this pleasing tale is probably a fish story. Anyhow, the Pascals once more found a job and security at Rouen. There young Pascal met the tragic dramatist Corneille, who was duly impressed with the boy's genius. At the time Pascal was all mathematician, so probably Corneille did not suspect that his young friend was to become one of the great creators of French prose.

All this time Pascal was studying incessantly. Before the age of sixteen (about 1639)* he had proved one of the most beautiful theorems in the whole range of geometry. Fortunately it can be described in terms comprehensible to anyone. Sylvester, a mathematician of the nineteenth century whom we shall meet later, called Pascal's great theorem a sort of "cat's cradle." We state first a special form of the general theorem that can be constructed with the use of a ruler only.

Label two intersecting straight lines l and l'. On l take any three distinct points A, B, C, and on l' any three distinct points A', B', C'. Join up these points by straight lines, crisscross, as follows: A and B', A' and B, B and C', B' and C, C and A', C' and A. The two lines in each of these pairs intersect in a point. We thus get three points. The special case of Pascal's theorem which we are now describing states that these three points lie on one straight line.

---

*Authorities differ on Pascal's age when this work was done, the estimate varying from fifteen to seventeen. The 1819 edition of Pascal's works contains a brief résumé of the statements of certain propositions on conics, but this is not the *completed* essay which Leibniz saw.

Before giving the general form of the theorem we mention another result like the preceding. This is due to Desargues (1593–1662). If the three straight lines joining corresponding vertices of two triangles

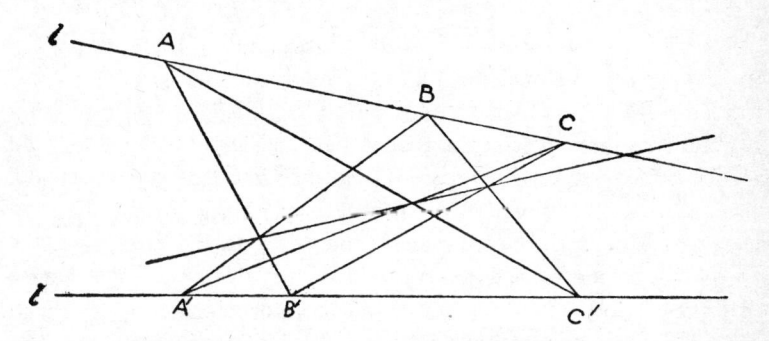

XYZ and xyz meet in a point, then the three intersections of pairs of corresponding sides lie on one straight line. Thus, *if* the straight lines joining X and x, Y and y, Z and z meet in a point, *then* the intersections of XY and xy, YZ and yz, ZX and zx lie in one straight line.

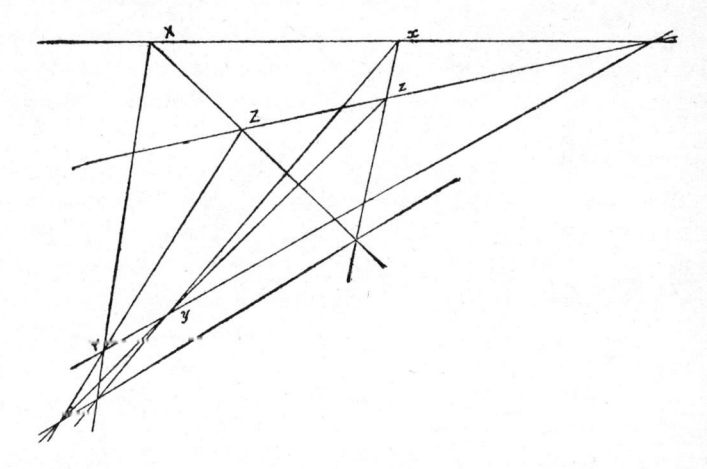

In Chapter 2 we stated what a conic section is. Imagine any conic section, for definiteness say an ellipse. On it mark any six points, A, B, C, D, E, F, and join them up, in this order, by straight lines. We thus have a six-sided figure inscribed in the conic section, in which AB and DE, BC and EF, CD and FA are pairs of opposite sides. The two lines in each of these three pairs intersect in a point; the three points of intersection lie on one straight line (see figure in Chapter 13, page 217). This is Pascal's theorem; the figure which it furnishes is what he called the "mystic hexagram." He probably first proved it true for a circle and then passed by projection to any conic section. Only a straightedge and a pair of compasses are required if the reader wishes to see what the figure looks like for a circle.

There are several amazing things about this wonderful proposition, not the least of which is that it was discovered and proved by a boy of sixteen. Again, in his *Essai pour les Coniques* (Essay on Conics), written around his great theorem by this extraordinarily gifted boy, no fewer than 400 propositions on conic sections, including the work of Apollonius and others, were systematically deduced as corollaries, by letting pairs of the six points move into coincidence, so that a chord became a tangent, and other devices. The full *Essai* itself was never published and is apparently lost irretrievably, but Leibniz saw and inspected a copy of it. Further, the *kind* of geometry which Pascal is doing here differs fundamentally from that of the Greeks; it is not *metrical*, but *descriptive*, or *projective*. Magnitudes of lines or angles cut no figure in either the statement or the proof of the theorem. This one theorem in itself suffices to abolish the stupid definition of mathematics, inherited from Aristotle and still sometimes reproduced in dictionaries, as the science of "quantity." There are no "quantities" in Pascal's geometry.

To see what the *projectivity* of the theorem means, imagine a (circular) cone of light issuing from a point and pass a flat sheet of glass through the cone in varying positions. The boundary curve of the figure in which the sheet cuts the cone is a *conic section*. If Pascal's "mystic hexagram" be drawn on the glass for any given position, and another flat sheet of glass be passed through the cone so that the shadow of the hexagram falls on it, *the shadow will be another "mystic hexagram"* with its three points of intersection of opposite pairs of sides lying on one straight line, the shadow of the "three-point-line" in the original hexagram. That is, Pascal's theorem is *invariant* (unchanged) *under conical projection*. The metrical properties of figures studied in common elementary geometry are *not* invariant under projection; for example, the shadow of a right angle is not a right angle for all positions of the second sheet. It is obvious that this kind of *projective*, or *descriptive* geometry, is one of the geometries naturally adapted to some of the problems of perspective. The *method* of projection was used by Pascal in proving his theorem, but had been applied previously by Desargues in deducing the result stated above concerning two triangles "in perspective." Pascal gave Desargues full credit for his great invention.

All this brilliance was purchased at a price. From the age of seventeen to the end of his life at thirty nine, Pascal passed but few days without pain. Acute dyspepsia made his days a torment and chronic insomnia his nights half-waking nightmares. Yet he worked incessantly. At the age of eighteen he invented and made the first calculating machine in history—the ancestor of all the arithmetical machines that have displaced armies of clerks from their jobs in our own generation. We shall see farther on what became of this ingenious device. Five years later, in 1646, Pascal suffered his first "conversion." It did not take deeply, possibly because Pascal was only twenty three and still absorbed in his mathematics. Up to this time the family had been decently enough devout; now they all seem to have gone mildly insane.

It is difficult for a modern to recreate the intense religious passions which inflamed the seventeenth century, disrupting families and hurling professedly Christian countries and sects at one another's throats. Among the would-be religious reformers of the age was Cornelius Jansen (1585–1638), a flamboyant Dutchman who became bishop of Ypres. A cardinal point of his dogma was the necessity for "conversion" as a means to "grace," somewhat in the manner of certain flourishing sects today. Salvation, however, at least to an unsympathetic eye, appears to have been the lesser of Jansen's ambitions. God, he was convinced, had especially elected him to blast the Jesuits in this life and toughen them for eternal damnation in the next. This was his call, his mission. His creed was neither Catholicism nor Protestantism, although it leaned rather toward the latter. Its moving spirit was, first, last and all the time, a rabid hatred of those who disputed its dogmatic bigotries. The Pascal family now (1646) ardently —but not too ardently at first—embraced this unlovely creed of Jansenism. Thus Pascal, at the early age of twenty three, began to die off at the top. In the same year his whole digestive tract went bad and he suffered a temporary paralysis. But he was not yet dead intellectually.

His scientific greatness flared up again in 1648 in an entirely new direction. Carrying on the work of Torricelli (1608–1647) on atmospheric pressure, Pascal surpassed him and demonstrated that he understood the scientific method which Galileo, the teacher of Torricelli, had shown the world. By experiments with the barometer, which he suggested, Pascal proved the familiar facts now known to every beginner in physics regarding the pressure of the atmosphere. Pascal's sister Gilberte had married a Mr. Périer. At Pascal's suggestion, Périer performed the experiment of carrying a barometer up the Puy de Dôme in Auvergne and noting the fall of the column of mercury as the atmospheric pressure decreased. Later Pascal, when he moved to Paris with his sister Jacqueline, repeated the experiment on his own account.

Shortly after Pascal and Jacqueline had returned to Paris they were joined by their father, now fully restored to favor as a state councillor. Presently the family received a somewhat formal visit from Descartes. He and Pascal talked over many things, including the barometer. There was little love lost between the two. For one thing, Descartes had openly refused to believe the famous *Essai pour les coniques* had been written by a boy of sixteen. For another, Descartes suspected Pascal of having filched the idea of the barometric experiments from himself, as he had discussed the possibilities in letters to Mersenne. Pascal, as has been mentioned, had been attending the weekly meetings at Father Mersenne's since he was fourteen. A third ground for dislike on both sides was furnished by their religious antipathies. Descartes, having received nothing but kindness all his life from the Jesuits, loved them; Pascal, following the devoted Jansen, hated a Jesuit worse than the devil is alleged to hate holy water. And finally, according to the candid Jacqueline, both her brother and Descartes were intensely jealous, each of the other. The visit was rather a frigid success.

The good Descartes however did give his young friend some excellent advice in a truly Christian spirit. He told Pascal to follow his own example and lie in bed every day till eleven. For poor Pascal's awful stomach he prescribed a diet of nothing but beef tea. But Pascal ignored the kindly meant advice, possibly because it came from Descartes. Among other things which Pascal totally lacked was a sense of humor.

Jacqueline now began to drag her genius of a brother down—or up; it all depends upon the point of view. In 1648, at the impressionable age of twenty three, Jacqueline declared her intention of moving to Port Royal, near Paris, the main hangout of the Jansenists in France, to become a nun. Her father sat down heavily on the project, and the devoted Jacqueline concentrated her thwarted efforts on her erring brother. She suspected he was not yet so thoroughly converted as he might have been, and apparently she was right. The family now returned to Clermont for two years.

During these two swift years Pascal seems to have become almost half human, in spite of sister Jacqueline's fluttering admonitions that he surrender himself utterly to the Lord. Even the recalcitrant stomach submitted to rational discipline for a few blessed months.

It is said by some and hotly denied by others that Pascal during this sane interlude and later for a few years discovered the predestined uses of wine and women. He did not sing. But these rumors of a basely human humanity may, after all, be nothing more than rumors. For after his death Pascal quickly passed into the Christian hagiocracy, and any attempts to get at the facts of his life as a human being were quietly but rigidly suppressed by rival factions, one of which strove to prove that he was a devout zealot, the other, a skeptical atheist, but both of which declared that Pascal was a saint not of this earth.

During these adventurous years the morbidly holy Jacqueline continued to work on her frail brother. By a beautiful freak of irony Pascal was presently to be converted—for good, this time—and it was to be *his* lot to turn the tables on his too pious sister and drive *her* into the nunnery which now, perhaps, seemed less desirable. This, of course, is not the orthodox interpretation of what happened; but to anyone other than a blind partisan of one sect or the other—Christian or Atheist—it is a more rational account of the unhealthy relationship between Pascal and his unmarried sister than that which is sanctioned by tradition.

Any modern reader of the *Pensées* must be struck by a certain something or another which either completely escaped our more reticent ancestors or was ignored by them in their wiser charity. The letters, too, reveal a great deal which should have been decently buried. Pascal's ravings in the *Pensées* about "lust" give him away completely, as do also the well-attested facts of his unnatural frenzies at the sight of his married sister Gilberte naturally caressing her children.

Modern psychologists, no less than the ancients with ordinary common sense, have frequently remarked the high correlation between sexual repression and morbid religious fervor. Pascal suffered from both, and his immortal *Pensées* is a brilliant if occasionally incoherent testimonial to his purely physiological eccentricities. If only the man could have been human enough to let himself go when his whole nature told him to cut loose, he might have lived out everything that was in him, instead of smothering the better half of it under a mass of meaningless mysticism and platitudinous observations on the misery and dignity of man.

Always shifting about restlessly the family returned to Paris in 1650. The next year the father died. Pascal seized the occasion to write Gilberte and her husband a lengthy sermon on death in general. This letter has been much admired. We need not reproduce any of it here; the reader who wishes to form his own opinion of it can easily locate it. Why this priggish effusion of pietistic and heartless moralizing on the death of a presumably beloved parent should ever have excited admiration instead of contempt for its author is, like the love of God which the letter in part dwells upon *ad nauseam*, a mystery that passeth all understanding. However, there is no arguing about tastes, and those who like the sort of thing that Pascal's much-quoted letter is, may be left to their undisturbed enjoyment of what is, after all, one of the masterpieces of self-conscious self-revelation in French literature.

A more practical result of Pascal senior's death was the opportunity which it offered Pascal, as administrator of the estate, of returning to normal intercourse with his fellow men. Encouraged by her brother, sister Jacqueline now joined Port Royal, her father being no longer capable of objecting. Her sweet concern over her brother's soul was now spiced by a quite human quarrel over the division of the estate.

A letter of the preceding year (1650) reveals another facet of Pascal's reverent character, or possibly his envy of Descartes. Dazzled by the transcendent brilliance of the Swedish Christine, Pascal humbly begged to lay his calculating machine at the feet of "the greatest princess in the world," who, he declares in liquid phrases dripping strained honey and melted butter, is as eminent intellectually as she is socially. What Christine did with the machine is not known. She did not invite Pascal to replace the Descartes whom she had done in.

At last, on November 23, 1654, Pascal was really converted. According to some accounts he had been living a fast life for three years. The best authorities seem to agree that there is not much in this tradition and that his life was not so fast after all. He had merely been doing his poor suffering best to live like a normal human being and to get something more than mathematics and piety out of life. On the day of his conversion he was driving a four-in-hand when the horses bolted. The leaders plunged over the parapet of the bridge at Neuilly, but the traces broke, and Pascal remained on the road.

To a man of Pascal's mystical temperament this lucky escape from a violent death was a direct warning from Heaven to pull himself up sharply on the brink of the moral precipice over which he, the victim of his morbid self-analysis, imagined he was about to plunge. He took a small piece of parchment, inscribed on it some obscure sentiments of mystical devotion, and thenceforth wore it next to his heart as an amulet to protect him from temptation and remind him of the goodness of God which had snatched him, a miserable sinner, from the very mouth of hell. Only once thereafter did he fall from grace (in his own pitiable opinion), although all the rest of his life he was haunted by hallucinations of a precipice before his feet.

Jacqueline, now a postulant for the nunnery at Port Royal, came to her brother's aid. Partly on his own account, partly because of his sister's persuasive pleadings, Pascal turned his back on the world and took up his residence at Port Royal, to bury his talent thenceforth in contemplation on "the greatness and misery of man." This was in 1654, when Pascal was thirty one. Before forever quitting things of the flesh and the mind, however, he had completed his most important contribution to mathematics, the joint creation, with Fermat, of the mathematical theory of probability. Not to interrupt the story of his life we shall defer an account of this for the moment.

His life at Port Royal was at least sanitary if not exactly as sane as might have been wished, and the quiet, orderly routine benefited his precarious health considerably. It was while at Port Royal that he composed the famous *Provincial Letters*, which were inspired by Pascal's desire to aid in acquitting Arnauld, the leading light of the institution, of the charge of heresy. These famous letters (there were eighteen, the first of which was printed on January 23, 1656) are masterpieces of controversial skill, and are said to have dealt the Jesuits a blow from which their Society has never fully recovered. However, as a commonplace of objective observation which anyone with eyes in his head can verify for himself, the Society of Jesus still flourishes; so it may be reasonably doubted whether the *Provincial*

Letters had in them the deadly potency ascribed to them by sympathetic critics.

In spite of his intense preoccupation with matters pertaining to his salvation and the misery of man, Pascal was still capable of doing excellent mathematics, although he regarded the pursuit of all science as a vanity to be eschewed for its derogatory effects on the soul. Nevertheless he did fall from grace once more, but only once. The occasion was the famous episode of the cycloid.

This beautifully proportioned curve (it is traced out by the motion of a fixed point on the circumference of a wheel rolling along a straight line on a flat pavement) seems to have turned up first in mathematical literature in 1501, when Charles Bouvelles described it in connection

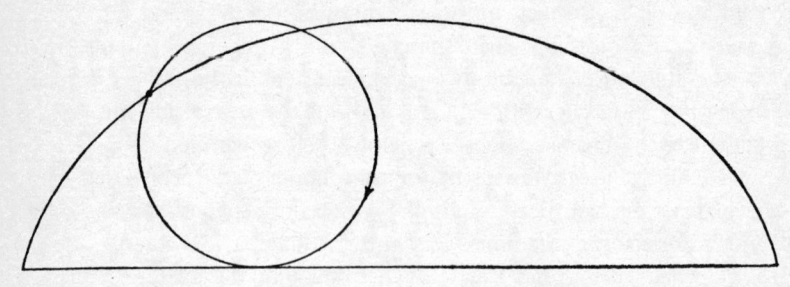

with the squaring of the circle. Galileo and his pupil Viviani studied it and solved the problem of constructing a tangent to the curve at any point (a problem which Fermat solved at once when it was proposed to him), and Galileo suggested its use as an arch for bridges. Since reinforced concrete has become common, cycloidal arches are frequently seen on highway viaducts. For mechanical reasons (unknown to Galileo) the cycloidal arch is superior to any other in construction. Among the famous men who investigated the cycloid was Sir Christopher Wren, the architect of St. Paul's Cathedral, who determined the length of any arc of the curve and its center of gravity, while Huygens, for mechanical reasons, introduced it into the construction of pendulum clocks. One of the most beautiful of all the discoveries of Huygens (1629–1695) was made in connection with the cycloid. He proved that it is the *tautochrone*, that is, the curve (when turned upside down like a bowl) down which beads placed *anywhere* on it will all slide to the lowest point under the influence of gravity *in the same time*. On account of its singular beauty, elegant properties, and the endless rows which it stirred up between quarrelsome mathematicians challenging one another to solve this or that problem in connection with it, the cycloid has been called "the Helen of Geometry," after the Graeco-Trojan lady whose mere face is said to have "launched a thousand ships."

Among other miseries which afflicted the wretched Pascal were persistent insomnia and bad teeth—in a day when such dentistry as was practised was done by the barber with a strong pair of forceps and brute force. Lying awake one night (1658) in the tortures of toothache, Pascal began to think furiously about the cycloid to take his mind off the excruciating pain. To his surprise he noticed presently that the pain had stopped. Interpreting this as a signal from Heaven that he was not sinning in thinking about the cycloid rather than his soul, Pascal let himself go. For eight days he gave himself up to the geometry of the cycloid and succeeded in solving many of the main problems in connection with it. Some of the things he discovered were issued under the pseudonym of Amos Dettonville as challenges to the French and English mathematicians. In his treatment of his rivals in this matter Pascal was not always as scrupulous as he might have been. It was his last flicker of mathematical activity and his only contribution to science after his entry to Port Royal.

The same year (1658) he fell more seriously ill than he had yet been in all his tormented life. Racking and incessant headaches now deprived him of all but the most fragmentary snatches of sleep. He suffered for four years, living ever more ascetically. In June, 1662, he gave up his own house to a poor family suffering from smallpox, as an act of self-denial, and went to live with his married sister. On August 19, 1662, his tortured existence came to an end in convulsions. He died at the age of thirty nine.

The post mortem revealed what had been expected regarding the stomach and vital organs; it also disclosed a serious lesion of the brain. Yet in spite of all this Pascal had done great work in mathematics and science and had left a name in literature that is still respected after nearly three centuries.

The beautiful things Pascal did in geometry, with the possible exception of the "mystic hexagram," would all have been done by other men had he not done them. This holds in particular for the investigations on the cycloid. After the invention of the calculus all such things became incomparably easier than they had been before and in time passed into the textbooks as mere exercises for young students. But in the joint creation with Fermat of the mathematical theory of probabilities Pascal made a new world. It seems quite likely that Pascal will be remembered for his part in this great and ever increasingly more important invention long after his fame as a writer has been forgotten. The *Pensées* and the *Provincial Letters*, apart from their literary excellences, appeal principally to a type of mind that is rapidly becoming extinct. The arguments for or against a particular point strike a modern mind as either trivial or unconvincing, and the very questions to which Pascal addressed himself with such fervent zeal now seem strangely ridiculous. If the problems which he discussed on the greatness and misery of man are indeed as profoundly important as enthusiasts have claimed, and not mere pseudo-problems mystically stated and incapable of solution, it seems unlikely that they will ever be solved by platitudinous moralizing. But in his theory of probabilities Pascal stated and solved a genuine problem, that of bringing the superficial lawlessness of pure chance under the domination of law, order, and regularity, and today this subtle theory appears to be at the very roots of human knowledge no less than at the foundation of physical science. Its ramifications are everywhere, from the quantum theory to epistemology.

The true founders of the mathematical theory of probability were Pascal and Fermat, who developed the fundamental principles of the subject in an intensely interesting correspondence during the year 1654. This correspondence is now readily available in the *Oeuvres de Fermat* (edited by P. Tannery and C. Henry, vol. 2, 1904). The letters show that Pascal and Fermat participated equally in the creation of the theory. Their correct solutions of problems differ in details but not in fundamental principles. Because of the tedious enumeration of possible cases in a certain problem on "points" Pascal tried to take a short cut and fell into error. Fermat pointed out the mistake, which Pascal acknowledged. The first letter of the series has been lost but the occasion of the correspondence is well attested.

The initial problem which started the whole vast theory was proposed to Pascal by the Chevalier de Méré, more or less of a professional gambler. The problem was that of "points": each of two players (at dice, say) requires a certain number of points to win the game; if they quit the game before it is finished, how should the stakes be divided between them? The score (number of points) of each player is given at the time of quitting, and the problem amounts to determining the probability which each player has at a given stage of the game of winning the game. It is assumed that the players have equal chances of winning a single point. The solution demands nothing more than sound common sense; the *mathematics* of probability enters when we seek a method for enumerating possible cases without actually counting them off. For example, how many possible different hands each consisting of three deuces and three other cards, none a deuce, are there in a common deck of fifty two? Or, in how many ways can a throw of three aces, five twos, and two sixes occur when ten dice are tossed? A third trifle of the same sort: how many different bracelets can be made by stringing ten pearls, seven rubies, six emeralds, and eight sapphires, if stones of one kind are considered as undistinguishable?

This detail of finding the number of ways in which a prescribed thing can be done or in which a completely specified event can happen, belongs to what is called *combinatorial analysis*. Its application to probability is obvious. Suppose, for example, we wish to know the probability of throwing two aces and one deuce in a single throw with three dice. If we know the *total* number of ways ($6 \times 6 \times 6$ or 216) in which the three dice can fall, and also the number of ways (say n, which the reader may find for himself) in which two aces and one deuce can fall, the required probability is $n/216$. (Here n is three, so the probability is $3/216$.) Antoine Gombaud, Chevalier de Méré, who instigated all this, is described by Pascal as a man having a very good mind but no mathematics, while Leibniz, who seems to have disliked the gay Chevalier, dubs him a man of penetrating mind, a philosopher, and a gambler—quite an unusual combination.

In connection with problems in combinatorial analysis and probability Pascal made extensive use of the arithmetical triangle

$$
\begin{array}{c}
1 \\
1 \; 1 \\
1 \; 2 \; 1 \\
1 \; 3 \; 3 \; 1 \\
1 \; 4 \; 6 \; 4 \; 1 \\
1 \; 5 \; 10 \; 10 \; 5 \; 1 \\
\cdots \quad \cdots \quad \cdots,
\end{array}
$$

in which the numbers in any row after the first two are obtained from those in the preceding row by copying down the terminal 1's and adding together the successive pairs of numbers from left to right to give the new row; thus $5 = 1 + 4$, $10 = 4 + 6$, $10 = 6 + 4$, $5 = 4 + 1$. The numbers in the nth row, after the 1, are the number of different selections of one thing, two things, three things, . . . that can be chosen from n distinct things. For example, 10 is the number of different pairs of things that can be selected from five distinct things. The numbers in the nth row are also the coefficients in the expansion of $(1 + x)^n$ by the binomial theorem, thus for $n = 4$, $(1 + x)^4 = 1 + 4x + 6x^2 + 4x^3 + x^4$. The triangle has numerous other interesting properties. Although it was known before the time of Pascal, it is usually named after him on account of the ingenious use he made of it in probabilities.

The theory which originated in a gamblers' dispute is now at the base of many enterprises which we consider more important than gambling, including all kinds of insurance, mathematical statistics and their application to biology and educational measurements, and much of modern theoretical physics. We no longer think of an electron being "at" a given place at a given instant, but we do calculate its probability of being in a given region. A little reflection will show that even the simplest measurements we make (when we attempt to measure anything accurately) are statistical in character.

The humble origin of this extremely useful mathematical theory is typical of many: some apparently trivial problem, first solved perhaps out of idle curiosity, leads to profound generalizations which, as in the case of the new statistical theory of the atom in the quantum theory, may cause us to revise our whole conception of the physical universe or, as has happened with the application of statistical methods to intelligence tests and the investigation of heredity, may induce us to modify our traditional beliefs regarding the "greatness and misery of man." Neither Pascal nor Fermat of course foresaw what was to issue from their disreputable child. The whole fabric of mathematics is so closely interwoven that we cannot unravel and eliminate any particular thread which happens to offend our individual taste without danger of destroying the whole pattern.

Pascal however did make one application of probabilities (in the *Pensées*) which for his time was strictly practical. This was his famous "wager." The "expectation" in a gamble is the value of the prize multiplied by the probability of winning the prize. According to Pascal the value of eternal happiness is infinite. He reasoned that even if the probability of winning eternal happiness by leading a religious life is very small indeed, nevertheless, since the expectation is infinite (*any* finite fraction of infinity is itself infinite) it will pay anyone to lead such a life. Anyhow, he took his own medicine. But just as if to show that he had not swallowed the bottle too, he jots down in another place in the *Pensées* this thoroughly skeptical query, "Is probability probable?" "It is annoying," as he says in another place, "to dwell upon such trifles; but there is a time for trifling." Pascal's difficulty was that he did not always see clearly when he was trifling, as in his wager against God, or when, as in the clearing up of the Chevalier de Méré's gambling difficulties for him, he was being profound.

CHAPTER SIX

On the Seashore

NEWTON

The method of Fluxions [the calculus] is the general key by help whereof the modern mathematicians unlock the secrets of Geometry, and consequently of Nature.—BISHOP BERKELEY

I do not frame hypotheses.—ISAAC NEWTON

"I DO NOT KNOW what I may appear to the world; but to myself I seem to have been only like a boy playing on the seashore, and diverting myself in now and then finding a smoother pebble or a prettier shell than ordinary, whilst the great ocean of truth lay all undiscovered before me."

Such was Isaac Newton's estimate of himself toward the close of his long life. Yet his successors capable of appreciating his work almost without exception have pointed to Newton as the supreme intellect that the human race has produced—"he who in genius surpassed the human kind."

Isaac Newton, born on Christmas Day ("old style" of dating), 1642, the year of Galileo's death, came of a family of small but independent farmers, living in the manor house of the hamlet of Woolsthorpe, about eight miles south of Grantham in the county of Lincoln, England. His father, also named Isaac, died at the age of thirty seven before the birth of his son. Newton was a premature child. At birth he was so frail and puny that two women who had gone to a neighbor's to get "a tonic" for the infant expected to find him dead on their return. His mother said he was so undersized at birth that a quart mug could easily have contained all there was of him.

Not enough of Newton's ancestry is known to interest students of heredity. His father was described by neighbors as "a wild, extravagant, weak man"; his mother, Hannah Ayscough, was thrifty, industrious, and a capable manageress. After her husband's death Mrs. Newton was recommended as a prospective wife to an old bachelor as "an extraordinary good woman." The cautious bachelor, the Reverend Barnabas Smith, of the neighboring parish of North Witham, married the widow on this testimonial. Mrs. Smith left her three-year-old son to the care of his grandmother. By her second marriage she had three children, none of whom exhibited any remarkable ability. From the property of his mother's second marriage and his father's estate Newton ultimately acquired an income of about £80 a year, which of course meant much more in the seventeenth century than it would now. Newton was not one of the great mathematicians who had to contend with poverty.

As a child Newton was not robust and was forced to shun the rough games of boys his own age. Instead of amusing himself in the usual way, Newton invented his own diversions, in which his genius first showed up. It is sometimes said that Newton was not precocious. This may be true so far as mathematics is concerned, but if it is so in other respects a new definition of precocity is required. The unsurpassed experimental genius which Newton was to exhibit as an explorer in the mysteries of light is certainly evident in the ingenuity of his boyish amusements. Kites with lanterns to scare the credulous villagers at night, perfectly constructed mechanical toys which he made entirely by himself and which worked—waterwheels, a mill that ground wheat into snowy flour, with a greedy mouse (who devoured most of the profits) as both miller and motive power, workboxes and toys for his many little girl friends, drawings, sundials, and a wooden clock (that went) for himself—such were some of the things with which this "un-precocious" boy sought to divert the interests of his playmates into "more philosophical" channels. In addition to these more noticeable evidences of talent far above the ordinary, Newton read extensively and jotted down all manner of mysterious recipes and out-of-the-way observations in his notebook. To rate such a boy as merely the normal, wholesome lad he appeared to his village friends is to miss the obvious.

The earliest part of Newton's education was received in the common village schools of his vicinity. A maternal uncle, the Reverend William Ayscough, seems to have been the first to recognize that Newton was something unusual. A Cambridge graduate himself, Ayscough finally persuaded Newton's mother to send her son to Cambridge instead of keeping him at home, as she had planned, to help her manage the farm on her return to Woolsthorpe after her husband's death when Newton was fifteen.

Before this, however, Newton had crossed his Rubicon on his own initiative. On his uncle's advice he had been sent to the Grantham Grammar School. While there, in the lowest form but one, he was tormented by the school bully who one day kicked Newton in the stomach, causing him much physical pain and mental anguish. Encouraged by one of the schoolmasters, Newton challenged the bully to a fair fight, thrashed him, and, as a final mark of humiliation, rubbed his enemy's cowardly nose on the wall of the church. Up till this young Newton had shown no great interest in his lessons. He now set out to prove his head as good as his fists and quickly rose to the distinction of top boy in the school. The Headmaster and Uncle Ayscough agreed that Newton was good enough for Cambridge, but the decisive die was thrown when Ayscough caught his nephew reading under a hedge when he was supposed to be helping a farmhand to do the marketing.

While at the Grantham Grammar School, and subsequently while preparing for Cambridge, Newton lodged with a Mr. Clarke, the village apothecary. In the apothecary's attic Newton found a parcel of old books, which he devoured, and in the house generally, Clarke's stepdaughter, Miss Storey, with whom he fell in love and to whom he became engaged before leaving Woolsthorpe for Cambridge in June, 1661, at the age of nineteen. But although Newton cherished a warm affection for his first and only sweetheart all her life, absence and growing absorption in his work thrust romance into the background, and Newton never married. Miss Storey became Mrs. Vincent.

Before going on to Newton's student career at Trinity College we may take a short look at the England of his times and some of the scientific knowledge to which the young man fell heir. The bullheaded and bigoted Scottish Stuarts had undertaken to rule England according to the divine rights they claimed were vested in them, with the not uncommon result that mere human beings resented the assumption of celestial authority and rebelled against the sublime conceit, the stupidity, and the incompetence of their rulers. Newton grew up in an atmosphere of civil war—political and religious—in which Puritans and Royalists alike impartially looted whatever was needed to keep their ragged armies fighting. Charles I (born in 1600, beheaded in 1649) had done everything in his power to suppress Parliament; but in spite of his ruthless extortions and the villainously able backing of his own Star Chamber through its brilliant perversions of the law and common justice, he was no match for the dour Puritans under Oliver Cromwell, who in his turn was to back his butcheries and his roughshod march over Parliament by an appeal to the divine justice of his holy cause.

All this brutality and holy hypocrisy had a most salutary effect on young Newton's character: he grew up with a fierce hatred of tyranny, subterfuge, and oppression, and when King James later sought to meddle repressively in University affairs, the mathematician and natural philosopher did not need to learn that a resolute show of backbone and a united front on the part of those whose liberties are endangered is the most effective defense against a coalition of unscrupulous politicians; he knew it by observation and by instinct.

To Newton is attributed the saying "If I have seen a little farther than others it is because I have stood on the shoulders of giants." He had. Among the tallest of these giants were Descartes, Kepler, and Galileo. From Descartes, Newton inherited analytic geometry, which he found difficult at first; from Kepler, three fundamental laws of planetary motion, discovered empirically after twenty two years of inhuman calculation; while from Galileo he acquired the first two of the three laws of motion which were to be the cornerstone of his own dynamics. But bricks do not make a building; Newton was the architect of dynamics and celestial mechanics.

As Kepler's laws were to play the rôle of hero in Newton's development of his law of universal gravitation they may be stated here.

I. *The planets move round the Sun in ellipses; the Sun is at one focus of these ellipses.*

[If S, S' are the foci, P any position of a planet in its orbit, SP + S'P is always equal to AA', the major axis of the ellipse: fig., page 94.]

II. *The line joining the Sun and a planet sweeps out equal areas in equal times.*

III. *The square of the time for one complete revolution of each planet is proportional to the cube of its mean* [or average] *distance from the Sun.*

These laws can be proved in a page or two by means of the calculus applied to Newton's law of universal gravitation:

Any two particles of matter in the universe attract one another with a force which is directly proportional to the product of their masses and in-

versely proportional to the square of the distance between them. Thus if m, M are the masses of the two particles and d the distance between them (all measured in appropriate units), the force of attraction between them is $\frac{k \times m \times M}{d^2}$, where k is some constant number (by suitably choosing the units of mass and distance k may be taken equal to 1, so that the attraction is simply $\frac{m \times M}{d^2}$).

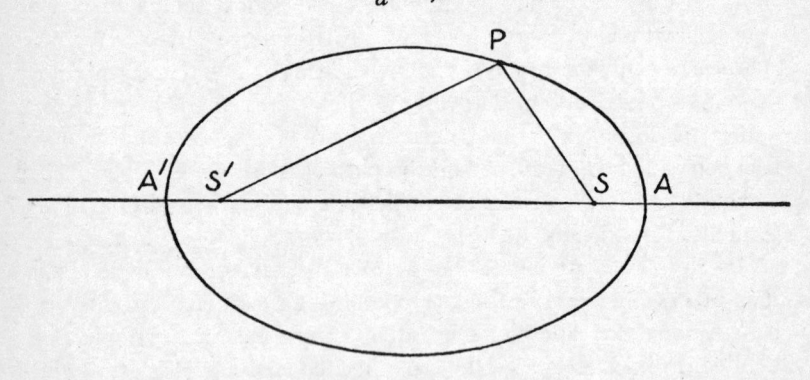

For completeness we state Newton's three laws of motion.

I. *Every body will continue in its state of rest or of uniform* [unaccelerated] *motion in a straight line except in so far as it is compelled to change that state by impressed force.*

II. *Rate of change of momentum* ["mass times velocity," mass and velocity being measured in appropriate units] *is proportional to the impressed force and takes place in the line in which the force acts.*

III. *Action and reaction* [as in the collision on a frictionless table of perfectly elastic billiard balls] *are equal and opposite* [the momentum one ball loses is gained by the other].

The most important thing for mathematics in all of this is the phrase opening the statement of the second law of motion, *rate of change*. What is a rate, and how shall it be measured? Momentum, as noted, is "mass times velocity." The masses which Newton discussed were assumed to remain constant during their motion—not like the electrons and other particles of current physics whose masses increase appreciably as their velocity approaches a measurable fraction of that of light. Thus, to investigate "rate of change of momentum," it sufficed Newton to clarify *velocity*, which is rate of change of position. His solution of this problem—giving a workable mathematical method for investigating the velocity of any particle moving in any continuous manner, no matter how erratic—gave him the master key to the whole mystery of rates and their measurement, namely, the *differential* calculus.

A similar problem growing out of rates put the *integral* calculus into his hands. How shall the total distance passed over in a given time by a moving particle whose velocity is varying continuously from instant to instant be calculated? Answering this or similar problems, some phrased geometrically, Newton came upon the integral calculus. Finally, pondering the two types of problem together, Newton made a capital discovery: he saw that the differential calculus and the integral calculus are intimately and reciprocally related by what is today called "the fundamental theorem of the calculus"—which will be described in the proper place.

In addition to what Newton inherited from his predecessors in science and mathematics he received from the spirit of his age two further gifts, a passion for theology and an unquenchable thirst for the mysteries of alchemy. To censure him for devoting his unsurpassed intellect to these things, which would now be considered unworthy of his serious effort, is to censure oneself. For in Newton's day alchemy *was* chemistry and it had *not* been shown that there was nothing much in it—except what was to come out of it, namely modern chemistry; and Newton, as a man of inborn scientific spirit, undertook to find out *by experiment* exactly what the claims of the alchemists amounted to.

As for theology, Newton was an unquestioning believer in an all-wise Creator of the universe and in his own inability—like that of the boy on the seashore—to fathom the entire ocean of truth in all its depths. He therefore believed that there were not only many things in heaven beyond his philosophy but plenty on earth as well, and he made it his business to understand for himself what the majority of intelligent men of his time accepted without dispute (to them it was as natural as common sense)—the traditional account of creation.

He therefore put what he considered his really serious efforts on attempts to prove that the prophecies of Daniel and the poetry of the Apocalypse make sense, and on chronological researches whose object was to harmonize the dates of the Old Testament with those of history. In Newton's day theology was still queen of the sciences and she sometimes ruled her obstreperous subjects with a rod of brass and a head of cast iron. Newton however did permit his rational science to influence his beliefs to the extent of making him what would now be called a Unitarian.

In June, 1661 Newton entered Trinity College, Cambridge, as a subsizar—a student who (in those days) earned his expenses by menial service. Civil war, the restoration of the monarchy in 1661, and uninspired toadying to the Crown on the part of the University had all brought Cambridge to one of the low-water marks in its history as an educational institution when Newton took up his residence. Nevertheless young Newton, lonely at first, quickly found himself and became absorbed in his work.

In mathematics Newton's teacher was Dr. Isaac Barrow (1630–1677), a theologian and mathematician of whom it has been said that brilliant and original as he undoubtedly was in mathematics, he had the misfortune to be the morning star heralding Newton's sun. Barrow gladly recognized that a greater than himself had arrived, and when (1669) the strategic moment came he resigned the Lucasian Professorship of Mathematics (of which he was the first holder) in favor of his incomparable pupil. Barrow's geometrical lectures dealt among other things with his own methods for finding areas and drawing tangents to curves—essentially the key problems of the integral and the differential calculus respectively, and there can be no doubt that these lectures inspired Newton to his own attack.

The record of Newton's undergraduate life is disappointingly meager. He seems to have made no very great impression on his fellow students, nor do his brief, perfunctory letters home tell anything of interest. The first two years were spent mastering elementary mathematics. If there is any reliable account of Newton's sudden maturity as a discoverer, none of his modern biographers seems to have located it. Beyond the fact that in the three years 1664–66 (age twenty one to twenty three) he laid the foundation of all his subsequent work in science and mathematics, and that incessant work and late hours brought on an illness, we know nothing definite. Newton's tendency to secretiveness about his discoveries has also played its part in deepening the mystery.

On the purely human side Newton was normal enough as an undergraduate to relax occasionally, and there is a record in his account book of several sessions at the tavern and two losses at cards. He took his B.A. degree in January, 1664.

The Great Plague (bubonic plague) of 1664–65, with its milder recurrence the following year, gave Newton his great if forced opportunity. The University was closed, and for the better part of two years Newton retired to meditate at Woolsthorpe. Up till then he had done nothing remarkable—except make himself ill by too assiduous observation of a comet and lunar halos—or, if he had, it was a secret. In these two years he invented the method of fluxions (the calculus), discovered the law of universal gravitation, and proved experimentally that white light is composed of light of all the colors. All this before he was twenty five.

A manuscript dated May 20, 1665, shows that Newton at the age of twenty three had sufficiently developed the principles of the calculus to be able to find the tangent and curvature at any point of any continuous curve. He called his method "fluxions"—from the idea of "flowing" or variable quantities and their rates of "flow" or "growth." His discovery of the binomial theorem, an essential step toward a fully developed calculus, preceded this.

The binomial theorem generalizes the simple results like

$$(a+b)^2 = a^2 + 2ab + b^2, (a+b)^3 = a^3 + 3a^2b + 3ab^2 + b^3,$$

and so on, which are found by direct calculation; namely,

$$(a+b)^n = a^n + \frac{n}{1}a^{n-1}b + \frac{n(n-1)}{1 \times 2}a^{n-2}b^2 + \frac{n(n-1)(n-2)}{1 \times 2 \times 3}a^{n-3}b^3 + \ldots,$$

where the dots indicate that the series is to be continued according to the same law as that indicated for the terms written; the next term is

$$\frac{n(n-1)(n-2)(n-3)}{1 \times 2 \times 3 \times 4}a^{n-4}b^4.$$

If n is one of the positive integers $1, 2, 3 \ldots$, the series automatically terminates after precisely $n+1$ terms. This much is easily proved (as in the school algebras) by mathematical induction.

But if n is not a positive integer, the series does not terminate, and this method of proof is inapplicable. As a proof of the binomial theorem for fractional and negative values of n (also for more general values), with a statement of the necessary restrictions on a,b, came only in the nineteenth century, we need merely state here that in extending the theorem to these values of n Newton satisfied himself that the theorem was correct for such values of a,b as he had occasion to consider in his work.

If all modern refinements are similarly ignored in the manner of the seventeenth century it is easy to see how the calculus finally got itself invented. The underlying notions are those of *variable*, *function*, and *limit*. The last took long to clarify.

A letter, say s, which can take on several different values during the course of a mathematical investigation is called a *variable*; for example s is a variable if it denotes the height of a falling body above the earth.

The word *function* (or its Latin equivalent) seems to have been introduced into mathematics by Leibniz in 1694; the concept now dominates much of mathematics and is indispensable in science. Since Leibniz' time the concept has been made precise. If y and x are two variables so related that whenever a numerical value is assigned to x there is determined a numerical value of y, then y is called a (one-valued, or *uniform*) function of x, and this is symbolized by writing $y = f(x)$.

Instead of attempting to give a modern definition of a *limit* we shall content ourselves with one of the simplest examples of the sort which led the followers of Newton and Leibniz (the former especially) to the use of limits in discussing rates of change. To the early developers of the calculus the notions of variables and limits were intuitive; to us they are extremely subtle concepts hedged about with thickets of semi-metaphysical mysteries concerning the nature of numbers, both rational and irrational.

Let y be a function of x, say $y = f(x)$. *The rate of change of y with respect to x*, or, as it is called, the *derivative of y with respect to x*, is defined as follows. To x is given any increment, say Δx (read, "increment of x"), so that x becomes $x + \Delta x$, and $f(x)$, or y, becomes $f(x + \Delta x)$. The corresponding increment, Δy, of y is its *new* value *minus* its initial value; namely, $\Delta y = f(x + \Delta x) - f(x)$. As a crude approximation to the rate of change of y with respect to x we may take, by our intuitive notion of a rate as an "average," the result of dividing the increment of y by the increment of x, that is, $\dfrac{\Delta y}{\Delta x}$.

But this obviously is too crude, as both of x and y are varying and we cannot say that this average represents the rate for *any particular* value of x. Accordingly, we decrease the increment Δx *indefinitely*, till, "in the limit" Δx approaches zero, and follow the "average" $\frac{\Delta y}{\Delta x}$ all through the process: Δy similarly decreases indefinitely and ultimately approaches zero; but $\frac{\Delta y}{\Delta x}$ does not, thereby, present us with the meaningless symbol $\frac{0}{0}$, but with a definite *limiting value*, which is the required rate of change of y with respect to x.

To see how it works out, let $f(x)$ be the particular function x^2, so that $y = x^2$. Following the above outline we get first

$$\frac{\Delta y}{\Delta x} = \frac{(x + \Delta x)^2 - x^2}{\Delta x}.$$

Nothing is yet said about limits. Simplifying the algebra we find

$$\frac{\Delta y}{\Delta x} = 2x + \Delta x.$$

Having simplified the algebra as far as possible, we *now* let Δx approach zero and see that the limiting value of $\frac{\Delta y}{\Delta x}$ is $2x$. Quite generally, in the same way, if $y = x^n$, the limiting value of $\frac{\Delta y}{\Delta x}$ is nx^{n-1}, as may be proved with the aid of the binomial theorem.

Such an argument would not satisfy a student today, but something not much better was good enough for the inventors of the calculus and it will have to do for us here. If $y = f(x)$, the *limiting value* of $\frac{\Delta y}{\Delta x}$ (provided such a value exists) is called the *derivative of y with respect to x*, and is denoted by $\frac{dy}{dx}$. This symbolism is due (essentially) to Leibniz and is the one in common use today; Newton used another ($\dot{y}$) which is less convenient.

The simplest instances of rates in physics are velocity and acceleration, two of the fundamental notions of dynamics. Velocity is rate of change of *distance* (or "position," or "space") with respect to *time*; *acceleration* is rate of change of *velocity* with respect to *time*.

If s denotes the distance traversed in the time t by a moving particle (it being assumed that the distance is a function of the time), the velocity at the time t is $\dfrac{ds}{dt}$. Denoting this velocity by v, we have the corresponding acceleration, $\dfrac{dv}{dt}$.

This introduces the idea of a *rate of a rate*, or of a *second derivative*. For in accelerated motion the velocity is not constant but variable, and hence it has a rate of change: the acceleration is the rate of change of the rate of change of distance (both rates with respect to time); and to indicate this *second* rate, or "rate of a rate," we write $\dfrac{d^2s}{dt^2}$ for the acceleration. This itself may have a rate of change with respect to the time; this *third* rate is written $\dfrac{d^3s}{dt^3}$. And so on for fourth, fifth, . . . rates, namely for fourth, fifth, . . . derivatives. The most important derivatives in the applications of the calculus to science are the first and second.

If now we look back at what was said concerning Newton's second law of motion and compare it with the like for acceleration, we see that "forces" are proportional to the accelerations they produce. With this much we can "set up" the *differential equation* for a problem which is by no means trivial—that of "central forces": a particle is attracted toward a fixed point by a force whose direction always passes through the fixed point. Given that the force varies as some function of the distance s, say as $F(s)$, where s is the distance of the particle at the time t from the fixed point O,

```
O              F(s)                        t
|---------------◄---------------------------|
                                            s
```

it is required to describe the motion of the particle. A little consideration will show that

$$\frac{d^2s}{dt^2} = -F(s),$$

the minus sign being taken because the attraction diminishes the velocity. This is the *differential equation* of the problem, so called because it involves a rate (the acceleration), and rates (or derivatives) are the object of investigation in the *differential* calculus.

Having translated the problem into a differential equation we are now required to solve this equation, that is, to find the relation between s and t, or, in mathematical language, to solve the differential equation by expressing s as a function of t. This is where the difficulties begin. It may be quite easy to translate a given physical situation into a set of differential equations which no mathematician can solve. In general every essentially new problem in physics leads to types of differential equations which demand the creation of new branches of mathematics for their solution. The particular equation above can however be solved quite simply in terms of elementary functions if $F(s) = \dfrac{1}{s^2}$ as in Newton's law of gravitational attraction. Instead of bothering with this particular equation, we shall consider a much simpler one which will suffice to bring out the point of importance:

$$\frac{dy}{dx} = x.$$

We are given that y is a function of x whose derivative is equal to x; it is required to express y as a function of x. More generally, consider in the same way

$$\frac{dy}{dx} = f(x).$$

This asks, what is the function y (of x) whose derivative (rate of change) with respect to x is equal to $f(x)$? Provided we can find the function required (or provided such a function exists), we call it the *anti-derivative* of $f(x)$ and denote it by $\int f(x)dx$—for a reason that will appear presently. For the moment we need note only that $\int f(x)dx$ symbolizes a function (if it exists) *whose derivative* is equal to $f(x)$.

By inspection we see that the first of the above equations has the solution $\frac{1}{2}x^2 + c$, where c is a constant (number not depending on the variable x); thus $\int x\,dx = \frac{1}{2}x^2 + c$.

Even this simple example may indicate that the problem of evaluating $\int f(x)dx$ for comparatively innocent looking functions $f(x)$ may be beyond our powers. It does not follow that an "answer" exists at all *in terms of known functions* when an $f(x)$ is chosen at random—the odds against such a chance are an infinity of the worst sort ("non-denumerable") to one. When a physical problem leads to one of these nightmares approximate methods are applied which give the result within the desired accuracy.

With the two basic notions, $\dfrac{dy}{dx}$ and $\int f(x)dx$, of the calculus we can now describe the *fundamental theorem of the calculus* connecting them. For simplicity we shall use a diagram, although this is not necessary and is undesirable in an exact account.

Consider a continuous, unlooped curve whose equation is $y = f(x)$ in Cartesian coordinates. It is required to find the area included between the curve, the x-axis and the two perpendiculars AA', BB' drawn to the x-axis from any two points A, B on the curve. The distances OA', OB' are a, b respectively—namely, the coordinates of A', B' are $(a, 0)$, $(b, 0)$. We proceed as Archimedes did, cutting the re-

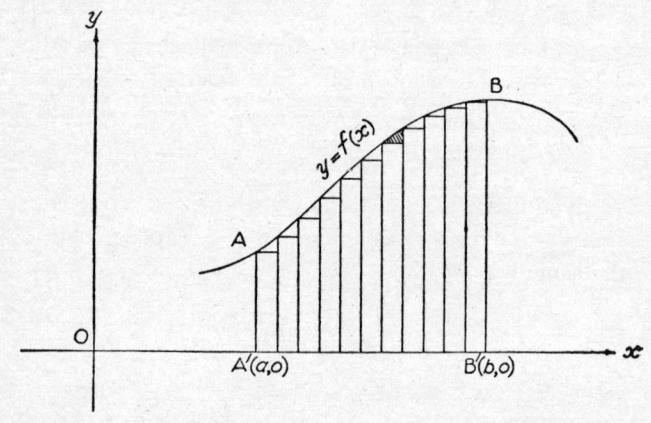

quired area into parallel strips of equal breadth, treating these strips as rectangles by disregarding the top triangular bits (one of which is shaded in the figure), adding the areas of all these rectangles, and finally evaluating *the limit of this sum* as the number of rectangles is increased indefinitely. This is all very well, but how are we to calculate the limit? The answer is surely one of the most astonishing things a mathematician ever discovered.

First, find $\int f(x) dx$. Say the result is $F(x)$. In this substitute a and b, getting $F(a)$ and $F(b)$. Then subtract the first from the second, $F(b) - F(a)$. *This is the required area.*

Notice the connection between $y = f(x)$, the equation of the given curve; $\dfrac{dy}{dx}$, which (as seen in the chapter on Fermat) gives the *slope* of the tangent line to the curve at the point (x, y); and $\int f(x) dx$, or $F(x)$, which is the function whose *rate of change* with respect to x is equal to $f(x)$. We have just stated that the *area* required, which is a *limiting sum* of the kind described in connection with Archimedes, is given by $F(b) - F(a)$. Thus we have connected *slopes*, or *derivatives*, with *limiting sums*, or, as they are called, *definite integrals*. The symbol $\int$ is an old-fashioned S, the first letter of the word *Summa*.

Summing all this up in symbols, we write for the area in question $\int_a^b f(x)dx$; a is the *lower limit* of the sum, b the *upper limit*; and

$$\int_a^b f(x)dx = F(b) - F(a),$$

in which $F(b)$, $F(a)$ are calculated by evaluating the *"indefinite integral"* $\int f(x)dx$, namely, by finding that function $F(x)$ such that its derivative with respect to x, $\frac{dF(x)}{dx}$, is equal to $f(x)$. This is the fundamental theorem of the calculus as it presented itself (in its geometrical form) to Newton and independently also to Leibniz. As a caution we repeat that numerous refinements demanded in a modern statement have been ignored.

Two simple but important matters may conclude this sketch of the leading notions of the calculus as they appeared to the pioneers. So far only functions of a single variable have been considered. But nature presents us with functions of several variables and even of an infinity of variables.

To take a very simple example, the volume, V, of a gas is a function of its temperature, T, and the pressure, P, on it; say $V = F(T,P)$ —the actual form of the function F need not be specified here. As T, P vary, V varies. But suppose *only one* of T, P varies while the other is held constant. We are then back essentially with a function of *one* variable, and the derivative of $F(T,P)$ can be calculated with respect to this variable. If T varies while P is held constant, the derivative of $F(T,P)$ with respect to T is called the *partial derivative* (with respect to T), and to show that the variable P is being held constant, a different symbol, ∂, is used for this partial derivative, $\frac{\partial F(T,P)}{\partial T}$. Similarly, if P varies while T is held constant, we get $\frac{\partial F(T,P)}{\partial P}$. Precisely as in the case of ordinary second, third, . . . derivatives, we have the like for partial derivatives; thus $\frac{\partial^2 F(T,P)}{\partial T^2}$ signifies the partial derivative of $\frac{\partial F(T,P)}{\partial T}$ with respect to T.

The great majority of the important equations of mathematical physics are *partial differential equations*. A famous example is Laplace's equation, or the "equation of continuity," which appears in the theory of Newtonian gravitation, electricity and magnetism, fluid motion, and elsewhere:

$$\frac{\partial^2 u}{\partial x^2} + \frac{\partial^2 u}{\partial y^2} + \frac{\partial^2 u}{\partial z^2} = 0.$$

In fluid motion this is the mathematical expression of the fact that a "perfect" fluid, in which there are no vortices, is indestructible. A derivation of this equation would be out of place here, but a statement of what it signifies may make it seem less mysterious. If there are no vortices in the fluid, the three component velocities parallel to the axes of x,y,z of any particle in the fluid are calculable as the partial derivatives

$$-\frac{\partial u}{\partial x}, \quad -\frac{\partial u}{\partial y}, \quad -\frac{\partial u}{\partial z}$$

of the *same* function u—which will be determined by the particular type of motion. Combining this fact with the obvious remark that if the fluid is incompressible and indestructible, as much fluid must flow out of any small volume in one second as flows into it; and noting that the amount of flow in one second across any small area is equal to the rate of flow multiplied by the area; we see (on combining these remarks and calculating the total inflow and total outflow) that Laplace's equation is more or less of a platitude.

The really astonishing thing about this and some other equations of mathematical physics is that a physical platitude, when subjected to mathematical reasoning, should furnish unforeseen information which is anything but platitudinous. The "anticipations" of physical phenomena mentioned in later chapters arose from such commonplaces treated mathematically.

Two very real difficulties, however, arise in this type of problem. The first concerns the physicist, who must have a feeling for what complications can be lopped off his problem, without mutilating it beyond all recognition, so that he can state it mathematically at all. The second concerns the mathematician, and this brings us to a matter of great importance—the last we shall mention in this sketch of the calculus—that of what are called *boundary-value problems*.

Science does not fling an equation like Laplace's at a mathematician's head and ask him to find the *general* solution. What it wants is something (usually) much more difficult to obtain, a *particular* solution which will not only satisfy the equation but which *in addition will satisfy certain auxiliary conditions* depending on the particular problem to be solved.

The point may be simply illustrated by a problem in the conduction of heat. There is a *general* equation (Fourier's) for the "motion" of heat in a conductor similar to Laplace's for fluid motion. Suppose it is required to find the final distribution of temperature in a cylindrical rod whose ends are kept at one constant temperature and whose curved surface is kept at another; "final" here means that there is a "steady state"—no further change in temperature—at all points of the rod. The solution must not only satisfy the *general* equation, it must also fit the *surface-temperatures*, or the *initial boundary conditions*.

The second is the harder part. For a cylindrical rod the problem is quite different from the corresponding problem for a bar of rectangular cross section. The theory of *boundary-value problems* deals with the fitting of solutions of differential equations to prescribed initial conditions. It is largely a creation of the past eighty years. In a sense mathematical physics is co-extensive with the theory of boundary-value problems.

The second of Newton's great inspirations which came to him as a youth of twenty two or three in 1666 at Woolsthorpe was his law of universal gravitation (already stated). In this connection we shall not repeat the story of the falling apple. To vary the monotony of the classical account we shall give Gauss' version of the legend when we come to him.

Most authorities agree that Newton did make some rough calculations in 1666 (he was then twenty three) to see whether his law of universal gravitation would account for Kepler's laws. Many years later (in 1684) when Halley asked him what law of attraction would account for the elliptical orbits of the planets Newton replied at once the inverse square.

"How do you know?" Halley asked—he had been prompted by Sir Christopher Wren and others to put the question, as a great argument over the problem had been going on for some time in London.

"Why, I have calculated it," Newton replied. On attempting to restore his calculation (which he had mislaid) Newton made a slip, and believed he was in error. But presently he found his mistake and verified his original conclusion.

Much has been made of Newton's twenty years' delay in the publication of the law of universal gravitation as an undeserved setback due to inaccurate data. Of three explanations a less romantic but more mathematical one than either of the others is to be preferred here.

Newton's delay was rooted in his inability to solve a certain problem in the integral calculus which was crucial for the whole theory of universal gravitation as expressed in the Newtonian law. Before he could account for the motion of both the apple and the Moon Newton had to find the total attraction of a solid homogeneous sphere on any mass particle outside the sphere. For *every* particle of the sphere attracts the mass particle outside the sphere with a force varying directly as the product of the masses of the two particles and inversely as the square of the distance between them: how are all these separate attractions, infinite in number, to be compounded or added into one resultant attraction?

This evidently is a problem in the integral calculus. Today it is given in the textbooks as an example which young students dispose of in twenty minutes or less. Yet it held Newton up for twenty years. He finally solved it, of course: the attraction is the same as if the entire mass of the sphere were concentrated in *a single point* at its centre. The problem is thus reduced to finding the attraction between two mass particles at a given distance apart, and the immediate solution of this is as stated in Newton's law. If this is the correct explanation for the twenty years' delay, it may give us some idea of the enormous amount of labor which generations of mathematicians since Newton's day have expended on developing and simplifying the calculus to the point where very ordinary boys of sixteen can use it effectively.

Although our principal interest in Newton centers about his greatness as a mathematician we cannot leave him with his undeveloped masterpiece of 1666. To do so would be to give no idea of his magnitude, so we shall go on to a brief outline of his other activities without entering into detail (for lack of space) on any of them.

On his return to Cambridge Newton was elected a Fellow of Trinity in 1667 and in 1669, at the age of twenty six, succeeded Barrow as Lucasian Professor of Mathematics. His first lectures were on optics. In these he expounded his own discoveries and sketched his corpuscular theory of light, according to which light consists in an emission of corpuscles and is not a wave phenomenon as Huygens and Hooke asserted. Although the two theories appear to be contradictory both are useful today in correlating the phenomena of light and are, in a purely mathematical sense, reconciled in the modern quantum theory. Thus it is not now correct to say, as it may have been a few years ago, that Newton was entirely wrong in his corpuscular theory.

The following year, 1668, Newton constructed a reflecting telescope with his own hands and used it to observe the satellites of Jupiter. His object doubtless was to see whether universal gravitation really was universal by observations on Jupiter's satellites. This year is also memorable in the history of the calculus. Mercator's calculation by means of infinite series of an area connected with a hyperbola was brought to Newton's attention. The method was practically identical with Newton's own, which he had not published, but which he now wrote out, gave to Dr. Barrow, and permitted to circulate among a few of the better mathematicians.

On his election to the Royal Society in 1672 Newton communicated his work on telescopes and his corpuscular theory of light. A commission of three, including the cantankerous Hooke, was appointed to report on the work on optics. Exceeding his authority as a referee Hooke seized the opportunity to propagandize for the undulatory theory and himself at Newton's expense. At first Newton was cool and scientific under criticism, but when the mathematician Lucas and the physician Linus, both of Liège, joined Hooke in adding suggestions and objections which quickly changed from the legitimate to the carping and the merely stupid, Newton gradually began to lose patience.

A reading of his correspondence in this first of his irritating controversies should convince anyone that Newton was not by nature secretive and jealous of his discoveries. The tone of his letters gradually changes from one of eager willingness to clear up the difficulties which others found, to one of bewilderment that scientific men should regard science as a battleground for personal quarrels. From bewilderment he quickly passes to cold anger and a hurt, somewhat childish resolution to play by himself in future. He simply could not suffer malicious fools gladly.

At last, in a letter of November 18, 1676, he says, "I see I have made myself a slave to philosophy, but if I get free of Mr. Lucas's business, I will resolutely bid adieu to it eternally, excepting what I do for my private satisfaction, or leave to come out after me; for I see a man must either resolve to put out nothing new, or become a slave to defend it." Almost identical sentiments were expressed by Gauss in connection with non-Euclidean geometry.

Newton's petulance under criticism and his exasperation at futile controversies broke out again after the publication of the *Principia*. Writing to Halley on June 20, 1688, he says, "Philosophy [science] is such an impertinently litigious Lady, that a man had as good be engaged to lawsuits, as to have to do with her. I found it so formerly, and now I am no sooner come near her again, but she gives me warning." Mathematics, dynamics, and celestial mechanics were in fact—we may as well admit it—secondary interests with Newton. His heart was in his alchemy, his researches in chronology, and his theological studies.

It was only because an inner compulsion drove him that he turned as a recreation to mathematics. As early as 1679, when he was thirty seven (but when also he had his major discoveries and inventions securely locked up in his head or in his desk), he writes to the pestiferous Hooke: "I had for some years last been endeavoring to bend myself from philosophy to other studies in so much that I have long grutched the time spent in that study unless it be perhaps at idle hours sometimes for diversion." These "diversions" occasionally cost him more incessant thought than his professed labors, as when he made himself seriously ill by thinking day and night about the motion of the Moon, the only problem, he says, that ever made his head ache.

Another side of Newton's touchiness showed up in the spring of 1673 when he wrote to Oldenburg resigning his membership in the Royal Society. This petulant action has been variously interpreted. Newton gave financial difficulties and his distance from London as his reasons. Oldenburg took the huffy mathematician at his word and told him that under the rules he could retain his membership without paying. This brought Newton to his senses and he withdrew his resignation, having recovered his temper in the meantime. Nevertheless Newton thought he was about to be hard pressed. However, his finances presently straightened out and he felt better. It may be noted here that Newton was no absent-minded dreamer when it came to a question of money. He was extremely shrewd and he died a rich man for his times. But if shrewd and thrifty he was also very liberal with his money and was always ready to help a friend in need as unobtrusively as possible. To young men he was particularly generous.

The years 1684–86 mark one of the great epochs in the history of all human thought. Skilfully coaxed by Halley, Newton at last consented to write up his astronomical and dynamical discoveries for publication. Probably no mortal has ever thought as hard and as continuously as Newton did in composing his *Philosophiae Naturalis Principia Mathematica* (Mathematical Principles of Natural Philosophy). Never careful of his bodily health, Newton seems to have forgotten that he had a body which required food and sleep when he gave himself up to the composition of his masterpiece. Meals were ignored or forgotten, and on arising from a snatch of sleep he would sit on the edge of the bed half-clothed for hours, threading the mazes of his mathematics. In 1686 the *Principia* was presented to the Royal Society, and in 1687 was printed at Halley's expense.

A description of the contents of the *Principia* is out of the question here, but a small handful of the inexhaustible treasures it contains may be briefly exhibited. The spirit animating the whole work is Newton's dynamics, his law of universal gravitation, and the application of both to the solar system—"the system of the world." Although the calculus has vanished from the synthetic geometrical demonstrations, Newton states (in a letter) that he used it to *discover* his results and, having done so, proceeded to rework the proofs furnished by the calculus into geometrical shape so that his contemporaries might the more readily grasp the main theme—the dynamical harmony of the heavens.

First, Newton deduced Kepler's empirical laws from his own law of gravitation, and he showed how the mass of the Sun can be calculated, also how the mass of any planet having a satellite can be determined. Second, he initiated the extremely important theory of *perturbations*: the Moon, for example, is attracted not only by the Earth but by the Sun also; hence the orbit of the Moon will be perturbed by the pull of the Sun. In this manner Newton accounted for two ancient observations due to Hipparchus and Ptolemy. Our own generation has seen the now highly developed theory of perturbations applied to electronic orbits, particularly for the helium atom. In addition to these ancient observations, seven other irregularities of the

Moon's motion observed by Tycho Brahe (1546–1601), Flamsteed (1646–1719), and others, were deduced from the law of gravitation.

So much for lunar perturbations. The like applies also to the planets. Newton began the theory of planetary perturbations, which in the nineteenth century was to lead to the discovery of the planet Neptune and, in the twentieth, to that of Pluto.

The "lawless" comets—still warnings from an angered heaven to superstitious eyes—were brought under the universal law as harmless members of the Sun's family, with such precision that we now calculate and welcome their showy return (unless Jupiter or some other outsider perturbs them unduly), as we did in 1910 when Halley's beautiful comet returned promptly on schedule after an absence of seventy four years.

He began the vast and still incomplete study of planetary evolution by calculating (from his dynamics and the universal law) the flattening of the earth at its poles due to diurnal rotation, and he proved that the shape of a planet determines the length of its day, so that if we knew accurately how flat Venus is at the poles, we could say how long it takes her to turn completely once round the axis joining her poles. He calculated the variation of weight with latitude. He proved that a hollow shell, bounded by concentric spherical surfaces, and homogeneous, exerts no force on a small body anywhere inside it. The last has important consequences in electrostatics—also in the realm of fiction, where it has been used as the motif for amusing fantasies.

The precession of the equinoxes was beautifully accounted for by the pull of the Moon and the Sun on the equatorial bulge of the Earth causing our planet to wobble like a top. The mysterious tides also fell naturally into the grand scheme—both the lunar and the solar tides were calculated, and from the observed heights of the spring and neap tides the mass of the Moon was deduced. The First Book laid down the principles of dynamics; the Second, the motion of bodies in resisting media, and fluid motion; the Third was the famous "System of the World."

Probably no other law of nature has so simply unified any such mass of natural phenomena as has Newton's law of universal gravitation in his *Principia*. It is to the credit of Newton's contemporaries that they recognized at least dimly the magnitude of what had been done, although but few of them could follow the reasoning by which the stupendous miracle of unification had been achieved, and made of the author of the *Principia* a demigod. Before many years had passed the Newtonian system was being taught at Cambridge (1699) and Oxford (1704). France slumbered on for half a century, still dizzy from the whirl of Descartes' angelic vortices. But presently mysticism gave way to reason and Newton found his greatest successor not in England but in France, where Laplace set himself the task of continuing and rounding out the *Principia*.

After the *Principia* the rest is anticlimax. Although the lunar theory continued to plague and "divert" him, Newton was temporarily sick of "philosophy" and welcomed the opportunity to turn to less celestial affairs. James II, obstinate Scot and bigoted Catholic that he was, had determined to force the University to grant a master's degree to a Benedictine over the protests of the academic authorities. Newton was one of the delegates who in 1687 went to London to present the University's case before the Court of High Commission presided over by that great and blackguardly lawyer the Lord High Chancellor George Jeffreys—"infamous Jeffreys" as he is known in history. Having insulted the leader of the delegates in masterly fashion, Jeffreys dismissed the rest with the injunction to go and sin no more. Newton apparently held his peace. Nothing was to be gained by answering a man like Jeffreys in his own kennel. But when the others would have signed a disgraceful compromise it was Newton who put backbone into them and kept them from signing. He won the day; nothing of any value was lost—not even honor. "An honest courage in these matters," he wrote later, "will secure all, having law on our sides."

Cambridge evidently appreciated Newton's courage, for in January, 1689, he was elected to represent the University at the Convention Parliament after James II had fled the country to make room for William of Orange and his Mary, and the faithful Jeffreys was burrowing into dunghills to escape the ready justice of the mob. Newton sat in Parliament till its dissolution in February, 1690. To his credit he never made a speech in the place. But he was faithful to his office and not averse to politics; his diplomacy had much to do with keeping the turbulent University loyal to the decent King and Queen.

Newton's taste of "real life" in London proved his scientific undoing. Influential and officious friends, including the philosopher John Locke (1632–1704) of *Human Understanding* fame, convinced Newton that he was not getting his share of the honors. The crowning imbecility of the Anglo-Saxon breed is its dumb belief in public office or an administrative position as the supreme honor for a man of intellect. The English finally (1699) made Newton Master of the Mint to reform and supervise the coinage of the Realm. For utter bathos this "elevation" of the author of the *Principia* is surpassed only by the jubilation of Sir David Brewster in his life of Newton (1860) over the "well-merited recognition" thus accorded Newton's genius by the English people. Of course if Newton really wanted anything of the sort there is nothing to be said; he had earned the right millions of times over to do anything he desired. But his busybody friends need not have egged him on.

It did not happen all at once. Charles Montagu, later Earl of Halifax, Fellow of Trinity College and a close friend of Newton, aided and abetted by the everlastingly busy and gossipy Samuel Pepys (1633–1703) of diary notoriety, stirred up by Locke and by Newton himself, began pulling wires to get Newton some recognition "worthy" of him.

The negotiations evidently did not always run smoothly and Newton's somewhat suspicious temperament caused him to believe that some of his friends were playing fast and loose with him—as they probably were. The loss of sleep and the indifference to food which had enabled him to compose the *Principia* in eighteen months took their revenge. In the autumn of 1692 (when he was nearly fifty and should have been at his best) Newton fell seriously ill. Aversion to all food and an almost total inability to sleep, aggravated by a temporary persecution mania, brought on something dangerously close to a total mental collapse. A pathetic letter of September 16, 1693 to Locke, written after his recovery, shows how ill he had been.

SIR,

Being of opinion that you endeavored to embroil me with women and by other means,* I was so much affected with it that when one told me you were sickly and would not live, I answered, 'twere better if you were dead. I desire you to forgive me for this uncharitableness. For I am now satisfied that what you have done is just, and I beg your pardon for having hard thoughts of you for it, and for representing that you struck at the root of morality, in

---

*There had been gossip that Newton's favorite niece had used her charms to further Newton's advancement.

a principle you laid down in your book of ideas, and designed to pursue in another book, and that I took you for a Hobbist. I beg your pardon also for saying or thinking that there was a design to sell me an office, or to embroil me.
>I am your most humble
>And unfortunate servant,
>Is. NEWTON

The news of Newton's illness spread to the Continent where, naturally, it was greatly exaggerated. His friends, including one who was to become his bitterest enemy, rejoiced at his recovery. Leibniz wrote to an acquaintance expressing his satisfaction that Newton was himself again. But in the very year of his recovery (1693) Newton heard for the first time that the calculus was becoming well known on the Continent and that it was commonly attributed to Leibniz.

The decade after the publication of the *Principia* was about equally divided between alchemy, theology, and worry, with more or less involuntary and headachy excursions into the lunar theory. Newton and Leibniz were still on cordial terms. Their respective "friends," ignorant as Kaffirs of all mathematics and of the calculus in particular, had not yet decided to pit one against the other with charges of plagiarism in the invention of the calculus, and even grosser dishonesty, in the most shameful squabble over priority in the history of mathematics. Newton recognized Leibniz' merits, Leibniz recognized Newton's, and at this peaceful stage of their acquaintance neither for a moment suspected that the other had stolen so much as a single idea of the calculus from the other.

Later, in 1712, when even the man in the street—the zealous patriot who knew nothing of the facts—realized vaguely that Newton had done something tremendous in mathematics (more, probably, as Leibniz said, than had been done in all history before him), the question as to who had invented the calculus became a matter of acute national jealousy, and all educated England rallied behind its somewhat bewildered champion, howling that his rival was a thief and a liar.

Newton at first was not to blame. Nor was Leibniz. But as the British sporting instinct presently began to assert itself, Newton acquiesced in the disgraceful attack and himself suggested or consented to shady schemes of downright dishonesty designed to win the international championship at any cost—even that of national honor. Leibniz and his backers did likewise. The upshot of it all was that the obstinate British practically rotted mathematically for all of a century after Newton's death, while the more progressive Swiss and French, following the lead of Leibniz, and developing his incomparably better way of merely *writing* the calculus, perfected the subject and made it the simple, easily applied implement of research that Newton's immediate successors should have had the honor of making it.

In 1696, at the age of fifty four, Newton became Warden of the Mint. His job was to reform the coinage. Having done so, he was promoted in 1699 to the dignity of Master. The only satisfaction mathematicians can take in this degradation of the supreme intellect of ages is the refutation which it afforded of the silly superstition that mathematicians have no practical sense. Newton was one of the best Masters the Mint ever had. He took his job seriously.

In 1701-2 Newton again represented Cambridge University in Parliament, and in 1703 was elected President of the Royal Society, an honorable office to which he was reëlected time after time till his death in 1727. In 1705 he was knighted by good Queen Anne. Probably this honor was in recognition of his services as a money-changer rather than in acknowledgment of his preëminence in the temple of wisdom. This is all as it should be: if "a riband to stick in his coat" is the reward of a turncoat politician, why should a man of intellect and integrity feel flattered if his name appears in the birthday list of honors awarded by the King? Caesar may be rendered the things that are his, ungrudgingly; but when a man of science, *as* a man of science, snaps up the droppings from the table of royalty he joins the mangy and starved dogs licking the sores of the beggars at the feast of Dives. It is to be hoped that Newton was knighted for his services to the money-changers and not for his science.

Was Newton's mathematical genius dead? Most emphatically no. He was still the equal of Archimedes. But the wiser old Greek, born aristocrat that he was—fortunately, cared nothing for the honors of a position which had always been his; to the very last minute of his long life he mathematicized as powerfully as he had in his youth. But for the accidents of preventable disease and poverty, mathematicians are a long-lived race intellectually; their creativeness outlives that of poets, artists, and even of scientists, by decades. Newton was still as virile of intellect as he had ever been. Had his officious friends but let him alone Newton might easily have created the calculus of variations, an instrument of physical and mathematical discovery second only to the calculus, instead of leaving it for the Bernoullis, Euler, and Lagrange to initiate. He had already given a hint of it in the *Principia* when he determined the shape of the surface of revolution which would cleave through a fluid with the least resistance. He had it in him to lay down the broad lines of the whole method. Like Pascal when he forsook this world for the mistier if more satisfying kingdom of heaven, Newton was still a mathematician when he turned his back on his Cambridge study and walked into a more impressive sanctum at the Mint.

In 1696 Johann Bernoulli and Leibniz between them concocted two devilish challenges to the mathematicians of Europe. The first is still of importance; the second is not in the same class. Suppose two points to be fixed at random in a vertical plane. What is the shape of the curve down which a particle must slide (without friction) under the influence of gravity so as to pass from the upper point to the lower in the *least time*? This is the problem of the *brachistochrone* (= "shortest time"). After the problem had baffled the mathematicians of Europe for six months, it was proposed again, and Newton heard of it for the first time on January 29, 1696, when a friend communicated it to him. He had just come home, tired out, from a long day at the Mint. After dinner he solved the problem (and the second as well), and the following day communicated his solutions to the Royal Society anonymously. But for all his caution he could not conceal his identity—while at the Mint Newton resented the efforts of mathematicians and scientists to entice him into discussions of scientific interest. On seeing the solution Bernoulli at once exclaimed, "Ah! I recognize the lion by his paw." (This is not an exact translation of B's Latin.) They all knew Newton when they saw him, even if he did have a moneybag over his head and did not announce his name.

A second proof of Newton's vitality was to come in 1716 when he was seventy four. Leibniz had rashly proposed what appeared to him a difficult problem as a challenge to the mathematicians of Europe and aimed at Newton in particular.* Newton received this at five o'clock one afternoon on returning exhausted from the blessed Mint.

---

*The problem was to find the orthogonal trajectories of any one-parameter family of curves (in modern language).

He solved it that evening. This time Leibniz somewhat optimistically thought he had trapped the Lion. In all the history of mathematics Newton has had no superior (and perhaps no equal) in the ability to concentrate all the forces of his intellect on a difficulty at an instant's notice.

The story of the honors that fall to a man's lot in his lifetime makes but trivial reading to his successors. Newton got all that were worth having to a living man. On the whole Newton had as fortunate a life as any great man has ever had. His bodily health was excellent up to his last years; he never wore glasses and he lost only one tooth in all his life. His hair whitened at thirty but remained thick and soft till his death.

The record of his last days is more human and more touching. Even Newton could not escape suffering. His courage and endurance under almost constant pain during the last two or three years of his life add but another laurel to his crown as a human being. He bore the tortures of "the stone" without flinching, though the sweat rolled from him, and always with a word of sympathy for those who waited on him. At last, and mercifully, he was seriously weakened by "a persistent cough," and finally, after having been eased of pain for some days, died peacefully in his sleep between one and two o'clock on the morning of March 20, 1727, in his eighty fifth year. He is buried in Westminster Abbey.

CHAPTER SEVEN

Master of All Trades

LEIBNIZ

I have so many ideas that may perhaps be of some use in time if others more penetrating than I go deeply into them some day and join the beauty of their minds to the labor of mine.—G. W. LEIBNIZ

"JACK OF ALL TRADES, master of none" has its spectacular exceptions like any other folk proverb, and Gottfried Wilhelm Leibniz (1646–1716) is one of them.

Mathematics was but one of the many fields in which Leibniz showed conspicuous genius: law, religion, statecraft, history, literature, logic, metaphysics, and speculative philosophy all owe to him contributions, any one of which would have secured his fame and have preserved his memory. "Universal genius" can be applied to Leibniz without hyperbole, as it cannot to Newton, his rival in mathematics and his infinite superior in natural philosophy.

Even in mathematics Leibniz' universality contrasts with Newton's undeviating direction to a unique end, that of applying mathematical reasoning to the phenomena of the physical universe: Newton imagined one thing of absolutely the first magnitude in mathematics; Leibniz, two. The first of these was the calculus, the second, combinatorial analysis. The calculus is the natural language of the *continuous*; combinatorial analysis does for the *discrete* (see Chapter 1) what the calculus does for the continuous. In combinatorial analysis we are confronted with an assemblage of distinct things, each with an individuality of its own, and we are asked, in the most general situation, to state what relations, if any, subsist between these completely heterogeneous individuals. Here we look, not at the smoothed-out resemblances of our mathematical population, but at whatever it may be that the individuals, *as individuals*, have in common—obviously not much. In fact it seems as if, in the end, all that we can say *combinatorially*, comes down to a matter of counting off the individuals in different ways, and comparing the results. That this apparently abstract and seemingly barren procedure should lead to anything of importance is in the nature of a miracle, but it is a fact. Leibniz was a pioneer in this field, and he was one of the first to perceive that the anatomy of logic—"the laws of thought"—is a matter of combinatorial analysis. In our own day the entire subject is being arithmetized.

In Newton the mathematical spirit of his age took definite form and substance. It was inevitable after the work of Cavalieri (1598–1647), Fermat (1601–1665), Wallis (1616–1703), Barrow (1630–1677), and others that the calculus should presently get itself organized as an autonomous discipline. Like a crystal being dropped into a saturated solution at the critical instant, Newton solidified the suspended ideas of his time, and the calculus took definite shape. Any mind of the first rank might equally well have served as the crystal. Leibniz was the other first-rate mind of the age, and he too crystallized the calculus. But he was more than an agent for the expression of the spirit of his times, which Newton, in mathematics, was not. In his dream of a "universal characteristic" Leibniz was well over two centuries ahead of his age, again only as concerns mathematics and logic. So far as historical research has yet shown, Leibniz was alone in his second great mathematical dream.

The union in one mind of the highest ability in the two broad, antithetical domains of mathematical thought, the analytical and the combinatorial, or the continuous and the discrete, was without precedent before Leibniz and without sequent after him. He is the one man in the history of mathematics who has had both qualities of thought in a superlative degree. His combinatorial side was reflected in the work of his German successors, largely in trivialities, and it was only in the twentieth century, when the work of Whitehead and Russell, following that of Boole in the nineteenth, partly realized the Leibnizian dream of a universal symbolic reasoning, that the supreme importance for all mathematical and scientific thought of the combinatorial side of mathematics became as significant as Leibniz had predicted that it must. Today Leibniz' combinatorial method, as developed in symbolic logic and its extensions, is as important for the analysis that he and Newton started toward its present complexity as analysis itself is; for the symbolic method offers the only prospect in sight of clearing mathematical analysis of the paradoxes and antinomies that have infested its foundations since Zeno.

Combinatorial analysis has already been mentioned in connection with the work of Fermat and Pascal in the mathematical theory of probability. This, however, is only a detail in the "universal characteristic" which Leibniz had in mind and toward which (as will appear) he took a considerable first step. But the development and applications of the calculus offered an irresistible attraction to the mathematicians of the eighteenth century, and Leibniz' program was not taken up seriously till the 1840's. Thereafter it was again ignored except by a few nonconformists to mathematical fashion until 1910, when the modern movement in symbolic reasoning originated in another *Principia*, that of Whitehead and Russell, *Principia Mathematica*.

Since 1910 the program has become one of the major interests of modern mathematics. By a curious sort of "eternal recurrence" the theory of probability, where combinatorial analysis in the narrow sense (as applied by Pascal, Fermat, and their successors) first appeared, has recently come under Leibniz' program in the fundamental revision of the basic concepts of probability which experience, partly in the new quantum mechanics, has shown to be desirable; and today the theory of probability is on its way to becoming a province in the empire of symbolic logic—"combinatoric" in the broad sense of Leibniz.

The part Leibniz played in the creation of the calculus was noted in the preceding chapter, also the disastrous controversy to which that part gave rise. For long after both Newton and Leibniz were dead and buried (Newton in Westminster Abbey, a relic to be reverenced by the whole English-speaking race; Leibniz, indifferently cast off by his own people, in an obscure grave where only the men with shovels and his own secretary heard the dirt thudding down on the coffin), Newton carried off all the honors—or dishonors, at least wherever English is spoken.

Leibniz did not himself elaborate his great project of reducing all exact reasoning to a symbolical technique. Nor, for that matter, has it been done yet. But he did imagine it all, and he did make a significant start. Servitude to the princelings of his day to earn worthless honors and more money than he needed, the universality of his mind, and exhausting controversies during his last years, all militated against the whole creation of a masterpiece such as Newton achieved in his

Principia. In the bare summary of what Leibniz accomplished, his multifarious activities and his restless curiosity, we shall see the familiar tragedy of frustration which has prematurely withered more than one mathematical talent of the highest order—Newton, pursuing a popular esteem not worthy his spitting on, and Gauss seduced from his greater work by his necessity to gain the attention of men who were his intellectual inferiors. Only Archimedes of all the greatest mathematicians never wavered. He alone was born into the social class to which the others strove to elevate themselves; Newton crudely and directly; Gauss indirectly and no doubt subconsciously, by seeking the approbation of men of established reputation and recognized social standing, although he himself was the simplest of the simple. So there may after all be something to be said for aristocracy: its possession by birthright or other social discrimination is the one thing that will teach its fortunate possessor its worthlessness.

In the case of Leibniz the greed for money which he caught from his aristocratic employers contributed to his intellectual dalliance: he was forever disentangling the genealogies of the semi-royal bastards whose descendants paid his generous wages, and proving with his unexcelled knowledge of the law their legitimate claims to duchies into which their careless ancestors had neglected to fornicate them. But more disastrously than his itch for money his universal intellect, capable of anything and everything had he lived a thousand years instead of a meager seventy, undid him. As Gauss blamed him for doing, Leibniz squandered his splendid talent for mathematics on a diversity of subjects in all of which no human being could hope to be supreme, whereas—according to Gauss—he had in him supremacy in mathematics. But why censure him? He was what he was, and willy-nilly he had to "dree his weird." The very diffusion of his genius made him capable of the dream which Archimedes, Newton, and Gauss missed—the "universal characteristic." Others may bring it to realization; Leibniz did his part in dreaming it to be possible.

Leibniz may be said to have lived not one life but several. As a diplomat, historian, philosopher, and mathematician he did enough in each field to fill one ordinary working life. Younger than Newton by about four years, he was born at Leipzig on July 1, 1646, and living only seventy years against Newton's eighty five, died in Hanover on November 14, 1716. His father was a professor of moral philosophy and came of a good family which had served the government of

Saxony for three generations. Thus young Leibniz' earliest years were passed in an atmosphere of scholarship heavily charged with politics.

At the age of six he lost his father, but not before he had acquired from him a passion for history. Although he attended a school in Leipzig, Leibniz was largely self-taught by incessant reading in his father's library. At eight he began the study of Latin and by twelve had mastered it sufficiently to compose creditable Latin verse. From Latin he passed on to Greek which he also learned largely by his own efforts.

At this stage his mental development parallels that of Descartes: classical studies no longer satisfied him and he turned to logic. From his attempts as a boy of less than fifteen to reform logic as presented by the classicists, the scholastics, and the Christian fathers, developed the first germs of his *Characteristica Universalis* or Universal Mathematics, which, as has been shown by Couturat, Russell, and others, is the clue to his metaphysics. The symbolic logic invented by Boole in 1847–54 (to be discussed in a later chapter) is only that part of the *Characteristica* which Leibniz called *calculus raticinator*. His own description of the universal characteristic will be quoted presently.

At the age of fifteen Leibniz entered the University of Leipzig as a student in law. The law, however, did not occupy all his time. In his first two years he read widely in philosophy and for the first time became aware of the new world which the modern, or "natural" philosophers, Kepler, Galileo, and Descartes had discovered. Seeing that this newer philosophy could be understood only by one acquainted with mathematics, Leibniz passed the summer of 1663 at the University of Jena, where he attended the mathematical lectures of Erhard Weigel, a man of considerable local reputation but scarcely a mathematician.

On returning to Leipzig he concentrated on law. By 1666, at the age of twenty, he was thoroughly prepared for his doctor's degree in law. This is the year, we recall, in which Newton began the rustication at Woolsthorpe that gave him the calculus and his law of universal gravitation. The Leipzig faculty, bilious with jealousy, refused Leibniz his degree, officially on account of his youth, actually because he knew more about law than the whole dull lot of them.

Before this he had taken his bachelor's degree in 1663 at the age of seventeen with a brilliant essay foreshadowing one of the cardinal doctrines of his mature philosophy. We shall not take space to go into this, but it may be mentioned that one possible interpretation of Leibniz' essay is the doctrine of "the organism as a whole," which one progressive school of biologists and another of psychologists has found attractive in our own time.

Disgusted at the pettiness of the Leipzig faculty Leibniz left his native town for good and proceeded to Nuremberg where, on November 5, 1666, at the affiliated University of Altdorf, he was not only granted his doctor's degree at once for his essay on a new method (the historical) of teaching law, but was begged to accept the University professorship of law. But, like Descartes refusing the offer of a lieutenant-generalship because he knew what he wanted out of life, Leibniz declined, saying he had very different ambitions. What these may have been he did not divulge. It seems unlikely that they could have been the higher pettifogging for princelets into which fate presently kicked him. Leibniz' tragedy was that he met the lawyers before the scientists.

His essay on the teaching of the law and its proposed recodification was composed on the journey from Leipzig to Nuremberg. This illustrates a lifelong characteristic of Leibniz, his ability to work anywhere, at any time, under any conditions. He read, wrote, and thought incessantly. Much of his mathematics, to say nothing of his other wonderings on everything this side of eternity and beyond, was written out in the jolting, draughty rattletraps that bumped him over the cow trails of seventeenth century Europe as he sped hither and thither at his employers' erratic bidding. The harvest of all this ceaseless activity was a mass of papers, of all sizes and all qualities, as big as a young haystack, that has never been thoroughly sorted, much less published. Today most of it lies baled in the royal Hanover library waiting the patient labors of an army of scholars to winnow the wheat from the straw.

It seems incredible that one head could have been responsible for all the thoughts, published and unpublished, that Leibniz committed to paper. As an item of interest to phrenologists and anatomists it has been stated (whether reliably or not I don't know) that Leibniz' skull was dug up, measured, and found to be markedly under the normal adult size. There may be something in this, as many of us have seen perfect idiots with noble brows bulging from heads as big as broth pots.

Newton's miraculous year 1666 was also the great year for Leibniz. In what he called a "schoolboy's essay," *De arte combinatoria*, the young man of twenty aimed to create *"a general method in which all truths of the reason would be reduced to a kind of calculation. At the same time this would be a sort of universal language or script, but infinitely different from all those projected hitherto; for the symbols and even the words in it would direct the reason; and errors, except those of fact, would be mere mistakes in calculation. It would be very difficult to form or invent this language or characteristic, but very easy to understand it without any dictionaries."* In a later description he confidently (and optimistically) estimates how long it would take to carry out his project: "I think a few chosen men could turn the trick within five years." Toward the end of his life Leibniz regretted that he had been too distracted by other things ever to work out his idea. If he were younger himself or had competent young assistants, he says, he could still do it—a common alibi for a talent squandered on snobbery, greed, and intrigue.

To anticipate slightly, it may be said that Leibniz' dream struck his mathematical and scientific contemporaries as a dream and nothing more, to be politely ignored as the fixed idea of an otherwise sane and universally gifted genius. In a letter of September 8, 1679, Leibniz (speaking of geometry in particular but of all reasoning in general) tells Huygens of a "new characteristic, entirely different from Algebra, which will have great advantages for representing exactly and naturally to the mind, and without figures, everything that depends on the imagination."

Such a direct, symbolic way of handling geometry was invented in the nineteenth century by Hermann Grassmann (whose work in algebra generalized that of Hamilton). Leibniz goes on to discuss the difficulties inherent in the project, and presently emphasizes what he considers its superiority over the Cartesian analytic geometry.

"But its principal utility consists in the consequences and reasonings which can be performed by the operations of characters [symbols], which could not be expressed by diagrams (or even by models) without too great elaboration, or without confusing them by an excessive number of points and lines, so that one would be obliged to make an infinity of useless trials: in contrast this method would lead surely and simply [to the desired end]. I believe mechanics could be handled by this method almost like geometry."

Of the definite things that Leibniz did in that part of his universal characteristic which is now called symbolic logic, we may cite his formulation of the principal properties of logical addition and logical multiplication, negation, identity, the null class, and class inclusion. For an explanation of what some of these terms mean and the postulates of the algebra of logic we must refer ahead to the chapter on Boole. All this fell by the wayside. Had it been picked up by able men when Leibniz scattered it broadcast, instead of in the 1840's, the history of mathematics might now be quite a different story from what it is. Almost as well never as too soon.

Having dreamed his universal dream at the age of twenty, Leibniz presently turned to something more practical, and he became a sort of corporation lawyer and glorified commercial traveller for the Elector of Mainz. Taking one last spree in the world of dreams before plunging up to his chin into more or less filthy politics, Leibniz devoted some months to alchemy in the company of the Rosicrucians infesting Nuremberg.

It was his essay on a new method of teaching law that undid him. The essay came to the attention of the Elector's right-hand statesman, who urged Leibniz to have it printed so that a copy might be laid before the august Elector. This was done, and Leibniz, after a personal interview, was appointed to revise the code. Before long he was being entrusted with important commissions of all degrees of delicacy and shadiness. He became a diplomat of the first rank, always pleasant, always open and aboveboard, but never scrupulous, even when asleep. To his genius is due, at least partly, that unstable formula known as the "balance of power." And for sheer cynical brilliance, it would be hard to surpass, even today, Leibniz' great dream of a holy war for the conquest and civilization of Egypt. Napoleon was quite chagrined when he discovered that Leibniz had anticipated him in this sublime vision.

Up till 1672 Leibniz knew but little of what in his time was modern mathematics. He was then twenty six when his real mathematical education began at the hands of Huygens, whom he met in Paris in the intervals between one diplomatic plot and another. Christian Huygens (1629–1695), while primarily a physicist, some of whose best work went into horology and the undulatory theory of light, was an accomplished mathematician. Huygens presented Leibniz with a copy of his mathematical work on the pendulum. Fascinated by the power of the mathematical method in competent hands, Leibniz begged Huygens to give him lessons, which Huygens, seeing that Leibniz had a first-class mind, gladly did. Leibniz had already drawn up an impressive list of discoveries he had made by means of his own methods—phases of the universal characteristic. Among these was a calculating machine far superior to Pascal's, which handled only addition and subtraction; Leibniz' machine did also multiplication, division, and the extraction of roots. Under Huygens' expert guidance Leibniz quickly found himself. He was a born mathematician.

The lessons were interrupted from January to March, 1673, during Leibniz' absence in London as an attaché for the Elector. While in London, Leibniz met the English mathematicians and showed them some of his work, only to learn that it was already known. His English friends told him of Mercator's quadrature of the hyperbola—one of the clues which Newton had followed to his invention of the calculus. This introduced Leibniz to the method of infinite series, which he carried on. One of his discoveries (sometimes ascribed to the Scotch mathematician James Gregory, 1638-1675) may be noted: if π is the ratio of the circumference of a circle to its diameter,

$$\frac{\pi}{4} = 1 - \frac{1}{3} + \frac{1}{5} - \frac{1}{7} + \frac{1}{9} - \frac{1}{11} + \ldots,$$

the series continuing in the same way indefinitely. This is not a practical way of calculating the numerical value of π (3.1415926 . . .), but the simple connection between π and *all* the odd numbers is striking.

During his stay in London Leibniz attended meetings of the Royal Society, where he exhibited his calculating machine. For this and his other work he was elected a foreign member of the Society before his return to Paris in March, 1673. He and Newton subsequently (1700) became the first foreign members of the French Academy of Sciences.

Greatly pleased with what Leibniz had done while away, Huygens urged him to continue. Leibniz devoted every spare moment to his mathematics, and before leaving Paris for Hanover in 1676 to enter the service of the Duke of Brunswick-Lüneburg, had worked out some of the elementary formulas of the calculus and had discovered "the fundamental theorem of the calculus" (see preceding chapter)—that is, if we accept his own date, 1675. This was not published till July 11, 1677, eleven years after Newton's unpublished discovery, which was not made public by Newton till after Leibniz' work had appeared. The controversy started in earnest, when Leibniz, diplomatically shrouding himself in editorial omniscience and anonymity, wrote a severely critical review of Newton's work in the *Acta Eruditorum*, which Leibniz himself had founded in 1682 and of which he was editor in chief. In the interval between 1677 and 1704 the Leibnizian calculus had been developed into an instrument of real power and easy applicability on the Continent, largely through the efforts of the Swiss Bernoullis, Jacob and his brother Johann, while in England, owing to Newton's reluctance to share his mathematical discoveries freely, the calculus was still a relatively untried curiosity.

One specimen of things that are now easy for beginners in the calculus, but which cost Leibniz (and possibly also Newton) much thought and many trials before the right way was found, may indicate how far mathematics has travelled since 1675. Instead of the infinitesimals of Leibniz we shall use the rates discussed in the preceding chapter. If u, v are functions of x, how shall the rate of change of uv with respect to x be expressed in terms of the respective rates of change of u and v with respect to x? In symbols, what is $\frac{d(uv)}{dx}$ in terms of $\frac{du}{dx}$ and $\frac{dv}{dx}$? Leibniz once thought it should be $\frac{du}{dx} \times \frac{dv}{dx}$, which is nothing like the correct

$$\frac{d(uv)}{dx} = u\frac{dv}{dx} + v\frac{du}{dx}.$$

The Elector died in 1673 and Leibniz was more or less free during the last of his stay in Paris. On leaving Paris in 1676 to enter the service of the Duke John Frederick of Brunswick-Lüneburg, Leibniz proceeded to Hanover by way of London and Amsterdam. It was while in the latter city that he engineered one of the shadiest transactions in all his long career as a philosophic diplomat. The history of Leibniz' commerce with "the God-intoxicated Jew" Benedict de Spinoza (1632-1677) may be incomplete, but as the account now stands it seems that for once Leibniz was grossly unethical over a matter—of all things—of ethics. Leibniz seems to have believed in applying his ethics to practical ends. He carried off copious extracts from Spinoza's unpublished masterpiece *Ethica* (*Ordina Geometrica Demonstrata*)—a treatise on ethics developed in the manner of

Euclid's geometry. When Spinoza died the following year Leibniz appears to have found it convenient to mislay his souvenirs of the Amsterdam visit. Scholars in this field seem to agree that Leibniz' own philosophy wherever it touches ethics was appropriated without acknowledgment from Spinoza.

It would be rash for anyone not an expert in ethics to doubt that Leibniz was guilty, or to suggest that his own thoughts on ethics were independent of Spinoza's. Nevertheless there are at least two similar instances in mathematics (elliptic functions, non-Euclidean geometry) where all the evidence at one time was sufficient to convict several men of dishonesty grosser than that attributed to Leibniz. When unsuspected diaries and correspondence were brought to light years after the death of all the accused it was seen that all were entirely innocent. It may pay occasionally to believe the best of human beings instead of the worst until all the evidence is in—which it can never be for a man who is tried after his death.

The remaining forty years of Leibniz' life were spent in the trivial service of the Brunswick family. In all he served three masters as librarian, historian, and general brains of the family. It was a matter of great importance to such a family to have an exact history of all its connections with other families as highly favored by heaven as itself. Leibniz was no mere cataloguer of books in his function as family librarian, but an expert genealogist and searcher of mildewed archives as well, whose function it was to confirm the claims of his employers to half the thrones of Europe or, failing confirmation, to manufacture evidence by judicious suppression. His historical researches took him all through Germany and thence to Austria and Italy in 1687–90.

During his stay in Italy Leibniz visited Rome and was urged by the Pope to accept the position of librarian at the Vatican. But as a prerequisite to the job was that Leibniz become a Catholic he declined—for once scrupulous. Or was he? His reluctance to throw up one good post for another may have started him off on the next application of his "universal characteristic," the most fantastically ambitious of all his universal dreams. Had he pulled this off he could have moved into the Vatican without leaving his face outside.

His grand project was no less than that of reuniting the Protestant and Catholic churches. It was then not so long since the first had split off from the second, so the project was not so insane as it now sounds.

In his wild optimism Leibniz overlooked a law which is as fundamental for human nature as the second law of thermodynamics is for the physical universe—indeed it is of the same kind: all creeds tend to split into two, each of which in turn splits into two more, and so on, until after a certain finite number of generations (which can be easily calculated by logarithms) there are fewer human beings in any given region, no matter how large, than there are creeds, and further attenuations of the original dogma embodied in the first creed dilute it to a transparent gas too subtle to sustain faith in any human being, no matter how small.

A quite promising conference at Hanover in 1683 failed to effect a reconciliation as neither party could decide which was to be swallowed by the other, and both welcomed the bloody row of 1688 in England between Catholics and Protestants as a legitimate ground for adjourning the conference *sine die*.

Having learned nothing from this farce Leibniz immediately organized another. His attempt to unite merely the two Protestant sects of his day succeeded only in making a large number of excellent men more obstinate and sorer at one another than they were before. The Protestant Conference dissolved in mutual recriminations and curses.

It was about this time that Leibniz turned to philosophy as his major consolation. In an endeavor to assist Pascal's old Jansenist friend Arnauld, Leibniz composed a semi-casuistical treatise on metaphysics destined to be of use to Jansenists and others in need of something more subtle than the too subtle logic of the Jesuits. His philosophy occupied the remainder of Leibniz' life (when he was not engaged on the unending history of the Brunswick family for his employers), in all about a quarter of a century. That a mind like Leibniz' evolved a vast cloud of philosophy in twenty five years need hardly be stated. Doubtless every reader has heard something of the ingenious theory of monads—miniature replicas of the universe out of which *everything* in the universe is composed, as a sort of one in all, all in one—by which Leibniz explained everything (except the monads) in this world and the next.

The power of Leibniz' method when applied to philosophy cannot be denied. As a specimen of the theorems *proved* by Leibniz in his philosophy, that concerning the existence of God may be mentioned. In his attempt to prove the fundamental theorem of optimism—

"everything is for the best in this best of all possible worlds"—Leibniz was less successful, and it was only in 1759, forty three years after Leibniz had died neglected and forgotten, that a conclusive demonstration was published by Voltaire in his epoch-making treatise *Candide*. One further isolated result may be mentioned. Those familiar with general relativity will recall that "empty space"—space totally devoid of matter—is no longer respectable. Leibniz rejected it as nonsensical.

The list of Leibniz' interests is still far from complete. Economics, philology, international law (in which he was a pioneer), the establishment of mining as a paying industry in certain parts of Germany, theology, the founding of academies, and the education of the young Electress Sophie of Brandenburg (a relative of Descartes' Elisabeth), all shared his attention, and in each of them he did something notable. Possibly his least successful ventures were in mechanics and physical science, where his occasional blunders show up glaringly against the calm, steady light of men like Galileo, Newton, and Huygens, or even Descartes.

Only one item in this list demands further attention here. On being called to Berlin in 1700 as tutor to the young Electress, Leibniz found time to organize the Berlin Academy of Sciences. He became its first president. The Academy was still one of the three or four leading learned bodies in the world till the Nazis "purged" it. Similar ventures in Dresden, Vienna, and St. Petersburg came to nothing during Leibniz' lifetime, but after his death the plans for the St. Petersburg Academy of Sciences which he had drawn up for Peter the Great were carried out. The attempt to found a Viennese Academy was frustrated by the Jesuits when Leibniz visited Austria for the last time, in 1714. Their opposition was only to have been expected after what Leibniz had done for Arnauld. That they got the better of the master diplomat in an affair of petty academic politics shows how badly Leibniz had begun to slip at the age of sixty eight. He was no longer himself, and indeed his last years were but a wasted shadow from his former glory.

Having served princes all his life he now received the usual wages of such service. Ill, fast ageing, and harassed by controversy, he was kicked out.

Leibniz returned to Brunswick in September, 1714, to learn that his employer the Elector George Louis—"the honest blockhead," as he is known in English history—having packed up his duds and his snuff, had left for London to become the first German King of England. Nothing would have pleased Leibniz better than to follow George to London, although his enemies at the Royal Society and elsewhere in England were now numerous and vicious enough owing to the controversy with Newton. But the boorish George, now socially a gentleman, had no further use for Leibniz' diplomacy, and curtly ordered the brains that had helped to lift him into civilized society to stick in the Hanover library and get on with their everlasting history of the illustrious Brunswick family.

When Leibniz died two years later (1716) the diplomatically doctored history was still incomplete. For all his hard labor Leibniz had been unable to bring the history down beyond the year 1005, and at that had covered less than three hundred years. The family was so very tangled in its marital adventures that even the universal Leibniz could not supply them all with unblemished scutcheons. The Brunswickers showed their appreciation of this immense labor by forgetting all about it till 1843, when it was published, but whether complete or expurgated will not be known until the rest of Leibniz' manuscripts have been sifted.

Today, over three hundred years after his death, Leibniz' reputation as a mathematician is higher than it was for many, many years after his secretary followed him to the grave, and it is still rising.

As a diplomat and statesman Leibniz was as good as the cream of the best of them in any time or any place, and far brainier than all of them together. There is but one profession in the world older than his, and until that is made respectable it would be premature to try any man for choosing diplomacy as his means to a livelihood.

CHAPTER EIGHT

Nature or Nurture?

THE BERNOULLIS

These men certainly accomplished much and admirably attained the goal they had set themselves.—JOHANNES BERNOULLI

SINCE THE GREAT DEPRESSION began deflating western civilization eugenists, geneticists, psychologists, politicians, and dictators—for very different reasons—have taken a renewed interest in the still unsettled controversy of heredity versus environment. At one extreme the hundred-percenter proletarians hold that anyone can be a genius given the opportunity; while at the other, equally positive Tories assert that genius is inborn and will out even in a London slum. Between the two stretches a whole spectrum of belief. The average opinion holds that nature, not nurture, is the determining factor in the emergence of genius, but that without deliberate or accidental assistance genius perishes. The history of mathematics offers abundant material for a study of this interesting problem. Without taking sides —to do so at present would be premature—we may say that the evidence furnished by the life histories of mathematicians seems to favor the average opinion.

Probably the most striking case history is that of the Bernoulli family, which in three generations produced eight mathematicians, several of them outstanding, who in turn produced a swarm of descendants about half of whom were gifted above the average and nearly all of whom, down to the present day, have been superior human beings. No fewer than 120 of the descendants of the mathematical Bernoullis have been traced genealogically, and of this considerable posterity the majority achieved distinction—sometimes amounting to eminence —in the law, scholarship, science, literature, the learned professions, administration, and the arts. None were failures. The most significant thing about a majority of the mathematical members of this family in the second and third generations is that they did not deliberately choose mathematics as a profession but drifted into it in spite of themselves as a dipsomaniac returns to alcohol.

As the Bernoulli family played a leading part in developing the calculus and its applications in the seventeenth and eighteenth centuries, they must be given more than a passing mention in even the briefest account of the evolution of modern mathematics. The Bernoullis and Euler were in fact the leaders above all others who perfected the calculus to the point where quite ordinary men could use it for the discovery of results which the greatest of the Greeks could never have found. But the mere volume of the Bernoulli family's work is too vast for detailed description in an account like the present, so we shall treat them briefly together.

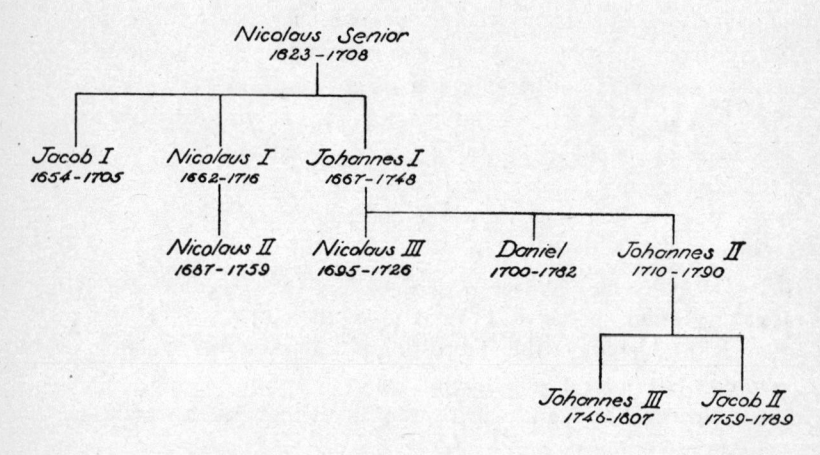

```
                    Nicolaus Senior
                      1623-1708
        ┌──────────────┼──────────────┐
     Jacob I       Nicolaus I      Johannes I
    1654-1705      1662-1716       1667-1748
                       │      ┌────────┼────────┬──────────┐
                  Nicolaus II  Nicolaus III   Daniel     Johannes II
                   1687-1759   1695-1726    1700-1782    1710-1790
                                                    ┌────────┴────────┐
                                                Johannes III       Jacob II
                                                 1746-1807         1759-1789
```

The Bernoullis were one of many Protestant families who fled from Antwerp in 1583 to escape massacre by the Catholics (as on St. Bartholomew's Eve) in the prolonged persecution of the Huguenots. The family sought refuge first in Frankfort, moving on presently to Switzerland, where they settled at Basle. The founder of the Bernoulli dynasty married into one of the oldest Basle families and became a great merchant. Nicolaus senior, who heads the genealogical table, was also a great merchant, as his grandfather and great-grandfather had been. All these men married daughters of merchants, and with one exception—the great-grandfather mentioned—accumulated large fortunes. The exception showed the first departure from the family tradition of trade by following the profession of medicine. Mathematical talent was probably latent for generations in this shrewd mercantile family, but its actual emergence was explosively sudden.

Referring now to the genealogical table we shall give a very brief summary of the chief scientific activities of the eight mathematicians descended from Nicolaus senior before continuing with the heredity.

Jacob I mastered the Leibnizian form of the calculus by himself. From 1687 to his death he was professor of mathematics at Basle. Jacob I was one of the first to develop the calculus significantly beyond the state in which Newton and Leibniz left it and to apply it to new problems of difficulty and importance. His contributions to analytic geometry, the theory of probability, and the calculus of variations were of the highest importance. As the last will recur frequently (in the work of Euler, Lagrange, and Hamilton), we may describe the nature of some of the problems attacked by Jacob I in this subject. We have already seen a specimen of the type of problem handled by the calculus of variations in Fermat's principle of least time.

The calculus of variations is of very ancient origin. According to one legend,* when Carthage was founded the city was granted as much land as a man could plow a furrow completely around in a day. What shape should the furrow be, given that a man can plow a straight furrow of a certain length in a day? Mathematically stated, what is the figure which has the greatest area of all figures having perimeters of the same length? This is an *isoperimetrical* problem; the answer here is a circle. This seems obvious, but it is by no means easy to prove. (The elementary "proofs" sometimes given in school geometries are rankly fallacious.) The mathematics of the problem comes down to making a certain integral a maximum subject to one restrictive condition. Jacob I solved this problem and generalized it.†

The discovery that the brachistochrone is a cycloid has been noted in previous chapters. This fact, that the cycloid is the curve of quickest descent, was discovered by the brothers Jacob I and Johannes I in 1697, and almost simultaneously by several others. But the cycloid is also the tautochrone. This struck Johannes I as something wonder-

---

*Actually, here, I have combined *two* legends. Queen Dido was given a bull's hide to "enclose" the greatest area. She cut it into one thong and enclosed a semicircle.

†Historical notes on this and other problems of the calculus of variations will be found in the book by G. A. Bliss, *Calculus of Variations*, Chicago, 1925. The Anglicized form of Jacob I is James.

ful and admirable: "With justice we may admire Huygens because he first discovered that a heavy particle falls on a cycloid in the same time always, no matter what the starting-point may be. But you will be petrified with astonishment when I say that exactly this same cycloid, the tautochrone of Huygens, is the brachistochrone we are seeking." (Bliss, loc. cit., p. 54.) Jacob also waxes enthusiastic. These again are instances of the sort of problem attacked by the calculus of variations. Lest they seem trivial, we repeat once more that a whole province of mathematical physics is frequently mapped into a simple *variational principle*—like Fermat's of least time in optics, or Hamilton's in dynamics.

After Jacob's death his great treatise on the theory of probability, the *Ars Conjectandi*, was published in 1713. This contains much that is still of the highest usefulness in the theory of probabilities and its applications to insurance, statistics, and the mathematical study of heredity.

Another research of Jacob's shows how far he had developed the differential and integral calculus: continuing the work of Leibniz, Jacob made a fairly exhaustive study of catenaries—the curves in which a uniform chain hangs suspended between two points, or in which loaded chains hang. This was no mere curiosity. Today the mathematics developed by Jacob I in this connection finds its use in applications to suspension bridges and high-voltage transmission lines. When Jacob I worked all this out it was new and difficult; today it is an exercise in the first course in the calculus or mechanics.

Jacob I and his brother Johannes I did not always get on well together. Johannes seems to have been the more quarrelsome of the two, and it is certain that he treated his brother with something pretty close to dishonesty in the matter of isoperimetrical problems. The Bernoullis took their mathematics in deadly earnest. Some of their letters about mathematics bristle with strong language that is usually reserved for horse thieves. For his part Johannes I not only attempted to steal his brother's ideas but threw his own son out of the house for having won a prize from the French Academy of Sciences for which Johannes himself had competed. After all, if rational human beings get excited about a game of cards, why should they not blow up over mathematics which is infinitely more exciting?

Jacob I had a mystical strain which is of some significance in the study of the heredity of the Bernoullis. It cropped out once in an interesting way toward the end of his life. There is a certain spiral (the logarithmic or equiangular) which is reproduced in a similar spiral after each of many geometrical transformations. Jacob was fascinated by this recurrence of the spiral, several of whose properties he discovered, and directed that a spiral be engraved on his tombstone with the inscription *Eadem mutata resurgo* (Though changed I shall arise the same).

Jacob's motto was *Invito patre sidera verso* (Against my father's will I study the stars)—in ironic memory of his father's futile opposition to Jacob's devoting his talents to mathematics and astronomy. This detail favors the "nature" view of genius over the "nurture." If his father had prevailed Jacob would have been a theologian.

Johannes I, brother of Jacob I, did not start as a mathematician but as a doctor of medicine. His dispute with the brother who had generously taught him mathematics has already been mentioned. Johannes was a man of violent likes and dislikes: Leibniz and Euler were his gods; Newton he positively hated and greatly underestimated, as a bigoted champion of Leibniz was almost bound to do from envy or spite. The obstinate father attempted to cramp his younger son into the family business, but Johannes I, following the lead of his brother Jacob I, rebelled and went in for medicine and the humanities, unaware that he was fighting against his heredity. At the age of eighteen he took his M.A. degree. Before long he realized his mistake in choosing medicine and turned to mathematics. His first academic appointment was at Groningen in 1695 as professor of mathematics; on the death of Jacob I in 1705 Johannes I succeeded to the professorship at Basle.

Johannes I was even more prolific than his brother in mathematics and did much to spread the calculus in Europe. His range included physics, chemistry, and astronomy in addition to mathematics. On the applied side Johannes I contributed extensively to optics, wrote on the theory of the tides and the mathematical theory of ship sails, and enunciated the principle of virtual displacements in mechanics. Johannes I was a man of unusual physical and intellectual vigor, remaining active till within a few days of his death at the age of eighty.

Nicolaus I, the brother of Jacob I and Johannes I, was also gifted in mathematics. Like his brothers he made a false start. At the age of sixteen he took his doctor's degree in philosophy at the University of Basle, and at twenty earned the highest degree in law. He was first a professor of law at Bern before becoming one of the mathematical faculty at the Academy of St. Petersburg. At the time of his death he was so highly thought of that the Empress Catherine gave him a public funeral at state expense.

Heredity came out curiously in the second generation. Johannes I tried to force his second son, Daniel, into business. But Daniel thought he preferred medicine and became a physician before landing, in spite of himself, in mathematics. At the age of eleven Daniel began taking lessons in mathematics from his elder brother Nicolaus III, only five years older than himself. Daniel and the great Euler were intimate friends and at times friendly rivals. Like Euler, Daniel Bernoulli has the distinction of having won the prize of the French Academy ten times (on a few occasions the prize was shared with other successful competitors). Some of Daniel's best work went into hydrodynamics, which he developed uniformly from the single principle that later came to be called the conservation of energy. All who work today in pure or applied fluid motion know the name of Daniel Bernoulli.

In 1725 (at the age of twenty five) Daniel became professor of mathematics at St. Petersburg, where the comparative barbarity of the life irked him so greatly that he returned at the first opportunity, eight years later, to Basle, where he became professor of anatomy and botany, and finally of physics. His mathematical work included the calculus, differential equations, probability, the theory of vibrating strings, an attempt at a kinetic theory of gases, and many other problems in applied mathematics. Daniel Bernoulli has been called the founder of mathematical physics.

From the standpoint of heredity it is interesting to note that Daniel had a marked vein of speculative philosophy in his nature—possibly a refined sublimation of the Huguenot religion of his ancestors. The like cropped out in numerous later descendants of the illustrious refugees from religious intolerance.

The third mathematician in the second generation, Johannes II, brother of Nicolaus III and Daniel, also made a false start and was pulled back into line by his heredity—or possibly by his brothers. Starting out in law he became professor of eloquence at Basle before succeeding his father in the chair of mathematics. His work was principally in physics and was sufficiently distinguished to capture the Paris prize on three occasions (once is usually enough to satisfy a good mathematician—provided he is good enough).

Johannes III, a son of Johannes II, repeated the family tradition of making a wrong start, and like his father began with law. At the age of thirteen he took his doctor's degree in philosophy. By nineteen Johannes III had found his true vocation and was appointed astronomer royal at Berlin. His interests embraced astronomy, geography, and mathematics.

Jacob II, another son of Johannes II, carried on the family blunder by starting in law, only to change over at twenty one to experimental physics. He also turned to mathematics, becoming a member of the St. Petersburg Academy in the section of mathematics and physics. His early death (at the age of thirty) by accidental drowning cut short a very promising career, and we do not know what Jacob II really had in him. He was married to a granddaughter of Euler.

The list of Bernoullis who showed mathematical talent is not yet exhausted, but the rest were less distinguished. It is sometimes asserted that the strain had worn thin. Quite the contrary seems to be the case. When mathematics was the most promising field for superior talent to cultivate, as it was immediately after the invention of the calculus, the gifted Bernoullis cultivated mathematics. But mathematics and science are only two of innumerable fields of human endeavor, and for gifted men to swarm into either when both are overcrowded with high ability indicates a lack of practical sense. The Bernoulli talent was not expended; it merely spent itself on things of equal—or perhaps greater—social importance than mathematics when that field began to resemble Epsom Downs on Derby Day.

Those interested in the vagaries of heredity will find plenty of material in the history of the Darwin and Galton families. The case of Francis Galton (a cousin of Charles Darwin) is particularly interesting, as the mathematical study of heredity was founded by him. To rail at the descendants of Charles Darwin because some of them have achieved eminence in mathematics or mathematical physics rather than in biology is slightly silly. The genius is still there, and one expression of it is not necessarily "better" or "higher" than another—unless we are the sort of bigots who insist that everything should be mathematics, or biology, or sociology, or bridge and golf. It may be that the abandonment of mathematics as the family trade by the Bernoullis was just one more instance of their genius.

Many legends and anecdotes have grown up round the famous Bernoullis, as is only natural in the case of a family as gifted and as violent in their language as the Bernoullis sometimes were. One of these ripe old chestnuts may be retailed again as it is one of the comparatively early authentic instances of a story which must be at least as old as ancient Egypt, and of which we daily see variants pinned onto all sorts of prominent characters from Einstein down. Once when travelling as a young man Daniel modestly introduced himself to an interesting stranger with whom he had been conversing: "I am Daniel Bernoulli." "And I," said the other sarcastically, "am Isaac Newton." This delighted Daniel to the end of his days as the sincerest tribute he had ever received.

CHAPTER NINE

Analysis Incarnate

EULER

History shows that those heads of empires who have encouraged the cultivation of mathematics, the common source of all the exact sciences, are also those whose reigns have been the most brilliant and whose glory is the most durable.
—MICHEL CHASLES

"EULER CALCULATED WITHOUT APPARENT EFFORT, as men breathe, or as eagles sustain themselves in the wind" (as Arago said), is not an exaggeration of the unequalled mathematical facility of Léonard Euler (1707–1783), the most prolific mathematician in history, and the man whom his contemporaries called "analysis incarnate." Euler wrote his great memoirs as easily as a fluent writer composes a letter to an intimate friend. Even total blindness during the last seventeen years of his life did not retard his unparalleled productivity; indeed, if anything, the loss of his eyesight sharpened Euler's perceptions in the inner world of his imagination.

The extent of Euler's work was not accurately known even in 1936, but it has been estimated that sixty to eighty large quarto volumes will be required for the publication of his collected works. In 1909 the Swiss Association for Natural Science undertook the collection and publication of Euler's scattered memoirs, with financial assistance from many individuals and mathematical societies throughout the world—rightly claiming that Euler belongs to the whole civilized world and not only to Switzerland. The careful estimates of the probable expense (about $80,000 in the money of 1909) were badly upset by the discovery in St. Petersburg (Leningrad) of an unsuspected mass of Euler's manuscripts.

Euler's mathematical career opened in the year of Newton's death. A more propitious epoch for a genius like that of Euler's could not have been chosen. Analytic geometry (made public in 1637) had been in use ninety years, the calculus about fifty, and Newton's law of universal gravitation, the key to physical astronomy, had been before the mathematical public for forty years. In each of these fields a vast number of isolated problems had been solved, with here and there notable attempts at unification; but no systematic attack had yet been launched against the whole of mathematics, pure and applied, as it then existed. In particular the powerful analytical methods of Descartes, Newton, and Leibniz had not yet been exploited to the limit of what they were then capable, especially in mechanics and geometry.

On a lower level algebra and trigonometry were then in shape for systematization and extension; the latter particularly was ready for essential completion. In Fermat's domain of Diophantine analysis and the properties of the common whole numbers no such "temporary perfection" was possible (it is not even yet); but even here Euler proved himself the master. In fact one of the most remarkable features of Euler's universal genius was its equal strength in both of the main currents of mathematics, the continuous and the discrete.

As an algorist Euler has never been surpassed, and probably never even closely approached, unless perhaps by Jacobi. An algorist is a mathematician who devises "algorithms" (or "algorisms") for the solution of problems of special kinds. As a very simple example, we assume (or prove) that every positive real number has a real square root. How shall the root be calculated? There are many ways known; an algorist devises practicable methods. Or again, in Diophantine analysis, also in the integral calculus, the solution of a problem may not be forthcoming until some ingenious (often simple) replacement of one or more of the variables by functions of other variables has been made; an algorist is a mathematician to whom such ingenious tricks come naturally. There is no uniform mode of procedure—algorists, like facile rhymesters, are born, not made.

It is fashionable today to despise the "mere algorist"; yet, when a truly great one like the Hindu Ramanujan arrives unexpectedly out of nowhere, even expert analysts hail him as a gift from Heaven: his all but supernatural insight into apparently unrelated formulas reveals hidden trails leading from one territory to another, and the analysts have new tasks provided for them in clearing the trails. An algorist is a "formalist" who loves beautiful formulas for their own sake.

Before going on to Euler's peaceful but interesting life we must mention two circumstances of his times which furthered his prodigious activity and helped to give it a direction.

In the eighteenth century the universities were not the principal centers of research in Europe. They might have become such sooner than they did but for the classical tradition and its understandable hostility to science. Mathematics was close enough to antiquity to be respectable, but physics, being more recent, was suspect. Further, a mathematician in a university of the time would have been expected to put much of his effort on elementary teaching; his research, if any, would have been an unprofitable luxury, precisely as in the average American institution of higher learning today. The Fellows of the British universities could do pretty well as they chose. Few, however, chose to do anything, and what they accomplished (or failed to accomplish) could not affect their bread and butter. Under such laxity or open hostility there was no good reason why the universities should have led in science, and they did not.

The lead was taken by the various royal academies supported by generous or farsighted rulers. Mathematics owes an undischargeable debt to Frederick the Great of Prussia and Catherine the Great of Russia for their broadminded liberality. They made possible a full century of mathematical progress in one of the most active periods in scientific history. In Euler's case Berlin and St. Petersburg furnished the sinews of mathematical creation. Both of these foci of creativity owed their inspiration to the restless ambition of Leibniz. The academies for which Leibniz had drawn up the plans gave Euler his chance to be the most prolific mathematician of all time; so, in a sense, Euler was Leibniz' grandson.

The Berlin Academy had been slowly dying of brainlessness for forty years when Euler, at the instigation of Frederick the Great, shocked it into life again; and the St. Petersburg Academy, which Peter the Great did not live to organize in accordance with Leibniz' program, was firmly founded by his successor.

These Academies were not like some of those today, whose chief function is to award membership in recognition of good work well done; they were research organizations which *paid* their leading members *to produce scientific research*. Moreover the salaries and perquisites were ample for a man to support himself and his family in decent comfort. Euler's household at one time consisted of no fewer than eighteen persons; yet he was given enough to support them all adequately. As a final touch of attractiveness to the life of an academician in the eighteenth century, his children, if worth anything at all, were assured of a fair start in the world.

This brings us to a second dominant influence on Euler's vast mathematical output. The rulers who paid the bills naturally wanted something in addition to abstract culture for their money. But it must be emphasized that when once the rulers had obtained a reasonable return on their investment, they did not insist that their employees spend the rest of their time on "productive" labor; Euler, Lagrange, and the other academicians were free to do as they pleased. Nor was any noticeable pressure brought to bear to squeeze out the few immediately practical results which the state could use. Wiser in their generation than many a director of a research institute today, the rulers of the eighteenth century merely suggested occasionally what they needed at once, and let science take its course. They seem to have felt instinctively that so-called "pure" research would throw off as by-products the instantly practical things they desired if given a hint of the right sort now and then.

To this general statement there is one important exception, which neither proves nor disproves the rule. It so happened that in Euler's time the outstanding problem in mathematical research chanced also to coincide with what was probably the first practical problem of the age—control of the seas. That nation whose technique in navigation surpassed that of all its competitors would inevitably rule the waves. But navigation is ultimately an affair of accurately determining one's position at sea hundreds of miles from land, and of doing it so much better than one's competitors that they can be outsailed to the scene, unfavorable only for them, of a naval battle. Britannia, as everyone knows, rules the waves. That she does so is due in no small measure to the practical application which her navigators were able to make of purely mathematical investigations in celestial mechanics during the eighteenth century.

One such application concerned Euler directly—if we may anticipate slightly. The founder of modern navigation is of course Newton, although he himself never bothered his head about the subject and never (so far as seems to be known) planted his shoe on the deck of a ship. Position at sea is determined by observations on the heavenly bodies (sometimes including the satellites of Jupiter in really fancy navigation); and after Newton's universal law had suggested that with sufficient patience the positions of the planets and the phases of the Moon could be calculated for a century in advance if necessary, those who wished to govern the seas set their computers on the nautical almanac to grinding out tables of future positions.

In such a practical enterprise the Moon offers a particularly vicious problem, that of three bodies attracting one another according to the Newtonian law. This problem will recur many times as we proceed to the twentieth century; Euler was the first to evolve a *calculable* solution for the problem of the Moon ("the lunar theory"). The three bodies concerned are the Moon, the Earth, and the Sun. Although we shall defer what little can be said here on this problem to later chapters, it may be remarked that the problem is one of the most difficult in the whole range of mathematics. Euler did not solve it, but his method of approximative calculation (superseded today by better methods) was sufficiently practical to enable an English computer to calculate the lunar tables for the British Admiralty. For this the computer received £5000 (quite a sum for the time), and Euler was voted a bonus of £300 for the method.

Léonard (or Leonhard) Euler, a son of Paul Euler and his wife Marguerite Brucker, is probably the greatest man of science that Switzerland has produced. He was born at Basle on April 15, 1707, but moved the following year with his parents to the nearby village of Riechen, where his father became the Calvinist pastor. Paul Euler himself was an accomplished mathematician, having been a pupil of Jacob Bernoulli. The father intended Léonard to follow in his footsteps and succeed him in the village church, but fortunately made the mistake of teaching the boy mathematics.

Young Euler knew early what he wanted to do. Nevertheless he dutifully obeyed his father, and on entering the University of Basle studied theology and Hebrew. In mathematics he was sufficiently advanced to attract the attention of Johannes Bernoulli, who generously gave the young man one private lesson a week. Euler spent the rest of the week preparing for the next lesson so as to be able to meet his teacher with as few questions as possible. Soon his diligence and marked ability were noticed by Daniel and Nicolaus Bernoulli, who became Euler's fast friends.

Léonard was permitted to enjoy himself till he took his master's degree in 1724 at the age of seventeen, when his father insisted that he abandon mathematics and put all his time on theology. But the father gave in when the Bernoullis told him that his son was destined to be a great mathematician and not the pastor of Riechen. Although the prophecy was fulfilled Euler's early religious training influenced him all his life and he never discarded a particle of his Calvinistic faith. Indeed as he grew older he swung round in a wide orbit toward the calling of his father, conducting family prayers for his whole household and usually finishing off with a sermon.

Euler's first independent work was done at the age of nineteen. It has been said that this first effort reveals both the strength and the weakness of much of Euler's subsequent work. The Paris Academy had proposed the masting of ships as a prize problem for the year 1727; Euler's memoir failed to win the prize but received an honorable mention. He was later to recoup this loss by winning the prize twelve times. The strength of the work was the analysis—the technical mathematics—it contained; its weakness the remoteness of the connection, if any, with practicality. The last is not very surprising when we remember the traditional jokes about the nonexistent Swiss navy. Euler might have seen a boat or two on the Swiss lakes, but he had not yet seen a ship. He has been criticized, sometimes justly, for letting his mathematics run away with his sense of reality. The physical universe was an occasion for mathematics to Euler, scarcely a thing of much interest in itself; and if the universe failed to fit his analysis it was the universe which was in error.

Knowing that he was a born mathematician, Euler applied for the professorship at Basle. Failing to get the position, he continued his studies, buoyed up by the hope of joining Daniel and Nicolaus Bernoulli at St. Petersburg. They had generously offered to find a place for Euler in the Academy and kept him well posted.

At this stage of his career Euler seems to have been curiously indifferent as to what he should do, provided only it was something scientific. When the Bernoullis wrote of a prospective opening in the medical section of the St. Petersburg Academy, Euler flung himself into physiology at Basle and attended the lectures on medicine. But even in this field he could not keep away from mathematics: the physiology of the ear suggested a mathematical investigation of sound, which in turn led out into another on the propagation of waves, and so on—this early work kept branching out like a tree gone mad in a nightmare all through Euler's career.

The Bernoullis were fast workers. Euler received his call to St.

Petersburg in 1727, officially as an associate of the medical section of the Academy. By a wise provision every imported member was obliged to take with him two pupils—actually apprentices to be trained. Poor Euler's joy was quickly dashed. The very day he set foot on Russian soil the liberal Catherine I died.

Catherine, Peter the Great's mistress before she became his wife, seems to have been a broadminded woman in more ways than one, and it was she who in her reign of only two years carried out Peter's wishes in establishing the Academy. On Catherine's death the power passed into the hands of an unusually brutal faction during the minority of the boy czar (who perhaps fortunately for himself died before he could begin his reign). The new rulers of Russia looked upon the Academy as a dispensable luxury and for some anxious months contemplated suppressing it and sending all the foreign members home. Such was the state of affairs when Euler arrived in St. Petersburg. Nothing was said in the confusion about the medical position to which he had been called, and he slipped into the mathematical section, after having almost accepted a naval lieutenancy in desperation.

Thereafter things went better and Euler settled down to work. For six years he kept his nose to the grindstone, not wholly because he was absorbed in his mathematics but partly because he dared not lead a normal social life on account of the treacherous spies everywhere.

In 1733 Daniel Bernoulli returned to free Switzerland, having had enough of holy Russia, and Euler, at the age of twenty six, stepped into the leading mathematical position in the Academy. Feeling that he was to be stuck in St. Petersburg for the rest of his life, Euler decided to marry, settle down, and make the best of things. The lady was Catharina, a daughter of the painter Gsell, whom Peter the Great had taken back to Russia with him. Political conditions became worse, and Euler longed more desperately than ever to escape. But with the rapid arrival of one child after another Euler felt more securely tied than before and took refuge in incessant work. Some biographers trace Euler's unmatched productivity to this first sojourn in Russia; common prudence forced him into an unbreakable habit of industry.

Euler was one of several great mathematicians who could work anywhere under any conditions. He was very fond of children (he had thirteen of his own, all but five of whom died very young), and would often compose his memoirs with a baby in his lap while the older children played all about him. The ease with which he wrote the most difficult mathematics is incredible.

Many legends of his constant outflow of ideas have survived. Some no doubt are exaggerations, but it is said that Euler would dash off a mathematical paper in the half hour or so before the first and second calls to dinner. As soon as a paper was finished it was laid on top of the growing stack awaiting the printer. When material to fill the transactions of the Academy was needed, the printer would gather up a sheaf from the top of the pile. Thus it happened that the dates of publication frequently ran counter to those of composition. The crazy effect was heightened by Euler's habit of returning many times to a subject in order to clarify or extend what he had already done, so that occasionally a sequence of papers on a given topic is seen in print through the wrong end of the telescope.

When the boy czar died, Anna Ivanovna (niece of Peter) became Empress in 1730, and so far as the Academy was concerned, things brightened up considerably. But under the indirect rule of Anna's paramour, Ernest John de Biron, Russia suffered one of the bloodiest reigns of terror in its history, and Euler settled down to a spell of silent work that was to last ten years. Halfway through he suffered his first great misfortune. He had set himself to win the Paris prize for an astronomical problem for which some of the leading mathematicians had asked several months' time. (As a similar problem occurs in connection with Gauss we shall not describe it here.) Euler solved it in three days. But the prolonged effort brought on an illness in which he lost the sight of his right eye.

It should be noted that the modern higher criticism which has been so effective in discrediting all the interesting anecdotes in the history of mathematics has shown that the astronomical problem was in no way responsible for the loss of Euler's eye. But how the scholarly critics (or anyone else) come to know so much about the so-called law of cause and effect is a mystery for David Hume's (a contemporary of Euler) ghost to resolve. With this caution we shall tell once more the famous story of Euler and the atheistic (or perhaps only pantheistic) French philosopher Denis Diderot (1713–1784). This is slightly out of its chronological order, as it happened during Euler's second stay in Russia.

Invited by Catherine the Great to visit her Court, Diderot earned his keep by trying to convert the courtiers to atheism. Fed up, Catherine commissioned Euler to muzzle the windy philosopher. This was easy because all mathematics was Chinese to Diderot. De Morgan tells what happened (in his classic *Budget of Paradoxes*, 1872): "Diderot was informed that a learned mathematician was in possession of an algebraical demonstration of the existence of God, and would give it before all the Court, if he desired to hear it. Diderot gladly consented. ... Euler advanced toward Diderot, and said gravely, and in a tone of perfect conviction:

"Sir, $\dfrac{a + b^n}{n} = x$, hence God exists; reply!"

It sounded like sense to Diderot. Humiliated by the unrestrained laughter which greeted his embarrassed silence, the poor man asked Catherine's permission to return at once to France. She graciously gave it.

Not content with this masterpiece, Euler in all seriousness painted his lily with solemn proofs, in deadly earnest, that God exists and that the soul is not a material substance. It is reported that both proofs passed into the treatises on theology of his day. These are probably the choicest flowers of the mathematically unpractical side of his genius.

Mathematics alone did not absorb all of Euler's energies during his stay in Russia. Wherever he was called upon to exercise his mathematical talents in ways not too far from pure mathematics he gave the government its full money's worth. Euler wrote the elementary mathematical textbooks for the Russian schools, supervised the government department of geography, helped to reform the weights and measures, and devised practical means for testing scales. These were but some of his activities. No matter how much extraneous work he did, Euler continued to pour out mathematics.

One of the most important works of this period was the treatise of 1736 on mechanics. Note that the date of publication lacks but a year of marking the centenary of Descartes' publication of analytic geometry. Euler's treatise did for mechanics what Descartes' had done for geometry—freed it from the shackles of synthetic demonstration and made it analytical. Newton's *Principia* might have been written by Archimedes; Euler's mechanics could not have been written by any Greek. For the first time the full power of the calculus was directed against mechanics and the modern era in that basic science began. Euler was to be surpassed in this direction by his friend Lagrange, but the credit for having taken the decisive step is Euler's.

On the death of Anna in 1740 the Russian government became more liberal, but Euler had had enough and was glad to accept the invitation of Frederick the Great to join the Berlin Academy. The Dowager Queen took a great fancy to Euler and tried to draw him out. All she got was monosyllables.

"Why don't you want to speak to me?" she asked.

"Madame," Euler replied, "I come from a country where, if you speak, you are hanged."

The next twenty four years of his life were spent in Berlin, not altogether happily, as Frederick would have preferred a polished courtier instead of the simple Euler. Although Frederick felt it his duty to encourage mathematics he despised the subject, being no good at it himself. But he appreciated Euler's talents sufficiently to engage them in practical problems—the coinage, water conduits, navigation canals, and pension systems, among others.

Russia never let go of Euler completely and even while he was in Berlin paid part of his salary. In spite of his many dependents Euler was prosperous, owning a farm near Charlottenburg in addition to his house in Berlin. During the Russian invasion of the March of Brandenburg in 1760 Euler's farm was pillaged. The Russian general, declaring that he was "not making war on the sciences," indemnified Euler for considerably more than the actual damage. When the Empress Elizabeth heard of Euler's loss she sent him a handsome sum in addition to the more than sufficient indemnity.

One cause of Euler's unpopularity at Frederick's court was his inability to keep out of arguments on philosophical questions about which he knew nothing. Voltaire, who spent much of his time toadying to Frederick, delighted with the other brilliant verbalists surrounding Frederick in tying the hapless Euler into metaphysical knots. Euler took it all good-naturedly and joined the others in roaring with laughter at his own ridiculous blunders. But Frederick gradually became irritated and cast about for a more sophisticated philosopher to head his Academy and entertain his Court.

D'Alembert (whom we shall meet later) was invited to Berlin to look over the situation. He and Euler had had a slight coolness over mathematics. But D'Alembert was not the man to let a personal difference cloud his judgment, and he told Frederick bluntly that it would be an outrage to put any other mathematician over Euler. This only made Frederick more stubborn and angrier than ever, and conditions became intolerable for Euler. His sons, he felt, would have no chance in Prussia. At the age of fifty nine (in 1766) he pulled up his stakes once more and migrated back to St. Petersburg at the cordial invitation of Catherine the Great.

Catherine received the mathematician as if he were royalty, setting aside a fully furnished house for Euler and his eighteen dependents, and donating one of her own cooks to run the kitchen.

It was at this time that Euler began to lose the sight of his remaining eye (by a cataract), and before long he was totally blind. The progress of his oncoming darkness is followed with alarm and sympathy in the correspondence of Lagrange, D'Alembert, and other leading mathematicians of the time. Euler himself watched the approach of blindness with equanimity. There can be no doubt that his deep religious faith helped him to face what was ahead of him. But he did not "resign" himself to silence and darkness. He immediately set about repairing the irreparable. Before the last light faded he accustomed himself to writing his formulas with chalk on a large slate. Then, his sons (particularly Albert) acting as amanuenses, he would dictate the words explaining the formulas. Instead of diminishing, his mathematical productivity increased.

All his life Euler had been blessed with a phenomenal memory. He knew Virgil's *Aeneid* by heart, and although he had seldom looked at the book since he was a youth, could always tell the first and last lines on any page of his copy. His memory was both visual and aural. He also had a prodigious power for mental calculation, not only of the arithmetical kind but also of the more difficult type demanded in higher algebra and the calculus. All the leading formulas of the whole range of mathematics as it existed in his day were accurately stowed away in his memory.

As one instance of his prowess, Condorcet tells how two of Euler's students had summed a complicated convergent series (for a particular value of the variable) to seventeen terms, only to disagree by a unit in the fiftieth place of the result. To decide which was right Euler performed the whole calculation *mentally*; his answer was found to be correct. All this now came to his aid and he did not greatly miss the light. But even at that, one feat of his seventeen blind years almost passes belief. The lunar theory—the motion of the Moon, the only problem which had ever made Newton's head ache—received its first thorough workout at Euler's hands. All the complicated analysis was done entirely in his head.

Five years after Euler's return to St. Petersburg another disaster overtook him. In the great fire of 1771 his house and all its furnishings were destroyed, and it was only by the heroism of his Swiss servant (Peter Grimm, or Grimmon) that Euler escaped with his life. At the risk of his own life Grimm carried his blind and ailing master through the flames to safety. The library was burned, but thanks to the energy of Count Orloff all of Euler's manuscripts were saved. The Empress Catherine promptly made good all the loss and soon Euler was back at work again.

In 1776 (when he was sixty nine) Euler suffered a greater loss in the death of his wife. The following year he married again. The second wife, Salome Abigail Gsell, was a half-sister of the first. His greatest tragedy was the failure (through surgical carelessness, possibly) of an operation to restore the sight of his left eye—the only one for which there was any hope. The operation was "successful" and Euler's joy passed all bounds. But presently infection set in, and after prolonged suffering which he described as hideous, he lapsed back into darkness.

In looking back over Euler's enormous output we may be inclined at the first glance to believe that any gifted man could have done a large part of it almost as easily as Euler. But an inspection of mathematics as it exists today soon disabuses us. For the present state of mathematics with its jungles of theories is relatively no more complicated, when we consider the power of the methods now at our disposal, than what Euler faced. Mathematics is ripe for a second Euler. In his day he systematized and unified vast tracts cluttered with partial results and isolated theorems, clearing the ground and binding up the valuable things by the easy power of his analytical machinery. Even today much of what is learned in a college course in mathematics is practically as Euler left it—the discussion of conic sections and quadrics in three-space from the unified point of view provided by the general equation of the second degree, for example, is Euler's. Again, the subject of annuities and all that grows out of it (insurance, old-age pensions, and so on) were put into the shape now familiar to students of the "mathematical theory of investment" by Euler.

As Arago points out, one source of Euler's great and immediate success as a teacher through his writings was his total lack of false pride. If certain works of comparatively low intrinsic merit were demanded to clarify earlier and more impressive works, Euler did not hesitate to write them. He had no fear of lowering his reputation.

Even on the creative side Euler combined instruction with discovery. His great treatises of 1748, 1755 and 1768–70 on the calculus (*Introductio in analysin infinitorum; Institutiones calculi differentialis; Institutiones calculi integralis*) instantly became classic and continued for three-quarters of a century to inspire young men who were to become great mathematicians. But it was in his work on the calculus of variations (*Methodus inveniendi lineas curvas maximi minimive proprietate gaudentes*, 1744) that Euler first revealed himself as a mathematician of the first rank. The importance of this subject has been noted in previous chapters.

Euler's great step forward when he made mechanics *analytical* has already been remarked; every student of rigid dynamics is familiar with Euler's analysis of rotations, to cite but one detail of this advance. Analytical mechanics is a branch of pure mathematics, so that Euler was not tempted here, as in some of his other flights toward the practical, to fly off on the first tangent he saw leading into the infinite blue of pure calculation. The severest criticism which Euler's contemporaries made of his work was his uncontrollable impulse to calculate merely for the sake of the beautiful analysis. He may occasionally have lacked a sufficient understanding of the physical situations he attempted to reduce to calculation without seeing what they were all about. Nevertheless, the fundamental equations of fluid motion, in use today in hydrodynamics, are Euler's. He could be practical enough when it was worth his trouble.

One peculiarity of Euler's analysis must be mentioned in passing, as it was largely responsible for one of the main currents of mathematics in the nineteenth century. This was his recognition that unless an infinite series is *convergent* it is unsafe to use. For example, by long division we find

$$\frac{1}{x-1} = \frac{1}{x} + \frac{1}{x^2} + \frac{1}{x^3} + \frac{1}{x^4} + \dots,$$

the series continuing indefinitely. In this put $x = \frac{1}{2}$. Then

$$-2 = 2 + 2^2 + 2^3 + 2^4 + \dots,$$
$$= 2 + 4 + 8 + 16 + \dots.$$

The study of *convergence* (to be discussed in the chapter on Gauss) shows us how to avoid absurdities like this. (See also the chapter on

Cauchy.) The curious thing is that although Euler recognized the necessity for caution in dealing with *infinite* processes, he failed to observe it in much of his own work. His faith in analysis was so great that he would sometimes seek a preposterous "explanation" to make a patent absurdity respectable.

But when all this is said, we must add that few have equalled or approached Euler in the mass of sound and novel work of the first importance which he put out. Those who love arithmetic—not a very "important" subject, possibly—will vote Euler a palm in Diophantine analysis of the same size and freshness as those worn by Fermat and Diophantus himself. Euler was the first and possibly the greatest of the mathematical universalists.

Nor was he merely a narrow mathematician: in literature and all of the sciences, including the biologic, he was at least well read. But even while he was enjoying his *Aeneid* Euler could not help seeing a problem for his mathematical genius to attack. The line "The anchor drops, the rushing keel is stay'd" set him to working out the ship's motion under such circumstances. His omnivorous curiosity even swallowed astrology for a time, but he showed that he had not digested it by politely declining to cast the horoscope of Prince Ivan when ordered to do so in 1740, pointing out that horoscopes belonged in the province of the court astronomer. The poor astronomer had to do it.

One work of the Berlin period revealed Euler as a graceful (if somewhat too pious) writer, the celebrated *Letters to a German Princess*, composed to give lessons in mechanics, physical optics, astronomy, sound, etc., to Frederick's niece, the Princess of Anhalt-Dessau. The famous letters became immensely popular and circulated in book form in seven languages. Public interest in science is not the recent development we are sometimes inclined to imagine it is.

Euler remained virile and powerful of mind to the very second of his death, which occurred in his seventy seventh year, on September 18, 1783. After having amused himself one afternoon calculating the laws of ascent of balloons—on his slate, as usual—he dined with Lexell and his family. "Herschel's Planet" (Uranus) was a recent discovery; Euler outlined the calculation of its orbit. A little later he asked that his grandson be brought in. While playing with the child and drinking tea he suffered a stroke. The pipe dropped from his hand, and with the words "I die," "Euler ceased to live and calculate."*

---

*The quotation is from Condorcet's *Éloge*.

CHAPTER TEN

A Lofty Pyramid

LAGRANGE

I do not know.—J. L. LAGRANGE

"LAGRANGE IS THE LOFTY PYRAMID of the mathematical sciences." This was Napoleon Bonaparte's considered estimate of the greatest and most modest mathematician of the eighteenth century, Joseph-Louis Lagrange (1736–1813), whom he had made a Senator, a Count of the Empire, and a Grand Officer of the Legion of Honor. The King of Sardinia and Frederick the Great had also honored Lagrange, but less lavishly than the imperial Napoleon.

Lagrange was of mixed French and Italian blood, the French predominating. His grandfather, a French cavalry captain, had entered the service of Charles Emmanuel II, King of Sardinia, and on settling at Turin had married into the illustrious Conti family. Lagrange's father, once Treasurer of War for Sardinia, married Marie-Thérèse Gros, the only daughter of a wealthy physician of Cambiano, by whom he had eleven children. Of this numerous brood only the youngest, Joseph-Louis, born on January 25, 1736, survived beyond infancy. The father was rich, both in his own right and his wife's. But he was also an incorrigible speculator, and by the time his son was ready to inherit the family fortune there was nothing worth inheriting. In later life Lagrange looked back on this disaster as the luckiest thing that had ever happened to him: "If I had inherited a fortune I should probably not have cast my lot with mathematics."

At school Lagrange's first interests were in the classics, and it was more or less of an accident that he developed a passion for mathematics. In line with his classical studies he early became acquainted with the geometrical works of Euclid and Archimedes. These do not seem to have impressed him greatly. Then an essay by Halley (Newton's friend) extolling the superiority of the calculus over the synthetic geometrical methods of the Greeks fell into young Lagrange's hands. He was captivated and converted. In an incredibly short time he had mastered entirely by himself what in his day was modern analysis. At the age of sixteen (according to Delambre there may be a slight inaccuracy here) Lagrange became professor of mathematics at the Royal Artillery School in Turin. Then began one of the most brilliant careers in the history of mathematics.

From the first Lagrange was an analyst, never a geometer. In him we see the first conspicuous example of that specialization which was to become almost a necessity in mathematical research. Lagrange's analytical preferences came out strongly in his masterpiece, the *Mécanique analytique* (Analytical Mechanics), which he had projected as a boy of nineteen at Turin, but which was published in Paris only in 1788 when Lagrange was fifty two. "No diagrams will be found in this work," he says in the preface. But with a half-humorous libation to the gods of geometry he remarks that the science of mechanics may be considered as the geometry of a space of four dimensions—three Cartesian coordinates with one time-coordinate sufficing to locate a moving particle in both space and time, a way of looking at mechanics that has become popular since 1915 when Einstein exploited it in his general relativity.

Lagrange's analytical attack on mechanics marks the first complete break with the Greek tradition. Newton, his contemporaries, and his immediate successors found diagrams helpful in their study of mechanical problems; Lagrange showed that greater flexibility and incomparably greater power are attained if general analytical methods are employed from the beginning.

At Turin the boyish professor lectured to students all older than himself. Presently he organized the more able into a research society from which the Turin Academy of Sciences developed. The first volume of the Academy's memoirs was published in 1759, when Lagrange was twenty three. It is usually supposed that the modest and unobtrusive Lagrange was responsible for much of the fine mathematics in these early works published by others. One paper by Foncenex was so good that the King of Sardinia put the supposed author in charge of the Department of the Navy. Historians of mathematics have sometimes wondered why Foncenex never lived up to his first mathematical success.

Lagrange himself contributed a memoir on maxima and minima (the calculus of variations, described in Chapters 4, 8) in which he promises to treat the subject in a work from which he will deduce

A Lofty Pyramid

the whole of mechanics, of both solids and fluids. Thus at twenty three—actually earlier—Lagrange had imagined his masterpiece, the *Mécanique analytique*, which does for general mechanics what Newton's law of universal gravitation did for celestial mechanics. Writing ten years later to the French mathematician D'Alembert (1717–1783), Lagrange says he regards his early work, the calculus of variations, thought out when he was nineteen, as his masterpiece. It was by means of this calculus that Lagrange unified mechanics and, as Hamilton said, made of it "a kind of scientific poem."

When once understood the Lagrangian method is almost a platitude. As some have remarked the Lagrangian equations dominating mechanics are the finest example in all science of the art of getting something out of nothing. But if we reflect a moment we see that any scientific principle which is general to the extent of uniting a whole vast universe of phenomena *must* be simple: only a principle of the utmost simplicity can dominate a multitude of diverse problems which on even a close inspection appear to be individual and distinct.

In the same volume of Turin memoirs Lagrange took another long step forward: he applied the differential calculus to the theory of probability. As if this were not enough for the young giant of twenty three he advanced beyond Newton with a radical departure in the mathematical theory of sound, bringing that theory under the sway of the mechanics of systems of elastic particles (rather than of the mechanics of fluids), by considering the behavior of all the air particles in one straight line under the action of a shock transmitted along the line from particle to particle. In the same general direction he also settled a vexed controversy that had been going on for years between the leading mathematicians over the correct mathematical formulation of the problem of a vibrating string—a problem of fundamental importance in the whole theory of vibrations. At twenty three Lagrange was acknowledged the equal of the greatest mathematicians of the age—Euler and the Bernoullis.

Euler was always generously appreciative of the work of others. His treatment of his young rival Lagrange is one of the finest pieces of unselfishness in the history of science. When as a boy of nineteen Lagrange sent Euler some of his work the famous mathematician at once recognized its merit and encouraged the brilliant young beginner to continue. When four years later Lagrange communicated to Euler the true method for attacking the isoperimetrical problems (the calculus of variations, described in connection with the Bernoullis), which had baffled Euler with his semi-geometrical methods for many years, Euler wrote to the young man saying that the new method had enabled him to overcome his difficulties. And instead of rushing into print with the long-sought solution, Euler held it back till Lagrange could publish his first, "so as not to deprive you of any part of the glory which is your due."

Private letters, however flattering, could not have helped Lagrange. Realizing this, Euler went out of his way when he published his work (after Lagrange's) to say how he had been held up by difficulties which, till Lagrange showed the way over them, were insuperable. Finally, to clinch the matter, Euler got Lagrange elected as a foreign member of the Berlin Academy (October 2, 1759) at the unusually early age of twenty three. This official recognition abroad was a great help to Lagrange at home. Euler and D'Alembert schemed to get Lagrange to Berlin. Partly for personal reasons they were eager to see their brilliant young friend installed as court mathematician at Berlin. After lengthy negotiations they succeeded, and the great Frederick, slightly outwitted in the whole transaction, was childishly (but justifiably) delighted.

Something must be said in passing about D'Alembert, Lagrange's devoted friend and generous admirer, if only for the grateful contrast one aspect of his character offers to that of the snobbish Laplace, whom we shall meet later.

Jean le Rond d'Alembert took his name from the little chapel of St. Jean le Rond hard by Notre-Dame in Paris. An illegitimate son of the Chevalier Destouches, D'Alembert had been abandoned by his mother on the steps of St. Jean le Rond. The parish authorities turned the foundling over to the wife of a poor glazier, who reared the child as if he were her own. The Chevalier was forced by law to pay for his bastard's education. D'Alembert's real mother knew where he was, and when the boy early gave signs of genius, sent for him, hoping to win him over.

"You are only my stepmother," the boy told her (a good pun in English, but not in French); "the glazier's wife is my true mother." And with that he abandoned his own flesh and blood as she had abandoned hers.

When he became famous and a great figure in French science D'Alembert repaid the glazier and his wife by seeing that they did not fall into want (they preferred to keep on living in their humble quarters), and he was always proud to claim them as his parents. Although we shall not have space to consider him apart from Lagrange, it must be mentioned that D'Alembert was the first to give a complete solution of the outstanding problem of the precession of the equinoxes. His most important purely mathematical work was in partial differential equations, particularly in connection with vibrating strings.

D'Alembert encouraged his modest young correspondent to attack difficult and important problems. He also took it upon himself to make Lagrange take reasonable care of his health—his own was not good. Lagrange had in fact seriously impaired his digestion by quite unreasonable application between the ages of sixteen and twenty six, and all his life thereafter he was forced to discipline himself severely, especially in the matter of overwork. In one of his letters D'Alembert lectures the young man for indulging in tea and coffee to keep awake; in another he lugubriously calls Lagrange's attention to a recent medical book on the diseases of scholars. To all of which Lagrange blithely replies that he is feeling fine and working like mad. But in the end he paid his tax.

In one respect Lagrange's career is a curious parallel to Newton's. By middle age prolonged concentration on problems of the first magnitude had dulled Lagrange's enthusiasm, and although his mind remained as powerful as ever, he came to regard mathematics with indifference. When only forty five he wrote to D'Alembert, "I begin to feel the pull of my inertia increasing little by little, and I cannot say that I shall still be doing mathematics ten years from now. It also seems to me that the mine is already too deep, and that unless new veins are discovered it will have to be abandoned."

When he wrote this Lagrange was ill and melancholic. Nevertheless it expressed the truth so far as he was concerned. D'Alembert's last letter (September, 1783), written a month before his death, reverses his early advice and counsels work as the only remedy for Lagrange's psychic ills: "In God's name do not renounce work, for you the strongest of all distractions. Goodbye, perhaps for the last time. Keep some memory of the man who of all in the world cherishes and honors you the most."

Happily for mathematics Lagrange's blackest depression, with its inescapable corollary that no human knowledge is worth striving for, was twenty glorious years in the future when D'Alembert and Euler were scheming to get Lagrange to Berlin. Among the great problems Lagrange attacked and solved before going to Berlin was that of the libration of the Moon. Why does the Moon always present the same face to the Earth—within certain slight irregularities that can be accounted for? It was required to deduce this fact from the Newtonian law of gravitation. The problem is an instance of the famous "Problem of Three Bodies"—here the Earth, Sun, and Moon—mutually attracting one another according to the law of the inverse square of the distances between their centers of gravity. (More will be said on this problem when we come to Poincaré.)

For his solution of the problem of libration Lagrange was awarded the Grand Prize of the French Academy of Sciences in 1764—he was then only twenty eight.

Encouraged by this brilliant success the Academy proposed a yet more difficult problem, for which Lagrange again won the prize in 1766. In Lagrange's day only four satellites of Jupiter had been discovered. Jupiter's system (himself, the Sun, and his satellites) thus made a six-body problem. A *complete* mathematical solution is beyond our powers even today (1936) in a shape adapted to practical computation. But by using methods of approximation Lagrange made a notable advance in explaining the observed inequalities.

Such applications of the Newtonian theory were one of Lagrange's major interests all his active life. In 1772 he again captured the Paris prize for his memoir on the three-body problem, and in 1774 and 1778 he had similar successes with the motion of the Moon and cometary perturbations.

The earlier of these spectacular successes induced the King of Sardinia to pay Lagrange's expenses for a trip to Paris and London in 1766. Lagrange was then thirty. It had been planned that he was to accompany Caraccioli, the Sardinian minister to England, but on reaching Paris Lagrange fell dangerously ill—the result of an overgenerous banquet of rich Italian dishes in his honor—and he was forced to remain in Paris. While there he met all the leading intellectuals, including the Abbé Marie, who was later to prove an invaluable friend. The banquet cured Lagrange of his desire to live in Paris and he eagerly returned to Turin as soon as he was able to travel.

At last, on November 6, 1766, Lagrange was welcomed, at the

age of thirty, to Berlin by Frederick, "the greatest King in Europe," as he modestly styled himself, who would be honored to have at his court "the greatest mathematician." The last, at least, was true. Lagrange became director of the physico-mathematical division of the Berlin Academy, and for twenty years crowded the transactions of the Academy with one great memoir after another. He was not required to lecture.

At first the young director found himself in a somewhat delicate position. Naturally enough the Germans rather resented foreigners being brought in over their heads and were inclined to treat Frederick's importations with a little less than cool civility. In fact they were frequently quite insulting. But in addition to being a mathematician of the first rank Lagrange was a considerate, gentle soul with the rare gift of knowing when to keep his mouth shut. In letters to trusted friends he could be outspoken enough, even about the Jesuits, whom he and D'Alembert seem to have disliked, and in his official reports to academies on the scientific work of others he could be quite blunt. But in his social contacts he minded his own business and avoided giving even justifiable offense. Until his colleagues got used to his presence he kept out of their way.

Lagrange's constitutional dislike of all disputes stood him in good stead at Berlin. Euler had blundered from one religious or philosophical controversy to another; Lagrange, if cornered and pressed, would always preface his replies with his sincere formula "I do not know." Yet when his own convictions were attacked he knew how to put up a spirited, reasoned defense.

On the whole Lagrange was inclined to sympathize with Frederick who had sometimes been irritated by Euler's tilting at philosophical problems about which he knew nothing. "Our friend Euler," he wrote to D'Alembert, "is a great mathematician, but a bad enough philosopher." And on another occasion, referring to Euler's effusion of pious moralizing in the celebrated *Letters to a German Princess*, he dubs the classic "Euler's commentary on the Apocalypse"—incidentally a backhand allusion to the indiscretion which Newton permitted himself when he had lost his taste for natural philosophy. "It is incredible," Lagrange said of Euler, "that he could have been so flat and childish in metaphysics." And for himself, "I have a great aversion to disputes." When he did philosophize in his letters it was with an unexpected touch of cynicism which is wholly absent from the works he published, as when he remarks, "I have always observed that the pretensions of all people are in exact inverse ratio to their merits; this is one of the axioms of morals." In religious matters Lagrange was, if anything at all, agnostic.

Frederick was delighted with his prize and spent many friendly hours with Lagrange, expounding the advantages of a regular life. The contrast Lagrange offered to Euler was particularly pleasing to Frederick. The King had been irritated by Euler's too obvious piety and lack of courtly sophistication. He had even gone so far as to call poor Euler a "lumbering cyclops of a mathematician," because Euler at the time was blind in only one of his eyes. To D'Alembert the grateful Frederick overflowed in both prose and verse. "To your trouble and to your recommendation," he wrote, "I owe the replacement in my Academy of a mathematician blind in one eye by a mathematician with two eyes, which will be especially pleasing to the anatomical section." In spite of sallies like this Frederick was not a bad sort.

Shortly after settling in Berlin Lagrange sent to Turin for one of his young lady relatives and married her. There are two accounts of how this happened. One says that Lagrange had lived in the same house with the girl and her parents and had taken an interest in her shopping. Having an economical streak in his cautious nature, Lagrange was scandalized by what he considered the girl's extravagance and bought her ribbons himself. From there on he was dragooned into marrying her.

The other version can be inferred from one of Lagrange's letters —certainly the strangest confession of indifference ever penned by a supposedly doting young husband. D'Alembert had joked his friend: "I understand that you have taken what we philosophers call the fatal plunge. . . . A great mathematician should know above all things how to calculate his happiness. I do not doubt then that after having performed this calculation you found the solution in marriage."

Lagrange either took this in deadly earnest or set out to beat D'Alembert at his own game—and succeeded. D'Alembert had expressed surprise that Lagrange had not mentioned his marriage in his letters.

"I don't know whether I calculated ill or well," Lagrange replied, "or rather, I don't believe I calculated at all; for I might have done as Leibniz did, who, compelled to reflect, could never make up his mind. I confess to you that I never had a taste for marriage, . . . but circumstances decided me to engage one of my young kinswomen to take care of me and all my affairs. If I neglected to inform you it was because the whole thing seemed to me so inconsequential in itself that it was not worth the trouble of informing you of it."

The marriage was turning out happily for both when the wife declined in a lingering illness. Lagrange gave up his sleep to nurse her himself and was heartbroken when she died.

He consoled himself in his work. "My occupations are reduced to cultivating mathematics, tranquilly and in silence." He then tells D'Alembert the secret of the perfection of all his work which has been the despair of his hastier successors. "As I am not pressed and work more for my pleasure than from duty, I am like the great lords who build: I make, unmake, and remake, until I am passably satisfied with my results, which happens only rarely." And on another occasion, after complaining of illness brought on by overwork, he says it is impossible for him to rest: "My bad habit of rewriting my memoirs several times till I am passably satisfied is impossible for me to break."

Not all of Lagrange's main efforts during his twenty years at Berlin went into celestial mechanics and the polishing of his masterpiece. One digression—into Fermat's domain—is of particular interest as it may suggest the inherent difficulty of simple-looking things in arithmetic. We see even the great Lagrange puzzled over the unexpected effort his arithmetical researches cost him.

"I have been occupied these last few days," he wrote to D'Alembert on August 15, 1768, "in diversifying my studies a little with certain problems of Arithmetic, and I assure you I found many more difficulties than I had anticipated. Here is one, for example, at whose solution I arrived only with great trouble. Given any positive integer n which is not a square, to find a square integer, x^2, such that $nx^2 + 1$ shall be a square. This problem is of great importance in the theory of squares [today, *quadratic forms*, to be described in connection with Gauss] which [squares] are the principal object in Diophantine analysis. Moreover I found on this occasion some very beautiful theorems of Arithmetic, which I will communicate to you another time if you wish."

The problem Lagrange describes has a long history going back to Archimedes and the Hindus. Lagrange's classic memoir on making $nx^2 + 1$ a square is a landmark in the theory of numbers. He was also the first to prove some of Fermat's theorems and that of John Wilson (1741-1793), who had stated that if p is any prime number, then if all the numbers $1, 2, \ldots$ up to $p - 1$ are multiplied together and 1 be added to the result, the sum is divisible by p. The like is not true if p is not prime. For example, if $p = 5$, $1 \times 2 \times 3 \times 4 + 1 = 25$. This can be proved by elementary reasoning and is another of those arithmetical super-intelligence tests.*

In his reply D'Alembert states his belief that Diophantine analysis may be useful in the integral calculus, but does not go into detail. Curiously enough, the prophecy was fulfilled in the 1870's by the Russian mathematician, G. Zolotareff.

Laplace also became interested in arithmetic for a while and told Lagrange that the existence of Fermat's unproved theorems, while one of the greatest glories of French mathematics, was also its most conspicuous blemish, and it was the duty of French mathematicians to remove the blemish. But he prophesied tremendous difficulties. The root of the trouble, in his opinion, is that *discrete* problems (those dealing ultimately with $1, 2, 3, \ldots$) are not yet attackable by any general weapon such as the calculus provides for the continuous. D'Alembert also remarks of arithmetic that he found it "more difficult than it seems at first." These experiences of mathematicians like Lagrange and his friends may imply that arithmetic really is hard.

Another letter of Lagrange's (February 28, 1769) records the conclusion of the matter. "The problem I spoke of has occupied me much more than I anticipated at first; but finally I am happily finished and I believe I have left practically nothing to be desired in the subject of indeterminate equations of the second degree in two unknowns." He was too optimistic here; Gauss had yet to be heard from—his father and mother had still seven years to go before meeting one another. Two years before the birth of Gauss (in 1777), Lagrange looked back over his work in a pessimistic mood: "The arithmetical

---

*A ridiculous "proof" by a Spanish gentleman is funny enough to be quoted. The customary abbreviation for $1 \times 2 \times \ldots \times n$ is $n!$ Now $p - 1 + 1 = p$, which is divisible by p. Put exclamation points throughout: $(p - 1)! + 1! = p!$. The right side is again divisible by p; hence $(p - 1)! + 1$ is divisible by p. Unfortunately this works equally well if p is not prime.

researches are those which have cost me most trouble and are perhaps the least valuable."

When he was feeling well Lagrange seldom lapsed into the error of estimating the "importance" of his work. "I have always regarded mathematics," he wrote to Laplace in 1777, "as an object of amusement rather than of ambition, and I can assure you that I enjoy the works of others much more than my own, with which I am always dissatisfied. You will see by that, if you are exempt from jealousy by your own success, I am none the less so by my disposition." This was in reply to a somewhat pompous declaration by Laplace that he worked at mathematics only to appease his own sublime curiosity and did not give a hang for the plaudits of "the multitude"—which, in his case, was partly balderdash.

A letter of September 15, 1782, to Laplace is of great historical interest as it tells of the finishing of the *Mécanique analytique*: "I have almost completed a Treatise on Analytical Mechanics, founded solely on the principle or formula in the first section of the accompanying memoir; but as I do not know when or where I can get it printed, I am not hurrying with the finishing touches."

Legendre undertook the editing of the work for the press and Lagrange's old friend the Abbé Marie finally persuaded a Paris publisher to risk his reputation. This canny individual consented to proceed with the printing only when the Abbé agreed to purchase all stock remaining unsold after a certain date. The book did not appear until 1788, after Lagrange had left Berlin. A copy was delivered into his hands when he had grown so indifferent to all science and all mathematics that he did not even bother to open the book. For all he knew at the time the printer might have got it out in Chinese. He did not care.

One investigation of Lagrange's Berlin period is of the highest importance in the development of modern algebra, the memoir of 1767 *On the Solution of Numerical Equations* and the subsequent additions dealing with the general question of the algebraic solvability of equations. Possibly the greatest importance of Lagrange's researches in the theory and solution of equations is the inspiration they proved to be to the leading algebraists of the early nineteenth century. Time after time we shall see the men who finally disposed of a problem which had baffled algebraists for three centuries or more returning to Lagrange for ideas and inspiration. Lagrange himself did not resolve the central difficulty—that of stating necessary and sufficient conditions that a given equation shall be solvable algebraically, but the germ of the solution is to be found in his work.

As the problem is one of those major things in all algebra which can be simply described we may glance at it in passing; it will recur many times as a leading motive in the work of some of the great mathematicians of the nineteenth century—Cauchy, Abel, Galois, Hermite, and Kronecker, among others.

First it may be emphasized that there is no difficulty whatever in solving an algebraic equation with numerical coefficients. The labor may be excessive if the equation is of high degree, say

$$3x^{101} - 17.3x^{70} + x - 11 = 0,$$

but there are many straightforward methods known whereby a root of such a *numerical* equation can be found to any prescribed degree of accuracy. Some of these are part of the regular school course in algebra. But in Lagrange's day uniform methods for solving numerical equations to a preassigned degree of accuracy were not commonplace —if known at all. Lagrange provided such a method. Theoretically it did what was required, but it was not practical. No engineer faced with a numerical equation today would dream of using Lagrange's method.

The really significant problem arises when we seek an *algebraic* solution of an equation with *literal* coefficients, say $ax^2 + bx + c = 0$, or $ax^3 + bx^2 + cx + d = 0$, and so on for degrees higher than the third. What is required is a set of formulas expressing the *unknown* x in terms of the *given* $a, b, c, \ldots$, such that if any one of these expressions for x be put in the lefthand side of the equation, that side shall reduce to zero. For an equation of degree n the unknown x has precisely n values. Thus for the above quadratic (second degree) equation,

$$\frac{1}{2a}(-b + \sqrt{b^2 - 4ac}), \frac{1}{2a}(-b - \sqrt{b^2 - 4ac})$$

are the two values which when substituted for x will reduce $ax^2 + bx + c$ to zero. *The required values of x in any case are to be expressed in terms of $a, b, c, \ldots$ by means of only a* FINITE *number of additions, subtractions, multiplications, divisions, and extractions of roots.* This is the problem. Is it solvable? The answer to this was not given till about twenty years after Lagrange's death, but the clue is easily traced to his work.

As a first step toward a comprehensive theory Lagrange made an exhaustive study of all the solutions given by his predecessors for the general equations of the first four degrees, and succeeded in showing that all of the dodges by which solutions had been obtained could be replaced by a uniform procedure. A detail in this general method contains the clue mentioned. Suppose we are given an algebraic expression involving letters a, b, c, ... : how many *different* expressions can be derived from the given one if the letters in it are interchanged in all possible ways? For example, from $ab + cd$ we get $ad + cb$ by interchanging b and d. This problem suggests another closely related one, also part of the clue Lagrange was seeking. What interchanges of letters will leave the given expression *invariant* (unaltered)? Thus $ab + cd$ becomes $ba + cd$ under the interchange of a and b, which is the same as $ab + cd$ since $ab = ba$. From these questions the *theory of finite groups* originated. This was found to be the key to the question of algebraic solvability. It will reappear when we consider Cauchy and Galois.

Another significant fact showed up in Lagrange's investigation. For degrees 2, 3, and 4 the general algebraic equation is solved by making the solution depend upon that of an equation of *lower degree* than the one under discussion. This works beautifully and uniformly for equations of degrees 2, 3, and 4, but when a precisely similar process is attempted on the general equation of degree 5,

$$ax^5 + bx^4 + cx^3 + dx^2 + ex + f = 0,$$

the *resolvent equation*, instead of being of degree *less than* 5 turns out to be of degree 6. This has the effect of replacing the given equation by a harder one. *The method which works for 2, 3, 4 breaks down for 5*, and unless there is some way round the awkward 6 the road is blocked. As a matter of fact we shall see that there is no way of avoiding the difficulty. We might as well try to square the circle or trisect an angle by Euclidean methods.

After the death of Frederick the Great (August 17, 1786) resentment against non-Prussians and indifference to science made Berlin an uncomfortable spot for Lagrange and his foreign associates in the

Academy, and he sought his release. This was granted on condition that he continue to send memoirs to the proceedings of the Academy for a period of years, to which Lagrange agreed. He gladly accepted the invitation of Louis XVI to continue his mathematical work in Paris as a member of the French Academy. On his arrival in Paris in 1787 he was received with the greatest respect by the royal family and the Academy. Comfortable quarters were assigned him in the Louvre, where he lived till the Revolution, and he became a special favorite of Marie Antoinette—then less than six years from the guillotine. Marie was about nineteen years younger than Lagrange, but she seemed to understand him and did what she could to lighten his overwhelming depression.

At the age of fifty one Lagrange felt that he was through. It was a clear case of nervous exhaustion from long-continued and excessive overwork. The Parisians found him gentle and agreeable in conversation, but he never took the lead. He spoke but little and appeared distrait and profoundly melancholy. At Lavoisier's gatherings of scientific men Lagrange would stand staring absently out of a window, his back to the guests who had come to do him honor, a picture of sad indifference. He said himself that his enthusiasm was extinct and that he had lost the taste for mathematics. If he were told that some mathematician was engaged on an important research he would say "So much the better; I began it; I shall not have to finish it." The *Mécanique analytique* lay unopened on his desk for two years.

Sick of everything smelling of mathematics Lagrange now turned to what he considered his real interests—as Newton had done after the *Principia*: metaphysics, the evolution of human thought, the history of religions, the general theory of languages, medicine, and botany. In this strange miscellany he surprised his friends with his extensive knowledge and the penetrating quality of his mind on matters alien to mathematics. Chemistry at the time was fast becoming a science—in distinction to the alchemy which preceded it, largely through the efforts of Lagrange's close friend Lavoisier (1743–1794). In a sense which any student of elementary chemistry will appreciate Lagrange declared that Lavoisier had made chemistry "as easy as algebra."

As for mathematics, Lagrange considered that it was finished or at least passing into a period of decadence. Chemistry, physics, and science generally he foresaw as the future fields of greatest interest to first-class minds, and he even predicted that the chairs of mathematics in academies and universities would presently sink to the undistinguished level of those for Arabic. In a sense he was right. Had not Gauss, Abel, Galois, Cauchy, and others injected new ideas into mathematics the surge of the Newtonian impulse would have spent itself by 1850. Happily Lagrange lived long enough to see Gauss well started on his great career and to realize that his own forebodings had been unfounded. We may smile at Lagrange's pessimism today, thinking of the era before 1800 at its brightest as only the dawn of the modern mathematics in the first hour of whose morning we now stand, wondering what the noon will be like—if there is to be any; and we may learn from his example to avoid prophecy.

The Revolution broke Lagrange's apathy and galvanized him once more into a living interest in mathematics. As a convenient point of reference we may remember July 14, 1789, the day on which the Bastille fell.

When the French aristocrats and men of science at last realized what they were in for, they urged Lagrange to return to Berlin where a welcome awaited him. No objection would have been raised to his departure. But he refused to leave Paris, saying he would prefer to stay and see the "experiment" through. Neither he nor his friends foresaw the Terror, and when it came Lagrange bitterly regretted having stayed until it was too late to escape. He had no fear for his own life. In the first place as a half-foreigner he was reasonably safe, and in the second he did not greatly value his life. But the revolting cruelties sickened him and all but destroyed what little faith he had left in human nature and common sense. *"Tu l'as voulu"* ("You wished it," or "You *would* do it"), he would keep reminding himself as one atrocity after another shocked him into a realization of his error in staying to witness the inevitable horrors of a revolution.

The grandiose schemes of the revolutionists for the regeneration of mankind and the reform of human nature left him cold. When Lavoisier went to the guillotine—as he no doubt would have deserved had it been merely a question of social justice—Lagrange expressed his indignation at the stupidity of the execution: "It took them only a moment to cause this head to fall, and a hundred years perhaps will not suffice to produce its like." But the outraged and oppressed citizens had assured the tax-farmer Lavoisier that "the people have no need of science" when the great chemist's contributions to science were urged as a common-sense reason that his head be left on his shoulders. They may have been right. Without the science of chemistry soap is impossible.

Although practically the whole of Lagrange's working life had been spent under the patronage of royalty his sympathies were not with the royalists. Nor were they with the revolutionists. He stood squarely and unequivocally on the middle ground of civilization which both sides had ruthlessly invaded. He could sympathize with the people who had been outraged beyond human endurance and wish them success in their struggle to gain decent living conditions. But his mind was too realistic to be impressed by any of the chimerical schemes put forth by the leaders of the people for the amelioration of human misery, and he refused to believe that the fabrication of such schemes was indubitable evidence of the greatness of the human mind as claimed by the enthusiastic guillotineers. "If you wish to see the human mind truly great," he said, "enter Newton's study when he is decomposing white light or unveiling the system of the world."

They treated him with remarkable tolerance. A special decree granted him his "pension," and when the inflation by paper money reduced the pension to nothing, they appointed him on the committee of inventions to eke out his pay, and again on the committee for the mint. When the École Normale was established in 1795 (for an ephemeral first existence), Lagrange was appointed professor of mathematics. When the Normale closed and the great École Polytechnique was founded in 1797, Lagrange mapped out the course in mathematics and was the first professor. He had never taught before he was called upon to lecture to ill-prepared students. Adapting himself to his raw material, Lagrange led his pupils through arithmetic and algebra to analysis, seeming more like one of his pupils than their teacher. The greatest mathematician of the age became a great teacher of mathematics—preparing Napoleon's fierce young brood of military engineers for their part in the conquest of Europe. The sacred superstition that a man who knows anything is incapable of teaching was shattered. Advancing far beyond the elements Lagrange developed new mathematics before his pupils' eyes and presently they were taking part in the development themselves.

Two works thus developed were to exercise a great influence on the analysis of the first three decades of the nineteenth century. Lagrange's pupils found difficulty with the concepts of the infinitely small and the infinitely great permeating the traditional form of the calculus. To remove these difficulties Lagrange undertook the development of the calculus without the use of Leibniz' "infinitesimals" and without Newton's peculiar conception of a limit. His own theory was published in two works, the *Theory of Analytic Functions* (1797), and the *Lessons on the Calculus of Functions* (1801). The importance of these works is not in their mathematics but in the impulse they gave Cauchy and others to construct a satisfactory calculus. Lagrange failed completely. But in saying this we must remember that even in our own day the difficulties with which Lagrange grappled unsuccessfully have not been completely overcome. His was a notable attempt and, for its epoch, satisfactory. If our own lasts as long as his did we shall have done well enough.

Lagrange's most important work during the period of the Revolution was his leading part in perfecting the metric system of weights and measures. It was due to Lagrange's irony and common sense that 12 was not chosen as a base instead of 10. The "advantages" of 12 are obvious and continue to the present day to be set forth in impressive treatises by earnest propagandists who escape the circle-squaring fraternity only by a hairsbreadth. A base of 12 superimposed on the 10 of our number-system would be a hexagonal peg in a pentagonal hole. To bring home the absurdity of 12 even to the cranks, Lagrange proposed 11 as better yet—*any prime number* would have the advantage of giving all fractions in the system the same denominator. The disadvantages are numerous and obvious enough to anyone who understands short division. The committee saw the point and stuck to 10.

Laplace and Lavoisier were members of the committee as first constituted, but after three months they were "purged" out of their seats with some others. Lagrange remained as president. "I do not know why they kept me," he remarked, modestly unaware that his gift for silence had saved not only his seat but his head.

In spite of all his interesting work Lagrange was still lonely and inclined to despondency. He was rescued from this twilight between life and death at the age of fifty six by a young girl nearly forty years his junior, the daughter of his friend the astronomer Lemonnier. She was touched by Lagrange's unhappiness and insisted on marrying him. Lagrange gave in, and contrary to all the laws of whatever it may be that governs the way of a man with a maid, the marriage turned out ideal. The young wife proved not only devoted but competent. She made it her life to draw her husband out and reawaken his desire to live. For his part Lagrange gladly made many concessions and accompanied his wife to balls where he would never have thought of going alone. Before long he could not bear to have her out of his sight for long, and during her brief absences—shopping—he was miserable.

Even in his new happiness Lagrange retained his curiously detached attitude to life and his perfect honesty about his own wishes. "I had no children by my first marriage," he said; "I don't know whether I shall have any by my second. I scarcely desire any." Of all his successes the one he prized most highly, he said simply and sincerely, was having found so tender and devoted a companion as his young wife.

Honors were showered on him by the French. The man who had been a favorite of Marie Antoinette now became an idol of the people who had put her to death. In 1796 when France annexed Piedmont, Talleyrand was ordered to wait in state on Lagrange's father, still living in Turin, to tell him that "Your son, whom Piedmont is proud to have produced and France to possess, has done honor to all mankind by his genius." When Napoleon turned to civil affairs between his campaigns he often talked with Lagrange on philosophical questions and the function of mathematics in a modern state, and conceived the highest respect for the gently-spoken man who always thought before he spoke and who was never dogmatic.

Beneath his calm reserve Lagrange concealed an ironic wit which flashed out unexpectedly on occasion. Sometimes it was so subtle that coarser men—Laplace, for one—missed the point when it was directed at themselves. Once in defense of experiment and observation against mere woolgathering and vague theorizing Lagrange remarked "These astronomers are queer; they won't believe in a theory unless it agrees with their observations." Noticing his rapt forgetfulness at a musicale, someone asked him why he liked music. "I like it because it isolates me," he replied. "I hear the first three measures; at the fourth I distinguish nothing; I give myself up to my thoughts; nothing interrupts me; and it is thus that I have solved more than one difficult problem." Even his sincere reverence for Newton has a faint flavor of the same gentle irony. "Newton," he declared, "was assuredly the man of genius *par excellence*, but we must agree that he was also the luckiest: one finds only once the system of the world to be established." And again: "How lucky Newton was that in his time the system of the world still remained to be discovered!"

Lagrange's last scientific effort was the revision and extension of the *Mécanique analytique* for a second edition. All his old power returned to him although he was past seventy. Resuming his former habits he worked incessantly, only to discover that his body would no longer obey his mind. Presently he began to have fainting spells, especially on getting out of bed in the morning. One day his wife found him unconscious on the floor, his head badly cut by a fall against the edge of a table. Thereafter he moderated his pace but kept on working. His illness, which he knew to be grave, did not disturb his serenity; all his life Lagrange lived as a philosopher would like to live, indifferent to his fate.

Two days before Lagrange died Monge and other friends called, knowing that he was dying and that he wished to tell them something of his life. They found him temporarily better, except for lapses of memory which obliterated what he had wished to tell them.

"I was very ill yesterday, my friends," he said. "I felt I was going to die; my body grew weaker little by little; my intellectual and physical faculties were extinguished insensibly; I observed the well-graduated progression of the diminution of my strength, and I came to the end without sorrow, without regrets, and by a very gentle decline. Oh, death is not to be dreaded, and when it comes without pain, it is a last function which is not unpleasant."

He believed that the seat of life is in all the organs, in the whole of the bodily machine, which, in his case, weakened equally in all its parts.

"In a few moments there will be no more functions anywhere, death will be everywhere; death is only the absolute repose of the body.

"I wish to die; yes, I wish to die, and I find a pleasure in it. But my wife did not wish it. In these moments I should have preferred a wife less good, less eager to revive my strength, who would have let me end gently. I have had my career; I have gained some celebrity in Mathematics. I never hated anyone, I have done nothing bad, and it would be well to end; but my wife did not wish it."

He soon had his wish. A fainting spell from which he never awoke came on shortly after his friends had left. He died early on the morning of April 10, 1813, in his seventy sixth year.

CHAPTER ELEVEN

From Peasant to Snob

LAPLACE

All the effects of nature are only the mathematical consequences of a small number of immutable laws.—P. S. LAPLACE

THE MARQUIS PIERRE-SIMON DE LAPLACE (1749–1827) was not born a peasant nor did he die a snob. Yet to within small quantities of the second order his illustrious career is comprised within the limits indicated, and it is from this approximate point of view that he is of greatest interest as a specimen of humanity.

As a mathematical astronomer Laplace has justly been called the Newton of France; as a mathematician he may be regarded as the founder of the modern phase of the theory of probability. On the human side he is perhaps the most conspicuous refutation of the pedagogical superstition that noble pursuits necessarily ennoble a man's character. Yet in spite of all his amusing foibles—his greed for titles, his political suppleness, and his desire to shine in the constantly changing spotlight of public esteem—Laplace had elements of true greatness in his character. We may not believe all that he said about his unselfish devotion to truth for truth's sake, and we may smile at the care with which he rehearsed his sententious last words—"What we know is not much; what we do not know is immense"—in an endeavor to telescope Newton's boy playing on the seashore into a neat epigram, but we cannot deny that Laplace in his generosity to unknown beginners was anything but a shifty and ungrateful politician. To give one young man a helping hand up Laplace once cheated himself.

Very little is known of Laplace's early years. His parents were peasants living in Beaumont-en-Auge, Department of Calvados, France, where Pierre-Simon was born on March 23, 1749. The obscurity surrounding Laplace's childhood and youth is due to his own snobbishness: he was thoroughly ashamed of his humble parents and did everything in his power to conceal his peasant origin.

Laplace got his chance through the friendly interest of wealthy neighbors on the occasion, presumably, of his having shown remarkable talent in the village school. It is said that his first success was in theological disputations. If this is true it is an interesting prelude to the somewhat aggressive atheism of his maturity. He took to mathematics early. There was a military academy at Beaumont, which Laplace attended as an externe, and in which he is said to have taught mathematics for a time. One dubious legend states that the young man's prodigious memory attracted more attention than his mathematical ability and was responsible for the cordial recommendations from influential people which he carried with him to Paris when, at the age of eighteen he wiped the mud of Beaumont off his boots forever and set out to seek his fortune. His own estimate of his powers was high, but not too high. With justified self-confidence young Laplace invaded Paris to conquer the mathematical world.

Arriving in Paris, Laplace called on D'Alembert and sent in his recommendations. He was not received. D'Alembert was not interested in young men who came recommended only by prominent people. With remarkable insight for so young a man Laplace sensed what the trouble was. He returned to his lodgings and wrote D'Alembert a wonderful letter on the general principles of mechanics. This did the trick. In his reply inviting Laplace to call, D'Alembert wrote: "Sir, you see that I paid little enough attention to your recommendations; you don't need any. You have introduced yourself better. That is enough for me; my support is your due." A few days later, thanks to D'Alembert, Laplace was appointed professor of mathematics at the Military School of Paris.

Laplace now threw himself into his life work—the detailed application of the Newtonian law of gravitation to the entire solar system. If he had done nothing else he would have been greater than he was. The kind of man Laplace would have liked to be is described in a letter of 1777, when he was twenty seven, to D'Alembert. The picture Laplace gives of himself is one of the strangest mixtures of fact and fancy a man ever perpetrated in the way of self-analysis.

"I have always cultivated mathematics by taste rather than from the desire for a vain reputation," he declares. "My greatest amusement is to study the march of the inventors, to see their genius at grips with the obstacles they have encountered and overcome. I then put myself in their place and ask myself how I should have gone about surmounting these same obstacles, and although this substitution in the great majority of instances has only been humiliating to my self-love, nevertheless the pleasure of rejoicing in their success has amply repaid me for this little humiliation. If I am fortunate enough to add something to their works, I attribute all the merit to their first efforts, well persuaded that in my position they would have gone much farther than I. . . . "

He may be granted the first sentence. But what about the rest of his smug little essay which might have been handed in by a priggish youngster of ten to his gullible Sunday-school teacher? Notice particularly the generous attribution of his own "modest" successes to the preliminary work of his predecessors. Nothing could be farther from the truth than this frank avowal of indebtedness. To call a spade a spade, Laplace stole outrageously, right and left, wherever he could lay his hands on anything of his contemporaries and predecessors which he could use. From Lagrange, for example, he lifted the fundamental concept of the potential (to be described presently); from Legendre he took whatever he needed in the way of analysis; and finally, in his masterpiece, the *Mécanique céleste*, he deliberately omits references to the work of others incorporated in his own, with the intention of leaving posterity to infer that he alone created the mathematical theory of the heavens. Newton, of course, he cannot avoid mentioning repeatedly. Laplace need not have been so ungenerous. His own colossal contributions to the dynamics of the solar system easily overshadow the works of others whom he ignores.

The complications and difficulties of the problem Laplace attacked cannot be conveyed to anyone who has never seen anything similar attempted. In discussing Lagrange we mentioned the problem of three bodies. What Laplace undertook was similar, but on a grander scale. He had to work out from the Newtonian law the combined effects of the perturbations—cross-pulling and hauling—of all the members of the Sun's family of planets on one another and on the Sun. Would Saturn, in spite of an apparently steady decrease of his mean motion, wander off into space, or would he continue as a member of the Sun's family? Or would the accelerations of Jupiter and the Moon ultimately cause one to fall into the Sun and the other to smash down on the Earth? Were the effects of these perturbations cumulative and dissipative, or were they periodic and conservative? These and similar riddles were details of the grand problem: is the solar system stable or is it unstable? It is assumed that the Newtonian law of gravitation is indeed universal and the only one controlling the motions of the planets.

Laplace's first important step toward the general problem was taken in 1773, when he was twenty four, in which he proved that the mean distances of the planets from the Sun are invariable to within certain slight periodic variations.

When Laplace attacked the problem of stability expert opinion was at best neutral. Newton himself believed that divine intervention might be necessary from time to time to put the solar system back in order and prevent it from destruction or dissolution. Others, like Euler, impressed by the difficulties of the lunar theory (motion of the Moon), rather doubted whether the motions of the planets and their satellites could be accounted for on the Newtonian hypothesis. The forces involved were too numerous, and their mutual interactions too complicated, for any reasonably fair guess. Until Laplace *proved* the stability of the solar system one man's guess was as good as another's.

To dispose here of an objection which the reader doubtless has already raised, it may be stated that Laplace's solution of the problem of stability is good only for the highly idealized solar system which Newton and he imagined. Tidal friction (acting like a brake on diurnal rotation) among other things was ignored. Since the *Mécanique céleste* was published we have learned a great deal about the solar system and everything in it of which Laplace was ignorant. It is probably not too radical to say that the problem of stability for the actual solar system—as opposed to Laplace's ideal—is still open. However, the experts on celestial mechanics might disagree, and a competent opinion can be obtained only from them.

As a matter of temperament some find the Laplacian conception of an eternally stable solar system repeating the complicated cycle of its motions time after time for ever and ever as depressing as an endless nightmare. For these there is the recent comfort that the Sun will probably explode some day as a nova. Then stability will cease to trouble us, for we shall all quite suddenly become perfect gases.

For this brilliant start Laplace was rewarded with the first substantial honor of his career when he was barely twenty four, associate membership in the Academy of Sciences. His subsequent scientific life is summarized by Fourier: "Laplace gave to all his works a fixed direction from which he never deviated; the imperturbable constancy of his views was always the principal feature of his genius. He was already [when he began his attack on the solar system] at the extreme of mathematical analysis, knowing all that is most ingenious in this, and no one was more competent than he to extend its domain. He had solved a capital problem of astronomy [that communicated to the Academy in 1773], and he decided to devote all his talents to mathematical astronomy, which he was destined to perfect. He meditated profoundly on his great project and passed his whole life perfecting it with a perseverance unique in the history of science. The vastness of the subject flattered the just pride of his genius. He undertook to compose the *Almagest* of his age—the *Mécanique céleste*; and his immortal work carries him as far beyond that of Ptolemy as the analytical science [mathematical analysis] of the moderns surpasses the *Elements* of Euclid."

This is no more than just. Whatever Laplace did in mathematics was designed as an aid to the solution of the grand problem. Laplace is the great example of the wisdom—for a man of genius—of directing all of one's efforts to a single central objective worthy of the best that a man has in him. Occasionally Laplace was tempted to turn aside, but not for long. Once he was strongly attracted by the theory of numbers, but quickly abandoned it on realizing that its puzzles were likely to cost him more time than he could spare from the solar system. Even his epochal work in the theory of probabilities, although at first sight off the main road of his interests, was inspired by his need for it in mathematical astronomy. Once well into the theory he saw that it is indispensable in all exact science and felt justified in developing it to the limit of his powers.

The *Mécanique céleste*, which bound all Laplace's astronomical work into a reasoned whole, was published in parts over a period of twenty six years. Two volumes appeared in 1799, dealing with the motions of the planets, their shapes (as rotating bodies), and the tides; two further volumes in 1802 and 1805 continued the investigation, which was finally completed in the fifth volume, 1823-25. The mathematical exposition is extremely concise and occasionally awkward. Laplace was interested in results, not in how he got them. To avoid condensing a complicated mathematical argument to a brief, intelligible form he frequently omits everything but the conclusion, with the optimistic remark *"Il est aisé à voir"* (It is easy to see). He himself would often be unable to restore the reasoning by which he had "seen" these easy things without hours—sometimes days—of hard labor. Even gifted readers soon acquired the habit of groaning whenever the famous phrase appeared, knowing that as likely as not they were in for a week's blind work.

A more readable account of the main results of the *Mécanique céleste* appeared in 1796, the classic *Exposition du système du monde* (Exposition of the System of the World), which has been described as Laplace's masterpiece with all the mathematics left out. In this work, as in the long nonmathematical introduction (153 quarto pages) to the treatise on probabilities (third edition, 1820), Laplace revealed himself as almost as great a writer as he was a mathematician. Anyone wishing to glimpse the scope and fascination of the theory of probability, without being held up by technicalities intelligible only to mathematicians, could not do better than to read Laplace's introduction. Much has been done since Laplace wrote, especially in recent years and particularly in the foundations of the theory of probability, but his exposition is still classic and a perfect expression of at least one philosophy of the whole subject. The theory, it need scarcely be said, is not yet complete. Indeed it is beginning to seem as if it has not yet been begun—the next generation may have it all to do over again.

One interesting detail of Laplace's astronomical work may be mentioned in passing, the famous nebular hypothesis of the origin of the solar system. Apparently unaware that Kant had anticipated him, Laplace (only half seriously) proposed the hypothesis in a note. His mathematics was inadequate for a systematic attack, and it was not till Jeans in the present century resumed the discussion that it had any scientific meaning.

Lagrange and Laplace, the two leading French men of science of the eighteenth century, offer an interesting contrast, and one typical of a difference which was to become increasingly sharp with the expansion of mathematics: Laplace belongs to the tribe of mathematical physicists, Lagrange to that of pure mathematicians. Poisson, himself a mathematical physicist, seems to favor Laplace as the more desirable type:

"There is a profound difference between Lagrange and Laplace in all their work, whether in a study of numbers or the libration of the Moon. Lagrange often appeared to see in the questions he treated only mathematics, of which the questions were the occasion—hence the high value he put upon elegance and generality. Laplace saw in mathematics principally a tool, which he modified ingeniously to fit every special problem as it arose. One was a great mathematician; the other a great philosopher who sought to know nature by making higher mathematics serve it."

Fourier (whom we shall consider later) was also struck by the radical difference between Lagrange and Laplace. Himself rather narrowly "practical" in his mathematical outlook, Fourier was yet capable—at one time—of estimating Lagrange at his true worth:

"Lagrange was no less a philosopher than he was a great mathematician. By his whole life he proved, in the moderation of his desires, his immovable attachment to the general interests of humanity, by the noble simplicity of his manners and the elevation of his character, and finally by the accuracy and the depth of his scientific works."

Coming from Fourier this statement is remarkable. It may smack of the bland rhetoric we are accustomed to expect in French funeral orations, yet it is true, at least today. Lagrange's great influence on modern mathematics is due to "the depth and accuracy of his scientific works," qualities which are sometimes absent from Laplace's masterpieces.

To the majority of his contemporaries and immediate followers Laplace ranked higher than Lagrange. This was due partly to the magnitude of the problem Laplace attacked—the grandiose project of demonstrating that the solar system is a gigantic perpetual motion machine. A sublime project in itself, no doubt, but essentially illusory: not enough about the actual physical universe was known in Laplace's day—or even in our own—to give the problem any real significance, and it will probably be many years before mathematics is sufficiently advanced to handle the complicated mass of data we now have. Mathematical astronomers will doubtless continue to play with idealized models of "the universe," or even of the infinitely less impressive solar system, and will continue to flood us with inspiring or depressing bulletins regarding the destiny of mankind; but in the end the by-products of their investigations—the perfection of the purely mathematical tools they have devised—will be their fairly permanent contribution to the advancement of science (as opposed to the propagation of guessing), precisely as has happened in the case of Laplace.

If the foregoing seems too strong, consider what has happened to the *Mécanique céleste*. Does anyone but an academic mathematician really believe today that Laplace's conclusions about the stability of the solar system are a reliable verdict on the infinitely complicated situation which Laplace replaced by an idealized dream? Possibly many do; but no worker in mathematical physics doubts the power and utility of the mathematical methods developed by Laplace to attack his ideal.

To take but one instance, the theory of the potential is more significant today than Laplace ever dreamed it would become. Without the mathematics of this theory we should be halted almost at the beginning of our attempt to understand electromagnetism. Out of this theory grew one vigorous branch of the mathematics of boundary-value problems, today of greater significance for physical science than the whole Newtonian theory of gravitation. The concept of the potential was a mathematical inspiration of the first order—it made possible an attack on physical problems which otherwise would have been unapproachable.

The potential is merely the function u described in connection with fluid motion and Laplace's equation in the chapter on Newton. The function u is there a "velocity potential"; if it is a question of the force of Newtonian gravitational attraction, u is a "gravitational potential." The introduction of the potential into the theories of fluid motion, gravitation, electromagnetism, and elsewhere was one of the longest strides ever taken in mathematical physics. It had the effect of replacing partial differential equations in two or three unknowns by equations in one unknown.

In 1785, at the age of thirty six, Laplace was promoted to full membership in the Academy. Important as this honor was in the career of a man of science, the year 1785 stands out as a landmark of yet greater significance in Laplace's career as a public character. For in that year Laplace had the unique distinction of examining a singular candidate of sixteen at the Military School. This youth was destined to upset Laplace's plans and deflect him from his avowed devotion to mathematics into the muddy waters of politics. The young man's name was Napoleon Bonaparte (1769–1821).

Laplace rode through the Revolution on horseback, as it were, and saw everything in comparative safety. But no man of his prominence and restless ambition could escape danger entirely. If De Pastoret knew what he was talking about in his eulogy, both Lagrange and Laplace escaped the guillotine only because they were requisitioned to calculate trajectories for the artillery and to help in directing the manufacture of saltpetre for gunpowder. Neither was forced to eat grass as some less necessary savants were driven to do, nor was either so careless as to betray himself, as their unfortunate friend Condorcet did, by ordering an aristocrat's omelet. Not knowing how many eggs go into a normal omelet Cordorcet ordered a dozen. The good cook asked Condorcet his trade. "Carpenter."—"Let me see your hands. You're no carpenter." That was the end of Laplace's close friend Condorcet. They either poisoned him in prison or let him commit suicide.

After the Revolution Laplace went in heavily for politics, possibly in the hope of beating Newton's record. The French refer politely to Laplace's "versatility" as a politician. This is too modest. Laplace's alleged defects as a politician are his true greatness in the slippery game. He has been criticized for his inability to hold public office under successive regimes without changing his politics. It would seem that a man who is sharp enough to convince opposing parties that he is a loyal supporter of whichever one happens to be in power at the moment is a politician of no mean order. It was his patrons who played the game like amateurs, not Laplace. What would we think of a Republican Postmaster General who gave all the fattest jobs to undeserving Democrats? Or the other way about? Laplace got a better job every time the government flopped. It cost him nothing to switch overnight from rabid republicanism to ardent royalism.

Napoleon shoved everything Laplace's way, including the portfolio of the interior—about which more later. All the Napoleonic orders of any note adorned the versatile mathematician's chest—including the Grand Cross of the Legion of Honor and the Order of the Reunion, and he was made a Count of the Empire. Yet what did he do when Napoleon fell? Signed the decree which banished his benefactor.

After the restoration Laplace had no difficulty in transferring his loyalty to Louis XVIII, especially as he now sat in the Chamber of Peers as the Marquis de Laplace. Louis recognized his supporter's merits and in 1816 appointed Laplace president of the committee to reorganize the École Polytechnique.

Perhaps the most perfect expressions of Laplace's political genius are those to be found in his scientific writings. It takes real genius to doctor science according to fluctuating political opinion and get away with it. The first edition of the *Exposition du système du monde*, dedicated to the Council of Five Hundred, closes with these noble words: "The greatest benefit of the astronomical sciences is to have dissipated errors born of ignorance of our true relations with nature, errors all the more fatal since the social order must rest solely on these relations. *Truth* and *justice* are its immutable bases. Far from us be the dangerous maxim that it may sometimes be useful to deceive or to enslave men the better to insure their happiness! Fatal experiences have proved in all ages that these sacred laws are never infringed with impunity." In 1824 this is suppressed and the Marquis de Laplace substitutes: "Let us conserve with care and increase the store of this advanced knowledge, the delight of thinking beings. It has rendered important services to navigation and geography; but its greatest benefit is to have dissipated the fears produced by celestial phenomena and to have destroyed the errors born of ignorance of our true relations with nature, errors which will soon reappear if the torch of the sciences is extinguished." In loftiness of sentiment there is but little to choose between these two sublime maxima.

This is enough on the debit side of the ledger. The last extract does indeed suggest one trait in which Laplace overtopped all courtiers— his moral courage where his true convictions were questioned. The story of Laplace's encounter with Napoleon over the *Mécanique céleste* shows the mathematician as he really was. Laplace had presented Napoleon with a copy of the work. Thinking to get a rise out of Laplace, Napoleon took him to task for an apparent oversight. "You have written this huge book on the system of the world without once mentioning the author of the universe." "Sire," Laplace retorted, "I had no need of that *hypothesis.*" When Napoleon repeated this to Lagrange, the latter remarked "Ah, but that is a fine hypothesis. *It explains so many things.*"

It took nerve to stand up to Napoleon and tell him the truth. Once at a session of the Institut when Napoleon was in one of his most insultingly bad tempers he caused poor old Lamarck to burst into tears with his deliberate brutality.

Also on the credit side was Laplace's sincere generosity to beginners. Biot tells how as a young man he read a paper before the Academy when Laplace was present, and was drawn aside afterward by Laplace who showed him the identical discovery in a yellowed old manuscript of his own, still unpublished. Cautioning Biot to secrecy, Laplace told him to go ahead and publish his work. This was but one of several such acts. Beginners in mathematical research were his stepchildren, Laplace liked to say, but he treated them as well as he did his own son.

As it is often quoted as an instance of the unpracticality of mathematicians we shall give Napoleon's famous estimate of Laplace, of which he is reported to have delivered himself while he was a prisoner at St. Helena.

"A mathematician of the first rank, Laplace quickly revealed himself as only a mediocre administrator; from his first work we saw that we had been deceived. Laplace saw no question from its true point of view; he sought subtleties everywhere, had only doubtful ideas, and finally carried the spirit of the infinitely small into administration."

This sarcastic testimonial was inspired by Laplace's short tenure—only six weeks—of the Ministry of the Interior. However, as Lucien Bonaparte needed a job at the moment and succeeded Laplace, Napoleon may have been rationalizing his well-known inclination to nepotism. Laplace's testimonial for Napoleon has not been preserved. It might have run somewhat as follows.

"A soldier of the first rank, Napoleon quickly revealed himself as only a mediocre politician; from his first exploits we saw that he was deceived. Napoleon saw all questions from the obvious point of view; he suspected treachery everywhere but where it was, had only a childlike faith in his supporters, and finally carried the spirit of infinite generosity into a den of thieves."

Which, after all, was the more practical administrator? The man who could not hang onto his gains and who died a prisoner of his enemies, or the other who continued to gather wealth and honor to the day of his death?

Laplace spent his last days in comfortable retirement at his country estate at Arcueil, not far from Paris. After a short illness he died on March 5, 1827, in his seventy eighth year. His last words have already been reported.

CHAPTER TWELVE

Friends of an Emperor

MONGE AND FOURIER

I cannot tell you the efforts to which I was condemned to understand something of the diagrams of Descriptive Geometry, which I detest.
—CHARLES HERMITE

Fourier's Theorem is not only one of the most beautiful results of modern analysis, but it may be said to furnish an indispensable instrument in the treatment of nearly every recondite question in modern physics.
—WILLIAM THOMSON AND P. G. TAIT

THE CAREERS OF GASPARD MONGE (1746–1818) and Joseph Fourier (1768–1830) are curiously parallel and may be considered together. On the mathematical side each made one fundamental contribution: Monge invented descriptive geometry (not to be confused with the projective geometry of Desargues, Pascal, and others); Fourier started the current phase of mathematical physics with his classic investigations on the theory of heat-conduction.

Without Monge's geometry—originally invented for use in military engineering—the wholesale spawning of machinery in the nineteenth century would probably have been impossible. Descriptive geometry is the root of all the mechanical drawing and graphical methods that help to make mechanical engineering a fact.

The methods inaugurated by Fourier in his work on the conduction of heat are of a similar importance in boundary-value problems—a trunk nerve of mathematical physics.

Monge and Fourier between them are thus responsible for a considerable part of our own civilization, Monge on the practical and industrial side, Fourier on the purely scientific. But even on the practical side Fourier's methods are indispensable today; they are in fact a commonplace in all electrical and acoustical engineering (including wireless) beyond the rule of thumb and handbook stages.

A third man must be named with these mathematicians, although we shall not take space to tell his life: the chemist Count Claude-Louis

Berthollet, (1748-1822), a close friend of Monge, Laplace, Lavoisier, and Napoleon. With Lavoisier, Berthollet is regarded as one of the founders of modern chemistry. He and Monge became so thick that their admirers gave up trying to distinguish between them in their nonscientific labors and called them simply Monge-Berthollet.

Gaspard Monge, born on May 10, 1746, at Beaune, France, was a son of Jacques Monge, a peddler and knife grinder who had a tremendous respect for education and who sent his three sons through the local college. All the sons had successful careers; Gaspard was the genius of the family. At the college (run by a religious order) Gaspard regularly captured the first prize in everything and earned the unique distinction of having *puer aureus* inscribed after his name.

At the age of fourteen Monge's peculiar combination of talents showed up in the construction of a fire engine. "How could you, without a guide or a model, carry through such an undertaking successfully?" he was asked by the astonished citizens. Monge's reply is a summary of the mathematical part of his career and of much of the rest. "I had two infallible means of success: an invincible tenacity, and fingers which translated my thought with geometric fidelity." He was in fact a born geometer and engineer with an unsurpassed gift for visualizing complicated space-relations.

At the age of sixteen he made a wonderful map of Beaune entirely on his own initiative, constructing his own surveying instruments for the purpose. This map got him his first great chance.

Impressed by his obvious genius, Monge's teachers recommended him for the professorship of physics at the college in Lyon run by their order. Monge was appointed at the age of sixteen. His affability, patience, and lack of all affectation, added to his sound knowledge, made him a great teacher. The order begged him to take their vows and cast his lot for life with them. Monge consulted his father. The astute knife grinder advised caution.

Some days later, on a visit home, Monge met an officer of engineers who had seen the famous map. The officer begged Jacques to send his son to the military school at Mézières. Perhaps fortunately for Monge's future career the officer omitted to state that on account of his humble birth Monge could never get a commission. Not knowing this, Monge eagerly accepted and proceeded to Mézières.

Monge quickly learned where he stood at Mézières. There were only twenty pupils at the school, of whom ten were graduated each year as lieutenants in engineering. The rest were destined for the "practical" work—the dirty jobs. Monge did not complain. He rather enjoyed himself, as the routine work in surveying and drawing left him plenty of time for mathematics. An important part of the regular course was the theory of fortification, in which the problem was to design the works so that no part should be exposed to the direct fire of the enemy. The usual calculations demanded endless arithmetic. One day Monge handed in his solution of a problem of this sort. It was turned over to a superior officer for inspection.

Skeptical that anyone could have solved the problem in the time, the officer declined to check the solution. "Why should I give myself the trouble of subjecting a supposed solution to tedious verifications? The author has not even taken the time to group his figures. I can believe in a great facility in calculation, but not in miracles!" Monge persisted, saying he had not used arithmetic. His tenacity won; the solution was checked and found correct.

This was the beginning of descriptive geometry. Monge was at once given a minor teaching position to instruct the future military engineers in the new method. Problems which had been nightmares before—sometimes solved only by tearing down what had been built and beginning all over again—were now as simple as ABC. Monge was sworn not to divulge his method, and for fifteen years it was a jealously guarded military secret. Only in 1794 was he allowed to teach it publicly, at the École Normale in Paris, where Lagrange was among the auditors. Lagrange's reaction to descriptive geometry was like M. Jourdain's when he discovered that he had been talking prose all his life. "Before hearing Monge," Lagrange said after a lecture, "I did not know that I knew descriptive geometry."

The idea behind it all now seems as ridiculously simple to us as it did to Lagrange. Descriptive geometry is a method for representing solids and other figures in ordinary three-dimensional space on *one* plane. Imagine first two planes at right angles to one another, like two pages of a thin book opened at a ninety degree angle; one plane is horizontal, the other vertical. The figure to be represented is projected onto each of these planes by rays perpendicular to the plane. There are thus *two* projections of the figure; that on the horizontal plane is called a *plan* of the figure, that on the vertical plane an *elevation*. The vertical plane is now turned down ("rabbatted") till it and the horizontal plane lie in *one* plane (that of the horizontal plane)—as if the book were now opened out flat on a table.

The solid or other figure in space is now represented by two projections on one plane (that of the drawing board). A plane, for instance, is represented by its *traces*—the straight lines in which it cut the vertical and horizontal planes before the former was rabbatted; a solid, say a cube, is represented by the projections of its edges and vertices. Curved surfaces cut the vertical and horizontal planes in curves; these curves, or *traces* of the surface, represent the surface on the one plane.

When these and other equally simple remarks are developed we have a *descriptive* method which puts on one flat sheet of paper what we ordinarily visualize in space of three dimensions. A short training enables the draughtsman to read such representations as easily as others read good photographs—and to get a great deal more out of them. This was the simple invention that revolutionized military engineering and mechanical design. Like many of the first-rate things in applied mathematics its most conspicuous feature is its simplicity. There are many ways in which descriptive geometry can be developed or modified, but they all go back to Monge. The subject is now so thoroughly worked out that it is not of much interest to professional mathematicians.

To finish with Monge's contributions to mathematics before continuing with his life, we recall that his name is familiar to every student in the second course in the calculus today in connection with the geometry of surfaces. Monge's great step forward was a systematic (and brilliant) application of the calculus to the investigation of the curvature of surfaces. In his general theory of curvature Monge prepared the way for Gauss, who in his turn was to inspire Riemann, who again was to develop the geometry known by his name in the theory of relativity.

It seems rather a pity that a born geometer like Monge should have lusted after the fleshpots of Egypt, but so he did. His work in differential equations, closely connected with that in geometry, also showed what he had in him. Years after he left Mézières, where these great things were done, Monge lectured on his discoveries to his colleagues at the École Polytechnique. Lagrange again was an auditor. "My dear colleague," he told Monge after the lecture, "you have just explained some very elegant things; I should have liked to have done them myself." And on another occasion: "With his application of analysis to geometry this devil of a man will make himself immortal!" He did; and it is interesting to note that although more urgent calls on his genius distracted him from mathematics, he never lost his talent. Like all the great mathematicians Monge was a mathematician to the last.

In 1768, at the age of twenty two, Monge was promoted to the professorship of mathematics at Mézières, and three years later, on the death of the professor of physics, stepped into his place also. The double work did not bother him at all. Powerfully built and as strong of body as he was of mind, Monge was always capable of doing three or four men's work and frequently did.

His marriage had a touch of eighteenth century romance. At a reception Monge heard some noble bounder slandering a young widow to get even with her for having rejected him. Shouldering his way through the cackling crowd, Monge demanded to know whether he had heard aright. "What is it to you?" Monge demonstrated with a punch on the jaw. There was no duel. A few months later at another reception Monge was very much taken by a charming young woman. On being introduced he recognized her name—Madame Horbon—as that of the unknown lady he had tried to fight a duel for. She was the widow, only twenty, and somewhat reluctant to marry before her late husband's affairs were straightened out. "Never mind all that," Monge reassured her, "I've solved lots of more difficult problems in my time." Monge and she were married in 1777. She survived him and did what she could to perpetuate his memory—unaware that her husband had raised his own monument long before he ever met her. Monge's wife was the one human being who stuck to him through everything. Even Napoleon at the very last would have let him down on account of his age.

At about this time Monge began corresponding with D'Alembert and Condorcet. In 1780 these two had induced the Government to found an institute at the Louvre for the study of hydraulics. Monge was called to Paris to take charge, on the understanding that he spend half his time at Mézières. He was then thirty four. Three years later he was relieved of his duties at Mézières and appointed examiner of candidates for commissions in the navy, a position which he held till the outbreak of the Revolution in 1789.

In looking back over the careers of all these mathematicians of the Revolutionary period we cannot help noticing how blind they and everyone else were to what now seems so obvious to us. Not one of them suspected that he was sitting on a mine and that the train was already sputtering. Possibly our successors in 2036 will be saying the same about us.

For the six years he held the naval job Monge proved himself an incorruptible public servant. Disgruntled aristocrats threatened him with dire penalties when he unmercifully disqualified their incompetent sons, but Monge never gave in. "Get someone else to run the job if you don't like the way I am doing it." As a consequence the navy was ready for business in 1789.

His birth and his experiences with snobs seeking unmerited favors made Monge a natural revolutionist. By first-hand experience he knew the corruption of the old order and the economic disabilities of the masses, and he believed that the time had come for a new deal. But like the majority of early liberals Monge did not know that a mob which has once tasted blood is not satisfied till no more is forthcoming. The early revolutionists had more faith in Monge than he had in himself. Against his better judgment they forced him into the Ministry of the Navy and the Colonies on August 10, 1792. He was the man for the position, but it was not healthy to be a public official in the Paris of 1792.

The mob was already out of hand; Monge was put on the Provisional Executive Council to attempt some measure of control. A son of the people himself, Monge felt that he understood them better than did some of his friends—Condorcet, for instance, who had wisely declined the naval job to save his head.

But there are people and people, all of whom together comprise "the people." By February, 1793 Monge found himself suspect of being not quite radical enough, and on the 13th he resigned, only to be re-elected on the 18th to a job which stupid political interference, "liberty, equality, and fraternity" among the sailors, and approaching bankruptcy of the state had made impossible. Any day during this difficult time Monge might have found himself on the scaffold. But he never truckled to ignorance and incompetence, telling his critics to their faces that he knew what was what while they knew nothing. His only anxiety was that dissension at home would lay France open to an attack which would nullify all the gains of the Revolution.

At last, on April 10, 1793, Monge was allowed to resign in order to undertake more urgent work. The anticipated attack was now plainly visible.

With the arsenals almost empty the Convention began raising an army of 900,000 men for defense. Only a tenth of the necessary munitions existed and there was no hope of importing the requisite materials—copper and tin for the manufacture of bronze cannon, saltpetre for gunpowder, and steel for firearms. "Give us saltpetre from the earth and in three days we shall be loading our cannon," Monge told the Convention. All very well, they retorted, but where were they to get the saltpetre? Monge and Berthollet showed them.

The entire nation was mobilized. Under Monge's direction bulletins were sent to every town, farmstead, and village in France telling the people what to do. Led by Berthollet the chemists invented new and better methods for refining the raw material and simplified the manufacture of gunpowder. The whole of France became a vast powder factory. The chemists also showed the people where to find tin and copper—in clock metal and church bells. Monge was the soul of it all. With his prodigious capacity for work he spent his days supervising the foundries and arsenals, and his nights writing bulletins for the direction of the workers, and throve on it. His bulletin on *The Art of Manufacturing Cannon* became the factory handbook.

Monge was not without enemies as the Revolution continued to fester. One day Monge's wife heard that Berthollet and her husband were to be denounced. Frantic with fear she ran to the Tuileries to learn the truth. She found Berthollet sitting quietly under the chestnut trees. Yes; he had heard the rumor, but believed nothing would happen for a week. "Then," he added with his habitual composure, "we shall certainly be arrested, tried, condemned, and executed."

When Monge came home that evening his wife told him Berthollet's prediction. "My word!" Monge exclaimed; "I know nothing of all that. What I do know is that my cannon factories are going forward marvelouslly!"

Shortly after this Citizen Monge was denounced by the porter at his lodgings. This was too much, even for Monge. He prudently left Paris till the storm blew over.

The third stage of Monge's career opened in 1796 with a letter from Napoleon. The two had already met in 1792, but Monge was unaware of the fact. Monge at the time was fifty, Napoleon twenty three years younger.

"Permit me," Napoleon wrote, "to thank you for the cordial welcome that a young artillery officer, little in favor, received from the Minister of the Navy in 1792; he has preciously preserved its memory. You see this officer in the present general of the Army [of invasion] of Italy; he is happy to extend you a hand of recognition and friendship."

Thus began the long intimacy between Monge and Napoleon. Commenting on this singular alliance, Arago* reports Napoleon's words "Monge loved me as one loves a mistress." On the other side Monge seems to have been the only man for whom Napoleon ever had an unselfish and abiding friendship. Napoleon knew of course that Monge had helped to make his career possible; but that was not the root of his affection for the older man.

The "recognition" mentioned in Napoleon's letter was the appointment of Monge and Berthollet by the Directory as commissioners sent to Italy to select the paintings, sculpture, and other works of art "donated" by the Italians (after being bled white of money) as part of their contribution to the expenses of Napoleon's campaign. In picking over the loot Monge developed a keen appreciation of art and became quite a connoisseur.

The practical implications of the looting, however, disturbed him somewhat, and when enough to furnish the Louvre half a dozen times over had been lifted and shipped to Paris, Monge counselled moderation. It would not do, he said, in governing a people either for their own good or for that of the conquerors to beggar them completely. His advice was heeded, and the goose continued laying its golden eggs.

After the Italian adventure Monge joined Napoleon at his château near Udine. The two became great cronies, Napoleon revelling in Monge's conversation and inexhaustible fund of interesting information, and Monge basking in the commander-in-chief's genial humor. At public banquets Napoleon always ordered the band to strike up

---

*F. J. D. Arago, 1786–1853, astronomer, physicist, and scientific biographer.

the *Marseillaise*—"Monge is an enthusiast for it!" Indeed he was, shouting it at the top of his lungs before sitting down to meals,

> "*Allons, enfants de la patrie,*
> *Le jour de gloire est arrivé!*"

It will be our special privilege to see the day of glory arriving in the company of another great Napoleonic mathematician—Poncelet.

In December, 1797, Monge made a second trip to Italy, this time as a member of the commission to investigate the "great crime" of General Duphot's assassination. The General had been shot down in Rome while standing near Lucien Bonaparte. The commission (rudely anticipated by one of the martyred General's brothers in arms) somewhat lamely prescribed a republic modelled on the French for the obstreperous Italians. "There must be an end of everything, even of the rights of conquest," as one of the negotiators remarked when the matter of further extortions came up.

How right this canny diplomat was came out eight months later when the Italians scrapped their republic to the great embarrassment of Napoleon, then in Cairo, and to the greater embarrassment of Monge and Fourier who happened to be with him.

Monge was one of the dozen or so to whom Napoleon in 1798 confided his plan for the invasion, conquest, and civilization of Egypt. As Fourier enters naturally here we shall go back and pick him up.

Jean-Baptiste-Joseph Fourier, born on March 21, 1768, at Auxerre, France, was the son of a tailor. Orphaned at the age of eight, he was recommended to the Bishop of Auxerre by a charitable lady who had been captivated by the boy's good manners and serious deportment—little did she dream what he was to become. The Bishop got Fourier into the local military college run by the Benedictines, where the boy soon proved his genius. By the age of twelve he was writing magnificent sermons for the leading church dignitaries of Paris to palm off as their own. At thirteen he was a problem child, wayward, petulant, and full of the devil generally. Then, at his first encounter with mathematics, he changed as if by magic. He knew what had ailed him and cured himself. To provide light for his mathematical studies after he was supposed to be asleep he collected candle-ends in the kitchen and wherever he could find them in the college. His secret study was an inglenook behind a screen.

The good Benedictines prevailed upon the young genius to choose the priesthood as his profession, and he entered the abbey of Saint-Benoît to become a novitiate. But before Fourier could take his vows 1789 arrived. He had always wanted to be a soldier and had chosen the priesthood only because commissions were not given to sons of tailors. The Revolution set him free. His old friends at Auxerre were broadminded enough to see that Fourier would never make a monk. They took him back and made him professor of mathematics. This was the first step—a long one—toward his ambition. Fourier proved his versatility by teaching his colleagues' classes when they were ill, usually better than they did themselves, in everything from physics to the classics.

In December, 1789, Fourier (then twenty one) went to Paris to present his researches on the solution of numerical equations before the Academy. This work advanced beyond Lagrange, and is still of value, but as it is overshadowed by Fourier's methods in mathematical physics, we shall not discuss it further; it may be found in elementary texts on the theory of equations. The subject became one of his lifelong interests.

On returning to Auxerre Fourier joined the people's party and used his natural eloquence, which had enabled him as a small boy to compose stirring sermons, to stir up the people to put an end to mere sermonizers (among others).

From the first Fourier was an enthusiast for the Revolution—till it got out of hand. During the Terror, ignoring the danger to himself, he protested against the needless brutality. If he were living today Fourier would probably belong to the intelligentsia, blissfully unaware that such are among the first to be swept into the gutter when the real revolution begins. He was all for the masses and the renaissance of science and culture which the intellectuals imagined they foresaw. Instead of the generous encouragement of the sciences which he had predicted, Fourier presently saw men of science riding in the tumbrils or fleeing the country, and science itself fighting for its life in a rapidly rising tide of barbarism.

It is to Napoleon's everlasting credit that he was one of the first to see with cold-blooded clarity that ignorance of itself can do nothing but destroy. His own remedy in the end may not have been much better, but he did recognize that such a thing as civilization might be possible. To check the mere blood-letting Napoleon ordered or encouraged the creation of schools. But there were no teachers. All the brains that might have been pressed into immediate service had long since fallen into the buckets. It became imperative to train a new teaching corps of fifteen hundred, and for this purpose the École Normale was created in 1794. As a reward for his recruiting in Auxerre Fourier was called to the chair of mathematics.

With this appointment a new era in the teaching of French mathematics began. Remembering the deadly lectures of defunct professors, memorized and delivered verbatim the same year after dreary year, the Convention called in *creators* of mathematics to do the *teaching*, and forbade them to lecture from any notes at all. The lectures were to be delivered standing (not sitting half asleep behind a desk), and were to be a free interchange of questions and explanations between the professor and his class. It was up to the lecturer to prevent a session from degenerating into a profitless debate.

The success of this scheme even surpassed expectations and led to one of the most brilliant periods in the history of French mathematics and science. Both at the short-lived Normale and the enduring Polytechnique Fourier demonstrated his genius for teaching. At the Polytechnique he enlivened his lectures on mathematics by out-of-the-way historical allusions (many of which he was the first to trace to their sources), and he skilfully tempered abstractions with interesting applications.

Fourier was still turning out engineers and mathematicians at the Polytechnique when Napoleon in 1798 decided to take him along as one of the Legion of Culture to civilize Egypt—"to offer a succouring hand to unhappy peoples, to free them from the brutalizing yoke under which they have groaned for centuries, and finally to endow them without delay with all the benefits of European civilization."

Incredible as it may seem, the quotation is not from Signor Mussolini in 1935 justifying an invasion of Ethiopia, but from Arago in 1833 setting forth the lofty and humane aims of Napoleon's assault on Egypt. It will be interesting to see how the unregenerate inhabitants of Egypt received "all the benefits of European civilization" which Messrs. Monge, Berthollet, and Fourier strove to ram down their throats, and what those three musketeers of European culture themselves got out of their unselfish missionary work.

The French fleet of five hundred ships arrived at Malta on June 9,

1798, and three days later captured the place. As a first step toward civilizing the East, Monge started fifteen elementary schools and a higher school somewhat on the lines of the Polytechnique. A week later the fleet was on its way again, with Monge aboard Napoleon's flagship, *l'Orient*. Every morning Napoleon outlined a program for discussion after dinner in the evening. Needless to say, Monge was the star of these soirées. Among the topics solemnly debated were the age of the earth, the possibility of the world coming to an end by fire or water, and "Are the planets inhabited?" The last suggests that even at this comparatively early stage of his career Napoleon's ambitions outran Alexander's.

The fleet reached Alexandria on July 1, 1798. Monge was one of the first to leap ashore, and it was only by exercising his authority as Commander in Chief that Napoleon restrained the *Marseillaising* geometer from participating in the assault on the city. It would never do to have the Legion of Culture annihilated in the first skirmish before the work of civilization could begin; so Napoleon sent Monge and the rest of them up the Nile by boat to Cairo.

While Monge and company lolled like Cleopatra and her court under their sunshade, Napoleon marched resolutely along the bank, civilizing the uncultured (and poorly armed) inhabitants with shot and flame. Presently the intrepid General heard a devil of a cannonade from the direction of the river. Guessing the worst he abandoned the battle in which he was engaged at the moment and galloped to the rescue. The blessed boat was hard aground on a sand bar. There was Monge serving the cannon like a veteran. Napoleon arrived just in the nick of time to chase the attackers up the bank and give Monge his well-merited decoration for conspicuous bravery. So Monge after all had his way and got his sniff of powder. Napoleon was so overjoyed at having saved his friend that he did not regret the decisive victory Monge's rescue had cost him.

Following the victory of July 20, 1798, at the Battle of the Pyramids, the triumphant army whooped into Cairo. Everything went off like fireworks, precisely as that great idealist Napoleon had dreamed, but for one trifling fizzle. The obtuse Egyptians cared not a single curse for the cultural banquet which Messrs. Monge, Fourier, and Berthollet spread before them at the Egyptian Institute (founded, August 27, 1798, in parody of the *Institut de France*), but sat like mummies through the great chemist's scientific legerdemain, the enthusiastic

Monge's concerts, and the historical disquisitions of the scholarly Fourier on the glories of their own mummified civilization. The sweating savants shed their sangfroid, damning their prospective enlightenees as tasteless cattle incapable of relishing the rich hash of French erudition offered for their spiritual nourishment, but to no avail. Once more the wily, "unsophisticated" native made a complete ass of his determined uplifters by holding his peace and waiting for the plague of locusts to be blown away in the scavenging winds. To keep his self-respect till the breezes blew, the uncivilized Egyptian criticized the superior civilization of his conquerors in the one language they could understand. Three hundred of Napoleon's bravest had their hairy throats cut at one swipe in a street brawl. Monge himself saved his own windpipe and those of his beleaguered companions only by an exhibition of heroism for which any Boy Scout today in the English-speaking world might well receive a medal.

This ingratitude on the part of the unregenerate Egyptians cut Napoleon to the quick. His suspicion that it was his moral duty to desert his companions in arms was strengthened by disturbing news from Paris. During his absence things on the Continent had been going from purgatory to damnation; and now he must hurry back to preserve the honor of France and his own skin. Monge shared the General's confidence; the less beloved Fourier did not. Fourier, however, had the satisfaction of knowing that he was considerable enough in his commander's masterful eyes to be left in Cairo to educate Egypt or have his throat cut, when Napoleon, accompanied by the complaisant Monge, took secret passage for France without so much as an adieu to the troops who had suffered hell for him in the desert. Not being a Commander in Chief, Fourier was not entitled to take to his heels in the face of danger. He stayed, perforce. Only in 1801, when the French after Trafalgar finally acknowledged that the British, not they, were to regenerate the Egyptians, did the devoted—but disillusioned—Fourier return to France.

The return trip of Monge and Napoleon was less amusing for both of them than the voyage out. Instead of speculating about the end of the world Napoleon spent much anxious thought on his own probable end should the British sailors bag him. The reward for desertion in the field, he recalled, was a strictly private interview with a firing squad. Would the British treat him as a deserter for having run away from his army? If he must die he would die theatrically.

"Monge," he said one day, "if we are attacked by the British, our ship must be blown up the instant they board us. I charge you to carry it out."

The very next day a sail topped the horizon and all hands stood to their posts to repel the expected attack. But it turned out to be a French ship after all.

"Where's Monge?" somebody asked when all the excitement was over.

They found him in the powder magazine with a lighted lamp in his hand. If only that had been a British ship—. They always blow in fifteen minutes or fifteen years too late.

Berthollet and Monge arrived home looking like a pair of tramps. Neither had had a change of clothes since he left, and it was only with difficulty that Monge got by his wife's porter.

The friendship with Napoleon continued unmarred. Probably Monge was the only man in France who dared to stand up to Napoleon and tell him the truth in the days of his greatest arrogance. When Napoleon crowned himself Emperor the young men of the Polytechnique revolted. They were Monge's pride.

"Well, Monge," Napoleon remarked one day, "your pupils are nearly all in revolt against me; they have decidedly declared themselves my enemies."

"Sire," Monge replied, "we have had trouble enough to make republicans out of them; give them time to become imperialists. Moreover, permit me to say, you have turned rather abruptly!"

Little spats like this meant nothing between old lovers. In 1804 Napoleon showed his appreciation of Monge's merits by creating him Count of Péluse (Pelusium). For his part Monge accepted the honor gratefully and lived up to the title with all the usual trappings of nobility, forgetting that he had once voted for the abolition of all titles.

And so it went, in an ever more dazzling blaze of splendor till the year 1812, which was to have ushered in the day of glory, but which brought instead the retreat from Moscow. Too old (he was sixty six) to accompany Napoleon into Russia, Monge had stayed behind in France at his country estate, eagerly following the progress of the Grand Army through the official bulletins. When he read the fatal "Bulletin 29," announcing the disaster to French arms, Monge suffered a stroke of apoplexy. On recovering he said, "A little while ago I did not know something that I know now; I know how I shall die."

Monge was to be spared for the final curtain; Fourier helped to lower it. On his return from Egypt Fourier was appointed (January 2, 1802) prefect of the Department of Isère, with headquarters at Grenoble. The district was then in political turmoil; Fourier's first task was to restore order. He was met by a curious opposition which he subdued in a ludicrous fashion. While in Egypt Fourier had taken a leading part in administering the archaeological research of the Institute. The good citizens of Grenoble were much upset by the religious implications of some of the Institute's discoveries, particularly the great age assigned to the older monuments, which conflicted (they imagined) with the chronology of the Bible. They were quite satisfied however and took Fourier to their bosoms when, as the result of some further archaeological researches nearer home, he dug up a saint in his own family, the blessed Pierre Fourier, his great-uncle, whose memory was hallowed because he had founded a religious order. His respectability established, Fourier accomplished a vast amount of useful work, draining marshlands, stamping out malaria, and otherwise lifting his district out of the Middle Ages.

It was while at Grenoble that Fourier composed the immortal *Theorie analytique de la chaleur* (The Mathematical Theory of Heat), a landmark in mathematical physics. His first memoir on the conduction of heat was submitted in 1807. This was so promising that the Academy encouraged Fourier to continue by setting a contribution to the mathematical theory of heat as its problem for the Grand Prize in 1812. Fourier won the prize, but not without some criticism which he resented deeply but which was well taken.

Laplace, Lagrange, and Legendre were the referees. While admitting the novelty and importance of Fourier's work they pointed out that the mathematical treatment was faulty, leaving much to be desired in the way of rigor. Lagrange himself had discovered special cases of Fourier's main theorem but had been deterred from proceeding to the general result by the difficulties which he now pointed out. These subtle difficulties were of such a nature that their removal at the time would probably have been impossible. More than a century was to elapse before they were satisfactorily met.

In passing it is interesting to observe that this dispute typifies a radical distinction between pure mathematicians and mathematical physicists. The only weapon at the disposal of pure mathematicians is sharp and rigid proof, and unless an alleged theorem can withstand the severest criticism of which its epoch is capable, pure mathematicians have but little use for it.

The applied mathematician and the mathematical physicist, on the other hand, are seldom so optimistic as to imagine that the infinite complexity of the physical universe can be described fully by any mathematical theory simple enough to be understood by human beings. Nor do they greatly regret that Airy's beautiful (or absurd) picture of the universe as a sort of interminable, self-solving system of differential equations has turned out to be an illusion born of mathematical bigotry and Newtonian determinism; they have something more real to appeal to at their own back door—the physical universe itself. They can *experiment* and check the deductions of their purposely imperfect mathematics against the verdict of experience—which, by the very nature of mathematics, is impossible for a pure mathematician. If their mathematical predictions are contradicted by experiment they do not, as a mathematician might, turn their backs on the physical evidence, but throw their mathematical tools away and look for a better kit.

This indifference of scientists to mathematics for its own sake is as enraging to one type of *pure* mathematician as the omission of a doubtful iota subscript is to another type of pedant. The result is that but few *pure* mathematicians have ever made a significant contribution to science—apart, of course, from inventing many of the tools which scientists find useful (perhaps indispensable). And the curious part of it all is that the very purists who object to the boldly imaginative attack of the scientists are the loudest in their insistence that mathematics, contrary to a widely diffused belief, is not all an affair of grubbing, meticulous accuracy, but is as creatively imaginative, and sometimes as loose, as great poetry or music can be on occasion. Sometimes the physicists beat the mathematicians at their own game in this respect: ignoring the glaring lack of rigor in Fourier's classic on the analytical theory of heat, Lord Kelvin called it "a great mathematical poem."

As has already been stated Fourier's main advance was in the direction of boundary-value problems (described in the chapter on Newton)—the fitting of solutions of differential equations to prescribed initial conditions, probably the central problem of mathematical physics. Since Fourier applied this method to the mathematical theory of heat conduction a crowded century of splendidly gifted men has gone farther than he would ever have dreamed possible, but his step was decisive. One or two of the things he did are simple enough for description here.

In algebra we learn to plot the graphs of simple algebraic equations and soon notice that the curves we get, if continued sufficiently far, do not break off suddenly and end for good. What sort of an equation would result in a graph like that of the heavy line *segment* (finite length, terminated at both ends) repeated indefinitely as in the figure?

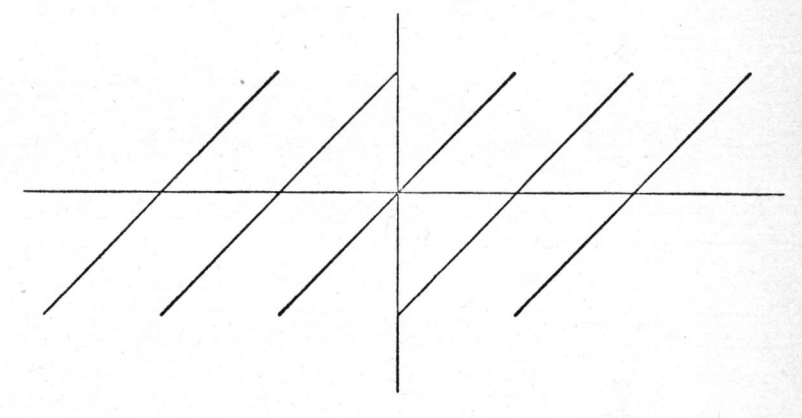

Such graphs, made up of disjointed fragments of straight or curved lines recur repeatedly in physics, for example in the theories of heat, sound, and fluid motion. It can be proved that it is impossible to represent them by finite, closed, mathematical expressions; *an infinity* of terms occur in their equations. "Fourier's Theorem" provides a means for representing and investigating such graphs mathematically: it expresses (within certain limitations) a given function continuous within a certain interval, or with only a finite number of discontinuities in the interval, and having in the interval only a finite number of turning-points, as an infinite sum of sines or cosines, or both. (This is only a rough description.)

Having mentioned sines and cosines we shall recall their most important property, *periodicity*. Let the radius of the circle in the figure be 1 unit in length. Through the center O draw rectangular axes as in Cartesian geometry, and mark off AB equal to 2π units of length; thus AB is equal in length to the circumference of the circle (since the radius is 1). Let the point P start from A and trace out the circle in the direction of the arrow. Drop PN perpendicular to OA. Then, for any position of P, the length of NP is called the *sine* of the angle AOP, and ON the *cosine*; NP and ON are to have their signs as in Cartesian geometry (NP is positive above OA, negative below; ON is positive to the right of OC, negative to the left).

For any position of P, the angle AOP will be that fraction of four right angles (360°) that the arc AP is of the whole circumference of the circle. So we may scale off these angles AOP by marking along AB the fractions of 2π which correspond to the arcs AP. Thus, when P is at C, ¼ the whole circumference has been traversed; hence, corresponding to the angle AOC we have the point K at ¼ of AB from A.

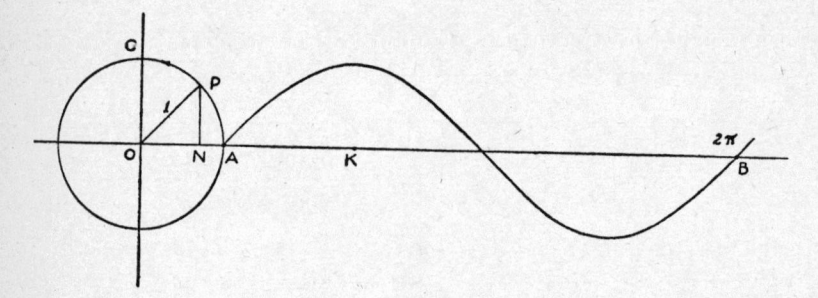

At each of these points on AB we erect a perpendicular equal in length to the sine of the corresponding angle, and above or below AB according as the sine is positive or negative. The ends of these perpendiculars not on AB lie on the continuous curve shown, the *sine curve*. When P returns to A and begins retracing the circle the curve is repeated beyond B, and so on indefinitely. If P revolves in the opposite direction, the curve is repeated to the left. After an interval of 2π the curve repeats: the sine of an angle (here AOP) is a *periodic function*, the *period* being 2π. The word "sine" is abbreviated to "sin"; and, if x is any angle, the equation

$$\sin(x + 2\pi) = \sin x$$

expresses the fact that sin x is a function of x having the period 2π.

It is easily seen that if the whole curve in the figure is shifted to the left a distance equal to AK, it now graphs the cosine of AOP. As before

$$\cos(x + 2\pi) = \cos x,$$

"cos" being the short for "cosine."

Inspection of the figure shows that sin $2x$ will go through its complete period "twice as fast" as sin x, and hence that the graph for a complete period will be one half as long as that for sin x. Similarly sin $3x$ will require only $2\pi/3$ for its complete period, and so on. The same holds for cos x, cos $2x$, cos $3x$, ...

Fourier's main mathematical result can now be described roughly. Within the restrictions already mentioned in connection with "broken" graphs, any function having a well-determined graph can be represented by an equation of the type

$$y = a_0 + a_1 \cos x + a_2 \cos 2x + a_3 \cos 3x + \ldots$$
$$+ b_1 \sin x + b_2 \sin 2x + b_3 \sin 3x + \ldots$$

where the dots indicate that the two series are to continue indefinitely according to the rule shown, and the coefficients $a_0, a_1, a_2, \ldots, b_1, b_2, b_3, \ldots$ are determinable when y, any given function of x, is known. In other words, any given function of x, say $f(x)$, can be expanded in a series of the type stated above, a *trigonometric* or *Fourier* series. To repeat, all this holds only within certain restrictions which, fortunately, are not of much importance in mathematical physics; the exceptions are more or less freak cases of little or no physical significance. Once more, Fourier's was the first great attack on boundary value problems. The specimens of such problems given in the chapter on Newton are solved by Fourier's method. In any given problem it is required to find the coefficients $a_0, a_1, \ldots, b_0, b_1, \ldots$ in a form adapted to computation. Fourier's analysis provides this.

The concept of periodicity (*simple* periodicity) as described above is of obvious importance for natural phenomena; the tides, the phases of the Moon, the seasons, and a multitude of other familiar things are periodic in character. Sometimes a periodic phenomenon, such for example as the recurrence of sunspots, can be closely approximated by superposition of a certain number of graphs having simple periodicity. The study of such situations can then be simplified by analysing the individual periodic phenomena of which the original is the resultant.

The process is the same mathematically as the analysis of a musical sound into its fundamental and successive harmonics. As a first very crude approximation to the "quality" of the sound only the fundamental is considered; the superposition of only a few harmonics usually suffices to produce a sound indistinguishable from the ideal (in which there is an infinity of harmonics). The like holds for phenomena attacked by "harmonic" or "Fourier" analysis. Attempts have even been made to detect long periods (the fundamentals) in the recurrence of earthquakes and annual rainfall. The notion of simple periodicity is as important in pure mathematics as it is in applied, and we shall see it being generalized to *multiple* periodicity (in connection with elliptic functions and others), which in its turn reacts on applied mathematics.

Fully aware that he had done something of the first magnitude Fourier paid no attention to his critics. They were right, he wrong, but he had done enough in his own way to entitle him to independence.

When the work begun in 1807 was completed and collected in the treatise on heat-conduction in 1822, it was found that the obstinate Fourier had not changed a single word of his original presentations, thus exemplifying the second part of Francis Galton's advice to all authors: "Never resent criticism, and never answer it." Fourier's resentment was rationalized in attacks on pure mathematicians for minding their own proper business and not blundering about in mathematical physics.

All was going well with Fourier and France in general when Napoleon, having escaped from Elba, landed on the French coast on March 1, 1815. Veterans and all were just getting comfortably over their headache when the cause of it popped up again to give them a worse one. Fourier was at Grenoble at the time. Fearing that the populace would welcome Napoleon back for another spree, Fourier hastened to Lyons to tell the Bourbons what was about to happen. With their usual stupidity they refused to believe him. On his way back Fourier learned that Grenoble had capitulated. Fourier himself was taken prisoner and brought before Napoleon at Bourgoin. He was confronted by the same old commander he had known so well in Egypt and had learned to distrust with his head but not with his viscera. Napoleon was bending over a map, a pair of compasses in his hand. He looked up.

"Well, Monsieur Prefect! You too; you have declared war against me?"

"Sire," Fourier stammered, "my oaths made it a duty."

"A duty, do you say? Don't you see that nobody in the country is of your opinion? And don't let yourself imagine that your plan of campaign frightens me much. I suffer only at seeing amongst my adversaries an *Egyptian*, a man who has eaten the bread of the bivouac with me, an old friend! How, moreover, Monsieur Fourier, have you been able to forget that I made you what you are?"

That Fourier, remembering Napoleon's callous abandonment of him in Egypt, could swallow such tripe and like it says a great deal for the goodness of his heart and the toughness of his stomach but precious little for the soundness of his head.

Some days later Napoleon asked the now loyal Fourier: "What do you think of my plan?"

"Sire, I believe you will fail. You will meet a fanatic on your road, and everything will be over."

"Bah! Nobody is for the Bourbons—not even a fanatic. As for that, you have read in the papers that they have put me outside the law. I myself will be more indulgent: I shall content myself with putting them outside the Tuileries!"

The leopard's spots and Napoleon's swellhead should be wedded in one proverb instead of pining apart in two.

The second restoration found Fourier in Paris pawning his effects to keep alive. But before he could starve to death old friends took pity on him and got him appointed director of the Bureau of Statistics for the Seine. The Academy tried to elect him to membership in 1816, but the Bourbon government ordered that no friend of their late kicker was to be honored in any way. The Academy stuck to its guns and elected Fourier the following year. This action of the Bourbons against Fourier may seem petty, but beside what they did to poor old Monge it was princely. *Noblesse oblige!*

Fourier's last years evaporated in clouds of talk. As Permanent Secretary of the Academy he was always able to find listeners. To say that he bragged of his achievements under Napoleon is putting it altogether too mildly. He became an insufferable, shouting bore. And instead of continuing with his scientific work he entertained his audience with boastful accounts of what he was *going* to do. However, he had done far more than his share for the advancement of science, and if any human work merits immortality, Fourier's does. He did not need to boast or bluff.

Fourier's experiences in Egypt were responsible for a curious habit which may have hastened his death. Desert heat, he believed, was the ideal condition for health. In addition to swathing himself like a mummy he lived in rooms which his uncooked friends said were hotter than hell and the Sahara desert combined. He died of heart disease (some say an aneurism) on May 16, 1830, in the sixty third year of his life. Fourier belongs to that select company of mathematicians whose work is so fundamental that their names have become adjectives in every civilized language.

Monge's decline was slower and more distressing. After the first restoration Napoleon felt embittered and vindictive toward the snobocracy of his own creation which, naturally, had let him down the moment his power waned. Once more in the saddle Napoleon was inclined to use the butt end of his crop on the skulls of the ungrateful. Monge, good old plebeian that he was, counselled mercy and common sense: Napoleon might some day find himself with his back to the wall (after an earthquake had cut off all means of flight), and be grateful for the support of the ingrates. Cooling off, Napoleon wisely tempered injustice with mercy. For this gracious dispensation Monge alone was responsible.

After Napoleon had run away from Waterloo, leaving his troops to get out of the mess as best they could, he returned to Paris. Fourier's devotion cooled then; Monge's boiled.

The school histories often tell of Napoleon's last dream—the conquest of America. The Mongian version differs and is on a much higher—in fact, incredibly high—plane. Hemmed in by enemies and appalled at the thought of enforced idleness for lack of further European conquest, Napoleon turned his eagle eye West, and in one flashing glance surveyed America from Alaska to Cape Horn. But, like the sick devil he was, Bonaparte longed to become a monk. The sciences alone could satisfy him, he declared; he would become a second and infinitely greater Alexander von Humboldt.

"I wish," he confessed to Monge, "in this new career to leave works, discoveries, worthy of me."

What, precisely, are the works which could be worthy of a Napoleon? Continuing, the fallen eagle outlined his dream.

"I need a companion," he admitted, "to first put me abreast of the present state of the sciences. Then you [Monge] and I will traverse the whole continent, from Canada to Cape Horn; and in this immense journey we shall study all those prodigious phenomena of terrestrial physics on which the scientific world has not pronounced its verdict." **Paranoia?**

"Sire," Monge exclaimed—he was nearly sixty seven—"your collaborator is already found; I will go with you!"

His old self once more, Napoleon curtly dismissed the thought of the willing veteran hampering his lightning marches from Baffin Bay to Patagonia.

"You are too old, Monge. I need a younger man."

Monge tottered off to find "a younger man." He approached the fiery Arago as the ideal travelling companion for his energetic master. But Arago, in spite of all his eloquent rhetoric on the gloriousness of glory, had learned his lesson. A general who could desert his troops as Napoleon had done at Waterloo, Arago pointed out, was no leader to follow anywhere, even in easy America.

Further negotiations were rudely halted by the British. By the middle of October Napoleon was exploring St. Helena. The hoard of money which had been put aside for the conquest of America found its way into deeper pockets than those of the scientists, and no "American Institute" rose on the banks of the Mississippi or the Amazon to match its fantastic twin overlooking the Nile.

Having enjoyed the bread of imperialism Monge now tasted the salt. His record as a revolutionist and favorite of the upstart Corsican made his head an extremely desirable object to the Bourbons, and Monge dodged from one slum to another in an endeavor to keep his head on his shoulders. For sheer human pettiness the treatment accorded Monge by the sanctified Bourbons would take a lot of beating. Small enough for anything they stripped the old man of his last honor —one with which the generosity of Napoleon had had nothing whatever to do. In 1816 they commanded that Monge be expelled from the Academy. The academicians, tame as rabbits now, obeyed.

The final touch of Bourbon pettiness graced the day of Monge's funeral. As he had foreseen he died after a prolonged stupor following a stroke. The young men at the Polytechnique, whom he had protected from Napoleon's domineering interference, were the pride of Monge's heart, and he was their idol. When Monge died on July 28, 1818, the Polytechnicians asked permission to attend the funeral. The King denied the request.

Well disciplined, the Polytechnicians observed the ban. But they were more resourceful or more courageous than the timid academicians. The King's order covered only the funeral. The following day they marched in a body to the cemetery and laid a wreath on the grave of their master and friend, Gaspard Monge.

CHAPTER THIRTEEN

The Day of Glory

PONCELET

Projective geometry has opened up for us with the greatest facility new territories in our science, and has rightly been called a royal road to its own particular field of knowledge.—FELIX KLEIN

MORE THAN ONCE during the World War when the French troops were hard pressed and reinforcements nonexistent, the high command saved the day by routing some prima donna out of her boudoir, rushing her to the front, draping her from neck to heels in the tricolor, and ordering her to sing the *Marseillaise* to the exhausted men. Having sung her piece the lady rolled back to Paris in her limousine; the heartened troops advanced, and the following morning a cynically censored press once more unanimously assured a gullible public that "the day of glory has arrived"—with unmentioned casualties.

In 1812 the day of glory was still on its way. Prima donnas did not accompany Napoleon Bonaparte's half-million troops on their triumphal march into Russia. The men did their own singing as the Russians retreated before the invincible Grand Army, and the endless plains rang to the stirring chant which had swept tyrants from their thrones and elevated Napoleon to their place.

All was going as gloriously as the most enthusiastic singer could have wished: six days before Napoleon crossed the Niemen his brilliant diplomatic strategy had indirectly exasperated President Madison into hurling the United States into a distracting war on England; the Russians were running harder than ever on their race back to Moscow, and the Grand Army was doing its valiant best to keep up with the reluctant enemy. At Borodino the Russians turned, fought, and, retired. Napoleon continued without opposition—except from the erratic weather—to Moscow, whence he notified the Czar of his willingness to consider an unconditional surrender of all the

Russian forces. The competent inhabitants of Moscow, led by the Governor, took matters into their own hands, fired their city, burned it to the ground, and smoked Napoleon and all his men out into the void. Chagrined but still master of the situation, Napoleon disregarded this broad hint—the second or third so far vouchsafed to his military obstinacy—that "who killeth with the sword must perish by the sword," presently ordered his driver to give the horses the lash, and dashed back post-haste over the now frozen plains to prepare for his rendezvous with Blücher at Leipzig, leaving the Grand Army to walk home or freeze as it should see fit.

With the deserted French army was a young officer of engineers, Jean-Victor Poncelet (July 1, 1788–December 23, 1867) who, as a student at the École Polytechnique in Paris, later at the military academy at Metz, had been inspired by the new descriptive geometry of Monge (1746–1818) and the *Géométrie de position* (published in 1803) of the elder Carnot (Lazare-Nicolas-Marguerite Carnot, May 13, 1753–August 2, 1823), whose revolutionary if somewhat reactionary program was devised "to free geometry from the hieroglyphics of analysis."

In the preface to his classic *Applications d'analyse et de géométrie* (second edition 1862, of the work first published in 1822), Poncelet recounts his experiences in the disastrous retreat from Moscow. On November 18, 1812, the exhausted remnant of the French army under Marshal Ney was overwhelmed at Krasnoï. Among those left for dead on the frozen battlefield was young Poncelet. His uniform as an officer of engineers saved his life. A searching party, discovering that he still breathed, took him before the Russian staff for questioning.

As a prisoner of war the young officer was forced to march for nearly five months across the frozen plains in the tatters of his uniform, subsisting on a meagre ration of black bread. In a cold so intense that the mercury of the thermometer frequently froze, many of Poncelet's companions in misery died in their tracks, but his ruggeder strength pulled him through, and in March, 1813 he entered his prison at Saratoff on the banks of the Volga. At first he was too exhausted to think. But when "the splendid April sun" restored his vitality, he remembered that he had received a good mathematical education, and to soften the rigors of his exile he resolved to reproduce as much as he could of what he had learned. It was thus that he created projective geometry.

Without books and with only the scantiest writing materials at first, he retraced all that he had known of mathematics from arithmetic to higher geometry and the calculus. These first labors were enlivened by Poncelet's efforts to coach his fellow officers for the examinations they must take should they ever see France again. One legend states that at first Poncelet had only scraps of charcoal, salvaged from the meager brazier which kept him from freezing to death, for drawing his diagrams on the wall of his cell. He makes the interesting observation that practically all details and complicated developments of the mathematics he had been taught had evaporated, while the general, fundamental principles remained as clear as ever in his memory. The same was true of physics and mechanics.

In September, 1814, Poncelet returned to France, carrying with him "the material of seven manuscript notebooks written at Saratoff in the prisons of Russia (1813 to 1814), together with divers other writings, old and new," in which he, as a young man of twenty four, had given projective geometry its strongest impulse since Desargues and Pascal initiated the subject in the seventeenth century. The first edition of his classic, as already mentioned, was published in 1822. It lacked the intimate "apology for his life" which has been used above, but it started a tremendous nineteenth century surge forward in projective geometry, modern synthetic geometry generally, and the geometric interpretation of the "imaginary" numbers that present themselves in algebraic manipulations, giving to such "imaginaries" geometrical interpretations as "ideal" elements of space. It also proposed the powerful and (for a time) controversial "doctrine of continuity," to be described presently, which greatly simplified the study of geometric configurations by unifying apparently unrelated properties of figures into uniform, self-contained complete wholes. Exceptions and awkward special cases appeared under Poncelet's broader point of view as merely different aspects of things already familiar. The classic treatise also made full use of the creative "principle of duality" and introduced the method of "reciprocation" devised by Poncelet himself. In short, a whole arsenal of new weapons was added to geometry by the young military engineer who had been left for dead on the field of Krasnoï, and who might indeed have died before morning had not his officer's uniform distinguished him as a likely candidate for questioning by the Russian staff.

For the next decade (1815–25) Poncelet's duties as a military engineer left him only odd moments for his real ambition—the exploitation of his new methods in geometry. Relief was not to come for many years. His high sense of duty and his fatal efficiency made Poncelet an easy prey for short-sighted superiors. Some of the tasks he was set could have been done only by a man of his calibre, for example the creation of the school of practical mechanics at Metz and the reform of mathematical education at the Polytechnique. But the reports on fortifications, his work on the Committee of Defense, and his presidency of the mechanical sections at the international expositions of London and Paris (1851–58), to mention only a few of his numerous routine jobs, could all have been done by lesser men. His high scientific merits, however, were not unappreciated. The Academy of Sciences elected him (1831) as successor to Laplace. For political reasons Poncelet declined the honor till three years later.

Poncelet's whole mature life was one long internal conflict between that half of him which was born to do lasting work and the other half which accepted all the odd or dirty jobs shortsighted politicians and obtuse militarists shoved in its way. Poncelet himself longed to escape, but a mistaken sense of duty, drilled into his very bones in Napoleon's armies, impelled him to serve the shadow and turn his back on the substance. That he did not suffer an early and permanent nervous breakdown is a remarkable testimonial to the ruggedness of his physique. And that he retained his creative abilities almost to his death at the age of seventy nine is a shining proof of his unquenchable genius. When they could think of nothing better for this splendidly endowed man to do with his time they sent him traipsing about France to inspect cotton mills, silk mills, and linen mills. They did not need a Poncelet to do that sort of thing, and he knew it. He would have been the last man in France to object had his unique talents been indispensable in such affairs, for he was anything but the sort of intellectual prude who holds that science loses her perennial virginity every time she shakes hands with industry. But he was not the only man available for the work, as possibly Pasteur was in the equally important matters of the respective diseases of beer, silkworms, and human beings.

We now glance at one or two of the weapons either devised or remodelled by Poncelet for the conquest of projective geometry. First there is his "principle of continuity," which refers to the permanence of geometrical properties as one figure shades, by projection or otherwise, into another. This no doubt is rather vague, but Poncelet's own statement of the principle was never very exact and, as a matter of fact, embroiled him in endless controversies with more conservative geometers whom he politely designated as old fossils—always in the dignified diction suitable to an officer and a gentleman, of course. With the caution that the principle is of great heuristic value but does not always of itself provide proofs of the theorems which it suggests, we may see something of its spirit from a few simple examples.

Imagine two intersecting circles. Say they intersect in the points A and B. Join A and B by a straight line. The figure presents ocular evidence of two *real* points A, B and the common chord AB of the two circles. Now imagine the two circles pulled gradually apart.

The common chord presently becomes a common tangent to the two circles at their point of contact. At any stage so far the following theorem (usually set as an exercise in school geometry) is true: if *any* point P be taken on the common chord, *four* tangent lines may be drawn from it to the two circles, and if the points in which these tangent lines touch the circles are T_1, T_2, T_3, T_4, then the segments PT_1, PT_2, PT_3, PT_4 are all equal in length. Conversely, if it is asked where do *all* the points P lie such that the four tangent-segments to the two circles shall all be equal, the answer is *on the common chord*. Stating all this briefly in the usual language, we say that the *locus* (which merely means *place*) of a point P which moves so that the lengths of the tangent-segments from it to two *intersecting* circles are equal, is the common chord of the two circles.* All this is familiar and straightforward; there is no element of mystery or incomprehensi-

---

*In what precedes the tangents are *real* (visible) if the point P lies *outside* the circles; if P is *inside*, the tangents are *"imaginary."*

bility as some may say there is in the next where the "principle of continuity" enters.

Pull the circles completely apart. Their two intersections (or in the last moment their one point of contact) are no longer visible on the paper and the "common chord" is left suspended between the two circles, cutting neither visibly. But it is known that there is still a *locus* of equal tangent-segments, and it is easily proved that this locus is a straight line perpendicular to the line joining the centres of the two circles, just as the original locus (the common chord) was. Merely as a manner of speaking, if we object to "imaginaries," we continue to *say* that the two circles intersect in two points in the infinite part of the plane, even when they have been pulled apart, and we *say* also that the new straight-line locus is still the common chord of the circles: the points of intersection are "imaginary" or "ideal," but the straight line joining them (the new "common chord") is "real"—we actually draw it on the paper.

If we write the equations of the circles and lines algebraically in the manner of Descartes, all that we do in the algebra of solving the equations for the intersections has its unique correlate in the enlarged geometry, whereas if we do not first expand our geometry—or at least increase its vocabulary, to take account of "ideal" elements—much of the meaningful algebra is geometrically meaningless.

All this of course requires logical justification. Such justification has been given so far as is necessary, that is, up to the stage which includes the applications of the "principle of continuity" useful in geometry.

A more important instance of the principle is furnished by parallel straight lines. Before describing this we may repeat the remark a venerable and distinguished judge relieved himself of a few days ago when the matter was revealed to him. The judge had been under the weather; an amateur mathematician, thinking to cheer the old fellow up, told him something of the geometrical concept of infinity. They were strolling through the judge's garden at the time. On being informed that "parallel lines meet at infinity," the judge stopped dead. "Mr. Blank," he said with great emphasis, "any man who says parallel lines meet at infinity, or anywhere else, simply hasn't got good sense." To obviate an argument we may say as before that it is all a way of speaking to avoid irritating exceptions and separations into exasperating distinct cases. But once the language has been agreed upon, logical consistency demands that it be followed to the end without traversing the rules of logical grammar and syntax, and this is what is done.

To see the reasonableness of the language, imagine a fixed straight line l and fixed point P not on l. Through P draw any straight line l' intersecting l in P', and imagine l' to rotate about P, so that P' recedes along l. When does P' stop receding? We say it stops when l, l' become parallel or, if we prefer, when the point of intersection P' is at infinity. For reasons already indicated this language is convenient and suggestive—not of a lunatic asylum, as the judge might think, but of interesting and sometimes highly practical things to do in geometry.

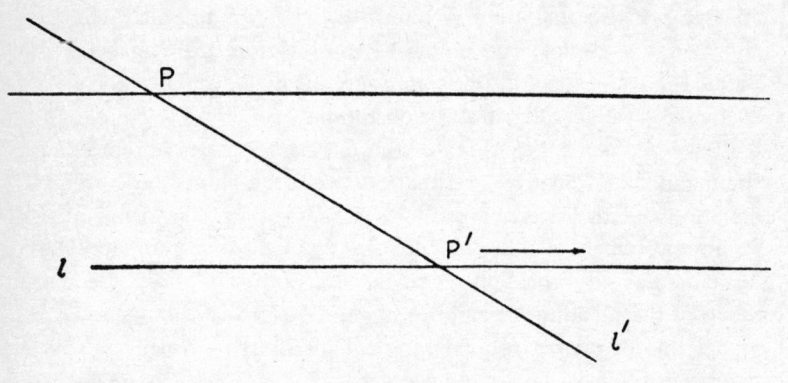

In a similar manner the visualizable, *finite* parts of lines, planes and three-dimensional space (also of higher space) are enriched by the adjunction of "ideal" points, lines, planes, or "regions" *at infinity*. If the judge happens to see this he may enjoy the following shocking example of the behavior of the infinite in geometry: *any two circles in a plane intersect in four points, two of which are imaginary and at infinity*. If the circles are concentric, they touch one another in two points lying on the line at infinity. Further, *all* circles in a plane go through *the same* two points at infinity—they are usually denoted by I and J, and are sometimes called Isaac and Jacob by irreverent students.

In the chapter on Pascal we described what is meant by projective properties in distinction to metrical properties in geometry. At this point we may glance back at Hadamard's remarks on Descartes' analytic geometry. Hadamard observed among other things that modern synthetic geometry repaid the debt of geometry in general to algebra by suggesting important researches in algebra and analysis. This modern synthetic geometry was the object of Poncelet's researches. Although all this may seem rather involved at the moment, we shall close the chain by taking a link from the 1840's, as the matter really is important, not only for the history of pure mathematics but for that of recent mathematical physics as well.

The link from the 1840's is the creation by Boole, Cayley, Sylvester and others, of the algebraic theory of invariance which (as will be explained in a later chapter) is of fundamental importance in current theoretical physics. The projective geometry of Poncelet and his school played a very important part in the development of the theory of invariance: the geometers had discovered a whole continent of properties of figures *invariant* under projection; the algebraists of the 1840's, notably Cayley, translated the geometrical *operations of projection* into analytical language, applied this translation to the *algebraic*, Cartesian mode of expressing geometric relationships, and were thus enabled to make phenomenally rapid progress in the elaboration of the theory of algebraic invariants. If Desargues, the daring pioneer of the seventeenth century, could have foreseen what his ingenious method of projection was to lead to, he might well have been astonished. He knew that he had done something good, but he probably had no conception of just how good it was to prove.

Isaac Newton was a young man of twenty when Desargues died. There is no evidence that Newton ever heard the name of Desargues. If he had, he also might have been astonished could he have foreseen that the humble link forged by his elderly contemporary was to form part of the strong chain which, in the twentieth century, was to pull his law of universal gravitation from its supposedly immortal pedestal. For without the mathematical machinery of the tensor calculus which developed naturally (as we shall see) from the algebraic work of Cayley and Sylvester, it is improbable that Einstein or anyone else could ever have budged the Newtonian theory of gravitation.

One of the useful ideas in projective geometry is that of *cross-ratio* or *anharmonic ratio*. Through a point O draw any four straight lines l, m, n, p. Across these four draw any straight line x, and label the points in which x cuts the others L, M, N, P respectively. We thus have on x the line segments LM, MN, LP, PN. From these form the ratios $\dfrac{LM}{MN}$ and $\dfrac{LP}{PN}$. Finally we take the ratio of these two ratios, and get the *cross-ratio* $\dfrac{LM \times PN}{MN \times LP}$. The remarkable thing about this cross-ratio is that it has the same numerical magnitude for *all* positions of the line x.

Later we shall refer to Felix Klein's unification of Euclidean geometry and the common non-Euclidean geometries into one comprehensive geometry. This unification was made possible by Cayley's revision of the usual notions of *distance* and *angle* on which *metrical* geometry is founded. In this revision, cross-ratio played the leading part, and through it, by the introduction of "ideal" elements of his own devising, Cayley was enabled to reduce *metrical* geometry to a species of *projective* geometry.

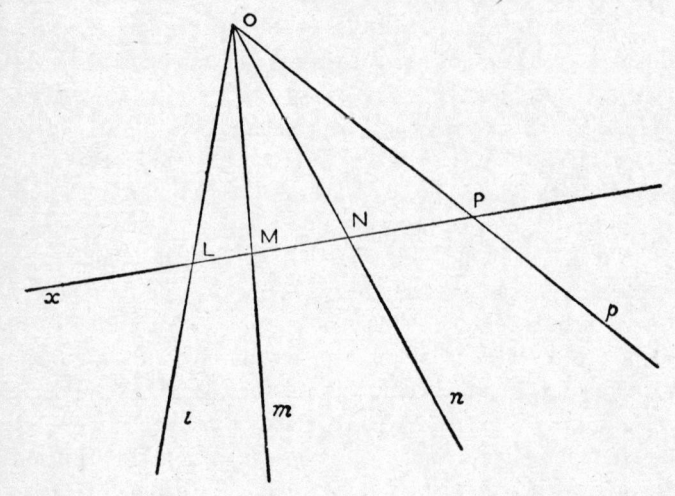

To close this inadequate description of the kind of weapons that Poncelet used we shall mention the extremely fruitful "principle of duality." For simplicity we consider only how the principle operates in plane geometry.

Note first that any continuous curve may be regarded in either of two ways: either as being generated by the motion of a point, or as being swept out by the turning motion of a straight line. To see the latter, imagine the tangent line drawn at each point of the curve. Thus *points* and *lines* are intimately and reciprocally associated with respect to the curve: *through* every *point* of the curve there is a *line* of the curve; *on* every line of the curve there is a point of the curve. Instead of "through" in the preceding sentence, write "on." Then the two assertions separated by ";" after the ":" are identical except that the words "point" and "line" are interchanged.

As a matter of terminology we say that a line (straight or curved) is *on* a point if the line passes through the point, and we note that if a line is *on* a point, then the point is *on* the line, and conversely. To make this correspondence universal we "adjoin" to the usual plane in which Euclidean geometry (common school geometry) is valid, a so-called *metric plane*, "ideal elements" of the kind already

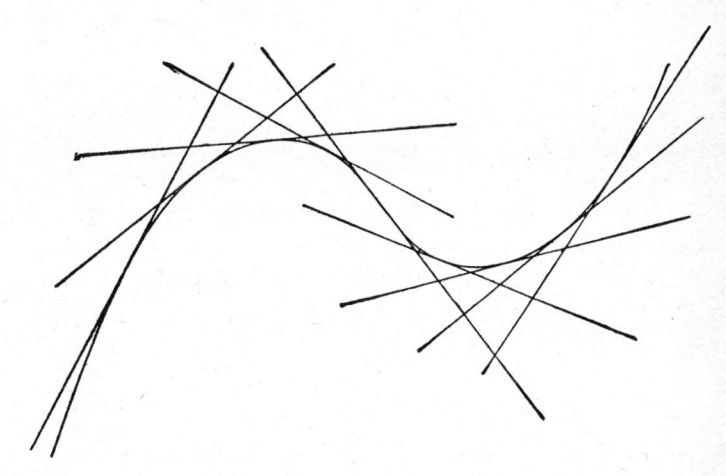

described. The result of this adjunction is a *projective plane*: a projective plane consists of all the ordinary points and straight lines of a metric plane and, in addition, of a set of ideal points all of which are assumed to lie on one ideal line and such that one such ideal point lies on every ordinary line.*

In Euclidean language we would say that two parallel lines have the same direction; in projective phraseology this becomes "two parallel lines have the same ideal point." Again, in the old, if two or more lines have the same direction, they are parallel; in the new, if two or more lines have the same ideal point they are parallel. Every

---

*This definition, and others of a similar character given presently, is taken from *Projective Geometry* (Chicago, 1930) by the late John Wesley Young. This little book is comprehensible to anyone who has had an ordinary school course in common geometry.

straight line in the projective plane is conceived of as having on it *one ideal point* ("at infinity"); *all* the ideal points are thought of as making up *one ideal line*, "the line at infinity."

The purpose of these conceptions is to avoid the exceptional statements of Euclidean geometry necessitated by the postulated existence of parallels. This has already been commented on in connection with Poncelet's formulation of the principle of continuity.

With these preliminaries the *principle of duality* in plane geometry can now be stated: All the propositions of plane projective geometry occur in dual pairs which are such that, from either proposition of a particular pair another can be immediately inferred by interchanging the parts played by the words *point* and *line*.

In his projective geometry Poncelet exploited this principle to the limit. Opening almost any book on projective geometry at random we note pages of propositions printed in double columns, a device introduced by Poncelet. Corresponding propositions in the two columns are duals of one another; if either has been proved, a proof of the other is superfluous, as implied by the principle of duality. Thus geometry at one stroke is doubled in extent with no expenditure of extra labor. As a specimen of dual propositions we give the following pair.

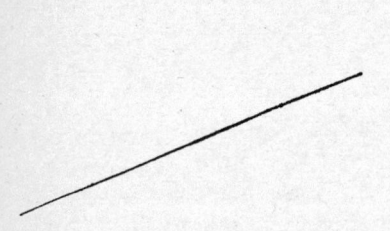

Two distinct points are on one, and only one, line.	Two distinct lines are on one, and only one, point.

It may be granted that this is not very exciting. The mountain has labored and brought forth a mouse. Can it do any better?

The proposition in the left-hand column (page 217) is Pascal's concerning his *Hexagrammum Mysticum* which we have already seen; that on the right is Brianchon's theorem, which was *discovered* by means of the principle of duality. Brianchon (1785-1864) discovered his theorem while he was a student at the École Polytechnique; it was printed in the *Journal* of that school in 1806. The figures for the two propositions look nothing alike. This may indicate the power of the methods used by Poncelet.

Brianchon's discovery was the one which put the principle of duality on the map of geometry. Far more spectacular examples of the power of the principle will be found in any textbook on projective geometry, particularly in the extension of the principle to ordinary three-dimensional space. In this extension the parts played by the words *point* and *plane* are interchangeable; *straight line* stays as it was.

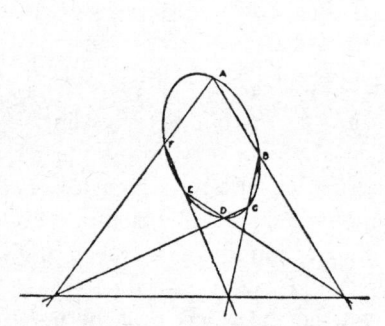

If A,B,C,D,E,F are any points on a conic section, the points of intersection of the pairs of lines AB and DE, BC and EF, CD and FA are on a straight line; and conversely.

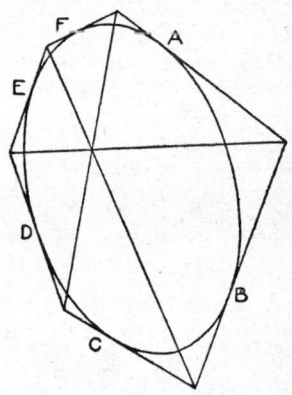

If A,B,C,D,E,F are tangent straight lines on a conic section, the lines joining the pairs of intersections of A with B and D with E, B with C and E with F, C with D and F with A, meet in one point; and conversely.

The conspicuous beauty of projective geometry and the supple elegance of its demonstrations made it a favorite study with the geometers of the nineteenth century. Able men swarmed into the new goldfield and quickly stripped it of its more accessible treasures. Today the majority of experts seem to agree that the subject is worked out so far as it is of interest to professional mathematicians. However, it is conceivable that there may yet be something in it as obvious as the principle of duality which has been overlooked. In any event it is an easy subject to acquire and one of fascinating delight to amateurs and even to professionals at some stage of their careers. Unlike some other fields of mathematics, projective geometry has been blessed with many excellent textbooks and treatises, some of them by master geometers, including Poncelet himself.

CHAPTER FOURTEEN

The Prince of Mathematicians

GAUSS

> *The further elaboration and development of systematic arithmetic, like nearly everything else which the mathematics of our [nineteenth] century has produced in the way of original scientific ideas, is knit to Gauss.*—LEOPOLD KRONECKER

ARCHIMEDES, NEWTON, AND GAUSS, these three, are in a class by themselves among the great mathematicians, and it is not for ordinary mortals to attempt to range them in order of merit. All three started tidal waves in both pure and applied mathematics: Archimedes esteemed his pure mathematics more highly than its applications; Newton appears to have found the chief justification for his mathematical inventions in the scientific uses to which he put them, while Gauss declared that it was all one to him whether he worked on the pure or the applied side. Nevertheless Gauss crowned the higher arithmetic, in his day the least practical of mathematical studies, the Queen of all.

The lineage of Gauss, Prince of Mathematicians, was anything but royal. The son of poor parents, he was born in a miserable cottage at Brunswick (Braunschweig), Germany, on April 30, 1777. His paternal grandfather was a poor peasant. In 1740 this grandfather settled in Brunswick, where he drudged out a meager existence as a gardener. The second of his three sons, Gerhard Diederich, born in 1744, became the father of Gauss. Beyond that unique honor Gerhard's life of hard labor as a gardener, canal tender, and bricklayer was without distinction of any kind.

The picture we get of Gauss' father is that of an upright, scrupulously honest, uncouth man whose harshness to his sons sometimes bordered on brutality. His speech was rough and his hand heavy. Honesty and persistence gradually won him some measure of comfort, but his circumstances were never easy. It is not surprising that such a man did everything in his power to thwart his young son and prevent him from acquiring an education suited to his abilities. Had the father prevailed, the gifted boy would have followed one of the family trades, and it was only by a series of happy accidents that Gauss was saved from becoming a gardener or a bricklayer. As a child he was respectful and obedient, and although he never criticized his poor father in later life, he made it plain that he had never felt any real affection for him. Gerhard died in 1806. By that time the son he had done his best to discourage had accomplished immortal work.

On his mother's side Gauss was indeed fortunate. Dorothea Benz's father was a stonecutter who died at the age of thirty of tuberculosis, the result of unsanitary working conditions in his trade, leaving two children, Dorothea and her younger brother Friederich.

Here the line of descent of Gauss' genius becomes evident. Condemned by economic disabilities to the trade of weaving, Friederich was a highly intelligent, genial man whose keen and restless mind foraged for itself in fields far from his livelihood. In his trade Friederich quickly made a reputation as a weaver of the finest damasks, an art which he mastered wholly by himself. Finding a kindred mind in his sister's child, the clever uncle Friederich sharpened his wits on those of the young genius and did what he could to rouse the boy's quick logic by his own quizzical observations and somewhat mocking philosophy of life.

Friederich knew what he was doing; Gauss at the time probably did not. But Gauss had a photographic memory which retained the impressions of his infancy and childhood unblurred to his dying day. Looking back as a grown man on what Friederich had done for him, and remembering the prolific mind which a premature death had robbed of its chance of fruition, Gauss lamented that "a born genius was lost in him."

Dorothea moved to Brunswick in 1769. At the age of thirty four (in 1776) she married Gauss' father. The following year her son was born. His full baptismal name was Johann Friederich Carl Gauss. In later life he signed his masterpieces simply Carl Friedrich Gauss. If a great genius was lost in Friederich Benz his name survives in that of his grateful nephew.

Gauss' mother was a forthright woman of strong character, sharp intellect, and humorous good sense. Her son was her pride from the day of his birth to her own death at the age of ninety seven. When the "wonder child" of two, whose astounding intelligence impressed all who watched his phenomenal development as something not of this earth, maintained and even surpassed the promise of his infancy as he grew to boyhood, Dorothea Gauss took her boy's part and defeated her obstinate husband in his campaign to keep his son as ignorant as himself.

Dorothea hoped and expected great things of her son. That she may sometimes have doubted whether her dreams were to be realized is shown by her hesitant questioning of those in a position to judge her son's abilities. Thus, when Gauss was nineteen, she asked his mathematical friend Wolfgang Bolyai whether Gauss would ever amount to anything. When Bolyai exclaimed "The greatest mathematician in Europe!" she burst into tears.

The last twenty two years of her life were spent in her son's house, and for the last four she was totally blind. Gauss himself cared little if anything for fame; his triumphs were his mother's life.* There was always the completest understanding between them, and Gauss repaid her courageous protection of his early years by giving her a serene old age. When she went blind he would allow no one but himself to wait on her, and he nursed her in her long last illness. She died on April 19, 1839.

Of the many accidents which might have robbed Archimedes and Newton of their mathematical peer, Gauss himself recalled one from his earliest childhood. A spring freshet had filled the canal which ran by the family cottage to overflowing. Playing near the water, Gauss was swept in and nearly drowned. But for the lucky chance that a laborer happened to be about his life would have ended then and there.

In all the history of mathematics there is nothing approaching the precocity of Gauss as a child. It is not known when Archimedes first gave evidence of genius. Newton's earliest manifestations of the highest mathematical talent may well have passed unnoticed. Although it seems incredible, Gauss showed his caliber before he was three years old.

---

* The legend of Gauss' relations to his parents has still to be authenticated. Although, as will be seen later, the *mother* stood by her son, the *father* opposed him; and, as was customary *then* (usually, also, *now*), in a German household, the *father* had the last word.—I allude later to legends from living persons who had known members of the Gauss family, particularly in respect to Gauss' treatment of his sons. These allusions refer to first-hand evidence; but I do not vouch for them, as the people were very old.

One Saturday Gerhard Gauss was making out the weekly payroll for the laborers under his charge, unaware that his young son was following the proceedings with critical attention. Coming to the end of his long computations, Gerhard was startled to hear the little boy pipe up, "Father, the reckoning is wrong, it should be . . . ." A check of the account showed that the figure named by Gauss was correct.

Before this the boy had teased the pronunciations of the letters of the alphabet out of his parents and their friends and had taught himself to read. Nobody had shown him anything about arithmetic, although presumably he had picked up the meanings of the digits 1,2, . . . along with the alphabet. In later life he loved to joke that he knew how to reckon before he could talk. A prodigious power for involved mental calculations remained with him all his life.

Shortly after his seventh birthday Gauss entered his first school, a squalid relic of the Middle Ages run by a virile brute, one Büttner, whose idea of teaching the hundred or so boys in his charge was to thrash them into such a state of terrified stupidity that they forgot their own names. More of the good old days for which sentimental reactionaries long. It was in this hell-hole that Gauss found his fortune.

Nothing extraordinary happened during the first two years. Then, in his tenth year, Gauss was admitted to the class in arithmetic. As it was the beginning class none of the boys had ever heard of an arithmetical progression. It was easy then for the heroic Büttner to give out a long problem in addition whose answer he could find by a formula in a few seconds. The problem was of the following sort, $81297 + 81495 + 81693 + \ldots + 100899$, where the step from one number to the next is the same all along (here 198), and a given number of terms (here 100) are to be added.

It was the custom of the school for the boy who first got the answer to lay his slate on the table; the next laid his slate on top of the first, and so on. Büttner had barely finished stating the problem when Gauss flung his slate on the table: "There it lies," he said—"*Ligget se'*" in his peasant dialect. Then, for the ensuing hour, while the other boys toiled, he sat with his hands folded, favored now and then by a sarcastic glance from Büttner, who imagined the youngest pupil in the class was just another blockhead. At the end of the period

Büttner looked over the slates. On Gauss' slate there appeared but a single number. To the end of his days Gauss loved to tell how the one number he had written was the correct answer and how all the others were wrong. Gauss had not been shown the trick for doing such problems rapidly. It is very ordinary once it is known, but for a boy of ten to find it instantaneously by himself is not so ordinary.

This opened the door through which Gauss passed on to immortality. Büttner was so astonished at what the boy of ten had done without instruction that he promptly redeemed himself and to at least one of his pupils became a humane teacher. Out of his own pocket he paid for the best textbook on arithmetic obtainable and presented it to Gauss. The boy flashed through the book. "He is beyond me," Büttner said; "I can teach him nothing more."

By himself Büttner could probably not have done much for the young genius. But by a lucky chance the schoolmaster had an assistant, Johann Martin Bartels (1769–1836), a young man with a passion for mathematics, whose duty it was to help the beginners in writing and cut their quill pens for them. Between the assistant of seventeen and the pupil of ten there sprang up a warm friendship which lasted out Bartels' life. They studied together, helping one another over difficulties and amplifying the proofs in their common textbook on algebra and the rudiments of analysis.

Out of this early work developed one of the dominating interests of Gauss' career. He quickly mastered the binomial theorem,

$$(1+x)^n = 1 + \frac{n}{1}x + \frac{n(n-1)}{1 \times 2}x^2 + \frac{n(n-1)(n-2)}{1 \times 2 \times 3}x^3 + \ldots,$$

in which n is not necessarily a positive integer, but may be any number. If n is not a positive integer, the series on the right is *infinite* (non-terminating), and in order to state when this series is actually equal to $(1+x)^n$, it is mandatory to investigate what restrictions must be imposed upon x and n in order that the infinite series shall *converge to a definite, finite limit*. Thus, if $x = -2$, and $n = -1$, we get the absurdity that $(1-2)^{-1}$, which is $(-1)^{-1}$ or $1/(-1)$, or finally -1, is equal to $1 + 2 + 2^2 + 2^3 + \ldots$ and so on *ad infinitum*; that is, -1 is equal to the "infinite number" $1 + 2 + 4 + 8 + \ldots$, which is nonsense.

Before young Gauss asked himself whether infinite series *converge* and really do enable us to calculate the mathematical expressions (functions) they are used to represent, the older analysts had not seriously troubled themselves to explain the mysteries (and nonsense) arising from an uncritical use of infinite processes. Gauss' early encounter with the binomial theorem inspired him to some of his greatest work and he became the first of the "rigorists." A *proof* of the binomial theorem when n is not an integer greater than zero is even today beyond the range of an elementary textbook. Dissatisfied with what he and Bartels found in their book, Gauss made a proof. This initiated him to mathematical analysis. The very essence of analysis is the correct use of infinite processes.

The work thus well begun was to change the whole aspect of mathematics. Newton, Leibniz, Euler, Lagrange, Laplace—all great analysts for their times—had practically no conception of what is now acceptable as a proof involving infinite processes. The first to see clearly that a "proof" which may lead to absurdities like "minus 1 equals infinity" is no proof at all, was Gauss. Even if in *some* cases a formula gives consistent results, it has no place in mathematics until the precise conditions under which it will continue to yield consistency have been determined.

The rigor which Gauss imposed on analysis gradually overshadowed the whole of mathematics, both in his own habits and in those of his contemporaries—Abel, Cauchy—and his successors—Weierstrass, Dedekind, and mathematics after Gauss became a totally different thing from the mathematics of Newton, Euler, and Lagrange.

In the constructive sense Gauss was a revolutionist. Before his schooling was over the same critical spirit which left him dissatisfied with the binomial theorem had caused him to question the demonstrations of elementary geometry. At the age of twelve he was already looking askance at the foundations of Euclidean geometry; by sixteen he had caught his first glimpse of a geometry other than Euclid's. A year later he had begun a searching criticism of the proofs in the theory of numbers which had satisfied his predecessors and had set himself the extraordinarily difficult task of filling up the gaps and *completing* what had been only half done. Arithmetic, the field of his earliest triumphs, became his favorite study and the locus of his masterpiece. To his sure feeling for what constitutes proof Gauss added a prolific mathematical inventiveness that has never been surpassed. The combination was unbeatable.

Bartels did more for Gauss than to induct him into the mysteries of algebra. The young teacher was acquainted with some of the influential men of Brunswick. He now made it his business to interest these men in his find. They in turn, favorably impressed by the obvious genius of Gauss, brought him to the attention of Carl Wilhelm Ferdinand, Duke of Brunswick.

The Duke received Gauss for the first time in 1791. Gauss was then fourteen. The boy's modesty and awkward shyness won the heart of the generous Duke. Gauss left with the assurance that his education would be continued. The following year (February, 1792) Gauss matriculated at the Collegium Carolinum in Brunswick. The Duke paid the bills and he continued to pay them till Gauss' education was finished.

Before entering the Caroline College at the age of fifteen, Gauss had made great headway in the classical languages by private study and help from older friends, thus precipitating a crisis in his career. To his crassly practical father the study of ancient languages was the height of folly. Dorothea Gauss put up a fight for her boy, won, and the Duke subsidized a two-years' course at the Gymnasium. There Gauss' lightning mastery of the classics astonished teachers and students alike.

Gauss himself was strongly attracted to philological studies, but fortunately for science he was presently to find a more compelling attraction in mathematics. On entering college Gauss was already master of the supple Latin in which many of his greatest works are written. It is an ever-to-be-regretted calamity that even the example of Gauss was powerless against the tides of bigoted nationalism which swept over Europe after the French Revolution and the downfall of Napoleon. Instead of the easy Latin which sufficed for Euler and Gauss, and which any student can master in a few weeks, scientific workers must now acquire a reading knowledge of two or three languages in addition to their own. Gauss resisted as long as he could, but even he had to submit when his astronomical friends in Germany pressed him to write some of his astronomical works in German.

Gauss studied at the Caroline College for three years, during which he mastered the more important works of Euler, Lagrange and, above all, Newton's *Principia*. The highest praise one great man can get is from another in his own class. Gauss never lowered the estimate which as a boy of seventeen he had formed of Newton. Others—Euler, Laplace, Lagrange, Legendre—appear in the flowing Latin of Gauss with the complimentary *clarissimus*; Newton is *summus*.

While still at the college Gauss had begun those researches in the higher arithmetic which were to make him immortal. His prodigious powers of calculation now came into play. Going directly to the numbers themselves he experimented with them, discovering by induction recondite general theorems whose proofs were to cost even him an effort. In this way he rediscovered "the gem of arithmetic," "*theorema aureum*," which Euler also had come upon inductively, which is known as the law of quadratic reciprocity, and which he was to be the first to prove. (Legendre's attempted proof slurs over a crux.)

The whole investigation originated in a simple question which many beginners in arithmetic ask themselves: How many digits are there in the period of a repeating decimal? To get some light on the problem Gauss calculated the decimal representations of all the fractions $1/n$ for $n = 1$ to 1000. He did not find the treasure he was seeking, but something infinitely greater—the law of quadratic reciprocity. As this is quite simply stated we shall describe it, introducing at the same time one of the revolutionary improvements in arithmetical nomenclature and notation which Gauss invented, that of *congruence*. All numbers in what follows are integers (common whole numbers).

If the *difference* $(a - b$ or $b - a)$ of two numbers a, b is exactly divisible by the number m, we say that a, b are *congruent* with respect to the modulus m, or simply *congruent modulo m*, and we symbolize this by writing $a \equiv b \pmod{m}$. Thus $100 \equiv 2 \pmod{7}$, $35 \equiv 2 \pmod{11}$.

The advantage of this scheme is that it recalls the way we write algebraic equations, traps the somewhat elusive notion of arithmetical divisibility in a compact notation, and suggests that we try to carry over to arithmetic (which is much harder than algebra) some of the manipulations that lead to interesting results in algebra. For example we can "add" equations, and we find that congruences also can be "added," provided the modulus is the same in all, to give other congruences.

Let x denote an unknown number, r and m given numbers, of which r is not divisible by m. Is there a number x such that

$$x^2 \equiv r \pmod{m}?$$

If there is, r is called a *quadratic residue of m*, if not, a *quadratic non-residue of m*.

If r *is* a quadratic residue of m, then it must be possible to find at least one x whose square when divided by m leaves the remainder r; if r is a quadratic non-residue of m, then there is no x whose square when divided by m leaves the remainder r. These are immediate consequences of the preceding definitions.

To illustrate: is 13 a quadratic residue of 17? If so, it must be possible to solve the *congruence*

$$x^2 \equiv 13 \pmod{17}$$

Trying 1, 2, 3, ..., we find that $x = 8, 25, 42, 59, \ldots$ are solutions ($8^2 = 64 = 3 \times 17 + 13$; $25^2 = 625 = 36 \times 17 + 13$; etc.,) so that 13 *is* a quadratic residue of 17. But there is *no* solution of $x^2 \equiv 5 \pmod{17}$, so 5 is a quadratic non-residue of 17.

It is now natural to ask what are the quadratic residues and non-residues of a given m? Namely, given m in $x^2 \equiv r \pmod{m}$, what numbers r can appear and what numbers r cannot appear as x runs through all the numbers 1, 2, 3, ...?

Without much difficulty it can be shown that it is sufficient to answer the question when both r and m are restricted to be primes. So we restate the problem: If p is a *given* prime, what primes q will make the congruence $x^2 \equiv q \pmod{p}$ solvable? This is asking altogether too much in the present state of arithmetic. However, the situation is not utterly hopeless.

There is a beautiful "reciprocity" between the *pair* of congruences

$$x^2 \equiv q \pmod{p}, \; x^2 \equiv p \pmod{q},$$

in which *both* of p, q are *primes*: *both* congruences are *solvable*, or *both* are *unsolvable, unless both* of p, q leave the remainder 3 when divided by 4, in which case *one* of the congruences *is* solvable and *the other* is *not*. This is the law of quadratic reciprocity.

It was not easy to prove. In fact it baffled Euler and Legendre. Gauss gave the first proof at the age of nineteen. As this reciprocity is of fundamental importance in the higher arithmetic and in many advanced parts of algebra, Gauss turned it over and over in his mind for many years, seeking to find its taproot, until in all he had given six distinct proofs, one of which depends upon the straightedge and compass construction of regular polygons.

A numerical illustration will illuminate the statement of the law. First, take $p = 5$, $q = 13$. Since both of 5, 13 leave the remainder 1 on division by 4, *both* of $x^2 \equiv 13 \pmod 5$, $x^2 \equiv 5 \pmod{13}$ must be *solvable*, or *neither* is solvable. The latter is the case for this pair. For $p = 13$, $q = 17$, both of which leave the remainder 1 on division by 4, we get $x^2 \equiv 17 \pmod{13}$, $x^2 \equiv 13 \pmod{17}$, and *both*, or *neither* again must be solvable. The former is the case here: the first congruence has the solutions $x = 2, 15, 28, \ldots$; the second has the solutions $x = 8, 25, 42, \ldots$. There remains to be tested only the case when *both* of p, q leave the remainder 3 on division by 4. Take $p = 11$, $q = 19$. Then, according to the law, *precisely one* of $x^2 \equiv 19 \pmod{11}$, $x^2 \equiv 11 \pmod{19}$ must be solvable. The first congruence has no solution; the second has the solutions $7, 26, 45, \ldots$.

The mere discovery of such a law was a notable achievement. That it was first proved by a boy of nineteen will suggest to anyone who tries to prove it that Gauss was more than merely competent in mathematics.

When Gauss left the Caroline College in October, 1795 at the age of eighteen to enter the University of Göttingen he was still undecided whether to follow mathematics or philology as his life work. He had already invented (when he was eighteen) the method of "least squares," which today is indispensable in geodetic surveying, in the reduction of observations and indeed in all work where the "most probable" value of anything that is measured is to be inferred from a large number of measurements. (The most probable value is furnished by making the sum of the squares of the "residuals"—roughly, divergences from assumed exactness—a minimum.) Gauss shares this honor with Legendre who published the method independently in 1806. This work was the beginning of Gauss' interest in the theory of errors of observation. The Gaussian law of normal distribution of errors and its accompanying bell-shaped curve is familiar today to all who handle statistics, from high-minded intelligence testers to unscrupulous market manipulators.

March 30, 1796, marks the turning point in Gauss' career. On that day, exactly a month before his twentieth year opened, Gauss definitely decided in favor of mathematics. The study of languages was to remain a lifelong hobby, but philology lost Gauss forever on that memorable day in March.

As has already been told in the chapter on Fermat the regular polygon of seventeen sides was the die whose lucky fall induced Gauss to cross his Rubicon. The same day Gauss began to keep his scientific diary (*Notizenjournal*). It is one of the most precious documents in the history of mathematics. The first entry records his great discovery.

The diary came into scientific circulation only in 1898, forty three years after the death of Gauss, when the Royal Society of Göttingen borrowed it from a grandson of Gauss for critical study. It consists of nineteen small octavo pages and contains 146 extremely brief statements of discoveries or results of calculations, the last of which is dated July 9, 1814. A facsimile reproduction was published in 1917 in the tenth volume (part 1) of Gauss' collected works, together with an exhaustive analysis of its contents by several expert editors. Not all of Gauss' discoveries in the prolific period from 1796 to 1814 by any means are noted. But many of those that are jotted down suffice to establish Gauss' priority in fields—elliptic functions, for instance —where some of his contemporaries refused to believe he had preceded them. (Recall that Gauss was born in 1777.)

Things were buried for years or decades in this diary that would have made half a dozen great reputations had they been published promptly. Some were never made public during Gauss' lifetime, and he never claimed in anything he himself printed to have anticipated others when they caught up with him. But the record stands. He did anticipate some who doubted the word of his friends. These anticipations were not mere trivialities. Some of them became major fields of nineteenth century mathematics.

A few of the entries indicate that the diary was a strictly private affair of its author's. Thus for July 10, 1796, there is the entry

$$\text{ΕΥΡΗΚΑ!} \quad \text{num} = \Delta + \Delta + \Delta.$$

Translated, this echoes Archimedes' exultant "Eureka!" and states that every positive integer is the sum of three triangular numbers— such a number is one of the sequence 0,1,3,6,10,15, . . . where each (after 0) is of the form $\frac{1}{2}n(n+1)$, n being any positive integer. Another way of saying the same thing is that every number of the form $8n + 3$ is a sum of three odd squares: $3 = 1^2 + 1^2 + 1^2$; $11 = 1 + 1 + 3^2$; $19 = 1^2 + 3^2 + 3^2$, etc. It is not easy to prove this from scratch.

Less intelligible is the cryptic entry for October 11, 1796, "Vicimus GEGAN." What dragon had Gauss conquered this time? Or what giant had he overcome on April 8, 1799, when he boxes REV. GALEN up in a neat rectangle? Although the meaning of these is lost forever the remaining 144 are for the most part clear enough. One in particular is of the first importance, as we shall see when we come to Abel and Jacobi: the entry for March 19, 1797, shows that Gauss had already discovered the double periodicity of certain elliptic functions. He was then not quite twenty. Again, a later entry shows that Gauss had recognized the double periodicity in the general case. This discovery of itself, had he published it, would have made him famous. But he never published it.

Why did Gauss hold back the great things he discovered? This is easier to explain than his genius—if we accept his own simple statements, which will be reported presently. A more romantic version is the story told by W. W. R. Ball in his well-known history of mathematics. According to this, Gauss submitted his first masterpiece, the *Disquisitiones Arithmeticae*, to the French Academy of Sciences, only to have it rejected with a sneer. The undeserved humiliation hurt Gauss so deeply that he resolved thenceforth to publish only what anyone would admit was above criticism in both matter and form. There is nothing in this defamatory legend. It was disproved once for all in 1935, when the officers of the French Academy ascertained by an exhaustive search of the permanent records that the *Disquisitiones* was never even submitted to the Academy, much less rejected.

Speaking for himself Gauss said that he undertook his scientific works only in response to the deepest promptings of his nature, and it was a wholly secondary consideration to him whether they were ever published for the instruction of others. Another statement which Gauss once made to a friend explains both his diary and his slowness in publication. He declared that such an overwhelming horde of new ideas stormed his mind before he was twenty that he could hardly control them and had time to record but a small fraction. The diary contains only the final brief statements of the outcome of elaborate investigations, some of which occupied him for weeks. Contemplating as a youth the close, unbreakable chains of synthetic proofs in which

Archimedes and Newton had tamed their inspirations, Gauss resolved to follow their great example and leave after him only finished works of art, severely perfect, to which nothing could be added and from which nothing could be taken away without disfiguring the whole. The work itself must stand forth, complete, simple, and convincing, with no trace remaining of the labor by which it had been achieved. A cathedral is not a cathedral, he said, till the last scaffolding is down and out of sight. Working with this ideal before him, Gauss preferred to polish one masterpiece several times rather than to publish the broad outlines of many as he might easily have done. His seal, a tree with but few fruits, bore the motto *Pauca sed matura* (Few, but ripe).

The fruits of this striving after perfection were indeed ripe but not always easily digestible. All traces of the steps by which the goal had been attained having been obliterated, it was not easy for the followers of Gauss to rediscover the road he had travelled. Consequently some of his works had to wait for highly gifted interpreters before mathematicians in general could understand them, see their significance for unsolved problems, and go ahead. His own contemporaries begged him to relax his frigid perfection so that mathematics might advance more rapidly, but Gauss never relaxed. Not till long after his death was it known how much of nineteenth-century mathematics Gauss had foreseen and anticipated before the year 1800. Had he divulged what he knew it is quite possible that mathematics would now be half a century or more ahead of where it is. Abel and Jacobi could have begun where Gauss left off, instead of expending much of their finest effort rediscovering things Gauss knew before they were born, and the creators of non-Euclidean geometry could have turned their genius to other things.

Of himself Gauss said that he was "all mathematician." This does him an injustice unless it is remembered that "mathematician" in his day included also what would now be termed a mathematical physicist. Indeed his second motto*

> *Thou, nature, art my goddess; to thy laws
> My services are bound . . . ,*

truly sums up his life of devotion to mathematics and the physical sciences of his time. The "all mathematician" aspect of him is to be

---

*Shakespeare's *King Lear*, Act I, Scene II, 1-2, with the essential change of "laws" for "law."

understood only in the sense that he did not scatter his magnificent endowment broadcast over all fields where he might have reaped abundantly, as he blamed Leibniz for doing, but cultivated his greatest gift to perfection.

The three years (October, 1795–September, 1798) at the University of Göttingen were the most prolific in Gauss' life. Owing to the generosity of the Duke Ferdinand the young man did not have to worry about finances. He lost himself in his work, making but few friends. One of these, Wolfgang Bolyai, "the rarest spirit I ever knew," as Gauss described him, was to become a friend for life. The course of this friendship and its importance in the history of non-Euclidean geometry is too long to be told here; Wolfgang's son Johann was to retrace practically the same path that Gauss had followed to the creation of a non-Euclidean geometry, in entire ignorance that his father's old friend had anticipated him. The ideas which had overwhelmed Gauss since his seventeenth year were now caught—partly—and reduced to order. Since 1795 he had been meditating a great work on the theory of numbers. This now took definite shape, and by 1798 the *Disquisitiones Arithmeticae* (Arithmetical Researches) was practically completed.

To acquaint himself with what had already been done in the higher arithmetic and to make sure that he gave due credit to his predecessors, Gauss went to the University of Helmstedt, where there was a good mathematical library, in September, 1798. There he found that his fame had preceded him. He was cordially welcomed by the librarian and the professor of mathematics, Johann Friedrich Pfaff (1765–1825), in whose house he roomed. Gauss and Pfaff became warm friends, although the Pfaff family saw but little of their guest. Pfaff evidently thought it his duty to see that his hard-working young friend took some exercise, for he and Gauss strolled together in the evenings, talking mathematics. As Gauss was not only modest but reticent about his own work, Pfaff probably did not learn as much as he might have. Gauss admired the professor tremendously (he was then the best-known mathematician in Germany), not only for his excellent mathematics, but for his simple, open character. All his life there was but one type of man for whom Gauss felt aversion and contempt, the pretender to knowledge who will not admit his mistakes when he knows he is wrong.

Gauss spent the autumn of 1798 (he was then twenty one) in

Brunswick, with occasional trips to Helmstedt, putting the finishing touches to the *Disquisitiones*. He had hoped for early publication, but the book was held up in the press owing to a Leipzig publisher's difficulties till September, 1801. In gratitude for all that Ferdinand had done for him, Gauss dedicated his book to the Duke—*"Serenissimo Principi ac Domino Carolo Guilielmo Ferdinando."*

If ever a generous patron deserved the homage of his protégé, Ferdinand deserved that of Gauss. When the young genius was worried ill about his future after leaving Göttingen—he tried unsuccessfully to get pupils—the Duke came to his rescue, paid for the printing of his doctoral dissertation (University of Helmstedt, 1799), and granted him a modest pension which would enable him to continue his scientific work unhampered by poverty. *"Your* kindness," Gauss says in his dedication, "freed me from all other responsibilities and enabled me to assume this exclusively."

Before describing the *Disquisitiones* we shall glance at the dissertation for which Gauss was awarded his doctor's degree *in absentia* by the University of Helmstedt in 1799: *Demonstratio nova theorematis omnem functionem algebraicam rationalem integram unius variabilis in factores reales primi vel secundi gradus revolvi posse* (A New Proof that Every Rational Integral Function of One Variable Can Be Resolved into Real Factors of the First or Second Degree).

There is only one thing wrong with this landmark in algebra. The first two words in the title would imply that Gauss had merely added a *new* proof to others already known. He should have omitted "nova." His was the *first* proof. (This assertion will be qualified later.) Some before him had published what they supposed were proofs of this theorem—usually called the fundamental theorem of algebra—but none had attained a proof. With his uncompromising demand for logical and mathematical rigor Gauss insisted upon a *proof*, and gave the first. Another, equivalent, statement of the theorem says that every algebraic equation in one unknown has a root, an assertion which beginners often take for granted as being true without having the remotest conception of what it means.

If a lunatic scribbles a jumble of mathematical symbols it does not follow that the writing means anything merely because to the inexpert eye it is indistinguishable from higher mathematics. It is just as doubtful whether the assertion that every algebraic equation has a root means anything until we say *what sort* of a root the equation has. Vaguely, we feel that a *number* will "satisfy" the equation but that half a pound of butter will not.

Gauss made this feeling precise by proving that all the roots of any algebraic equation are "numbers" of the form $a + bi$, where a,b are real numbers (the numbers that correspond to the distances, positive, zero, or negative, measured from a fixed point O on a given straight line, as on the x-axis in Descartes' geometry), and i is the square root of -1. The new sort of "number" $a + bi$ is called *complex*.

Incidentally, Gauss was one of the first to give a coherent account of complex numbers and to interpret them as labelling the points of a plane, as is done today in elementary textbooks on algebra.

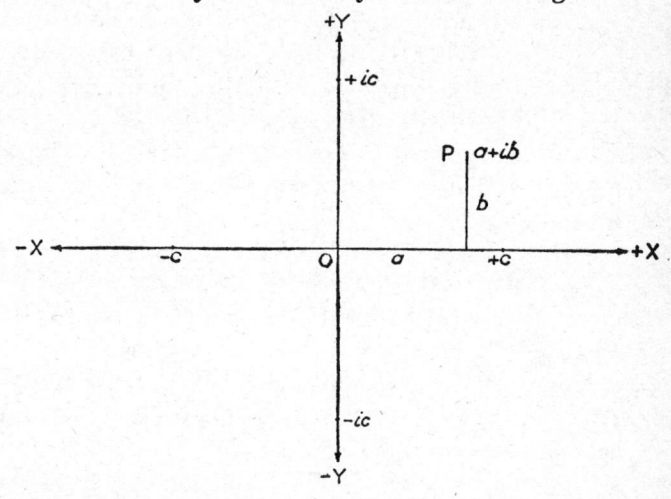

The Cartesian coordinates of P are (a,b); the point P is also labelled $a + bi$. Thus to every point of the plane corresponds precisely one complex number; the numbers corresponding to the points on XOX' are "real," those on YOY' "pure imaginary" (they are all of the type ic, where c is a real number).

The word "imaginary" is the great algebraical calamity, but it is too well established for mathematicians to eradicate. It should never have been used. Books on elementary algebra give a simple interpretation of imaginary numbers in terms of rotations. Thus if we interpret the multiplication $i \times c$, where c is real, as a rotation about O of the segment Oc through one right angle, Oc is rotated onto OY; another multiplication by i, namely $i \times i \times c$, rotates Oc through another right angle, and hence the total effect is to rotate Oc through two right angles, so that $+Oc$ becomes $-Oc$. As an operation, multiplication by $i \times i$ has the same effect as multiplication by -1; multiplication by i has the same effect as a rotation through a right angle, and these interpretations (as we have just seen) are consistent. If we like we may now write $i \times i = -1$, in operations, or $i^2 = -1$; so that the operation of rotation through a right angle is symbolized by $\sqrt{-1}$.

All this of course proves nothing. It is not meant to prove anything. *There is nothing to be proved;* we *assign* to the symbols and operations of algebra *any meanings whatever* that will lead to consistency. Although the *interpretation* by means of rotations *proves* nothing, it may suggest that there is no occasion for anyone to muddle himself into a state of mystic wonderment over nothing about the grossly misnamed "imaginaries." For further details we must refer to almost any schoolbook on elementary algebra.

Gauss thought the theorem that every algebraic equation has a root in the sense just explained so important that he gave four distinct proofs, the last when he was seventy years old. Today some would transfer the theorem from algebra (which restricts itself to processes that can be carried through in a finite number of steps) to analysis. Even Gauss *assumed* that the graph of a polynomial is a continuous curve and that if the polynomial is of odd degree the graph must cross the axis at least once. To any beginner in algebra this is obvious. But today it is *not obvious* without proof, and attempts to prove it again lead to the difficulties connected with continuity and the infinite. The roots of so simple an equation as $x^2 - 2 = 0$ cannot be computed exactly in any finite number of steps. More will be said about this when we come to Kronecker. We proceed now to the *Disquisitiones Arithmeticae*.

The *Disquisitiones* was the first of Gauss' masterpieces and by some considered his greatest. It was his farewell to pure mathematics as an exclusive interest. After its publication in 1801 (Gauss was then twenty four), he broadened his activity to include astronomy, geodesy, and electromagnetism in both their mathematical and practical aspects. But arithmetic was his first love, and he regretted in later life that he had never found the time to write the second volume he had planned as a young man. The book is in seven "sections." There was to have been an eighth, but this was omitted to keep down the cost of printing.

The opening sentence of the preface describes the general scope of the book. "The researches contained in this work appertain to that part of mathematics which is concerned with integral numbers, also fractions, surds [irrationals] being always excluded."

The first three sections treat the theory of congruences and give in particular an exhaustive discussion of the binomial *congruence* $x^n \equiv A$ (mod p), where the given integers n, A are arbitrary and p is prime; the unknown integer is x. This beautiful *arithmetical* theory has many resemblances to the corresponding *algebraic* theory of the binomial *equation* $x^n = A$, but in its peculiarly arithmetical parts is incomparably richer and more difficult than the algebra which offers no analogies to the arithmetic.

In the fourth section Gauss develops the theory of quadratic residues. Here is found the first published *proof* of the law of quadratic reciprocity. The proof is by an amazing application of mathematical induction and is as tough a specimen of that ingenious logic as will be found anywhere.

With the fifth section the theory of *binary quadratic forms* from the arithmetical point of view enters, to be accompanied presently by a discussion of *ternary* quadratic forms which are found to be necessary for the completion of the binary theory. The law of quadratic reciprocity plays a fundamental part in these difficult enterprises. For the first forms named the general problem is to discuss the solution in integers x, y of the indeterminate equation

$$ax^2 + 2bxy + cy^2 = m,$$

where a, b, c, m are any given integers; for the second, the integer solutions x, y, z of

$$ax^2 + 2bxy + cy^2 + 2dxz + 2eyz + fz^2 = m,$$

where a, b, c, d, e, f, m, are any given integers, are the subject of investigation. An easy-looking but hard question in this field is to impose necessary and sufficient restrictions upon a, c, f, m which will ensure the existence of a solution in integers x, y, z of the indeterminate equation

$$ax^2 + cy^2 + fz^2 = m.$$

The sixth section applies the preceding theory to various special cases, for example the integer solutions x, y of $mx^2 + ny^2 = A$, where m, n, A are any given integers.

In the seventh and last section, which many consider the crown of the work, Gauss applies the preceding developments, particularly the theory of binomial congruences, to a wonderful discussion of the algebraic equation $x^n = 1$, where n is any given integer, weaving together arithmetic, algebra, and geometry into one perfect pattern. The equation $x^n = 1$ is the *algebraic* formulation of the geometric problem to construct a regular polygon of n sides, or to divide the circumference of a circle into n equal parts (consult any secondary text book on algebra or trigonometry); the *arithmetical congruence* $x^m \equiv 1 \pmod{p}$, where m, p are given integers, and p is prime, is the thread which runs through the algebra and the geometry and gives the pattern its simple meaning. This flawless work of art is accessible to any student who has had the usual algebra offered in school, but the *Disquisitiones* is not recommended for beginners (Gauss' concise presentation has been reworked by later writers into a more readily assimilable form).

Many parts of all this had been done otherwise before—by Fermat, Euler, Lagrange, Legendre and others; but Gauss treated the whole from his individual point of view, added much of his own, and deduced the isolated results of his predecessors from his general formulations and solutions of the relevant problems. For example, Fermat's beautiful result that every prime of the form $4n + 1$ is a sum of two squares, and is such a sum in only one way, which Fermat proved by his difficult method of "infinite descent," falls out naturally from Gauss' general discussion of binary quadratic forms.

"The *Disquisitiones Arithmeticae* have passed into history," Gauss said in his old age, and he was right. A new direction was given to the higher arithmetic with the publication of the *Disquisitiones*, and the theory of numbers, which in the seventeenth and eighteenth centuries had been a miscellaneous aggregation of disconnected special results, assumed coherence and rose to the dignity of a mathematical science on a par with algebra, analysis, and geometry.

The work itself has been called a "book of seven seals." It is hard reading, even for experts, but the treasures it contains and (partly conceals) in its concise, synthetic demonstrations are now available to all who wish to share them, largely the result of the labors of Gauss' friend and disciple Peter Gustav Lejeune Dirichlet (1805-1859), who first broke the seven seals.

Competent judges recognized the masterpiece for what it was immediately. Legendre* at first may have been inclined to think that Gauss had done him but scant justice. But in the preface to the second edition of his own treatise on the theory of numbers (1808), which in large part was superseded by the *Disquisitiones*, he is enthusiastic. Lagrange also praised unstintedly. Writing to Gauss on May 31, 1804 he says "Your *Disquisitiones* have raised you at once to the rank of the first mathematicians, and I regard the last section as containing the most beautiful analytical discovery that has been made for a long time. . . . Believe, sir, that no one applauds your success more sincerely than I."

Hampered by the classic perfection of its style the *Disquisitiones* was somewhat slow of assimilation, and when finally gifted young men began studying the work deeply they were unable to purchase copies, owing to the bankruptcy of a bookseller. Even Eisenstein, Gauss' favorite disciple, never owned a copy. Dirichlet was more fortunate. His copy accompanied him on all his travels, and he slept with it under his pillow. Before going to bed he would struggle with some tough paragraph in the hope—frequently fulfilled—that he would wake up in the night to find that a re-reading made everything clear. To Dirichlet is due the marvellous theorem, mentioned in connection with Fermat, that every arithmetical progression

$$a, a + b, a + 2b, a + 3b, a + 4b, \ldots,$$

in which a, b are integers with no common divisor greater than 1, contains an infinity of primes. This was proved by analysis, in itself a miracle, for the theorem concerns integers, whereas analysis deals with the *continuous*, the *non-integral*.

Dirichlet did much more in mathematics than his amplification of the *Disquisitiones*, but we shall not have space to discuss his life. Neither shall we have space (unfortunately) for Eisenstein, one of the brilliant young men of the early nineteenth century who died before their time and, what is incomprehensible to most mathematicians, as the man of whom Gauss is reported to have said, "There have been but three epoch-making mathematicians, Archimedes, Newton, and Eisenstein." If Gauss ever did say this (it is impossible to check) it

---

* Adrien-Marie Legendre (1752–1833). Considerations of space preclude an account of his life; much of his best work was absorbed or circumvented by younger mathematicians.

deserves attention merely because he said it, and he was a man who did not speak hastily.

Before leaving this field of Gauss' activities we may ask why he never tackled Fermat's Last Theorem. He gives the answer himself. The Paris Academy in 1816 proposed the proof (or disproof) of the theorem as its prize problem for the period 1816–18. Writing from Bremen on March 7, 1816, Olbers tries to entice Gauss into competing: "It seems right to me, dear Gauss, that you should get busy about this."

But "dear Gauss" resisted the tempter. Replying two weeks later he states his opinion of Fermat's Last Theorem. "I am very much obliged for your news concerning the Paris prize. But I confess that Fermat's Theorem as an isolated proposition has very little interest for me, because I could easily lay down a multitude of such propositions, which one could neither prove nor dispose of."

Gauss goes on to say that the question has induced him to recall some of his old ideas for a great extension of the higher arithmetic. This doubtless refers to the theory of algebraic numbers (described in later chapters) which Kummer, Dedekind, and Kronecker were to develop independently. But the theory Gauss has in mind is one of those things, he declares, where it is impossible to foresee what progress shall be made toward a distant goal that is only dimly seen through the darkness. For success in such a difficult search one's lucky star must be in the ascendency, and Gauss' circumstances are now such that, what with his numerous distracting occupations, he is unable to give himself up to such meditations, as he did "in the fortunate years 1796–1798 when I shaped the main points of the *Disquisitiones Arithmeticae*. Still I am convinced that if I am as lucky as I dare hope, and if I succeed in taking some of the principal steps in that theory, then Fermat's Theorem will appear as only one of the least interesting corollaries."

Probably all mathematicians today regret that Gauss was deflected from his march through the darkness by "a couple of clods of dirt which we call planets"—his own words—which shone out unexpectedly in the night sky and led him astray. Lesser mathematicians than Gauss—Laplace for instance—might have done all that Gauss did in computing the orbits of Ceres and Pallas, even if the problem was of a sort which Newton said belonged to the most difficult in mathematical astronomy. But the brilliant success of Gauss in these matters brought him instant recognition as the first mathematician in Europe and thereby won him a comfortable position where he could work in comparative peace; so perhaps those wretched lumps of dirt were after all his lucky stars.

The second great stage in Gauss' career began on the first day of the nineteenth century, also a red-letter day in the histories of philosophy and astronomy. Since 1781 when Sir William Herschel (1738–1822) discovered the planet Uranus, thus bringing the number of planets then known up to the philosophically satisfying seven, astronomers had been diligently searching the heavens for further members of the Sun's family, whose existence was to be expected, according to Bode's law, between the orbits of Mars and Jupiter. The search was fruitless till Giuseppe Piazzi (1746–1826) of Palermo, on the opening day of the nineteenth century, observed what he at first mistook for a small comet approaching the Sun, but which was presently recognized as a new planet—later named Ceres, the first of the swarm of minor planets known today.

By one of the most ironic verdicts ever delivered in the agelong litigation of fact versus speculation, the discovery of Ceres coincided with the publication by the famous philosopher Georg Wilhelm Friedrich Hegel (1770–1831) of a sarcastic attack on astronomers for presuming to search for an eighth planet. Would they but pay some attention to philosophy, Hegel asserted, they must see immediately that there can be precisely seven planets, no more, no less. Their search therefore was a stupid waste of time. Doubtless this slight lapse on Hegel's part has been satisfactorily explained by his disciples, but they have not yet talked away the hundreds of minor planets which mock his Jovian ban.

It will be of interest here to quote what Gauss thought of philosophers who busy themselves with scientific matters they have not understood. This holds in particular for philosophers who peck at the foundations of mathematics without having first sharpened their dull beaks on some hard mathematics. Conversely, it suggests why Bertrand A. W. Russell (1872–   ), Alfred North Whitehead (1861–   ) and David Hilbert (1862–   ) in our own times have made outstanding contributions to the philosophy of mathematics: these men are mathematicians.

Writing to his friend Schumacher on November 1, 1844, Gauss says

"You see the same sort of thing [mathematical incompetence] in the contemporary philosophers Schelling, Hegel, Nees von Essenbeck, and their followers; don't they make your hair stand on end with their definitions? Read in the history of ancient philosophy what the big men of that day—Plato and others (I except Aristotle)—gave in the way of explanations. But even with Kant himself it is often not much better; in my opinion his distinction between analytic and synthetic propositions is one of those things that either run out in a triviality or are false." When he wrote this (1844) Gauss had long been in full possession of non-Euclidean geometry, itself a sufficient refutation of some of the things Kant said about "space" and geometry, and he may have been unduly scornful.

It must not be inferred from this isolated example concerning purely mathematical technicalities that Gauss had no appreciation of philosophy. He had. All philosophical advances had a great charm for him, although he often disapproved of the means by which they had been attained. "There are problems," he said once, "to whose solution I would attach an infinitely greater importance than to those of mathematics, for example touching ethics, or our relation to God, or concerning our destiny and our future; but their solution lies wholly beyond us and completely outside the province of science."

Ceres was a disaster for mathematics. To understand why she was taken with such devastating seriousness by Gauss we must remember that the colossal figure of Newton—dead for more than seventy years—still overshadowed mathematics in 1801. The "great" mathematicians of the time were those who, like Laplace, toiled to complete the Newtonian edifice of celestial mechanics. Mathematics was still confused with mathematical physics—such as it was then—and mathematical astronomy. The vision of mathematics as an autonomous science which Archimedes saw in the third century before Christ had been lost sight of in the blaze of Newton's splendor, and it was not until the youthful Gauss again caught the vision that mathematics was acknowledged as a science whose first duty is to itself. But that insignificant clod of dirt, the minor planet Ceres, seduced his unparalleled intellect when he was twenty four years of age, just as he was getting well into his stride in those untravelled wildernesses which were to become the empire of modern mathematics.

Ceres was not alone to blame. The magnificent gift for mental arithmetic whose empirical discoveries had given mathematics the

Disquisitiones Arithmeticae also played a fatal part in the tragedy. His friends and his father, too, were impatient with the young Gauss for not finding some lucrative position now that the Duke had educated him and, having no conception of the nature of the work which made the young man a silent recluse, thought him deranged. Here now at the dawn of the new century the opportunity which Gauss had lacked was thrust at him.

A new planet had been discovered in a position which made it extraordinarily difficult of observation. To compute an orbit from the meager data available was a task which might have exercised Laplace himself. Newton had declared that such problems are among the most difficult in mathematical astronomy. The mere arithmetic necessary to establish an orbit with accuracy sufficient to ensure that Ceres on her whirl round the sun should not be lost to telescopes might well deter an electrically-driven calculating machine even today; but to the young man whose inhuman memory enabled him to dispense with a table of logarithms when he was hard pressed or too lazy to reach for one, all this endless arithmetic—*logistica,* not *arithmetica*—was the sport of an infant.

Why not indulge his dear vice, calculate as he had never calculated before, produce the difficult orbit to the sincere delight and wonderment of the dictators of mathematical fashion and thus make it possible, a year hence, for patient astronomers to rediscover Ceres in the place where the Newtonian law of gravitation decreed that she *must* be found—*if* the law were indeed a law of nature? Why not do all this, turn his back on the insubstantial vision of Archimedes and forget his own unsurpassed discoveries which lay waiting for development in his diary? Why not, in short, be popular? The Duke's generosity, always ungrudged, had nevertheless wounded the young man's pride in its most secret place; honor, recognition, acceptance as a "great" mathematician in the fashion of the time with its probable sequel of financial independence—all these were now within his easy reach. Gauss, the mathematical god of all time, stretched forth his hand and plucked the Dead Sea fruits of a cheap fame in his own young generation.

For nearly twenty years the sublime dreams whose fugitive glimpses the boyish Gauss had pictured with unrestrained joy in his diary lay cold and all but forgotten. Ceres was rediscovered, precisely where the marvellously ingenious and detailed calculations of the young Gauss had predicted she must be found. Pallas, Vesta, and Juno, insignificant sister planets of the diminutive Ceres were quickly picked up by prying telescopes defying Hegel, and their orbits, too, were found to conform to the inspired calculations of Gauss. Computations which would have taken Euler three days to perform—one such is sometimes said to have blinded him—were now the simple exercises of a few laborious hours. Gauss had prescribed the *method*, the routine. The major part of his own time for nearly twenty years was devoted to astronomical calculations.

But even such deadening work as this could not sterilize the creative genius of a Gauss. In 1809 he published his second masterpiece, *Theoria motus corporum coelestium in sectionibus conicis solem ambientium* (Theory of the Motion of the Heavenly Bodies Revolving round the Sun in Conic Sections), in which an exhaustive discussion of the determination of planetary and cometary orbits from observational data, including the difficult analysis of perturbations, lays down the law which for many years is to dominate computational and practical astronomy. It was great work, but not as great as Gauss was easily capable of had he developed the hints lying neglected in his diary. No essentially new discovery was added to *mathematics* by the *Theoria motus*.

Recognition came with spectacular promptness after the rediscovery of Ceres. Laplace hailed the young mathematician at once as an equal and presently as a superior. Some time later when the Baron Alexander von Humboldt (1769-1859), the famous traveller and amateur of the sciences, asked Laplace who was the greatest mathematician in Germany, Laplace replied "Pfaff." "But what about Gauss?" the astonished Von Humboldt asked, as he was backing Gauss for the position of director at the Göttingen observatory. "Oh," said Laplace, "Gauss is the greatest mathematician in the world."

The decade following the Ceres episode was rich in both happiness and sorrow for Gauss. He was not without detractors even at that early stage of his career. Eminent men who had the ear of the polite public ridiculed the young man of twenty four for wasting his time on so useless a pastime as the computation of a minor planet's orbit. Ceres might be the goddess of the fields, but it was obvious to the merry wits that no corn grown on the new planet would ever find its way into the Brunswick market of a Saturday afternoon. No doubt they were right, but they also ridiculed him in the same way thirty years later when he laid the foundations of the mathematical theory of electromagnetism and invented the electric telegraph. Gauss let them enjoy their jests. He never replied publicly, but in private expressed his regret that men of honor and priests of science could stultify themselves by being so petty. In the meantime he went on with his work, grateful for the honors the learned societies of Europe showered on him but not going out of his way to invite them.

The Duke of Brunswick increased the young man's pension and made it possible for him to marry (October 9, 1805) at the age of twenty eight. The lady was Johanne Osthof of Brunswick. Writing to his old university friend, Wolfgang Bolyai, three days after he became engaged, Gauss expresses his unbelievable happiness. "Life stands still before me like an eternal spring with new and brilliant colors."

Three children were born of this marriage: Joseph, Minna, and Louis, the first of whom is said to have inherited his father's gift for mental calculations. Johanne died on October 11, 1809, after the birth of Louis, leaving her young husband desolate. His eternal spring turned to winter. Although he married again the following year (August 4, 1810) for the sake of his young children it was long before Gauss could speak without emotion of his first wife. By the second wife, Minna Waldeck, who had been a close friend of the first, he had two sons and a daughter.

According to gossip Gauss did not get on well with his sons, except possibly the gifted Joseph who never gave his father any trouble. Two are said to have run away from home and gone to the United States. As one of these sons is said to have left numerous descendants still living in America, it is impossible to say anything further here, except that one of the American sons became a prosperous merchant in St. Louis in the days of the river boats; both first were farmers in Missouri. With his daughters Gauss was always happy. An exactly contrary legend (vouched for forty years ago by old people whose memories of the Gauss family might be considered trustworthy) to that about the sons asserts that Gauss was never anything but kind to his boys, some of whom were rather wild and caused their distracted father endless anxiety. One would think that the memory of his own father would have made Gauss sympathetic with his sons.

In 1808 Gauss lost his father. Two years previously he had suffered an even severer loss in the death of his benefactor under tragic circumstances.

The Duke Ferdinand was not only an enlightened patron of learning and a kindly ruler but a first-rate soldier as well who had won the warm praise of Frederick the Great for his bravery and military brilliance in the Seven Years' War (1756–1763).

At the age of seventy Ferdinand was put in command of the Prussian forces in a desperate attempt to halt the French under Napoleon, after the Duke's mission to St. Petersburg in an effort to enlist the aid of Russia for Germany had failed. The battle of Austerlitz (December 2, 1805) was already history and Prussia found itself forsaken in the face of overwhelming odds. Ferdinand faced the French on their march toward the Saale at Auerstedt and Jena, was disastrously defeated and himself mortally wounded. He turned homeward.

Napoleon the Great here steps on the stage in person at his potbellied greatest. At the time of Ferdinand's defeat Napoleon was quartered at Halle. A deputation from Brunswick waited on the victorious Emperor of all the French to implore his generosity for the brave old man he had defeated. Would the mighty Emperor stretch a point of military etiquette and let his broken enemy die in peace by his own fireside? The Duke, they assured him, was no longer dangerous. He was dying.

It was the wrong time of the month and Napoleon was enjoying one of his womanish tantrums. He not only refused but did so with quite vulgar and unnecessary brutality. Revealing the true measure of himself as a man, Napoleon pointed his refusal with a superfluous vilification of his honorable opponent and a hysterical ridicule of the dying man's abilities as a soldier. There was nothing for the humiliated deputation to do but to try to save their gentle ruler from the disgrace of a death in prison. It does not seem surprising that these same Germans some nine years later fought like methodical devils at Waterloo and helped to topple the Emperor of the French into the ditch.

Gauss at the time was living in Brunswick. His house was on the main highway. One morning in late autumn he saw a hospital wagon hastening by. In it lay the dying Duke on his flight to Altona. With an emotion too deep for words Gauss saw the man who had been more than his own father to him hurried away to die in hiding like a hounded criminal. He said nothing then and but little afterwards, but his friends noticed that his reserve deepened and his always serious nature became more serious. Like Descartes in his earlier years Gauss had a horror of death, and all his life the passing of a close friend chilled him with a quiet, oppressive dread. Gauss was too vital to die or to witness death. The Duke died in his father's house in Altona on November 10, 1806.

His generous patron dead, it became necessary for Gauss to find some reliable livelihood to support his family. There was no difficulty about this as the young mathematician's fame had now spread to the farthest corners of Europe. St. Petersburg had been angling for him as the logical successor of Euler who had never been worthily replaced after his death in 1783. In 1807 a definite and flattering offer was tendered Gauss. Alexander von Humboldt and other influential friends, reluctant to see Germany lose the greatest mathematician in the world, bestirred themselves, and Gauss was appointed director of the Göttingen Observatory with the privilege—and duty, when necessary—of lecturing on mathematics to university students.

Gauss no doubt might have obtained a professorship of mathematics but he preferred the observatory as it offered better prospects for uninterrupted research. Although it may be too strong to say that Gauss hated teaching, the instruction of ordinary students gave him no pleasure, and it was only when a real mathematician sought him out that Gauss, sitting at a table with his students, let himself go and disclosed the secrets of his methods in his perfectly prepared lessons. But such incentives were regrettably rare and for the most part the students who took up Gauss' priceless time had better have been doing something other than mathematics. Writing in 1810 to his intimate friend the astronomer and mathematician Friedrich Wilhelm Bessel (1784-1846), Gauss says "This winter I am giving two courses of lectures to three students, of whom one is only moderately prepared, the other less than moderately, and the third lacks both preparation and ability. Such are the burdens of a mathematical calling."

The salary which Göttingen could afford to pay Gauss at the time —the French were then busy pillaging Germany in the interests of good government for the Germans by the French—was modest but sufficient for the simple needs of Gauss and his family. Luxury never attracted the Prince of Mathematicians whose life had been unaffectedly dedicated to science long before he was twenty. As his friend Sartorius von Waltershausen writes, "As he was in his youth, so he remained through his old age to his dying day, the unaffectedly simple Gauss. A small study, a little work table with a green cover, a standing-desk painted white, a narrow sopha and, after his seventieth year, an arm chair, a shaded lamp, an unheated bedroom, plain food, a dressing gown and a velvet cap, these were so becomingly all his needs."

If Gauss was simple and thrifty the French invaders of Germany in 1807 were simpler and thriftier. To govern Germany according to their ideas the victors of Auerstedt and Jena fined the losers for more than the traffic would bear. As professor and astronomer at Göttingen Gauss was rated by the extortionists to be good for an involuntary contribution of 2,000 francs to the Napoleonic war chest. This exorbitant sum was quite beyond Gauss' ability to pay.

Presently Gauss got a letter from his astronomical friend Olbers enclosing the amount of the fine and expressing indignation that a scholar should be subjected to such petty extortion. Thanking his generous friend for his sympathy, Gauss declined the money and sent it back at once to the donor.

Not all the French were as thrifty as Napoleon. Shortly after returning Olbers' money Gauss received a friendly little note from Laplace telling him that the famous French mathematician had paid the 2,000-franc fine for the greatest mathematician in the world and had considered it an honor to be able to lift this unmerited burden from his friend's shoulders. As Laplace had paid the fine in Paris, Gauss was unable to return him the money. Nevertheless he declined to accept Laplace's help. An unexpected (and unsolicited) windfall was presently to enable him to repay Laplace with interest at the current market rate. Word must have got about that Gauss disdained charity. The next attempt to help him succeeded. An admirer in Frankfurt sent 1,000 guilders anonymously. As Gauss could not trace the sender he was forced to accept the gift.

The death of his friend Ferdinand, the wretched state of Germany under French looting, financial straits, and the loss of his first wife all did their part toward upsetting Gauss' health and making his life miserable in his early thirties. Nor did a constitutional predisposition to hypochondria, aggravated by incessant overwork, help matters. His unhappiness was never shared with his friends, to whom he is always the serene correspondent, but is confided—only once—to a private mathematical manuscript. After his appointment to the directorship at Göttingen in 1807 Gauss returned occasionally for three years to one of the great things noted in his diary. In a manuscript on elliptic functions purely scientific matters are suddenly interrupted by the finely pencilled words "Death were dearer to me than such a life." His work became his drug.

The years 1811–12 (Gauss was thirty four in 1811) were brighter. With a wife again to care for his young children Gauss began to have some peace. Then, almost exactly a year after his second marriage, the great comet of 1811, first observed by Gauss deep in the evening twilight of August 22, blazed up unannounced. Here was a worthy foe to test the weapons Gauss had invented to subjugate the minor planets.

His weapons proved adequate. While the superstitious peoples of Europe, following the blazing spectacle with awestruck eyes as the comet unlimbered its flaming scimitar in its approach to the Sun, saw in the fiery blade a sharp warning from Heaven that the King of Kings was wroth with Napoleon and weary of the ruthless tyrant, Gauss had the satisfaction of seeing the comet follow the path he had quickly calculated for it to the last decimal. The following year the credulous also saw their own prediction verified in the burning of Moscow and the destruction of Napoleon's Grand Army on the icy plains of Russia.

This is one of those rare instances where the popular explanation fits the facts and leads to more important consequences than the scientific. Napoleon himself had a basely credulous mind—he relied on "hunches," reconciled his wholesale slaughters with a childlike faith in a beneficent, inscrutable Providence, and believed himself a Man of Destiny. It is not impossible that the celestial spectacle of a harmless comet flaunting its gorgeous tail across the sky left its impress on the subconscious mind of a man like Napoleon and fuddled his judgment. The almost superstitious reverence of such a man for mathematics and mathematicians is no great credit to either, although it has been frequently cited as one of the main justifications for both.

Beyond a rather crass appreciation of the value of mathematics in military affairs, where its utility is obvious even to a blind idiot, Napoleon had no conception of what mathematics as practised by masters like his contemporaries, Lagrange, Laplace, and Gauss, is all about. A quick student of trivial, elementary mathematics at school, Napoleon turned to other things too early to certify his promise and, mathematically, never grew up. Although it seems incredible that a man of Napoleon's demonstrated capacity could so grossly underestimate the difficulties of matters beyond his comprehension as to patronize Laplace, it is a fact that he had the ludicrous audacity to assure the author of the *Mécanique céleste* that he would read the book the *first free month* he could find. Newton and Gauss might have been equal to the task; Napoleon no doubt could have turned the pages in his month without greatly tiring himself.

It is a satisfaction to record that Gauss was too proud to prostitute mathematics to Napoleon the Great by appealing to the Emperor's vanity and begging him in the name of his notorious respect for all things mathematical to remit the 2,000-franc fine, as some of Gauss' mistaken friends urged him to do. Napoleon would probably have been flattered to exercise his clemency. But Gauss could not forget Ferdinand's death, and he felt that both he and the mathematics he worshipped were better off without the condescension of a Napoleon.

No sharper contrast between the mathematician and the military genius can be found than that afforded by their respective attitudes to a broken enemy. We have seen how Napoleon treated Ferdinand. When Napoleon fell Gauss did not exult. Calmly and with a detached interest he read everything he could find about Napoleon's life and did his best to understand the workings of a mind like Napoleon's. The effort even gave him considerable amusement. Gauss had a keen sense of humor, and the blunt realism which he had inherited from his hardworking peasant ancestors also made it easy for him to smile at heroics.

The year 1811 might have been a landmark in mathematics comparable to 1801—the year in which the *Disquisitiones Arithmeticae* appeared—had Gauss made public a discovery he confided to Bessel. Having thoroughly understood complex numbers and their geometrical representation as points on the plane of analytic geometry, Gauss proposed himself the problem of investigating what are today called *analytic functions* of such numbers.

The complex number $x + iy$, where i denotes $\sqrt{-1}$, represents the point (x,y). For brevity $x + iy$ will be denoted by the single letter z. As x,y independently take on real values in any prescribed continuous manner, the point z wanders about over the plane, obviously not at random but in a manner determined by that in which x, y assume their values. Any expression containing z, such as z^2, or $1/z$, etc., which takes on a *single* definite value when a value is assigned to z, is called a *uniform function* of z. We shall denote such a function by $f(z)$. Thus if $f(z)$ is the particular function z^2, so that here $f(z) = (x + iy)^2 = x^2 + 2ixy + i^2y^2, = x^2 - y^2 + 2ixy$ (because $i^2 = -1$), it is clear that when any value is assigned to z, namely to $x + iy$, for example $x = 2$, $y = 3$, so that $z = 2 + 3i$, precisely one value of this $f(z)$ is thereby determined; here, for $z = 2 + 3i$ we get $z^2 = -5 + 12i$.

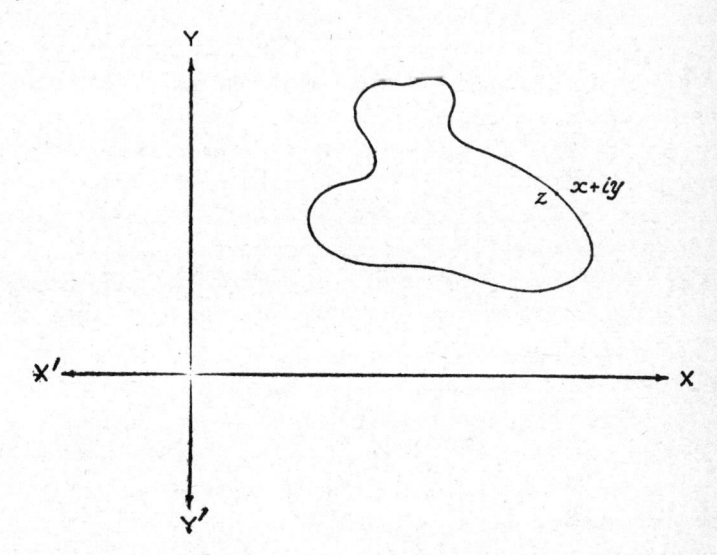

Not all uniform functions $f(z)$ are studied in the theory of functions of a complex variable; the *monogenic* functions are singled out for exhaustive discussion. The reason for this will be stated after we have described what "monogenic" means.

Let z move to another position, say to z'. The function $f(z)$ takes on another value, $f(z')$, obtained by substituting z' for z. The *difference* $f(z') - f(z)$ of the new and old values of the function is now divided by the difference of the new and old values of the variable, thus $[f(z') - f(z)]/(z' - z)$, and, precisely as is done in calculating the slope of a graph to find the derivative of the function the graph represents, we here let z' approach z indefinitely, so that $f(z')$ approaches $f(z)$ simultaneously. But here a remarkable new phenomenon appears.

There is not here a unique way in which z' can move into coincidence with z, for z' may wander about all over the plane of complex numbers by any of an infinity of different paths before coming into coincidence with z. We should not expect the limiting value of $[f(z') - f(z)]/(z' - z)$ when z' coincides with z to be *the same* for *all* of these paths, and in general it is *not*. But *if* $f(z)$ is such that the limiting value just described *is* the same for *all* paths by which z' moves into coincidence with z, then $f(z)$ is said to be monogenic at z (or at the point representing z). *Uniformity* (previously described) and *monogenicity* are distinguishing features of *analytic* functions of a complex variable.

Some idea of the importance of analytic functions can be inferred from the fact that vast tracts of the theories of fluid motion (also of mathematical electricity and representation by maps which do not distort angles) are naturally handled by the theory of *analytic* functions of a complex variable. Suppose such a function $f(z)$ is separated into its "real" part (that which does not contain the "imaginary unit" i) and its "imaginary" part, say $f(z) = U + iV$. For the special analytic function z^2 we have $U = x^2 - y^2$, $V = 2xy$. Imagine a film of fluid streaming over a plane. If the motion of the fluid is without vortices, a stream line of the motion is obtainable from *some* analytic function $f(z)$ by plotting the curve $U = a$, in which a is any real number, and likewise the equipotential lines are obtainable from $V = b$ (b any real number). Letting a, b range, we thus get a complete picture of the motion for as large an area as we wish. For a given situation, say that of a fluid streaming round an obstacle, the hard part of the problem is to find what analytic function to choose, and the whole matter has been gone at largely backwards: the simple analytic functions have been investigated and the physical problems which they fit have been sought. Curiously enough, many of these artificially prepared problems have proven of the greatest service in aerodynamics and other practical applications of the theory of fluid motion.

The theory of analytic functions of a complex variable was one of the greatest fields of mathematical triumphs in the nineteenth century. Gauss in his letter to Bessel states what amounts to the fundamental theorem in this vast theory, but he hid it away to be rediscovered by Cauchy and later Weierstrass. As this is a landmark in the history of mathematical analysis we shall briefly describe it, omitting all refinements that would be demanded in an exact formulation.

Imagine the complex variable z tracing out a closed curve of finite length without loops or kinks. We have an intuitive notion of what we mean by the "length" of a piece of this curve.

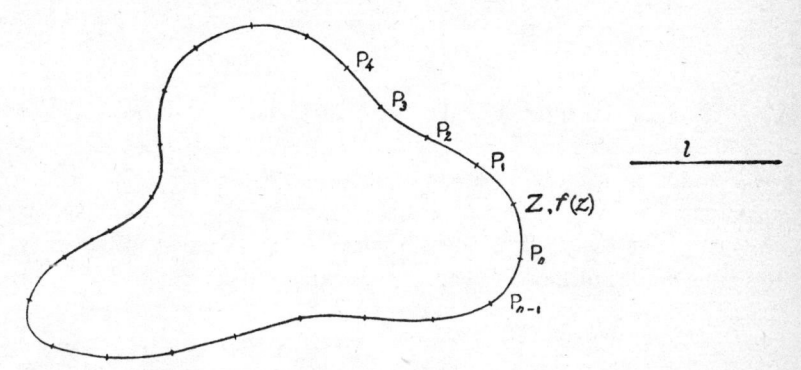

Mark n points $P_1, P_2, \ldots, P_n$ on the curve so that each of the pieces $P_1P_2, P_2P_3, P_3P_4, \ldots, P_nP_1$ is not greater than some preassigned finite length l. On each of these pieces choose a point, not at either end of the piece; form the value of $f(z)$ for the value of z corresponding to the point, and multiply this value by the length of the piece in which the point lies. Do the like for *all* the pieces, and add the results. Finally take the limiting value of this sum as the number of pieces is indefinitely increased. This gives the *"line integral"* of $f(z)$ for the curve.

When will this line integral be zero? In order that the line integral shall be zero it is sufficient that $f(z)$ be *analytic* (uniform and monogenic) at every point z on the curve and inside the curve.

Such is the great theorem which Gauss communicated to Bessel in 1811 and which, with another theorem of a similar kind, in the hands of Cauchy who rediscovered it independently, was to yield many of the important results of analysis as corollaries.

Astronomy did not absorb the whole of Gauss' prodigious energies in his middle thirties. The year 1812, which saw Napoleon's Grand Army fighting a desperate rear-guard action across the frozen plains, witnessed the publication of another great work by Gauss, that on the *hypergeometric series*

$$1 + \frac{ab}{c}x + \frac{a(a+1)b(b+1)x^2}{c(c+1)1 \times 2} + \cdots,$$

the dots meaning that the series continues indefinitely according to the law indicated; the next term is

$$\frac{a(a+1)(a+2)b(b+1)(b+2)}{c(c+1)(c+2)} \frac{x^3}{1 \times 2 \times 3}.$$

This memoir is another landmark. As has already been noted Gauss was the first of the modern rigorists. In this work he determined the restrictions that must be imposed on the numbers a, b, c, x in order that the series shall converge (in the sense explained earlier in this chapter). The series itself was no mere textbook exercise that may be investigated to gain skill in analytical manipulations and then be forgotten. It includes as special cases—obtained by assigning specific values to one or more of a, b, c, x—many of the most important series in analysis, for example those by which logarithms, the trigonometric functions, and several of the functions that turn up repeatedly in Newtonian astronomy and mathematical physics are calculated and tabulated; the general binomial theorem also is a special case. By disposing of this series in its general form Gauss slew a multitude at one smash. From this work developed many applications to the differential equations of physics in the nineteenth century.

The choice of such an investigation for a serious effort is characteristic of Gauss. He never published trivialities. When he put out anything it was not only finished in itself but was also so crammed with ideas that his successors were enabled to apply what Gauss had invented to new problems. Although limitations of space forbid discussion of the many instances of this fundamental character of Gauss' contributions to pure mathematics, one cannot be passed over in even the briefest sketch: the work on the law of biquadratic reciprocity. The importance of this was that it gave a new and totally unforeseen direction to the higher arithmetic.

Having disposed of *quadratic* (second degree) reciprocity, it was natural for Gauss to consider the general question of binomial congruences of any degree. If m is a given integer not divisible by the prime p, and if n is a given positive integer, and if further an integer x can be found such that $x^n \equiv m \pmod{p}$, m is called an *n-ic residue* of p; when $n = 4$, m is a *biquadratic residue* of p.

The case of *quadratic* binomial congruences ($n = 2$) suggests but little to do when n exceeds 2. One of the matters Gauss was to have included in the discarded eighth section (or possibly, as he told Sophie Germain, in the projected but unachieved second volume) of the *Disquisitiones Arithmeticae* was a discussion of these higher congruences and a search for the corresponding laws of reciprocity, namely the interconnections (as to solvability or non-solvability) of the pair $x^n \equiv p \pmod{q}$, $x^n \equiv q \pmod{p}$, where p, q are rational primes. In particular the cases $n = 3$, $n = 4$ were to have been investigated.

The memoir of 1825 breaks new ground with all the boldness of the great pioneers. After many false starts which led to intolerable complexity Gauss discovered the "natural" way to the heart of his problem. The *rational* integers 1, 2, 3, . . . are *not* those appropriate to the statement of the law of *biquadratic* reciprocity, as they are for *quadratic*; a totally new species of *integers* must be invented. These are called the *Gaussian complex integers* and are all those complex numbers of the form $a + bi$ in which a, b are *rational integers* and i denotes $\sqrt{-1}$.

To state the law of biquadratic reciprocity an exhaustive preliminary discussion of the laws of arithmetical divisibility for such *complex integers* is necessary. Gauss gave this, thereby inaugurating the theory of algebraic numbers—that which he probably had in mind when he gave his estimate of Fermat's Last Theorem. For *cubic* reciprocity ($n = 3$) he also found the right way in a similar manner. His work on this was found in his posthumous papers.

The significance of this great advance will become clearer when we follow the careers of Kummer and Dedekind. For the moment it is sufficient to say that Gauss' favorite disciple, Eisenstein, disposed of cubic reciprocity. He further discovered an astonishing connection between the law of biquadratic reciprocity and certain parts of the theory of elliptic functions, in which Gauss had travelled far but had refrained disclosing what he found.

Gaussian complex *integers* are of course a subclass of *all* complex *numbers*, and it might be thought that the *algebraic* theory of *all* the numbers would yield the *arithmetical* theory of the included *integers* as a trivial detail. Such is by no means the case. Compared to the arithmetical theory the algebraic is childishly easy. Perhaps a reason why this should be so is suggested by the *rational numbers* (numbers of the form a/b, where a, b are rational integers). We can *always* divide one rational number by another and get *another* rational number: a/b divided by c/d yields the rational number ad/bc. But a rational *integer* divided by another rational integer is not always another rational integer: 7 divided by 8 gives ⅞. Hence if we must restrict ourselves to *integers*, the case of interest for the theory of numbers, we have tied our hands and hobbled our feet before we start. This is one of the reasons why the higher arithmetic is harder than algebra, higher or elementary.

Equally significant advances in geometry and the applications of mathematics to geodesy, the Newtonian theory of attraction, and electromagnetism were also to be made by Gauss. How was it possible for one man to accomplish this colossal mass of work of the highest order? With characteristic modesty Gauss declared that "If others would but reflect on mathematical truths as deeply and as continuously as I have, they would make my discoveries." Possibly. Gauss' explanation recalls Newton's. Asked how he had made discoveries in astronomy surpassing those of all his predecessors, Newton replied, "By always thinking about them." This may have been plain to Newton; it is not to ordinary mortals.

Part of the riddle of Gauss is answered by his *involuntary* preoccupation with mathematical ideas—which itself of course demands explanation. As a young man Gauss would be "seized" by mathematics. Conversing with friends he would suddenly go silent, overwhelmed by thoughts beyond his control, and stand staring rigidly oblivious of his surroundings. Later he controlled his thoughts—or they lost their control over him—and he consciously directed all his energies to the solution of a difficulty till he succeeded. A problem once grasped was never released till he had conquered it, although several might be in the foreground of his attention simultaneously.

In one such instance (referring to the *Disquisitiones*, page 636) he relates how for four years scarcely a week passed that he did not spend some time trying to settle whether a certain sign should be plus or minus. The solution finally came of itself in a flash. But to imagine that it would have blazed out of itself like a new star without the "wasted" hours is to miss the point entirely. Often after spending days or weeks fruitlessly over some research Gauss would find on resuming work after a sleepless night that the obscurity had vanished and the whole solution shone clear in his mind. The capacity for intense and prolonged concentration was part of his secret.

In this ability to forget himself in the world of his own thoughts

Gauss resembles both Archimedes and Newton. In two further respects he also measures up to them, his gifts for precise observation and a scientific inventiveness which enabled him to devise the instruments necessary for his scientific researches. To Gauss geodesy owes the invention of the heliotrope, an ingenious device by which signals could be transmitted practically instantaneously by means of reflected light. For its time the heliotrope was a long step forward. The astronomical instruments he used also received notable improvements at Gauss' hands. For use in his fundamental researches in electromagnetism Gauss invented the bifilar magnetometer. And as a final example of his mechanical ingenuity it may be recalled that Gauss in 1833 invented the electric telegraph and that he and his fellow worker Wilhelm Weber (1804–1891) used it as a matter of course in sending messages. The combination of mathematical genius with first-rate experimental ability is one of the rarest in all science.

Gauss himself cared but little for the possible practical uses of his inventions. Like Archimedes he preferred mathematics to all the kingdoms of the earth; others might gather the tangible fruits of his labors. But Weber, his collaborator in electromagnetic researches, saw clearly what the puny little telegraph of Göttingen meant for civilization. The railway, we recall, was just coming into its own in the early 1830's. "When the globe is covered with a net of railroads and telegraph wires," Weber prophesied in 1835, "this net will render services comparable to those of the nervous system in the human body, partly as a means of transport, partly as a means for the propagation of ideas and sensations with the speed of lightning."

The admiration of Gauss for Newton has already been noted. Knowing the tremendous efforts some of his own masterpieces had cost him, Gauss had a true appreciation of the long preparation and incessant meditation that went into Newton's greatest work. The story of Newton and the falling apple roused Gauss' indignation. "Silly!" he exclaimed. "Believe the story if you like, but the truth of the matter is this. A stupid, officious man asked Newton how he discovered the law of gravitation. Seeing that he had to deal with a child in intellect, and wanting to get rid of the bore, Newton answered that an apple fell and hit him on the nose. The man went away fully satisfied and completely enlightened."

The apple story has its echo in our own times. When teased as to what led him to his theory of the gravitational field Einstein replied that he asked a workman who had fallen off a building, to land unhurt on a pile of straw, whether he noticed the tug of the "force" of gravity on the way down. On being told that no force had tugged, Einstein immediately saw that "gravitation" in a sufficiently small region of space-time can be replaced by an acceleration of the observer's (the falling workman's) reference system. This story, if true, is also probably all rot. What gave Einstein his idea was the hard labor he expended for several years mastering the tensor calculus of two Italian mathematicians, Ricci and Levi-Civita, themselves disciples of Riemann and Christoffel, both of whom in their turn had been inspired by the geometrical work of Gauss.

Commenting on Archimedes, for whom he also had a boundless admiration, Gauss remarked that he could not understand how Archimedes failed to invent the decimal system of numeration or its equivalent (with some base other than 10). The thoroughly un-Greek work of Archimedes in devising a scheme for writing and dealing with numbers far beyond the capacity of the Greek symbolism had—according to Gauss—put the decimal notation with its all-important principle of place-value ($325 = 3 \times 10^2 + 2 \times 10 + 5$) in Archimedes' hands. This oversight Gauss regarded as the greatest calamity in the history of science. "To what heights would science now be raised if Archimedes had made that discovery!" he exclaimed, thinking of his own masses of arithmetical and astronomical calculations which would have been impossible, even to him, without the decimal notation. Having a full appreciation of the significance for all science of improved methods of computation, Gauss slaved over his own calculations till pages of figures were reduced to a few lines which could be taken in almost at a glance. He himself did much of his calculating mentally; the improvements were intended for those less gifted than himself.

Unlike Newton in his later years, Gauss was never attracted by the rewards of public office, although his keen interest and sagacity in all matters pertaining to the sciences of statistics, insurance, and "political arithmetic" would have made him a good minister of finance. Till his last illness he found complete satisfaction in his science and his simple recreations. Wide reading in the literatures of Europe and the classics of antiquity, a critical interest in world politics, and the mastery of foreign languages and new sciences (including botany and mineralogy) were his hobbies.

English literature especially attracted him, although its darker aspect as in Shakespeare's tragedies was too much for the great mathematician's acute sensitiveness to all forms of suffering, and he tried to pick his way through the happier masterpieces. The novels of Sir Walter Scott (who was a contemporary of Gauss) were read eagerly as they came out, but the unhappy ending of *Kenilworth* made Gauss wretched for days and he regretted having read the story. One slip of Sir Walter's tickled the mathematical astronomer into delighted laughter, "the moon rises broad in the northwest," and he went about for days correcting all the copies he could find. Historical works in English, particularly Gibbon's *Decline and Fall of the Roman Empire* and Macaulay's *History of England* gave him special pleasure.

For his meteoric young contemporary Lord Byron, Gauss had almost an aversion. Byron's posturing, his reiterated world-weariness, his affected misanthropy, and his romantic good looks had captivated the sentimental Germans even more completely than they did the stolid British who—at least the older males—thought Byron somewhat of a silly ass. Gauss saw through Byron's histrionics and disliked him. No man who guzzled good brandy and pretty women as assiduously as Byron did could be so very weary of the world as the naughty young poet with the flashing eye and the shaking hand pretended to be.

In the literature of his own country Gauss' tastes were somewhat unusual for an intellectual German. Jean Paul was his favorite German poet; Goethe and Schiller, whose lives partly overlapped his own, he did not esteem very highly. Goethe, he said, was unsatisfying. Being completely at variance with Schiller's philosophical tenets, Gauss disliked his poetry. He called *Resignation* a blasphemous, corrupt poem and wrote "Mephistopheles!" on the margin of his copy.

The facility with which he mastered languages in his youth stayed with Gauss all his life. Languages were rather more to him than a hobby. To test the plasticity of his mind as he grew older he would deliberately acquire a new language. The exercise, he believed, helped to keep his mind young. At the age of sixty two he began an intensive study of Russian without assistance from anyone. Within two years he was reading Russian prose and poetical works fluently, and carrying on his correspondence with scientific friends in St. Petersburg wholly in Russian. In the opinion of Russians who visited him in Göttingen he also spoke the language perfectly. Russian literature he put on a par with English for the pleasure it gave him. He also tried Sanskrit but disliked it.

His third hobby, world politics, absorbed an hour or so of his time every day. Visiting the literary museum regularly, he kept abreast of events by reading all the newspapers to which the museum subscribed, from the London *Times* to the Göttingen local news.

In politics the intellectual aristocrat Gauss was conservative through and through, but in no sense reactionary. His times were turbulent, both in his own country and abroad. Mob rule and acts of political violence roused in him—as his friend Von Waltershausen reports—"an indescribable horror." The Paris revolt of 1848 filled him with dismay.

The son of poor parents himself, familiar from infancy with the intelligence and morality of "the masses," Gauss remembered what he had observed, and his opinion of the intelligence, morality, and political acumen of "the people"—taken in the mass, as demagogues find and take them—was extremely low. "*Mundus vult decepi*" he believed a true saying.

This disbelief in the innate morality, integrity, and intelligence of Rousseau's "natural man" when massed into a mob or when deliberating in cabinets, parliaments, congresses, and senates, was no doubt partly inspired by Gauss' intimate knowledge, as a man of science, of what "the natural man" did to the scientists of France in the early days of the French Revolution. It may be true, as the revolutionists declared, that "the people have no need of science," but such a declaration to a man of Gauss' temperament was a challenge. Accepting the challenge, Gauss in his turn expressed his acid contempt for all "leaders of the people" who lead the people into turmoil for their own profit. As he aged he saw peace and simple contentment as the only good things in any country. Should civil war break out in Germany, he said, he would as soon be dead. Foreign conquest in the grand Napoleonic manner he looked upon as an incomprehensible madness.

These conservative sentiments were not the nostalgia of a reactionary who bids the world defy the laws of celestial mechanics and stand still in the heavens of a dead and unchanging past. Gauss believed in reforms—when they were intelligent. And if brains are not to judge when reforms are intelligent and when they are not, what organ of the human body is? Gauss had brains enough to see where the ambitions of some of the great statesmen of his own reforming generation were taking Europe. The spectacle did not inspire his confidence.

His more progressive friends ascribed Gauss' conservatism to the closeness with which he stuck to his work. This may have had something to do with it. For the last twenty seven years of his life Gauss slept away from his observatory only once, when he attended a scientific meeting in Berlin to please Alexander von Humboldt who wished to show him off. But a man does not always have to be flying about all over the map to see what is going on. Brains and the ability to read newspapers (even when they lie) and government reports (especially when they lie) are sometimes better than any amount of sightseeing and hotel lobby gossip. Gauss stayed at home, read, disbelieved most of what he read, thought, and arrived at the truth.

Another source of Gauss' strength was his scientific serenity and his freedom from personal ambition. All his ambition was for the advancement of mathematics. When rivals doubted his assertion that he had anticipated them—not stated boastfully, but as a fact germane to the matter in hand—Gauss did not exhibit his diary to prove his priority but let his statement stand on its own merits.

Legendre was the most outspoken of these doubters. One experience made him Gauss' enemy for life. In the *Theoria motus* Gauss had referred to his early discovery of the method of least squares. Legendre published the method in 1806, before Gauss. With great indignation he wrote to Gauss practically accusing him of dishonesty and complaining that Gauss, so rich in discoveries, might have had the decency not to appropriate the method of least squares, which Legendre regarded as his own ewe lamb. Laplace entered the quarrel. Whether he believed the assurances of Gauss that Legendre had indeed been anticipated by ten years or more, he does not say, but he retains his usual suavity. Gauss apparently disdained to argue the matter further. But in a letter to a friend he indicates the evidence which might have ended the dispute then and there had Gauss not been "too proud to fight." "I communicated the whole matter to Olbers in 1802," he says, and if Legendre had been inclined to doubt this he could have asked Olbers, who had the manuscript.

The dispute was most unfortunate for the subsequent development of mathematics, as Legendre passed on his unjustified suspicions to Jacobi and so prevented that dazzling young developer of the theory of elliptic functions from coming to cordial terms with Gauss. The misunderstanding was all the more regrettable because Legendre was a man of the highest character and scrupulously fair himself. It was his fate to be surpassed by more imaginative mathematicians than himself in the fields where most of his long and laborious life was spent in toil which younger men—Gauss, Abel, and Jacobi—showed to have been superfluous. At every step Gauss strode far ahead of Legendre. Yet when Legendre accused him of unfair dealing Gauss felt that he himself had been left in the lurch. Writing to Schumacher (July 30, 1806), he complains that "It seems to be my fate to concur in nearly all my theoretical works with Legendre. So it is in the higher arithmetic, in the researches in transcendental functions connected with the rectification [the process for finding the length of an arc of a curve] of the ellipse, in the foundations of geometry and now again here [in the method of least squares, which] . . . is also used in Legendre's work and indeed right gallantly carried through."

With the detailed publication of Gauss' posthumous papers and much of his correspondence in recent years all these old disputes have been settled once for all in favor of Gauss. There remains another score on which he has been criticized, his lack of cordiality in welcoming the great work of others, particularly of younger men. When Cauchy began publishing his brilliant discoveries in the theory of functions of a complex variable, Gauss ignored them. No word of praise or encouragement came from the Prince of Mathematicians to the young Frenchman. Well, why should it have come? Gauss himself (as we have seen) had reached the heart of the matter years before Cauchy started. A memoir on the theory was to have been one of Gauss' masterpieces. Again, when Hamilton's work on quaternions (to be considered in a later chapter) came to his attention in 1852, three years before his death, Gauss said nothing. Why should he have said anything? The crux of the matter lay buried in his notes of more than thirty years before. He held his peace and made no claim for priority. As in his anticipations of the theory of functions of a complex variable, elliptic functions, and non-Euclidean geometry, Gauss was content to have done the work.

The gist of quaternions is the algebra which does for rotations in space of three dimensions what the algebra of complex numbers does for rotations in a plane. But in quaternions (Gauss called them mutations) one of the fundamental rules of algebra breaks down: it is no longer true that $a \times b = b \times a$, and it is impossible to make an algebra of rotations in three dimensions in which this rule *is* preserved. Hamilton, one of the great mathematical geniuses of the nineteenth century, records with Irish exuberance how he struggled for fifteen years to invent a consistent algebra to do what was required until a happy inspiration gave him the clue that $a \times b$ is not equal to $b \times a$ in the algebra he was seeking. Gauss does not state how long it took him to reach the goal; he merely records his success in a few pages of algebra that leave no mathematics to the imagination.

If Gauss was somewhat cool in his printed expressions of appreciation he was cordial enough in his correspondence and in his scientific relations with those who sought him out in a spirit of disinterested inquiry. One of his scientific friendships is of more than mathematical interest as it shows the liberality of Gauss' views regarding women scientific workers. His broadmindedness in this respect would have been remarkable for any man of his generation; for a German it was almost without precedent.

The lady in question was Mademoiselle Sophie Germain (1776–1831)—just a year older than Gauss. She and Gauss never met, and she died (in Paris) before the University of Göttingen could confer the honorary doctor's degree which Gauss recommended to the faculty. By a curious coincidence we shall see the most celebrated woman mathematician of the nineteenth century, another Sophie, getting her degree from the same liberal University many years later after Berlin had refused her on account of her sex. Sophie appears to be a lucky name in mathematics for women—provided they affiliate with broadminded teachers. The leading woman mathematician of our own times, Emmy Noether (1882–1935) also came from Göttingen.*

Sophie Germain's scientific interests embraced acoustics, the mathematical theory of elasticity, and the higher arithmetic, in all of which she did notable work. One contribution in particular to the study of Fermat's Last Theorem led in 1908 to a considerable advance in this direction by the American mathematician Leonard Eugene Dickson (1874–   ).

---

*"Came from" is right. When the sagacious Nazis expelled Fräulein Noether from Germany because she was a Jewess, Bryn Mawr College, Pennsylvania, took her in. She was the most creative abstract algebraist in the world. In less than a week of the new German enlightenment, Göttingen lost the liberality which Gauss cherished and which he strove all his life to maintain.

Entranced by the *Disquisitiones Arithmeticae*, Sophie wrote to Gauss some of her own arithmetical observations. Fearing that Gauss might be prejudiced against a woman mathematician, she assumed a man's name. Gauss formed a high opinion of the talented correspondent whom he addressed in excellent French as "Mr. Leblanc."

Leblanc dropped her—or his—disguise when she was forced to divulge her true name to Gauss on the occasion of her having done him a good turn with the French infesting Hanover. Writing on April 30, 1807, Gauss thanks his correspondent for her intervention on his behalf with the French General Pernety and deplores the war. Continuing, he pays her a high compliment and expresses something of his own love for the theory of numbers. As the latter is particularly of interest we shall quote from this letter which shows Gauss in one of his cordially human moods.

"But how describe to you my admiration and astonishment at seeing my esteemed correspondent Mr. Leblanc metamorphose himself into this illustrious personage [Sophie Germain] who gives such a brilliant example of what I would find it difficult to believe. A taste for the abstract sciences in general and above all the mysteries of numbers is excessively rare: one is not astonished at it; the enchanting charms of this sublime science reveal themselves only to those who have the courage to go deeply into it. But when a person of the sex which, according to our customs and prejudices, must encounter infinitely more difficulties than men to familiarize herself with these thorny researches, succeeds nevertheless in surmounting these obstacles and penetrating the most obscure parts of them, then without doubt she must have the noblest courage, quite extraordinary talents and a superior genius. Indeed nothing could prove to me in so flattering and less equivocal manner that the attractions of this science, which has enriched my life with so many joys, are not chimerical, as the predilection with which *you* have honored it." He then goes on to discuss mathematics with her. A delightful touch is the date at the end of the letter: "Bronsvic ce 30 Avril 1807 jour de ma naissance—Brunswick, this 30th of April, 1807, my birthday."

That Gauss was not merely being polite to a young woman admirer is shown by a letter of July 21, 1807 to his friend Olbers. " . . . Lagrange is warmly interested in astronomy and the higher arithmetic; the two test-theorems (for what primes 2 is a cubic or a biquadratic residue), which I also communicated to him some time ago, he considers 'among the most beautiful things and the most difficult to prove.' But Sophie Germain has sent me the proofs of these; I have not yet been able to go through them, but I believe they are good; at least she had attacked the matter from the right side, only somewhat more diffusely than would be necessary.... " The theorems to which Gauss refers are those stating for what odd primes p each of the congruences $x^3 \equiv 2 \pmod{p}$, $x^4 \equiv 2 \pmod{p}$ is solvable.

It would take a long book (possibly a longer one than would be required for Newton) to describe all the outstanding contributions of Gauss to mathematics, both pure and applied. Here we can only refer to some of the more important works that have not already been mentioned, and we shall select those which have added new techniques to mathematics or which rounded off outstanding problems. As a rough but convenient table of dates (from that adopted by the editors of Gauss' works) we summarize the principal fields of Gauss' interests after 1800 as follows: 1800–1820, astronomy; 1820–1830, geodesy, the theories of surfaces, and conformal mapping; 1830–1840, mathematical physics, particularly electromagnetism, terrestrial magnetism, and the theory of attraction according to the Newtonian law; 1841–1855, analysis situs, and the geometry associated with functions of a complex variable.

During the period 1821–1848 Gauss was scientific adviser to the Hanoverian (Göttingen was then under the government of Hanover) and Danish governments in an extensive geodetic survey. Gauss threw himself into the work. His method of least squares and his skill in devising schemes for handling masses of numerical data had full scope but, more importantly, the problems arising in the precise survey of a portion of the earth's surface undoubtedly suggested deeper and more general problems connected with all curved surfaces. These researches were to beget the mathematics of relativity. The subject was not new: several of Gauss' predecessors, notably Euler, Lagrange, and Monge, had investigated geometry on certain types of curved surfaces, but it remained for Gauss to attack the problem in all its generality, and from his investigations the first great period of *differential geometry* developed.

Differential geometry may be roughly described as the study of properties of curves, surfaces, etc., in the immediate neighborhood of a point, so that higher powers than the second of distances can be neglected. Inspired by this work, Riemann in 1854 produced his classic dissertation on the hypotheses which lie at the foundations of geometry, which, in its turn, began the second great period in differential geometry, that which is today of use in mathematical physics, particularly in the theory of general relativity.

Three of the problems which Gauss considered in his work on surfaces suggested general theories of mathematical and scientific importance: the measurement of *curvature*, the theory of *conformal representation* (or mapping), and the *applicability* of surfaces.

The unnecessarily mystical motion of a "curved" space-time, which is a purely mathematical extension of familiar, visualizable curvature to a "space" described by four coordinates instead of two, was a natural development of Gauss' work on curved surfaces. One of his definitions will illustrate the reasonableness of all. The problem is to devise some precise means for describing how the "curvature" of a surface varies from point to point of the surface; the description must satisfy our intuitive feeling for what "more curved" and "less curved" signify.

The total curvature of any part of a surface bounded by an unlooped closed curve C is defined as follows. The *normal* to a surface at a given

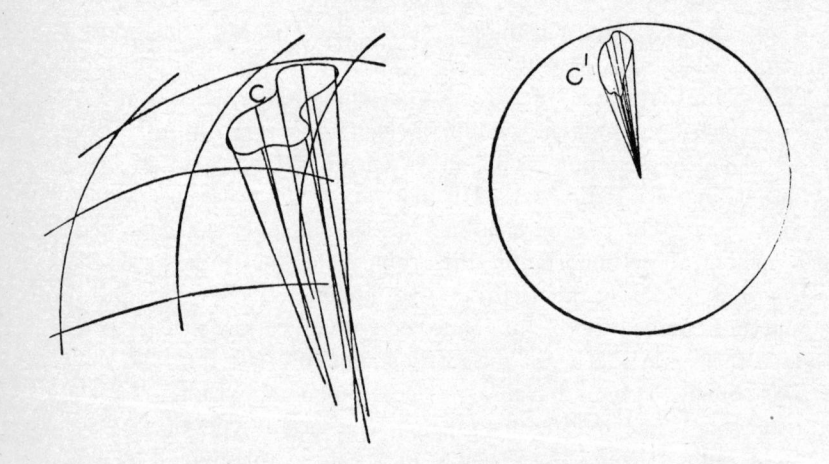

point is that straight line passing through the point which is perpendicular to the plane which touches the surface at the given point. At each point of C there is a normal to the surface. Imagine all these normals drawn. Now, from the center of a sphere (which may be anywhere with reference to the surface being considered), whose radius is equal to the unit length, imagine all the radii drawn which are parallel to the normals to C. These radii will cut out a curve, say C', on the sphere of unit radius. The *area* of that part of the spherical surface which is enclosed by C' is defined to be the *total curvature* of the part of the given surface which is enclosed by C. A little visualization will show that this definition accords with common notions as required.

Another fundamental idea exploited by Gauss in his study of surfaces was that of *parametric representation*.

It requires *two* coordinates to specify a particular point on a plane. Likewise on the surface of a sphere, or on a spheroid like the Earth: the coordinates in this case may be thought of as latitude and longitude. This illustrates what is meant by a *two-dimensional manifold*. Generally: if *precisely n* numbers are both necessary and sufficient to specify (individualize) each particular member of a class of things (points, sounds, colors, lines, etc.,) the class is said to be an *n-dimensional manifold*. In such specifications it is agreed that only certain characteristics of the members of the class shall be assigned numbers. Thus if we consider only the pitch of sounds, we have a one-dimensional manifold, because one number, the frequency of the vibration corresponding to the sound, suffices to determine the pitch; if we add loudness—measured on some convenient scale—sounds are now a two-dimensional manifold, and so on. If now we regard a *surface* as being made up of *points*, we see that it is a *two-dimensional manifold* (of points). Using the language of geometry we find it convenient to speak of *any* two-dimensional manifold as a "surface," and to apply to the manifold the reasoning of geometry—in the hope of finding something interesting.

The foregoing considerations lead to the parametric representation of surfaces. In Descartes' geometry *one* equation between *three* coordinates represents a surface. Say the coordinates (Cartesian) are x, y, z. Instead of using a single equation connecting x,y,z to represent the surface, we now seek *three*:

$$x = f(u,v),\ y = g(u,v),\ z = h(u,v),$$

where $f(u,v)$, $g(u,v)$, $h(u,v)$ are such functions (expressions) of the new variables u,v that when these variables are eliminated (got rid of—"put over the threshold," literally) there results between x,y,z the equation of the surface. The elimination is possible, because *two* of the equations can be used to solve for the *two* unknowns u,v; the results can then be substituted in the third. For example, if

$$x = u + v, \quad y = u - v, \quad z = uv,$$

we get $u = \frac{1}{2}(x + y)$, $v = \frac{1}{2}(x - y)$ from the first two, and hence $4z = x^2 - y^2$ from the third. Now as the variables u, v independently run through any prescribed set of numbers, the functions f, g, h will take on numerical values and x, y, z will move on the surface whose equations are the three written above. The variables u, v are called the *parameters* for the surface, and the three equations $x = f(u,v)$, $y = g(u,v)$, $z = h(u,v)$ their parametric equations. This method of representing surfaces has great advantages over the Cartesian when applied to the study of curvature and other properties of surfaces which vary rapidly from point to point.

Notice that the parametric representation is *intrinsic*; it refers to the surface itself for its coordinates, and not to an extrinsic, or extraneous, set of axes, not connected with the surface, as is the case in Descartes' method. Observe also that the *two* parameters u, v immediately show up the two-dimensionality of the surface. Latitude and longitude on the earth are instances of these intrinsic, "natural" coordinates; it would be most awkward to have to do all our navigation with reference to three mutually perpendicular axes drawn through the center of the Earth, as would be required for Cartesian sailing.

Another advantage of the method is its easy generalization to a space of any number of dimensions. It suffices to increase the number of parameters and proceed as before. When we come to Riemann we shall see how these simple ideas led naturally to a generalization of the metric geometry of Pythagoras and Euclid. The foundations of this generalization were laid down by Gauss, but their importance for mathematics and physical science was not fully appreciated till our own century.

Geodetic researches also suggested to Gauss the development of another powerful method in geometry, that of conformal mapping. Before a map can be drawn, say of Greenland, it is necessary to determine what is to be preserved. Are distances to be distorted, as they are on Mercator's projection, till Greenland assumes an exaggerated importance in comparison with North America? Or are distances to be preserved, so that one inch on the map, measured anywhere along the reference lines (say those for latitude and longitude) shall always correspond to the same distance measured on the surface of the earth? If so, one kind of mapping is demanded, and this kind will not preserve some other feature that we may wish to preserve; for example, if two roads on the earth intersect at a certain angle, the lines representing these roads on the map will intersect at a different angle. That kind of mapping which *preserves angles* is called conformal. In such mapping the theory of analytic functions of a complex variable, described earlier, is the most useful tool.

The whole subject of conformal mapping is of constant use in mathematical physics and its applications, for example in electrostatics, hydrodynamics and its offspring aerodynamics, in the last of which it plays a part in the theory of the airfoil.

Another field of geometry which Gauss cultivated with his usual thoroughness and success was that of the applicability of surfaces, in which it is required to determine what surfaces can be bent onto a given surface without stretching or tearing. Here again the methods Gauss invented were general and of wide utility.

To other departments of science Gauss contributed fundamental researches, for example in the mathematical theories of electromagnetism, including terrestrial magnetism, capillarity, the attraction of ellipsoids (the planets are special kinds of ellipsoids) where the law of attraction is the Newtonian, and dioptrics, especially concerning systems of lenses. The last gave him an opportunity to apply some of the purely abstract technique (continued fractions) he had developed as a young man to satisfy his curiosity in the theory of numbers.

Gauss not only mathematicized sublimely about all these things; he used his hands and his eyes, and was an extremely accurate observer. Many of the specific theorems he discovered, particularly in his researches on electromagnetism and the theory of attraction, have become part of the indispensable stock in trade of all who work seriously in physical science. For many years Gauss, aided by his friend Weber, sought a satisfying theory for all electromagnetic phenomena. Failing to find one that he considered satisfactory he abandoned his attempt. Had he found Clerk Maxwell's (1831–1879) equations of the electromagnetic field he might have been satisfied.

To conclude this long but still far from complete list of the great things that earned Gauss the undisputed title of Prince of Mathematicians we must allude to a subject on which he published nothing beyond a passing mention in his dissertation of 1799, but which he predicted would become one of the chief concerns of mathematics— *analysis situs*. A technical definition of what this means is impossible here (it requires the notion of a *continuous group*), but some hint of the type of problem with which the subject deals can be gathered from a simple instance. Any sort of a knot is tied in a string, and the ends of the string are then tied together. A "simple" knot is easily distinguishable by eye from a "complicated" one, but how are we to give an exact, *mathematical* specification of the difference between the two? And how are we to classify knots mathematically? Although he published nothing on this, Gauss had made a beginning, as was discovered in his posthumous papers. Another type of problem in this subject is to determine the least number of cuts on a given surface which will enable us to flatten the surface out on a plane. For a conical surface one cut suffices; for an anchor ring, two; for a sphere, no finite number of cuts suffices if no stretching is permitted.

These examples may suggest that the whole subject is trivial. But if it had been, Gauss would not have attached the extraordinary importance to it that he did. His prediction of its fundamental character has been fulfilled in our own generation. Today a vigorous school (including many Americans—J. W. Alexander, S. Lefschetz, O. Veblen, among others) is finding that analysis situs, or the "geometry of position" as it used sometimes to be called, has far-reaching ramifications in both geometry and analysis. What a pity it seems to us now that Gauss could not have stolen a year or two from Ceres to organize his thoughts on this vast theory which was to become the dream of his old age and a reality of our own young age.

His last years were full of honor, but he was not as happy as he had earned the right to be. As powerful of mind and as prolifically inventive as he had ever been, Gauss was not eager for rest when the first symptoms of his last illness appeared some months before his death.

A narrow escape from a violent death had made him more reserved than ever, and he could not bring himself to speak of the sudden passing of a friend. For the first time in more than twenty years he had left Göttingen on June 16, 1854, to see the railway under construction between his town and Cassel. Gauss had always taken a keen interest in the construction and operation of railroads; now he would see one being built. The horses bolted; he was thrown from his carriage, unhurt, but badly shocked. He recovered, and had the pleasure of witnessing the opening ceremonies when the railway reached Göttingen on July 31, 1854. It was his last day of comfort.

With the opening of the new year he began to suffer greatly from an enlarged heart and shortness of breath, and symptoms of dropsy appeared. Nevertheless he worked when he could, although his hand cramped and his beautifully clear writing broke at last. The last letter he wrote was to Sir David Brewster on the discovery of the electric telegraph.

Fully conscious almost to the end he died peacefully, after a severe struggle to live, early on the morning of February 23, 1855, in his seventy eighth year. He lives everywhere in mathematics.

CHAPTER FIFTEEN

Mathematics and Windmills

CAUCHY

A man may say even his pater noster *out of turn.*—SPANISH PROVERB

IN THE FIRST THREE DECADES of the nineteenth century mathematics quite suddenly became something noticeably different from what it had been in the heroic post-Newtonian age of the eighteenth. The change was in the direction of greater rigor in demonstration following an unprecedented generality and freedom of inventiveness. Something of the same sort is plainly visible again today, and he would be a rash prophet who would venture to forecast what mathematics will be like three-quarters of a century hence.

At the beginning of the nineteenth century only Gauss had any inkling of what was so soon to come, but his Newtonian reserve held him back from telling Lagrange, Laplace, and Legendre what he foresaw. Although the great French mathematicians lived well into the first third of the nineteenth century much of their work now appears to have been preparatory. Lagrange in the theory of equations prepared the way for Abel and Galois; Laplace, with his work on the differential equations of Newtonian astronomy—including the theory of gravitation—hinted at the phenomenal development of mathematical physics in the nineteenth century; while Legendre's researches in the integral calculus suggested to Abel and Jacobi one of the most fertile fields of investigation analysis has ever acquired. Lagrange's analytical mechanics is still modern; but even it was to receive magnificent additions at the hands of Hamilton and Jacobi and, later, Poincaré. Lagrange's work in the calculus of variations was also to remain classic and useful, but again the work of Weierstrass gave it a new direction under the rigorous, inventive spirit of the latter half of the nineteenth century, and this in its turn has been amplified and renovated in our own times (American and Italian mathematicians taking a leading part in the development).

Augustin-Louis Cauchy, the first of the great French mathematcians whose thought belongs definitely to the modern age, was born in Paris on August 21, 1789—a little less than six weeks after the fall of the Bastille. A child of the Revolution, he paid his tax to liberty and equality by growing up with an undernourished body. It was only by the diplomacy and good sense of his father that Cauchy survived at all in the midst of semi-starvation. Having outlived the Terror, he graduated from the Polytechnique into the service of Napoleon. After the downfall of the Napoleonic order Cauchy got his full share of deprivations from revolutions and counter-revolutions, and in a measure his work was affected by the social unrest of his times. If revolutions and the like do affect a scientific man's work, Cauchy should be the prize laboratory specimen for proving the fact. He had an extraordinary fertility in mathematical inventiveness and a fecundity that has been surpassed only twice—by Euler and Cayley. His work, like his times, was revolutionary.

Modern mathematics is indebted to Cauchy for two of its major interests, each of which marks a sharp break with the mathematics of the eighteenth century. The first was the introduction of rigor into mathematical analysis. It is difficult to find an adequate simile for the magnitude of this advance; perhaps the following will do. Suppose that for centuries an entire people has been worshipping false gods and that suddenly their error is revealed to them. Before the introduction of rigor mathematical analysis was a whole pantheon of false gods. In this Cauchy was one of the great pioneers with Gauss and Abel. Gauss might have taken the lead long before Cauchy entered the field, but did not, and it was Cauchy's habit of rapid publication and his gift for effective teaching which really got rigor in mathematical analysis accepted.

The second thing of fundamental importance which Cauchy added to mathematics was on the opposite side—the combinatorial. Seizing on the heart of Lagrange's method in the theory of equations, Cauchy made it abstract and began the systematic creation of the theory of groups. The nature of this will be described later; for the moment we note only the modernity of Cauchy's outlook.

Without enquiring whether the thing he invented had any application or not, even to other branches of mathematics, Cauchy developed it on its own merits as an abstract system. His predecessors, with the exception of the universal Euler who was as willing to write a memoir on a puzzle in numbers as on hydraulics or the "system of the world."

had found their inspiration growing out of the applications of mathematics. This statement of course has numerous exceptions, notably in arithmetic; but before the time of Cauchy few if any sought profitable discoveries in the mere manipulations of algebra. Cauchy looked deeper, saw the *operations* and their *laws of combination* beneath the symmetries of algebraic formulas, isolated them, and was led to the theory of groups. Today this elementary yet intricate theory is of fundamental importance in many fields of pure and applied mathematics, from the theory of algebraic equations to geometry and the theory of atomic structure. It is at the bottom of the geometry of crystals, to mention but one of its applications. Its later developments (on the analytical side) extend far into higher mechanics and the modern theory of differential equations.

Cauchy's life and character affect us like poor Don Quixote's—we sometimes do not know whether to laugh or to cry, and compromise by swearing. His father, Louis-François, was a paragon of virtue and piety, both excellent things in their way, but easily overdone. Heaven only knows how Cauchy senior escaped the guillotine; for he was a parliamentary lawyer, a cultured gentleman, an accomplished classical and biblical scholar, a bigoted Catholic, and a lieutenant of police in Paris when the Bastille fell. Two years before the Revolution broke he had married Marie-Madeleine Desestre, an excellent, not very intelligent woman who, like himself, was also a bigoted Catholic.

Augustin was the eldest of six children (two sons, four daughters). From his parents Cauchy inherited and acquired all the estimable qualities which make their lives read like one of those charming love stories, insipid as stewed cucumbers, concocted for French schoolgirls under sixteen, in which the hero and heroine are as pure and sexless as God's holy angels. With parents such as his it was perhaps natural that Cauchy should have grown up to be the obstinate Quixote of French Catholicism in the 1830's and 1840's when the Church was on the defensive. He suffered for his religion, and for that he deserves respect (possibly even if he was the smug hypocrite his colleagues accused him of being), but he also richly deserved to suffer on more than one occasion. His everlasting preaching about the beauty of holiness put peoples' backs up and engendered an opposition to his pious schemes which they did not always deserve. Abel, himself the son of a minister and a decent enough Christian, expressed the general disgust which some of Cauchy's antics inspired when he wrote home, "Cauchy is a bigoted Catholic—a strange thing for a man of science." The emphasis of course is on "bigoted," not on the word it qualifies. Two of the finest characters and greatest mathematicians we shall meet, Weierstrass and Hermite, were Catholics. They were devout but not bigoted.

Cauchy's childhood fell in the bloodiest period of the Revolution. The schools were closed. Having no need of science or culture at the moment, the Commune either left the cultured and men of science to starve or carted them off to the guillotine. To escape the obvious danger Cauchy senior moved his family to his country place in the village of Arcueil. There he sat out the Terror, half starved himself and feeding his wife and infant son largely from what scanty fruits and vegetables he could raise. As a consequence Cauchy grew up delicate and underdeveloped physically. He was nearly twenty before he began to recover from this early malnutrition, and all his life had to watch his health.

This retreat, gradually becoming less strict, lasted nearly eleven years, during which Cauchy senior undertook the education of his children. He wrote his own textbooks, several of them in the fluent verse of which he was master. Verse, he believed, made grammar, history and, above all, morals less repulsive to the juvenile mind. Young Cauchy thus acquired his own uncontrolled fluency in both French and Latin verse which he indulged all his life. His verse abounds in noble sentiments loftily expressed and admirably reflects the piety of his blameless life but is otherwise undistinguished. A large share of the lessons was devoted to narrow religious instruction, in which the mother assisted ably.

Arcueil adjoined the imposing estates of the Marquis Laplace and Count Claude-Louis Berthollet (1748–1822), the distinguished and eccentric chemist who kept his head in the Terror because he knew all about gunpowder. The two were great friends. Their gardens were separated by a common wall with a gate to which each had a key. In spite of the fact that both the mathematician and the chemist were anything but pious, Cauchy senior scraped an acquaintance with his distinguished and well-fed neighbors.

Berthollet never went anywhere. Laplace was more sociable and presently began dropping in at his friend's cottage, where he was struck by the spectacle of young Cauchy, too feeble physically to be tearing round like a properly nourished boy, poring over his books and papers like a penitent monk and seeming to enjoy it. Before long Laplace discovered that the boy had a phenomenal mathematical talent and advised him to husband his strength. Within a few years Laplace was to be listening apprehensively to Cauchy's lectures on infinite series, fearing that the bold young man's discoveries in convergence might have destroyed the whole vast edifice of his own celestial mechanics. "The system of the world" came within a hairsbreadth of going to smash that time; a slightly greater ellipticity of the Earth's almost circular orbit, and the infinite series on which Laplace had based his calculations would have diverged. Luckily his astronomical intuition had preserved him from disaster, as he discovered on rising with a sigh of infinite relief after a prolonged testing of the convergence of all his series by Cauchy's methods.

On January 1, 1800, Cauchy senior, who had kept discreetly in touch with Paris, was elected Secretary of the Senate. His office was in the Luxembourg Palace. Young Cauchy shared the office, using a corner as his study. Thus it came about that he frequently saw Lagrange—then Professor at the Polytechnique—who dropped in frequently to discuss business with Secretary Cauchy. Lagrange soon became interested in the boy and, like Laplace, was struck by his mathematical talent. On one occasion when Laplace and several other notables were present, Lagrange pointed to young Cauchy in his corner and said, "You see that little young man? Well! He will supplant all of us in so far as we are mathematicians."

To Cauchy senior Lagrange gave some sound advice, believing that the delicate boy might burn himself out: "Don't let him touch a mathematical book till he is seventeen." Lagrange meant higher mathematics. And on another occasion: "If you don't hasten to give Augustin a solid literary education his tastes will carry him away; he will be a great mathematician but he won't know how to write his own language." The father took this advice from the greatest mathematician of the age to heart and gave his son a sound literary education before turning him loose on advanced mathematics.

After his father had done all he could for him, Cauchy entered the Central School of the Panthéon, at about the age of thirteen. Napoleon had instituted several prizes in the school and a sort of grand sweepstakes prize for all the schools of France in the same class. From the first Cauchy was the star of the school, carrying off the first prizes in

Greek, Latin composition, and Latin verse. On leaving the school in 1804 he won the sweepstakes and a special prize in humanities. The same year Cauchy received his first communion, a solemn and beautiful occasion in the life of any Catholic and trebly so to him.

For the next ten months he studied mathematics intensively with a good tutor, and in 1805 at the age of sixteen passed second into the Polytechnique. There his experiences were not altogether happy among the ribald young skeptics who hazed him unmercifully for making a public exhibition of his religious observances. But Cauchy kept his temper and even tried to convert some of his scorners.

From the Polytechnique Cauchy passed to the civil engineering school (*Ponts et Chaussées*) in 1807. Although only eighteen he easily beat young men of twenty who had been two years in the school, and was early marked for special service. On completing his training in March, 1810, Cauchy was at once given an important commission. His ability and bold originality had singled him out as a man for whom red tape should be cut, even at the risk of lopping off some older man's head in the process. Whatever else may be said of Napoleon, he took ability wherever he found it.

In March, 1810, when Cauchy left Paris, "light of baggage, but full of hope," for Cherbourg on his first commission, the battle of Waterloo (June 18, 1815) was still over five years in the future, and Napoleon still confidently expected to take England by the neck and rub its nose in its own fragrant sod. Before an invasion could be launched an enormous fleet was necessary, and this had yet to be built. Harbors and fortifications to defend the shipyards from the seagoing British were the first detail to be disposed of in the glamorous dream. Cherbourg for many reasons was the logical point to begin all these grandiose operations which were to hasten "the day of glory" the French had been yelling about ever since the fall of the Bastille. Hence the gifted young Cauchy's assignment to Cherbourg to become a great military engineer.

In his light baggage Cauchy carried only four books, the *Mécanique céleste* of Laplace, the *Traité des fonctions analytiques* of Lagrange, Thomas à Kempis' *Imitation of Christ*, and a copy of Virgil's works— an unusual assortment for an ambitious young military engineer. Lagrange's treatise was to be the very book which caused its author's prophecy that "this young man will supplant all of us" to come true first, as it inspired Cauchy to seek some theory of functions free from the glaring defects of Lagrange's.

The third on the list was to occasion Cauchy some distress, for with it and his aggressive piety he rather got on the nerves of his practical associates who were anxious to get on with their job of killing. But Cauchy soon showed them by turning the other cheek that he had at least read the book. "You'll soon get over all that," they assured him. To which Cauchy replied by sweetly asking them to point out what was wrong in his conduct and he would gladly correct it. What answer this drew has not survived.

Rumors that her darling boy was fast becoming an infidel or worse reached the ears of his anxious mother. In a letter long enough and full enough of pious sentiments to calm all the mothers who ever sent their sons to the front or anywhere near it Cauchy reassured her, and she was happy once more. The conclusion of the letter shows that the holy Cauchy was quite capable of holding his own against his tormentors, who had hinted he was slightly cracked.

"It is therefore ridiculous to suppose that religion can turn anybody's head, and if all the insane were sent to insane asylums, more philosophers than Christians would be found there." Is this a slip on Cauchy's part, or did he really mean that no Christians are philosophers? He signs off with a flash from the other side of his head: "But enough of this—it is more profitable for me to work at certain Memoirs on Mathematics." Precisely; but every time he saw a windmill waving its gigantic arms against the sky he was off again full tilt.

Cauchy stayed approximately three years at Cherbourg. Outside of his heavy duties his time was well spent. In a letter of July 3, 1811, he describes his crowded life. "I get up at four and am busy from morning to night. My ordinary work is augmented this month by the arrival of the Spanish prisoners. We had only eight days' warning, and during those eight days we had to build barracks and prepare camp beds for 1200 men. . . . At last our prisoners are lodged and covered —since the last two days. They have camp beds, straw, food, and count themselves very fortunate. . . . Work doesn't tire me; on the contrary it strengthens me and I am in perfect health."

On top of all this good work *pour la gloire de la belle France* Cauchy found time for research. As early as December, 1810, he had begun "to go over again all the branches of Mathematics, beginning with Arithmetic and finishing with Astronomy, clearing up obscurities, applying [my own methods] to the simplification of proofs and the discovery of new propositions." And still on top of this the amazing young man found time to instruct others who begged for lessons so that they might rise in their profession, and he even assisted the mayor of Cherbourg by conducting school examinations. In this way he learned to teach. He still had time for hobbies.

The Moscow fiasco of 1812, war against Prussia and Austria, and the thorough drubbing he got at the battle of Leipzig in October, 1813, all distracted Napoleon's attention from the dream of invading England, and the works at Cherbourg languished. Cauchy returned to Paris in 1813, worn out by overwork. He was then only twenty four, but he had already attracted the attention of the leading mathematicians of France by his brilliant researches, particularly the memoir on polyhedra and that on symmetric functions. As the nature of both is easily understood, and each offers suggestions of the very first importance to the mathematics of today, we shall briefly describe them.

The first is of only minor interest in itself. What is significant regarding it today is the extraordinary acuteness of the criticism which Malus levelled at it. By a curious historical coincidence Malus was exactly one hundred years ahead of his times in objecting to Cauchy's reasoning in the precise manner in which he did. The Academy had proposed as its prize problem "To perfect in some essential point the theory of polyhedra," and Lagrange had suggested this as a promising research for young Cauchy to undertake. In February, 1811, Cauchy submitted his first memoir on the theory of polyhedra. This answered negatively a question asked by Poinsot (1777–1859): is it possible that regular polyhedra other than those having 4, 6, 8, 12, or 20 faces exist? In the second part of this memoir Cauchy extended the formula of Euler, given in the school books on solid geometry, connecting the number of edges (E), faces (F), and vertices (V) of a polyhedron, $E + 2 = F + V$.

This work was printed. Legendre thought highly of it and encouraged Cauchy to continue, which Cauchy did in a second memoir (January, 1812). Legendre and Malus (1775–1812) were the referees. Legendre was enthusiastic and predicted great things for the young author. But Malus was more reserved.

Étienne-Louis Malus was not a professional mathematician but an ex-officer of engineers in Napoleon's campaigns in Germany and

Egypt, who made himself famous by his accidental discovery of the polarization of light by reflexion. So possibly his objections struck young Cauchy as just the sort of captious criticisms to be expected from an obstinate physicist. In proving his most important theorems Cauchy had used the "indirect method" familiar to all beginners in geometry. It was to this method of proof that Malus objected.

In proving a proposition by the indirect method, a contradiction is deduced from the assumed falsity of the proposition; whence it follows, in Aristotelian logic, that the proposition is true. Cauchy could not meet the objection by supplying direct proofs, and Malus gave in— still unconvinced that Cauchy had proved anything. When we come to the conclusion of the whole story (in the last chapter) we shall see the same objection being raised in other connections by the intuitionists. If Malus failed to make Cauchy see the point in 1812, Malus was avenged by Brouwer in 1912 and thereafter when Brouwer succeeded in making some of Cauchy's successors in mathematical analysis at least see that there is a point to be seen. Aristotelian logic, as Malus was trying to tell Cauchy, is not always a safe method of reasoning in mathematics.

Passing to the *theory of substitutions*, begun systematically by Cauchy, and elaborated by him in a long series of papers in the middle 1840's, which developed into the *theory of finite groups*, we shall presently illustrate the underlying notions by a simple example. First, however, the leading properties of a *group of operations* may be described informally.

Operations will be denoted by capital letters, $A, B, C, D, \ldots$; and the performance of two operations *in succession*, say A *first*, B *second*, will be indicated by juxtaposition thus, AB. Note that BA, by what has just been said, means that B is performed first, A second; so that AB and BA are *not necessarily* the same operation. For example, if A is the operation "add 10 to a given number," and B is the operation "divide a given number by 10," AB applied to x gives $\frac{x+10}{10}$, while BA gives $\frac{x}{10} + 10$, or $\frac{x+100}{10}$, and the resulting fractions are unequal; hence AB and BA are distinct.

If the effects of two operations X, Y are the same, X and Y are said to be *equal* (or *equivalent*), and this is expressed by writing $X = Y$.

The next fundamental notion is that of *associativity*. If for *every* triple of operations, say U, V, W is any triple, in the set, $(UV)W = U(VW)$, the set is said to satisfy the *associative* law. By $(UV)W$ is meant that UV is performed first, then, on the result, W is performed; by $U(VW)$ is meant that U is performed first, then, on the result of this VW is performed.

The last fundamental notion is that of an *identical operation,* or an *identity*: an operation I which leaves unchanged whatever it operates on is called an *identity*.

With these notions we can now state the simple postulates which define a group of operations.

A set of operations $I, A, B, C, \ldots, X, Y, \ldots$ is said to form a *group* if the postulates (1) – (4) are satisfied.

(1) There is a rule of combination applicable to *any* pair X, Y of operations* in the set such that the result, denoted by XY, of combining X, Y, in this order, according to the rule of combination, is a uniquely determined operation in the set.

(2) For *any three* operations X, Y, Z in the set, the rule in (1) is associative, namely $(XY)Z = X(YZ)$.

(3) There is a unique identity I in the set, such that, for every operation X in the set, $IX = XI = X$.

(4) If X is *any* operation in the set, there is in the set a *unique* operation, say X', such that $XX' = I$ (it can be easily proved that $X'X = I$ also).

These postulates contain redundancies deducible from other statements in (1)–(4), but in the form given the postulates are easier to grasp. To illustrate a group we shall take a very simple example relating to *permutations* (arrangements) of letters. This may seem trivial, but such *permutation* or *substitution* groups were found to be the long-sought clue to the algebraic solvability of equations.

There are precisely 6 orders in which the 3 letters a,b,c can be written, namely $abc, acb, bca, bac, cab, cba$. Take any one of these, say the first abc, as the initial order. By what permutations of the letters can we pass from this to the remaining 5 arrangements? To pass from abc to acb it is sufficient to *interchange,* or *permute,* b and c. To indicate the *operation* of permuting b and c, we write (bc), which is read, "b into c, and c into b." From abc to bca we pass by a into b, b into c, and c into a, which is written (abc). The order abc itself is obtained from abc by *no* change, namely a into a, b into b, c into c, which

---

*The operations in a pair may be the same, thus X, X.

is the *identity* substitution and is denoted by I. Proceeding similarly with all 6 orders

$$abc,\ acb,\ bca,\ bac,\ cab,\ cba,$$

we get the corresponding *substitutions*,

$$I,\ (bc),\ (abc),\ (ab),\ (acb),\ (ac).$$

The "rule of combination" in the postulates is here as follows. Take any two of the substitutions, say (bc) and (acb), and consider the effect of these applied successively in the order stated, namely (bc) first and (acb) second: (bc) carries b into c, then (acb) carries c into b. Thus b is left as it was. Take the next letter, c, in (bc): by (bc), c is carried into b, which, by (acb) is carried into a; thus c is carried into a. Continuing, we see what a is now carried into: (bc) leaves a as it was, but (acb) carries a into c. Finally then the total effect of (bc) followed by (acb) is seen to be (ca), which we indicate by writing $(bc)(acb) = (ca) = (ac)$.

In the same way it is easily verified that

$$(acb)(abc) = (abc)(acb) = I;$$
$$(abc)(ac) = (ab);\ (bc)(ac) = (acb),$$

and so on for all possible pairs. Thus postulate (1) is satisfied for these 6 substitutions, and it can be checked that (2), (3), (4) are also satisfied.

All this is summed up in the "multiplication table" of the group, which we shall write out, denoting the substitutions by the letters under them (to save space),

$$I,\ (bc),\ (abc),\ (ab),\ (acb),\ (ac)$$
$$I,\ \ A,\ \ \ B,\ \ \ C,\ \ \ D,\ \ \ E.$$

In reading the table any letter, say C, is taken from the left-hand *column*, and any letter, say D, from the top *row*, and the entry, here A, where the corresponding *row* and *column* intersect is the result of CD. Thus $CD = A$, $DC = E$, $EA = B$, and so on.

As an example we may verify the associative law for $(AB)C$ and $A(BC)$, which should be equal. First, $AB = C$; hence $(AB)C = CC = I$. Again $BC = A$; hence $A(BC) = AA = I$. In the same way $A(DB) = AI = A$; $(AD)B = EB = A$; thus $(AD)B = A(DB)$.

	I	A	B	C	D	E
I	I	A	B	C	D	E
A	A	I	C	B	E	D
B	B	E	D	A	I	C
C	C	D	E	I	A	B
D	D	C	I	E	B	A
E	E	B	A	D	C	I

The total number of distinct operations in a group is called its *order*. Here 6 is the order of the group. By inspection of the table we pick out several *subgroups*, for example,

	I
I	I

	I	A
I	I	A
A	A	I

	I	B	D
I	I	B	D
B	B	D	I
D	D	I	B

which are of the respective orders 1, 2, 3. This illustrates one of the fundamental theorems proved by Cauchy: *the order of any subgroup is a divisor of the order of the group.*

The reader may find it amusing to try to construct groups of orders other than 6. For any given order the number of distinct groups (having different multiplication tables) is finite, but what this number may be for *any* given order (the *general* order n) is not known

—nor likely to be in our lifetime. So at the very beginning of a theory which on its surface is as simple as dominoes we run into unsolved problems.

Having constructed the "multiplication table" of a group, we forget about its derivation from substitutions (if that happens to be the way the table was made), and regard the table as defining an *abstract group*; that is, the symbols $I, A, B, \ldots$ are given no interpretation beyond that implied by the rule of combination, as in $CD = A$, $DC = E$, etc. This abstract point of view is that now current. It was not Cauchy's, but was introduced by Cayley in 1854. Nor were completely satisfactory sets of postulates for groups stated till the first decade of the twentieth century.

When the operations of a group are interpreted as substitutions, or as the rotations of a rigid body, or in any other department of mathematics to which groups are applicable, the interpretation is called a *realization* of the *abstract* group defined by the multiplication table. A given abstract group may have many diverse realizations. This is one of the reasons that groups are of *fundamental* importance in modern mathematics: one abstract, *underlying structure* (that summarized in the multiplication table) of one and the same group is the essence of several apparently unrelated theories, and by an intensive study of the properties of the abstract group, a knowledge of the theories in question and their mutual relationships is obtained by one investigation instead of several.

To give but one instance, the set of all rotations which twirl a regular icosahedron (twenty-sided regular solid) about its axes of symmetry, so that after any rotation of the set the volume of the solid occupies the same space as before, forms a group, and this group of rotations, when expressed abstractly, is the same group as that which appears, under permutations of the roots, when we attempt to solve the general equation of the fifth degree. Further, this same group turns up (to anticipate slightly) in the theory of elliptic functions. This suggests that although it is impossible to solve the general quintic algebraically, the equation may be—and in fact is—solvable in terms of the functions mentioned. Finally, all this can be pictured geometrically by describing the rotations of an icosahedron already mentioned. This beautiful unification was the work of Felix Klein (1849–1925) in his book on the icosahedron (1884).

Cauchy was one of the great pioneers in the theory of substitution groups. Since his day an immense amount of work has been done in the subject, and the theory itself has been vastly extended by the accession of *infinite groups*—groups having an infinity of operations which can be counted off 1,2,3, . . . , and further, to groups of *continuous* motions. In the latter an operation of the group shifts a body into another position by *infinitesimal* (arbitrarily small) displacements —not like the icosahedral group described above, where the rotations shift the whole body round by a finite amount. This is but one category of infinite groups (the terminology here is not exact, but is sufficient to bring out the point of importance—the distinction between *discrete* and *continuous* groups). Just as the theory of finite discrete groups is the structure underlying the theory of algebraic equations, so is the theory of infinite, continuous groups of great service in the theory of differential equations—those of the greatest importance in mathematical physics. So in playing with groups Cauchy was not idling.

To close this description of groups we may indicate how the groups of substitutions discussed by Cauchy have entered the modern theory of atomic structure. A substitution, say (xy), containing precisely two letters in its symbol, is called a *transposition*. It is easily proved that any substitution is a combination of transpositions. For example,

$$(abcdef) = (ab)(ac)(ad)(ae)(af),$$

from which the rule for writing out any substitution in terms of transpositions is evident.

Now, it is an entirely reasonable hypothesis to assume that the electrons in an atom are identical, that is, one electron is indistinguishable from another. Hence, if in an atom two electrons are interchanged, the atom will remain unchanged. Suppose for simplicity that the atom contains precisely three electrons, say a,b,c. To the *group of substitutions* on a,b,c (the one whose multiplication table we gave) will correspond all interchanges of electrons leaving the atom *invariant*—as it was. From this to the spectral lines in the light emitted by an excited gas consisting of atoms may seem a long step, but it has been taken, and one school of experts in quantum mechanics finds a satisfactory background for the elucidation of spectra (and other phenomena associated with atomic structure) in the theory of substitution groups. Cauchy of course foresaw no such applications of the theory which he developed for its own fascinations, nor did he foresee its application to the outstanding riddle of algebraic equations. That triumph was reserved for a boy in his teens whom we shall meet later.

By the age of twenty seven (in 1816) Cauchy had raised himself to the front rank of living mathematicians. His only serious rival was the reticent Gauss, twelve years older than himself. Cauchy's memoir of 1814 on definite integrals with complex-number limits inaugurated his great career as the independent creator and unequalled developer of the theory of functions of a complex variable. For the technical terms we must refer to the chapter on Gauss—who had reached the fundamental theorem in 1811, three years before Cauchy. Cauchy's luxuriantly detailed memoir on the subject was published only in 1827. The delay was due possibly to the length of the work—about 180 pages. Cauchy thought nothing of hurling massive works of from 80 to 300 pages at the Academy or the Polytechnique to be printed out of their stinted funds.

The following year (1815) Cauchy created a sensation by proving one of the great theorems which Fermat had bequeathed to a baffled posterity: every positive integer is a sum of three "triangles," four "squares," five "pentagons," six "hexagons," and so on, zero in each case being counted as a number of the kind concerned. A "triangle" is one of the numbers $0, 1, 3, 6, 10, 15, 21, \ldots$ got by building up *regular* (equilateral) triangles out of dots,

$$., \quad ..\,., \quad ...\,...\,., \quad ....\,....\,....\,., \quad \text{etc.};$$

"squares" are built up similarly,

$$., \quad ..\,., \quad ...\,...\,., \quad ....\,....\,....\,., \quad \text{etc.},$$

where the "bordering" by which one square is obtained from its predecessor is evident. Similarly "pentagons" are *regular* pentagons built up by dots; and so on for "hexagons" and the rest. This was not easy to prove. In fact it had been too much for Euler, Lagrange, and Legendre. Gauss had early proved the case of "triangles."

As if to show that he was not limited to first-rate work in pure mathematics Cauchy next captured the Grand Prize offered by the Academy in 1816 for a "theory of the propagation of waves on the surface of a heavy fluid of indefinite depth"—ocean waves are close enough to this type for mathematical treatment. This finally (when printed) ran to more than 300 pages. At the age of twenty seven Cauchy found himself being strongly "rushed" for membership in the Academy of Sciences—a most unusual honor for so young a man. The very first vacancy in the mathematical section would fall to him, he was assured on the quiet. So far as popularity is concerned this was the highwater mark of Cauchy's career.

In 1816, then, Cauchy was ripe for election to the Academy. But there were no vacancies. Two of the seats however might soon be expected to be empty owing to the age of the incumbents: Monge was seventy, L. M. N. Carnot sixty three. Monge we have already met; Carnot was a precursor of Poncelet. Carnot held his seat in the Academy on account of his researches which restored and extended the synthetic geometry of Pascal and Desargues, and for his heroic attempt to put the calculus on a firm logical foundation. Outside of mathematics Carnot made an enviable name for himself in French history, being the genius who in 1793 organized fourteen armies to defeat the half million troops hurled against France by the united antidemocratic reactionaries of Europe. When Napoleon seized the power for himself in 1796, Carnot was banished for opposing the tyrant: "I am an irreconcilable enemy of all kings," said Carnot. After the Russian campaign of 1812 Carnot offered his services as a soldier, but with one stipulation. He would fight for France, not for the French Empire of Napoleon.

In the reorganization of the Academy of Sciences during the political upheaval after Napoleon's glorious "Hundred Days" following his escape from Elba, Carnot and Monge were expelled. Carnot's successor took his seat without much being said, but when young Cauchy calmly sat down in Monge's chair the storm broke. The expulsion of Monge was sheer political indecency, and whoever profited by it showed at least that he lacked the finer sensibilities. Cauchy of course was well within his rights and his conscience.

The hippopotamus is said to have a tender heart by those who have eaten that delicacy baked, so a thick skin is not necessarily a reliable index to what is inside a man. Worshipping the Bourbons as he did, and believing the dynasty to be the direct representatives of Heaven sent to govern France—even when Heaven sent an incompetent clown like Charles X—Cauchy was merely doing his loyal duty to Heaven and to France when he slipped into Monge's chair. That he was sincere and not merely self-seeking will appear from his subsequent devotion to the sanctified Charles.

Honorable and important positions now came thick and fast to the greatest mathematician in France—still well under thirty. Since 1815 (when he was twenty six) Cauchy had been lecturing on analysis at the Polytechnique. He was now made Professor, and before long was appointed also at the Collège de France and the Sorbonne. Everything began coming his way. His mathematical activity was incredible; sometimes two full length papers would be laid before the Academy in the same week. In addition to his own research he drew up innumerable reports on the memoirs of others submitted to the Academy, and found time to emit an almost constant stream of short papers on practically all branches of mathematics, pure and applied. He became better known than Gauss to the mathematicians of Europe. Savants as well as students came to hear his beautifully clear expositions of the new theories he was developing, particularly in analysis and mathematical physics. His auditors included well-known mathematicians from Berlin, Madrid, and St. Petersburg.

In the midst of all this work Cauchy found time to do his courting. His fancy, Aloise de Bure, whom he married in 1818 and with whom he lived for nearly forty years, was the daughter of a cultured old family and, like himself, an ardent Catholic. They had two daughters, who were brought up as Cauchy had been.

One great work of this period may be noted. Encouraged by Laplace and others, Cauchy in 1821 wrote up for publication the course of lectures on analysis he had been giving at the Polytechnique. This is the work which for long set the standard in rigor. Even today Cauchy's definitions of limit and continuity, and much of what he wrote on the convergence of infinite series in this course of lectures, will be found in any carefully written book on the calculus. An extract from the preface will show what he had in mind and what he accomplished.

"I have sought to give to the methods [of analysis] all the rigor which is demanded in geometry, in such a way as never to refer to reasons drawn from the generality of algebra. [As it would be put today, the *formalism* of algebra.] Reasons of this kind, although commonly enough admitted, above all in the passage from convergent to divergent series, and from real quantities to imaginary, cannot be considered, it seems to me, as anything more than inductions which occasionally suggest the truth, but which agree but little with the boasted exactitude of mathematics. We must also observe that they tend to cause an indefinite validity to be attributed to algebraical formulas,* while, in reality, the majority of these formulas subsist only under certain conditions, and for certain values of the quantities which they contain. By determining these conditions and values, and by fixing precisely the meaning of the notations I make use of, I shall dispel all uncertainty."

Cauchy's productivity was so prodigious that he had to found a sort of journal of his own, the *Exercises de Mathématiques* (1826-30), continued in a second series as *Exercises d'Analyse Mathématique et de Physique*, for the publication of his expository and original work in pure and applied mathematics. These works were eagerly bought and studied, and did much to reform mathematical taste before 1860.

One aspect of Cauchy's terrific activity is rather amusing. In 1835 the Academy of Sciences began publishing its weekly bulletin (the *Comptes rendus*). Here was a virgin dumping ground for Cauchy, and he began swamping the new publication with notes and lengthy memoirs—sometimes more than one a week. Dismayed at the rapidly mounting bill for printing, the Academy passed a rule, in force today, prohibiting the publication of an article over four pages long. This cramped Cauchy's luxuriant style, and his longer memoirs, including a great one of 300 pages on the theory of numbers, were published elsewhere.

Happily married and as prolific in his research as a spawning salmon, Cauchy was ripe for the jester when the revolution of 1830 unseated his beloved Charles. Fate never enjoyed a heartier laugh than it did when it motioned Cauchy to rise from Monge's chair in the Academy and follow his anointed King into exile. Cauchy could

---

*For example, $\dfrac{1}{1-x} = 1 + x + x^2 + x^3 + \ldots$ to infinity, obtained by dividing 1 by $1 - x$, is nonsense if x is a positive number equal to or greater than 1.

not refuse; he had sworn a solemn oath of allegiance to Charles, and to Cauchy an oath was an oath, even if sworn to a deaf donkey. To his credit, Cauchy, at the age of forty, gave up all his positions and went into voluntary exile.

He was not sorry to go. The bloodied streets of Paris had turned his sensitive stomach. He firmly believed that good King Charles was in no way responsible for the gory mess.

Leaving his family in Paris, but not resigning his seat in the Academy, Cauchy went first to Switzerland, where he sought distraction in scientific conferences and research. He never asked the slightest favor from Charles and did not even know that the exiled king was aware of his voluntary sacrifice for a matter of principle. Shortly a more enlightened Charles, Charles Albert, King of Sardinia, heard that the renowned Cauchy was out of a job and made one for him as Professor of Mathematical Physics at Turin. Cauchy was perfectly happy. He quickly learned Italian and delivered his lectures in that language.

Presently overwork and excitement made him ill, and to his regret (as he wrote to his wife) he was forced to abandon evening work for a time. A vacation in Italy, with a visit to the Pope for good measure, completely restored him, and he returned to Turin, eagerly anticipating a long life devoted to teaching and research. But presently the obtuse Charles X butted into the retiring mathematician's life like a brainless goat and, in seeking to reward his loyal follower, did him a singular disservice. In 1833 Cauchy was entrusted with the education of Charles' heir, the thirteen-year-old Duke of Bordeaux. The job of male nurse and elementary tutor was the last thing on earth that Cauchy desired. Nevertheless he dutifully reported to Charles at Prague and took up the cross of loyalty. The following year he was joined by his family.

The education of the heir to the Bourbons proved no sinecure. From early morning to late evening, with barely time out for meals, Cauchy was pestered by the royal brat. Not only the elementary lessons of an ordinary school course had to be hammered somehow or another into the pampered boy, but Cauchy was detailed to see that his charge did not fall down and skin his knees on his gambols in the park. Needless to say the major part of Cauchy's instruction consisted in intimate talks on the peculiar brand of moral philosophy to which he was addicted; so perhaps it is as well that France finally decided not to take the Bourbons back to its heart, but to leave them and their innumerable descendants as prizes to be raffled off to the daughters of millionaires in the international marriage bureau.

In spite of almost constant attendance on his pupil Cauchy somehow managed to keep his mathematics going, dashing into his private quarters at odd moments to jot down a formula or scribble a hasty paragraph. The most impressive work of this period was the long memoir on the dispersion of light, in which Cauchy attempted to explain the phenomenon of dispersion (the separation of white light into colors owing to different refrangibilities of the colored lights composing the white) on the hypothesis that light is caused by the vibrations of an elastic solid. This work is of great interest in the history of physics, as it exemplified the tendency of the nineteenth century to try to account for physical phenomena in terms of mechanical models instead of merely constructing an abstract, mathematical theory to correlate observations. This was a departure from the prevailing practice of Newton and his successors—although there had been attempts to "explain" gravitation mechanically.

Today the tendency is in the opposite direction of a purely mathematical correlation and a complete abandonment of ethers, elastic solids, or other mechanical "explanations" more difficult to grasp than the thing explained. Physicists at present seem to have heeded Byron's query, "Who will then explain the explanation?" The elastic solid theory had a long and brilliant success, and even today some of the formulas Cauchy derived from his false hypothesis are in use. But the theory itself was abandoned when, as not infrequently happens, refined experimental technique and unsuspected phenomena (anomalous dispersion in this case) failed to accord with the predictions of the theory.

Cauchy escaped from his pupil in 1838 (he was then almost fifty). Friends in Paris had been urging him for some time to return, and Cauchy seized the excuse of his parents' golden wedding to bid adieu to Charles and all his entourage. By a special dispensation members of the Institut (of which the Academy of Sciences was, and is, a part) were not required to take an oath of allegiance to the Government, so Cauchy resumed his seat. His mathematical activity now became greater than ever. During the last nineteen years of his life he produced over 500 papers on all branches of mathematics, including mechanics, physics, and astronomy. Many of these works were long treatises.

His troubles were not yet over. When a vacancy occurred at the Collège de France Cauchy was unanimously elected to fill the place. But here there was no dispensation and before he could step into the position Cauchy would have to take the oath of allegiance. Believing the Government to be usurping the divine rights of his master, Cauchy stiffened his neck and refused to take the oath. Once more he was out of a job. But the Bureau des Longitudes could use a mathematician of his calibre. Again he was unanimously elected.

Then began an amusing tug of war between Baron Cauchy and the Bureau at one end of the rope and the unsanctified Government at the other. Conscious for once that it was making a fool of itself the Government let go and Cauchy was shot backwards into the Bureau without an oath. Defiance of the Government was grossly illegal, not to say treasonable, but Cauchy stuck to his job. His colleagues at the Bureau embarrassed the Government by politely ignoring its request to elect someone legally. For four years Cauchy turned his obstinate back on the Government and went on with his work.

To this period belong some of Cauchy's most important contributions to mathematical astronomy. Leverrier had unwittingly started Cauchy off with his memoir of 1840 on Pallas. This was a lengthy work packed with numerical calculations which it would take any referee as long to check as it had taken the author to perform them in the first place. When the memoir was presented to the Academy the officers began looking about for someone willing to undertake the inhuman task of verifying the correctness of the conclusions. Cauchy volunteered. Instead of following Leverrier's footsteps he quickly found shortcuts and invented new methods which enabled him to verify and extend the work in a remarkably short time.

The tussle with the Government reached its crisis in 1843 when Cauchy was fifty four. The Minister declined to be made a public laughing stock any longer and demanded that the Bureau hold an election to fill the position Cauchy refused to vacate. On the advice of his friends Cauchy laid his case before the people in an open letter. This letter is one of the finest things Cauchy ever wrote.

Whatever we may think of his quixotic championship of a cause which all but flyblown reactionaries knew had been well lost forever, we cannot help respecting Cauchy's fearlessness in stating his own case, with dignity and without passion, and in fighting for the freedom of his conscience. It was the old fight for free thought in a guise that was not very familiar then but is common enough now.

In the time of Galileo, Cauchy no doubt would have gone to the stake to maintain the freedom of his beliefs; under Louis Philippe he denied the right of any government to exact an oath of allegiance which traversed his conscience, and he suffered for his courage. His stand earned him the respect even of his enemies, and brought the Government into contempt, even in the eyes of its supporters. Presently the stupidity of repression was brought home to the Government in a way it could understand—street fighting, riots, strikes, civil war, and an unanswerable order to get out and stay out. Louis Philippe and all his gang were ousted in 1848. One of the first acts of the Provisional Government was to abolish the oath of allegiance. With rare good sense the politicians realized that all such oaths are either unnecessary or worthless.

In 1852, when Napoleon III took charge, the oath was restored. But by this time Cauchy had won his battle. Word was quietly passed to him that he might resume his lectures without taking the oath. It was understood on both sides that no fuss was to be made. The Government asked no thanks for its liberality, nor did Cauchy tender any, but went on with his lectures as if nothing had happened. From then to the end of his life he was the chief glory of the Sorbonne.

In the interim between official instability and unofficial stability Cauchy had taken time out to splinter a lance in defence of the Jesuits. The trouble was the old one—the State educational authorities insisting that the Jesuit training incurred a divided allegiance, the Jesuits defending religious instruction as the only sound basis for any education. It was a fight up Cauchy's own alley and he sailed into it with eloquent gusto. His defence of his friends was touching and sincere but unconvincing. Whenever Cauchy got off mathematics he substituted emotion for reason.

The Crimean War afforded Cauchy his last opportunity for getting himself disliked by his harder-headed colleagues. He became an enthusiastic propagandist for a singular enterprise known as Work of the Schools of the Orient. "Work" here is intended in the sense of a particular "good work."

"It was necessary," according to the sponsors of the Work in 1855, "to remedy the disorders of the past and at the same time impose a double check on Muscovite ambition and Mohammedan fanaticism: above all to prepare the regeneration of peoples brutalized by the Koran. . . ." In short the Crimean War had been the customary bayonet preparing the way for the Cross. Deeply impressed by the obvious necessity of replacing the brutalizing Koran by something more humane, Cauchy threw himself into the project, "completing and consolidating . . . the work of emancipation so admirably begun by the arms of France."

The Jesuit Council, grateful for Cauchy's expert help, gave him full credit for many of the details (including the collection of subscriptions) which were to accomplish "the moral regeneration of peoples enslaved to the law of the Koran, the triumph of the Gospel round the cradle and the sepulchre of Jesus Christ being the sole acceptable compensation for these billows of blood that have been shed" by the Christian French, English, Russians, Sardinians, and the Mohammedan Turks in the Crimean War.

It was good works of this character that caused some of Cauchy's friends, out of sympathy with the pious spirit of the orthodox religion of the time, to call him a smug hypocrite. The epithet was wholly undeserved. Cauchy was one of the sincerest bigots that ever lived.

The net result of the Work was the particularly revolting massacre of May, 1860. Cauchy did not live to see his labors crowned.

Reputations of great mathematicians are subject to the same vicissitudes as those of other great men. For long after his death—and even today—Cauchy was severely criticized for overproduction and hasty composition. His total output is 789 papers (many of them very extensive works) filling twenty four large quarto volumes. Criticism of this sort always seems rather beside the point if a man has put out a mass of first rate work in addition to some that is not of high quality, and is usually indulged in by men who themselves have done comparatively little, and that little not of the highest order of originality. Cauchy's part in modern mathematics is somewhere not far from the center of the stage. This is now almost universally admitted, if grudgingly in some quarters. Since his death, especially in recent decades, Cauchy's reputation as a mathematician has risen steadily. The methods he introduced, his whole program inaugurating the first period of modern rigor, and his almost unequalled inventiveness have made a mark on mathematics that is, so far as we can now see, destined to be visible for many years to come.

One apparently unimportant detail out of all the mass of new things Cauchy did may be mentioned as an illustration of his prophetic originality. Instead of using the "imaginary" $i \, (= \sqrt{-1})$ Cauchy proposed to accomplish all that complex numbers do in mathematics by operating with congruences to the modulus $i^2 + 1$. This was done in 1847. The paper—a short one—attracted but little attention. Yet it is the germ of something—Kronecker's program—that is on its way to revolutionizing some of the fundamental concepts of mathematics. This matter will reappear frequently in later chapters, so we may pass it here with this allusion.

In social contacts Cauchy was extremely polite, not to say oily on occasion as when, for example, he was soliciting subscriptions for one of his jousts. His habits were temperate and in all things except mathematics and religion he was moderate. On the last he lacked ordinary common sense. Everyone who came near him was a prospect for conversion. When William Thomson (Lord Kelvin) as a young man of twenty one called on Cauchy to discuss mathematics, Cauchy spent the time trying to convert his visitor—then a staunch adherent of the Scottish Free Church—to Catholicism.

Cauchy had his share of rows over priority in which his enemies accused him of greed and unfair play. His last year was marred by one such dispute wherein it would seem that Cauchy had no case. But with his usual stubbornness where a matter of principle was involved he braved the outcry and stuck to his point with invincible sweetness and pertinacity.

Another peculiarity added to Cauchy's unpopularity with his scientific colleagues. In scientific academies and societies a man is supposed to base his vote for a candidate only on the candidate's scientific merits; any other procedure is considered bad ethics. Whether rightly or wrongly Cauchy was accused of voting in accordance with his religious and political views. His last years were embittered by what he considered a lack of understanding among his colleagues on this and similar foibles. Neither side could get the point of view of the other.

Cauchy died rather unexpectedly in his sixty eighth year on May 23, 1857. Hoping to benefit a bronchial trouble, he retired to the country to recuperate, only to be smitten with a fever which proved fatal. A few hours before his death he was talking animatedly with the Archbishop of Paris of the charitable works he had in view—charity was one of Cauchy's lifelong interests. His last words were addressed to the Archbishop: "Men pass away but their deeds abide."

CHAPTER SIXTEEN

The Copernicus of Geometry

LOBATCHEWSKY

> *Lobatchewsky's theory was incomprehensible to his contemporaries, appearing as it did to contradict an axiom whose necessity is based only on a prejudice sanctified by thousands of years.*—The Editors of Lobatchewsky's Works

GRANTING THAT THE COMMONLY ACCEPTED ESTIMATE of the importance of what Copernicus did is correct, we shall have to admit that it is either the highest praise or the severest condemnation humanly possible to call another man the "Copernicus" of anything. When we understand what Lobatchewsky did in the creation of non-Euclidean geometry, and consider its significance for all human thought, of which mathematics is only a small if important part, we shall probably agree that Clifford (1845–1879), himself a great geometer and far more than a "mere mathematician," was not overpraising his hero when he called Lobatchewsky "The Copernicus of Geometry."

Nikolas Ivanovitch Lobatchewsky, the second son of a minor government official, was born on November 2, 1793 in the district of Makarief, government of Nijni Novgorod, Russia. The father died when Nikolas was seven, leaving his widow, Praskovia Ivanovna, the care of three young sons. As the father's salary had barely sufficed to keep his family going while he was alive the widow found herself in extreme poverty. She moved to Kazan, where she prepared her boys for school as best she could, and had the satisfaction of seeing them accepted, one after the other, as free scholars at the Gymnasium. Nikolas was admitted in 1802 at the age of eight. His progress was phenomenally rapid in both mathematics and the classics. At the age of fourteen he was ready for the university. In 1807 he entered the University of Kazan (founded in 1805), where he was to spend the next forty years of his life as student, assistant professor, professor, and finally rector.

Hoping to make Kazan ultimately the equal of any university in

Europe, the authorities had imported several distinguished professors from Germany. Among these was the astronomer Littrow, who later became director of the Observatory at Vienna, whom Abel mentioned as one of his excuses for seeing something of "the south." The German professors quickly recognized Lobatchewsky's genius and gave him every encouragement.

In 1811, at the age of eighteen, Lobatchewsky obtained his master's degree after a short tussle with the authorities, whose ire he had incurred through his youthful exuberance. His German friends on the faculty took his part and he got his degree with distinction. At this time his elder brother Alexis was in charge of the elementary mathematical courses for the training of minor government officials, and when Alexis presently took a sick-leave, Nikolas stepped into his place as substitute. Two years later, at the age of twenty one, Lobatchewsky received a probationary appointment as "Extraordinary Professor" or, as we should say in America, Assistant Professor.

Lobatchewsky's promotion to an ordinary professorship came in 1816 at the unusually early age of twenty three. His duties were heavy. In addition to his mathematical work he was charged with courses in astronomy and physics, the former to substitute for a colleague on leave. The fine balance with which he carried his heavy load made him a conspicuous candidate for yet more work, on the theory that a man who can do much is capable of doing more, and presently Lobatchewsky found himself University Librarian and curator of the chaotically disordered University Museum.

Students are often an unruly lot before life teaches them that generosity of spirit does not pay in the cut-throat business of earning a living. Among Lobatchewsky's innumerable duties from 1819 till the death of the Czar Alexander in 1825 was that of supervisor of all the students in Kazan, from the elementary schools to the men taking post-graduate courses in the University. The supervision was primarily over the political opinions of his charges. The difficulties of such a thankless job can easily be imagined. That Lobatchewsky contrived to send in his reports day after day and year after year to his suspicious superiors without once being called on the carpet for laxity in espionage, and without losing the sincere respect and affection of all the students, says more for his administrative ability than do all the gaudy orders and medals which a grateful Government showered on him and with which he delighted to adorn himself on state occasions.

The collections in the University Museum to all appearance had been tossed in with a pitchfork. A similar disorder made the extensive library practically unusable. Lobatchewsky was commanded to clean up these messes. In recognition of his signal services the authorities promoted him to the deanship of the Faculty of Mathematics and Physics, but omitted to appropriate any funds for hiring assistance in straightening out the library and the museum. Lobatchewsky did the work with his own hands, cataloguing, dusting and casing, or wielding a mop as the occasion demanded.

With the death of Alexander in 1825 things took a turn for the better. The particular official responsible for the malicious persecution of the University of Kazan was kicked out as being too corrupt for even a government post, and his successor appointed a professional curator to relieve Lobatchewsky of his endless tasks of cataloguing books, dusting mineral specimens, and deverminizing stuffed birds. Needing political and moral support for his work in the University, the new curator did some high politics on his own account and secured the appointment in 1827 of Lobatchewsky as Rector. The mathematician was now head of the University, but the new position was no sinecure. Under his able direction the entire staff was reorganized, better men were brought in, instruction was liberalized in spite of official obstruction, the library was built up to a higher standard of scientific sufficiency, a mechanical workshop was organized for making the scientific instruments required in research and instruction, an observatory was founded and equipped—a pet project of the energetic Rector's—and the vast mineralogical collection, representative of the whole of Russia, was put in order and constantly enriched.

Even the new dignity of his rectorship did not deter Lobatchewsky from manual labor in the library and museum when he felt that his help was necessary. The University was his life and he loved it. On the slightest provocation he would take off his collar and coat and go to work. Once a distinguished foreigner, taking the coatless Rector for a janitor or workman, asked to be shown through the libraries and museum collections. Lobatchewsky showed him the choicest treasures, explaining as he exhibited. The visitor was charmed and greatly impressed by the superior intelligence and courtesy of this obliging Russian worker. On parting from his guide he tendered a handsome tip. Lobatchewsky, to the foreigner's bewilderment, froze up in a cold rage and indignantly spurned the proffered coin. Thinking it but just one more eccentricity of the high-minded Russian janitor, the visitor bowed and pocketed his money. That evening he and Lobatchewsky met at the Governor's dinner table, where apologies were offered and accepted on both sides.

Lobatchewsky was a strong believer in the philosophy that in order to get a thing done to your own liking you must either do it yourself or understand enough about its execution to be able to criticize the work of another intelligently and constructively. As has been said, the University was his life. When the Government decided to modernize the buildings and add new ones, Lobatchewsky made it his business to see that the work was done properly and the appropriation not squandered. To fit himself for this task he learned architecture. So practical was his mastery of the subject that the buildings were not only handsome and suited for their purposes but, what must be almost unique in the history of governmental building, were constructed for less money than had been appropriated. Some years later (in 1842) a disastrous fire destroyed half Kazan and took with it Lobatchewsky's finest buildings, including the barely completed observatory—the pride of his heart. But due to his energetic cool-headedness the instruments and the library were saved. After the fire he set to work immediately to rebuild. Two years later not a trace of the disaster remained.

We recall that 1842, the year of the fire, was also the year in which, thanks to the good offices of Gauss, Lobatchewsky was elected a foreign correspondent of the Royal Society of Göttingen for his creation of non-Euclidean geometry. Although it seems incredible that any man so excessively burdened with teaching and administration as Lobatchewsky was, could find the time to do even one piece of mediocre scientific work, he had actually, somehow or another, made the opportunity to create one of the great masterpieces of all mathematics and a landmark in human thought. He had worked at it off and on for twenty years or more. His first public communication on the subject, to the Physical-Mathematical Society of Kazan, was made in 1826. He might have been speaking in the middle of the Sahara Desert for all the echo he got. Gauss did not hear of the work till about 1840.

Another episode in Lobatchewsky's busy life shows that it was not only in mathematics that he was far ahead of his time. The Russia of

1830 was probably no more sanitary than that of a century later, and it may be assumed that the same disregard of personal hygiene which filled the German soldiers in the World War with an amazed disgust for their unfortunate Russian prisoners, and which today causes the industrious proletariat to use the public parks and playgrounds of Moscow as vast and convenient latrines, distinguished the luckless inhabitants of Kazan in Lobatchewsky's day when the cholera epidemic found them richly prepared for a prolonged visitation. The germ theory of disease was still in the future in 1830, although progressive minds had long suspected that filthy habits had more to do with the scourge of the pestilence than the anger of the Lord.

On the arrival of the cholera in Kazan the priests did what they could for their smitten people, herding them into the churches for united supplication, absolving the dying and burying the dead, but never once suggesting that a shovel might be useful for any purpose other than digging graves. Realizing that the situation in the town was hopeless, Lobatchewsky induced his faculty to bring their families to the University and prevailed upon—practically ordered—some of the students to join him in a rational, human fight against the cholera. The windows were kept closed, strict sanitary regulations were enforced, and only the most necessary forays for replenishing the food supply were permitted. Of the 660 men, women and children thus sanely protected, only sixteen died, a mortality of less than 2.5 per cent. Compared to the losses under the traditional remedies practised in the town this was negligible.

It might be imagined that after all his distinguished services to the state and his European recognition as a mathematician, Lobatchewsky would be in line for substantial honors from his Government. To imagine anything of the kind would not only be extremely naïve but would also traverse the scriptural injunction "Put not your trust in princes." As a reward for all his sacrifices and his unswerving loyalty to the best in Russia, Lobatchewsky was brusquely relieved in 1846 of his Professorship and his Rectorship of the university. No explanation of this singular and unmerited double insult was made public. Lobatchewsky was in his fifty fourth year, vigorous of body and mind as ever, and more eager than he had ever been to continue with his mathematical researches. His colleagues to a man protested against the outrage, jeopardizing their own security, but were curtly informed that they as mere professors were constitutionally incapable of comprehending the higher mysteries of the science of government.

The ill-disguised disgrace broke Lobatchewsky. He was still permitted to retain his study at the University. But when his successor, hand-picked by the Government to discipline the disaffected faculty, arrived in 1847 to take up his ungracious task, Lobatchewsky abandoned all hope of ever being anybody again in the University which owed its intellectual eminence almost entirely to his efforts, and he appeared thereafter only occasionally to assist at examinations. Although his eyesight was failing rapidly he was still capable of intense mathematical thinking.

He still loved the University. His health broke when his son died, but he lingered on, hoping that he might still be of some use. In 1855 the University celebrated its semicentennial anniversary. To do honor to the occasion, Lobatchewsky attended the exercises in person to present a copy of his *Pangeometry*, the completed work of his scientific life. This work (in French and Russian) was not written by his own hand, but was dictated, as Lobatchewsky was now blind. A few months later he died, on February 24, 1856, at the age of sixty two.

To see what Lobatchewsky did we must first glance at Euclid's outstanding achievement. The name Euclid until quite recently was practically synonymous with elementary school geometry. Of the man himself very little is known beyond his doubtful dates, 330–275. B.C. In addition to a systematic account of elementary geometry his *Elements* contain all that was known in his time of the theory of numbers. Geometrical teaching was dominated by Euclid for over 2200 years. His part in the *Elements* appears to have been principally that of a coordinator and logical arranger of the scattered results of his predecessors and contemporaries, and his aim was to give a connected, reasoned account of elementary geometry such that every statement in the whole long book could be referred back to the postulates. Euclid did not attain this ideal or anything even distantly approaching it, although it was assumed for centuries that he had.

Euclid's title to immortality is based on something quite other than the supposed logical perfection which is still sometimes erroneously ascribed to him. This is his recognition that the fifth of his postulates (his Axiom XI) is a pure assumption. The fifth postulate can be stated in many equivalent forms, each of which is deducible from any one of the others by means of the remaining postulates of Euclid's geometry. Possibly the simplest of these equivalent statements is the following: Given any straight line l and a point P not on l, then in the plane determined by l and P it is possible to draw *precisely one* straight line l' through P such that l' never meets l no matter how far l' and l are extended (in either direction). Merely as a nominal definition we

say that two straight lines lying in one plane which never meet are *parallel*. Thus the fifth postulate of Euclid asserts that through P there is precisely one straight line parallel to l. Euclid's penetrating insight into the nature of geometry convinced him that this postulate had not, in his time, been deduced from the others, although there had been many attempts to *prove* the postulate. Being unable to deduce the postulate himself from his other assumptions, and wishing to use it in the proofs of many of his theorems, Euclid honestly set it out with his other postulates.

There are one or two simple matters to be disposed of before we come to Lobatchewsky's Copernican part in the extension of geometry. We have alluded to "equivalents" of the parallel postulate. One of these, "the hypothesis of the right angle," as it is called, will suggest two possibilities, neither equivalent to Euclid's assumption, one of which introduces Lobatchewsky's geometry, the other, Riemann's.

Consider a figure $AXYB$ which "looks like" a rectangle, consisting of four straight lines AX, XY, YB, BA, in which BA (or AB) is the base, AX and YB (or BY) are drawn equal and perpendicular to AB, and on the same side of AB. The essential things to be remembered about this figure are that each of the angles XAB, YBA (at the base) is a right angle, and that the sides AX, BY are equal in length. *Without using the parallel postulate*, it can be proved that the angles AXY, BYX, are *equal*, but, *without* using this postulate, *it is impossible to prove that AXY, BYX are right angles*, although they look it. If we *assume* the *parallel postulate* we can *prove* that AXY, BYX are *right angles* and, conversely, if we *assume* that AXY, BYX are *right* *angles*, we can *prove* the parallel postulate. Thus *the assumption that AXY, BYX are right angles* is equivalent *to the parallel postulate*. This assumption is today called *the hypothesis of the right angle* (since both angles are right angles the singular instead of the plural "angles" is used).

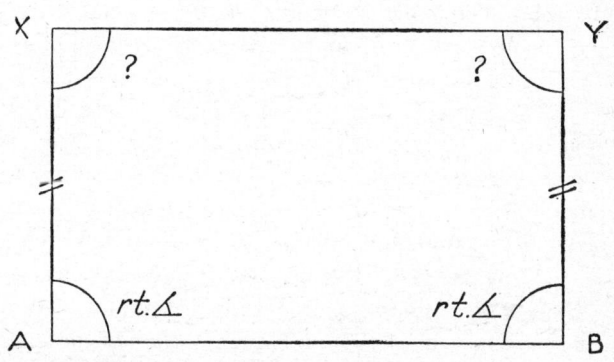

[Figure: quadrilateral $ABYX$ with A lower-left, B lower-right, X upper-left, Y upper-right; the angles at A and B are marked "rt. ∠", the angles at X and Y are marked "?", and the sides AX and BY are marked equal.]

It is known that the hypothesis of the right angle leads to a consistent, practically useful geometry, in fact to Euclid's geometry refurbished to meet modern standards of logical rigor. But the figure suggests two other possibilities: each of the equal angles AXY, BYX is *less* than a right angle—*the hypothesis of the acute angle*; each of the equal angles AXY, BYX is *greater* than a right angle —*the hypothesis of the obtuse angle*. Since any angle can satisfy one, and only one, of the requirements that it be *equal to*, *less than*, or *greater than* a right angle, the three hypotheses—of the right angle, acute angle, and obtuse angle respectively—exhaust the possibilities.

Common experience predisposes us in favor of the first hypothesis. To see that each of the others is not as unreasonable as might at first appear we shall consider something closer to actual human experience than the highly idealized "plane" in which Euclid imagined his figures drawn. But first we observe that neither the hypothesis of the acute angle nor that of the obtuse angle will enable us to prove Euclid's parallel postulate, because, as has been stated above, Euclid's postulate is *equivalent* to the hypothesis of the *right angle* (in the sense of interdeducibility; the hypothesis of the right angle is both necessary and sufficient for the deduction of the parallel postulate). Hence if we succeed in constructing geometries on either of the two new hypotheses, we shall not find in them parallels in Euclid's sense.

To make the other hypotheses less unreasonable than they may seem at first sight, suppose the Earth were a perfect sphere (without irregularities due to mountains, etc.). A plane drawn through the center of this ideal Earth cuts the surface in a *great circle*. Suppose we wish to go from one point A to another B on the surface of the Earth, keeping always *on* the surface in passing from A to B, and suppose further that we wish to make the journey by the shortest way possible. This is the problem of "great circle sailing." Imagine a plane passed through A, B, and the center of the Earth (there is one, and only one, such plane). This plane cuts the surface in a great circle. To make our shortest journey we go from A to B along the shorter of the two arcs of this great circle joining them. If A, B happen to lie at the extremities of a diameter, we may go by either arc.

The preceding example introduces an important definition, that of a *geodesic on a surface*, which will now be explained. It has just been seen that the *shortest* distance joining two points on a sphere, the distance itself being measured *on the surface*, is *an* arc of the great circle joining them. We have also seen that the *longest* distance joining the two points is the *other* arc of the same great circle, except in the case when the points are ends of a diameter, when shortest and longest are equal. In the chapter on Fermat "greatest" and "least" were subsumed under the common name "extreme," or "extremum." We recall now one usual definition of a straight-line segment joining two points in a plane—"the *shortest distance* between two points." Transferring this to the sphere, we say that to *straight line* in the *plane* corresponds *great circle* on the *sphere*. Since the Greek word for the Earth is the first syllable ge ($γῆ$) of geodesic we call *all extrema joining any two points on any surface the geodesics of that surface*. Thus in a plane the geodesics are Euclid's straight lines; on a sphere they are great circles. A geodesic can be visualized as the position taken by a string stretched as tight as possible between two points on a surface.

Now, in navigation at least, an ocean is not thought of as a flat surface (Euclidean plane) if even moderate distances are concerned; it is taken for what it very approximately is, namely a part of the surface of a sphere, and the geometry of great circle sailing is not Euclid's. Thus Euclid's is not the only geometry of human utility. On the plane two geodesics intersect in exactly *one* point *unless* they are parallel, when they do not intersect (in Euclidean geometry); but on the sphere *any* two geodesics always intersect in precisely *two* points. Again, on a plane, no two geodesics can enclose a space—as Euclid assumed in one of the postulates for his geometry; on a sphere, any two geodesics always enclose a space.

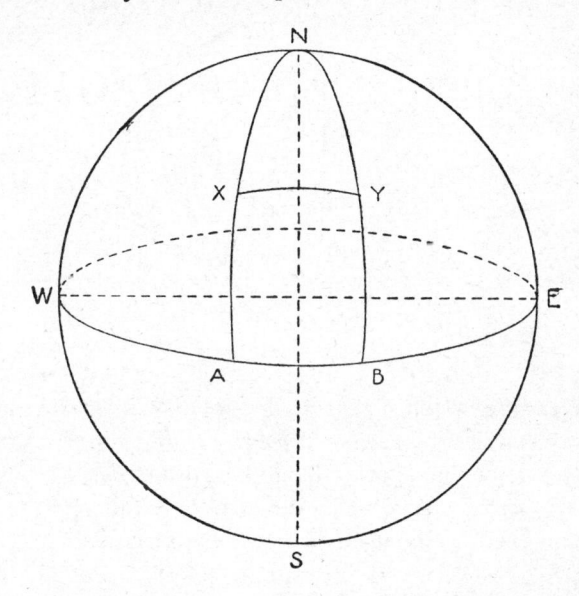

Imagine now the equator on the sphere and two geodesics drawn through the north pole perpendicular to the equator. In the northern hemisphere this gives a triangle with curved sides, two of which are equal. Each side of this triangle is an arc of a geodesic. Draw any other geodesic cutting the two equal sides so that the intercepted parts between the equator and the cutting line are equal. We now have, *on the sphere*, the four-sided figure corresponding to the $AXYB$ we had a few moments ago in the plane. The two angles at the base of this figure are right angles and the corresponding sides are equal, as before, *but each of the equal angles at X, Y is now greater than a right angle.* So, in the highly practical geometry of great circle sailing, which is closer to real human experience than the idealized diagrams of elementary geometry ever get, it is not Euclid's postulate which is true—or its equivalent in the hypothesis of the right angle—but the geometry which follows from the hypothesis of the obtuse angle.

In a similar manner, inspecting a less familiar surface, we can make reasonable the hypothesis of the acute angle. The surface looks like two infinitely long trumpets soldered together at their largest ends. To describe it more accurately we must introduce the plane curve called the *tractrix*, which is generated as follows. Let two lines XOX', YOY' be drawn in a horizontal plane intersecting at right angles in O, as in Cartesian geometry. Imagine an inextensible fiber lying along YOY', to one end of which is attached a small heavy pellet; the other end of the fiber is at O. Pull this end out along the line OX.

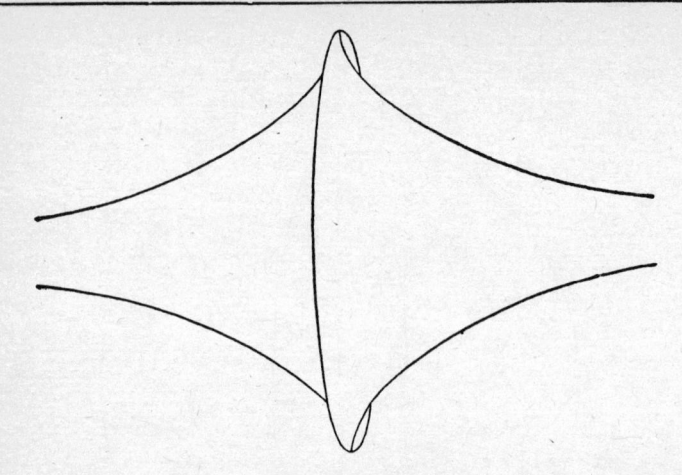

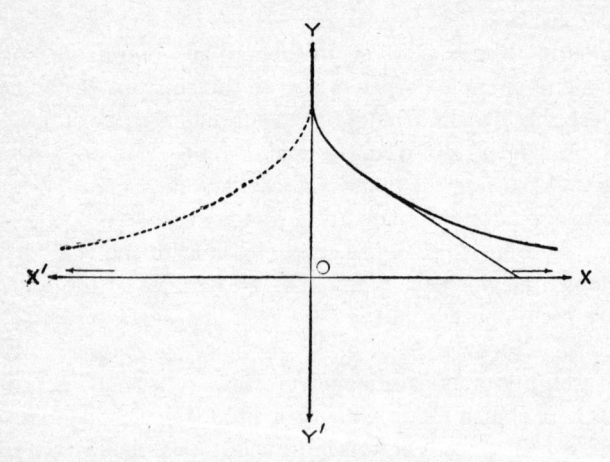

As the pellet follows, it traces out one half of the tractrix; the other half is traced out by drawing the end of the fiber along OX', and of course is merely the reflection or image in OY of the first half. The drawing out is supposed to continue indefinitely—"to infinity"—in each instance. Now imagine the tractrix to be revolved about the line XOX'. The double-trumpet surface is generated; for reasons we need not go into (it has constant negative curvature) it is called a *pseudo-sphere*. If on this surface we draw the four-sided figure with two equal sides and two right angles as before, using geodesics, we find that the hypothesis of the acute angle is realized.

Thus the hypotheses of the right angle, the obtuse angle, and the acute angle respectively are true on a Euclidean plane, a sphere, and a pseudosphere respectively, and in all cases "straight lines" are *geodesics* or *extrema*. Euclidean geometry is a limiting, or degenerate, case of geometry on a sphere, being attained when the radius of the sphere becomes infinite.

Instead of constructing a geometry to fit the Earth as human beings now know it, Euclid apparently proceeded on the assumption that the Earth is flat. If Euclid did not, his predecessors did, and by the time the theory of "space," or geometry, reached him the bald *assumptions* which he embodied in his postulates had already taken on the aspect of hoary and immutable necessary truths, revealed to mankind by a higher intelligence as the veritable essence of all material things. It took over two thousand years to knock the eternal truth out of geometry, and Lobatchewsky did it.

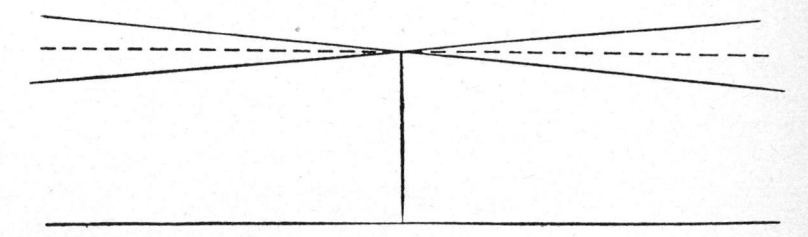

To use Einstein's phrase, Lobatchewsky *challenged an axiom*. Anyone who challenges an "accepted truth" that has seemed necessary or reasonable to the great majority of sane men for 2000 years or more takes his scientific reputation, if not his life, in his hands. Einstein himself challenged the axiom that two events can happen in *different places* at the *same time*, and by analyzing this hoary assumption was led to the invention of the special theory of relativity. Lobatchewsky challenged the assumption that Euclid's parallel postulate or, what is equivalent, the hypothesis of the right angle, is necessary to a consistent geometry, and he backed his challenge by producing a system of geometry based on the hypothesis of the acute angle in which there is not *one* parallel through a fixed point to a given straight line but *two*. Neither of Lobatchewsky's parallels meets the line to which both are parallel, nor does any straight line drawn through the fixed point and lying within the angle formed by the two parallels. This apparently bizarre situation is "realized" by the geodesics on a pseudosphere.

For any everyday purpose (measurements of distances, etc.), the differences between the geometries of Euclid and Lobatchewsky are too small to count, but this is not the point of importance: each is self-consistent and each is adequate for human experience. Lobatchewsky abolished the *necessary* "truth" of Euclidean geometry. His geometry was but the first of several constructed by his successors. Some of these substitutes for Euclid's geometry—for instance the Riemannian geometry of general relativity—are today at least as important in the still living and growing parts of physical science as Euclid's was, and is, in the comparatively static and classical parts. For some purposes Euclid's geometry is best or at least sufficient, for others it is inadequate and a non-Euclidean geometry is demanded.

Euclid in some sense was believed for 2200 years to have discovered an absolute truth or a necessary mode of human perception in his system of geometry. Lobatchewsky's creation was a pragmatic demonstration of the error of this belief. The boldness of his challenge and its successful outcome have inspired mathematicians and scientists in general to challenge other "axioms" or accepted "truths," for example the "law" of causality, which, for centuries, have seemed as necessary to straight thinking as Euclid's postulate appeared till Lobatchewsky discarded it.

The full impact of the Lobatchewskian method of challenging axioms has probably yet to be felt. It is no exaggeration to call Lobatchewsky the Copernicus of Geometry, for geometry is only a part of the vaster domain which he renovated; it might even be just to designate him as a Copernicus of all thought.

CHAPTER SEVENTEEN

Genius and Poverty

ABEL

I have finished a monument more lasting than bronze and loftier than the pyramids reared by kings, that neither corroding rain nor the uncontrolled north wind can dash apart, nor the countless succession of years and the flight of ages. I shall not wholly die; that greater part of me shall escape Death and ever shall I grow, still fresh in the praise of posterity.—HORACE (*Odes*, 3, XXX)

AN ASTROLOGER IN THE YEAR 1801 might have read in the stars that a new galaxy of mathematical genius was about to blaze forth inaugurating the greatest century of mathematical history. In all that galaxy of talent there was no brighter star than Niels Henrik Abel, the man of whom Hermite said, "He has left mathematicians something to keep them busy for five hundred years."

Abel's father was the pastor of the little village of Findö, in the diocese of Kristiansand, Norway, where his second son, Niels Henrik, was born on August 5, 1802. On the father's side several ancestors had been prominent in the work of the church and all, including Abel's father, were men of culture. Anne Marie Simonsen, Abel's mother, was chiefly remarkable for her great beauty, love of pleasure, and general flightiness—quite an exciting combination for a pastor's helpmeet. From her Abel inherited his striking good looks and a very human desire to get something more than everlasting hard work out of life, a desire he was seldom able to gratify.

The pastor was blessed with seven children in all at a time when Norway was desperately poor as the result of wars with England and Sweden, to say nothing of a famine thrown in for good measure between wars. Nevertheless the family was a happy one. In spite of pinching poverty and occasional empty stomachs they kept their chins up. There is a charming picture of Abel after his mathematical genius had seized him sitting by the fireside with the others chattering and laughing in the room while he researched with one eye on his mathematics and the other on his brothers and sisters. The noise never distracted him and he joined in the badinage as he wrote.

Like several of the first-rank mathematicians Abel discovered his talent early. A brutal schoolmaster unwittingly threw opportunity Abel's way. Education in the first decades of the nineteenth century was virile, at least in Norway. Corporal punishment, as the simplest method of toughening the pupils' characters and gratifying the sadistic inclinations of the masterful pedagogues, was generously administered for every trivial offense. Abel was not awakened through his own skin, as Newton is said to have been by that thundering kick donated by a playmate, but by the sacrifice of a fellow student who had been flogged so unmercifully that he died. This was a bit too thick even for the rugged schoolboard and they deprived the teacher of his job. A competent but by no means brilliant mathematician filled the vacancy, Bernt Michael Holmboë (1795–1850), who was later to edit the first edition of Abel's collected works in 1839.

Abel at the time was about fifteen. Up till now he had shown no marked talent for anything except taking his troubles with a sense of humor. Under the kindly, enlightened Holmboë's teaching Abel suddenly discovered what he was. At sixteen he began reading privately and thoroughly digesting the great works of his predecessors, including some of those of Newton, Euler, and Lagrange. Thereafter real mathematics was not only his serious occupation but his fascinating delight. Asked some years later how he had managed to forge ahead so rapidly to the front rank he replied, "By studying the masters, not their pupils"—a prescription some popular writers of textbooks might do well to mention in their prefaces as an antidote to the poisonous mediocrity of their uninspired pedagogics.

Holmboë and Abel soon became close friends. Although the teacher was himself no creative mathematician he knew and appreciated the masterpieces of mathematics, and under his eager suggestions Abel was soon mastering the toughest of the classics, including the *Disquisitiones Arithmeticae* of Gauss.

Today it is a commonplace that many fine things the old masters thought they had proved were not really proved at all. Particularly is this true of some of Euler's work on infinite series and some of Lagrange's on analysis. Abel's keen mind was one of the first to detect the gaps in his predecessors' reasoning, and he resolved to devote a fair share of his lifework to caulking the cracks and making the reasoning watertight. One of his classics in this direction is the first *proof* of the *general* binomial theorem, special cases of which had been stated by Newton and Euler. It is not easy to give a sound proof in the general case, so perhaps it is not astonishing to find alleged proofs still displayed in the schoolbooks as if Abel had never lived. This proof however was only a detail in Abel's vaster program of cleaning up the theory and application of infinite series.

Abel's father died in 1820 at the early age of forty eight. At the time Abel was eighteen. The care of his mother and six children fell on his shoulders. Confident of himself Abel assumed his sudden responsibilities cheerfully. Abel was a genial and optimistic soul. With no more than strict justice he foresaw himself as an honored and moderately prosperous mathematician in a university chair. Then he could provide for the lot of them in reasonable security. In the meantime he took private pupils and did what he could. In passing it may be noted that Abel was a very successful teacher. Had he been footloose poverty would never have bothered him. He could have earned enough for his own modest needs, somehow or other, at any time. But with seven on his back he had no chance. He never complained, but took it all in his stride as part of the day's work and kept at his mathematical researches in every spare moment.

Convinced that he had one of the greatest mathematicians of all time on his hands, Holmboë did what he could by getting subsidies for the young man and digging down generously into his own none too deep pocket. But the country was poor to the point of starvation and not nearly enough could be done. In those years of privation and incessant work Abel immortalized himself and sowed the seeds of the disease which was to kill him before he had half done his work.

Abel's first ambitious venture was an attack on the general equation of the fifth degree (the "quintic"). All of his great predecessors in algebra had exhausted their efforts to produce a solution, without success. We can easily imagine Abel's exultation when he mistakenly imagined he had succeeded. Through Holmboë the supposed solution was sent to the most learned mathematical scholar of the time in Denmark who, fortunately for Abel, asked for further particulars without committing himself to an opinion on the correctness of the solution. Abel in the meantime had found the flaw in his reasoning. The supposed solution was of course no solution at all. This failure gave him a most salutary jolt; it jarred him onto the right track and caused him to doubt whether an algebraic solution was possible. He *proved the impossibility*. At the time he was about nineteen. But he had been anticipated, at least in part, in the whole project.

As this question of the general quintic played a rôle in algebra similar to that of a crucial experiment to decide the fate of an entire scientific theory, it is worth a moment's attention. We shall quote presently a few things Abel himself says.

The nature of the problem is easily described. In early school algebra we learn to solve the *general* equations of the *first* and *second* degrees in the unknown x, say

$$ax + b = 0, \ ax^2 + bx + c = 0,$$

and a little later those of the *third* and *fourth* degrees, say

$$ax^3 + bx^2 + cx + d = 0, \ ax^4 + bx^3 + cx^2 + dx + e = 0.$$

That is, we produce *finite* (closed) formulas for each of these *general* equations of the first four degrees, expressing the unknown x in terms of the given coefficients a,b,c,d,e. A solution such as any one of these four which can be obtained by only a *finite number of additions, multiplications, subtractions, divisions, and extractions of roots*, all these operations being performed on the given coefficients, is called *algebraic*. The important qualification in this definition of an *algebraic solution* is "finite"; there is no difficulty in describing solutions for *any* algebraic equation which contain no extractions of roots at all, but which do imply an *infinity* of the other operations named.

After this success with algebraic equations of the first four degrees, algebraists struggled for nearly three centuries to produce a similar *algebraic solution* for the general quintic

$$ax^5 + bx^4 + cx^3 + dx^2 + ex + f = 0.$$

They failed. It is here that Abel enters.

The following extracts are given partly to show how a great inventive mathematician thought and partly for their intrinsic interest. They are from Abel's memoir *On the algebraic resolution of equations*.

"One of the most interesting problems of algebra is that of the algebraic solution of equations. Thus we find that nearly all mathematicians of distinguished rank have treated this subject. We arrive without difficulty at the general expression of the roots of equations of the first four degrees. A uniform method for solving these equations was discovered and it was believed to be applicable to an equation of any degree; but in spite of all the efforts of Lagrange and other distinguished mathematicians the proposed end was not reached. That led to the presumption that the solution of general equations was impossible algebraically; but this is what could not be decided, since the method followed could lead to decisive conclusions only in the case where the equations were solvable. In effect they proposed to solve equations without knowing whether it was possible. In this way one might indeed arrive at a solution, although that was by no means certain; but if by ill luck the solution was impossible, one might seek it for an eternity, without finding it. To arrive infallibly at something in this matter, we must therefore follow another road. We can give the problem such a form that it shall always be possible to solve it, as we can always do with any problem.* Instead of asking for a relation of which it is not known whether it exists or not, we must ask whether such a relation is indeed possible. . . . When a problem is posed in this way, the very statement contains the germ of the solution and indicates what road must be taken; and I believe there will be few instances where we shall fail to arrive at propositions of more or less importance, even when the complication of the calculations precludes a complete answer to the problem."

He goes on to say that this, the true scientific method to be followed, has been but little used owing to the extreme complication of the calculations (algebraic) which it entails; "but," he adds, "in many instances this complication is only apparent and vanishes after the first attack." He continues:

"I have treated several branches of analysis in this manner, and although I have often set myself problems beyond my powers, I have nevertheless arrived at a large number of general results which throw a strong light on the nature of those quantities whose elucidation is the object of mathematics. On another occasion I shall give the results at which I have arrived in these researches and the procedure which has led me to them. In the present memoir I shall treat the problem of the algebraic solution of equations in all its generality."

---

*". . . *ce qu'on peut toujours faire d'un problème quelconque*" is what Abel says. This seems a trifle too optimistic; at least for ordinary mortals. How would the method be applied to Fermat's Last Theorem?

Presently he states two general inter-related problems which he proposes to discuss:

"1. To find all the equations of any given degree which are solvable algebraically.

2. To determine whether a given equation is or is not solvable algebraically."

At bottom, he says, these two problems are the same, and although he does not claim a *complete* solution, he does *indicate* an infallible method (*des moyens sûrs*) for disposing of them fully.

Abel's irrepressible inventiveness hurried him on to vaster problems before he had time to return to these; their complete solution—the explicit statement of necessary and sufficient conditions that an algebraic equation be solvable algebraically—was to be reserved for Galois. When this memoir of Abel's was published in 1828, Galois was a boy of sixteen, already well started on his career of fundamental discovery. Galois later came to know and admire the work of Abel; it is probable that Abel never heard the name of Galois, although when Abel visited Paris he and his brilliant successor could have been only a few miles apart. But for the stupidity of Galois' teachers and the loftiness of some of Abel's mathematical "superiors," it is quite possible that he and Abel might have met.

Epoch-making as Abel's work in algebra was, it is overshadowed by his creation of a new branch of analysis. This, as Legendre said, is Abel's "time-outlasting monument." If the story of his life adds nothing to the splendor of his accomplishment it at least suggests what the world lost when he died. It is a somewhat discouraging tale. Only Abel's unconquerable cheerfulness and unyielding courage under the stress of poverty and lack of encouragement from the mathematical princes of his day lighten the story. He did however find one generous friend in addition to Holmboë.

In June, 1822, when Abel was nineteen, he completed his required work at the University of Kristiania. Holmboë had done everything possible to relieve the young man's poverty, convincing his colleagues that they too should subscribe to make it possible for Abel to continue his mathematical researches. They were immensely proud of him but they were also poor themselves. Abel quickly outgrew Scandinavia. He longed to visit France, then the mathematical queen of the world, where he could meet his great peers (he was in a class far above some of them, but he did not know it). He dreamed also of touring Germany and meeting Gauss, the undisputed prince of them all.

Abel's mathematical and astronomical friends persuaded the University to appeal to the Norwegian Government to subsidize the young man for a grand mathematical tour of Europe. To impress the authorities with his worthiness, Abel submitted an extensive memoir which, from its title, was probably connected with the fields of his greatest fame. He himself thought highly enough of it to believe its publication by the University would bring Norway honor, and Abel's opinion of his own work, never more than just, was probably as good as anyone's. Unfortunately the University was having a severe financial struggle of its own, and the memoir was finally lost. After undue deliberation the Government compromised—does any Government ever do anything else?—and instead of doing the only sensible thing, namely sending Abel at once to France and Germany, granted him a subsidy to continue his university studies at Kristiania in order that he might brush up his French and German. That is exactly the sort of decision he might have expected from any body of officials conspicuous for their good hearts and common sense. Common sense however has no business dictating to genius.

Abel dallied a year and a half at Kristiania, not wasting his time, but dutifully keeping his part of the contract by wrestling (not too successfully) with German, getting a fair start on French, and working incessantly at his mathematics. With his incurable optimism he had also got himself engaged to a young woman—Crelly Kemp. At last, on August 27, 1825, when Abel was twenty three, his friends overcame the last objection of the Government, and a royal decree granted him sufficient funds for a year's travel and study in France and Germany. They did not give him much, but the fact that they gave him anything at all in the straitened financial condition of the country says more for the state of civilization in Norway in 1825 than could a whole encyclopaedia of the arts and trades. Abel was grateful. It took him about a month to straighten out his dependents before leaving. But thirteen months before this, innocently believing that all mathematicians were as generous-minded as himself, he had burned one of his ladders before ever setting foot on it.

Out of his own pocket—God only knows how—Abel had paid for the printing of his memoir in which the impossibility of solving the general equation of the fifth degree algebraically is proved. It was a pretty poor job of printing but the best backward Norway could manage. This, Abel naïvely believed, was to be his scientific passport to the great mathematicians of the Continent. Gauss in particular, he hoped, would recognize the signal merits of the work and grant him more than a formal interview. He could not know that "the prince of mathematicians" sometimes exhibited anything but a princely generosity to young mathematicians struggling for just recognition.

Gauss duly received the paper. Through unimpeachable witnesses Abel heard how Gauss welcomed the offering. Without deigning to read it he tossed it aside with the disgusted exclamation "Here is another of those monstrosities!" Abel decided not to call on Gauss. Thereafter he disliked Gauss intensely and nicked him whenever he could. He said Gauss wrote obscurely and hinted that the Germans thought a little too much of him. It is an open question whether Gauss or Abel lost more by this perfectly understandable dislike.

Gauss has often been censured for his "haughty contempt" in this matter, but those are hardly the right words to describe his conduct. The problem of the general equation of the fifth degree had become notorious. Cranks as well as reputable mathematicians had been burrowing into it. Now, if a mathematician today receives an alleged squaring of the circle, he may or may not write a courteous note of acknowledgement to the author, but he is almost certain to file the author's manuscript in the wastebasket. For he knows that Lindemann in 1882 proved that it is impossible to square the circle by straightedge and compass alone—the implements to which cranks limit themselves, just as Euclid did. He knows also that Lindemann's proof is accessible to anyone. In 1824 the problem of the general quintic was almost on a par with that of squaring the circle. Hence Gauss' impatience. But it was not quite as bad; the impossibility had not yet been proved. Abel's paper supplied the proof; Gauss might have read something to interest him intensely had be kept his temper. It is a tragedy that he did not. A word from him and Abel would have been made. It is even possible that his life would have been lengthened, as we shall admit when we have his whole story before us.

After leaving home in September, 1825, Abel first visited the notable mathematicians and astronomers of Norway and Denmark and then, instead of hurrying to Göttingen to meet Gauss as he had intended, proceeded to Berlin. There he had the great good fortune to fall in with a man, August Leopold Crelle (1780–1856), who was to be a scientific Holmboë to him and who had far more weight in the mathematical world than the good Holmboë ever had. If Crelle helped to make Abel's reputation, Abel more than paid for the help by making Crelle's. Wherever mathematics is cultivated today the name of Crelle is a household word, indeed more; for "Crelle" has become a proper noun signifying the great journal he founded, the first three volumes of which contained twenty two of Abel's memoirs. The journal made Abel, or at least made him more widely known to Continental mathematicians than he could ever have been without it; Abel's great work started the journal off with a bang that was heard round the mathematical world; and finally the journal made Crelle. This self-effacing amateur of mathematics deserves more than a passing mention. His business ability and his sure instinct for picking collaborators who had real mathematics in them did more for the progress of mathematics in the nineteenth century than half a dozen learned academies.

Crelle himself was a self-taught lover of mathematics rather than a creative mathematician. By profession he was a civil engineer. He early rose to the top in his work, built the first railroad in Germany, and made a comfortable stake. In his leisure he pursued mathematics as something more than a hobby. He himself contributed to mathematical research before and after the great stimulus to German mathematics which his *Journal für die reine und angewandte Mathematik* (Journal for pure and applied Mathematics) gave on its foundation in 1826. This is Crelle's greatest contribution to the advancement of mathematics.

The Journal was the first periodical in the world devoted exclusively to mathematical *research*. Expositions of old work were not welcomed. Papers (except some of Crelle's own) were accepted from anyone, provided only the matter was new, true, and of sufficient "importance"—an intangible requirement—to merit publication. Regularly once every three months from 1826 to the present day "Crelle" has appeared with its sheaf of new mathematics. In the chaos after the World War "Crelle" tottered and almost went down, but was sustained by subscribers from all over the world who were unwilling to see this great monument to a more tranquil civilization than our own obliterated. Today hundreds of periodicals are devoted either wholly or in considerable part to the advancement of pure and applied mathematics. How many of them will survive our next outburst of epidemic insanity is anybody's guess.

When Abel arrived in Berlin in 1825 Crelle had just about made up his mind to start his great venture with his own funds. Abel played a part in clinching the decision. There are two accounts of the first meeting of Abel and Crelle, both interesting. Crelle at the time was holding down a government job for which he had but little aptitude and less liking, that of examiner at the Trade-School (*Gewerbe-Institut*) in Berlin. At third-hand (Crelle to Weierstrass to Mittag-Leffler) Crelle's account of that historic meeting is as follows.

"One fine day a fair young man, much embarrassed, with a very youthful and very intelligent face, walked into my room. Believing that I had to do with an examination-candidate for admission to the Trade-School, I explained that several separate examinations would be necessary. At last the young man opened his mouth and explained [in poor German], 'Not examination, only mathematics.'"

Crelle saw that Abel was a foreigner and tried him in French, in which Abel could make himself understood with some difficulty. Crelle then questioned him about what he had done in mathematics. Diplomatically enough Abel replied that he had read, among other things, Crelle's own paper of 1823, then recently published, on "analytical faculties" (now called "factorials" in English). He had found the work most interesting he said, but ——. Then, not so diplomatically, he proceeded to tell Crelle that parts of the work were quite wrong. It was here that Crelle showed his greatness. Instead of freezing or blowing up in a rage at the daring presumption of the young man before him, he pricked up his ears and asked for particulars, which he followed with the closest attention. They had a long mathematical talk, only parts of which were intelligible to Crelle. But whether he understood all that Abel told him or not, Crelle saw clearly what Abel was. Crelle never did understand a tenth of what Abel was up to, but his sure instinct for mathematical genius told him that Abel was a mathematician of the first water and he did everything in his power to gain recognition for his young protégé. Before the interview was ended Crelle had made up his mind that Abel must be one of the first contributors to the projected *Journal*.

Abel's account differs, but not essentially. Reading between the lines we may see that the differences are due to Abel's modesty. At first Abel feared his project of interesting Crelle was fated to go on the rocks. Crelle could not make out what the young man wanted, who he was, or anything about him. But at Crelle's question as to what Abel had read in mathematics things brightened up considerably. When Abel mentioned the works of the masters he had studied Crelle became instantly alert. They had a long talk on several outstanding unsettled problems, and Abel ventured to spring his proof of the impossibility of solving the general quintic algebraically on the unsuspecting Crelle. Crelle wouldn't hear of it; there must be something wrong with any such proof. But he accepted a copy of the paper, thumbed through it, admitted the reasoning was beyond him—and finally published Abel's amplified proof in his *Journal*. Although he was a limited mathematician with no pretensions to scientific greatness, Crelle was a broadminded man, in fact a great man.

Crelle took Abel everywhere, showing him off as the finest mathematical discovery yet made. The self-taught Swiss Steiner—"the greatest geometer since Apollonius"—sometimes accompanied Crelle and Abel on their rounds. When Crelle's friends saw him coming with his two geniuses in tow they would exclaim "Here comes Father Adam again with Cain and Abel."

The generous sociability of Berlin began to distract Abel from his work and he fled to Freiburg where he could concentrate. It was at Freiburg that he hewed his greatest work into shape, the creation of what is now called Abel's Theorem. But he had to be getting on to Paris to meet the foremost French mathematicians of the day—Legendre, Cauchy, and the rest.

It can be said at once that Abel's reception at the hands of the French mathematicians was as civil as one would expect from distinguished representatives of a very civil people in a very civil age. They were all very civil to him—damned civil, in fact, and that was about all that Abel got out of the visit to which he had looked forward with such ardent hopes. Of course they did not know who or what he was. They made only perfunctory efforts to find out. If Abel opened his mouth—when he got within talking distance of them—about his own work, they immediately began lecturing about their own greatness. But for his indifference the venerable Legendre might have learned something about his own lifelong passion (for elliptic integrals) which would have interested him beyond measure. But he was just stepping into his carriage when Abel called and had time for little more than a very civil good-day. Later he made handsome amends.

Late in July, 1826, Abel took up his lodgings in Paris with a poor but grasping family who gave him two bad meals a day and a vile room for a sufficiently outrageous rent. After four months of Paris Abel writes his impressions to Holmboë:

"*Paris, 24 October 1826.*

"To tell you the truth this noisiest capital of the Continent has for the moment the effect of a desert on me. I know practically nobody; this is the lovely season when everybody is in the country. . . . Up till now I have made the acquaintance of Mr. *Legendre,* Mr. *Cauchy* and Mr. *Hachette,* and some less celebrated but very able mathematicians: Mr. *Saigey,* editor of the Bulletin des Sciences, and Mr. *Lejeune-Dirichlet,* a Prussian who came to see me the other day believing me to be a compatriot of his. He is a mathematician of great penetration. With Mr. *Legendre* he has proved the impossibility of solving $x^5 + y^5 = z^5$ in whole numbers, and other very fine things. *Legendre* is extremely polite, but unfortunately very old. *Cauchy* is mad. . . . What he does is excellent but very muddled. At first I understood practically none of it; now I see some of it more clearly. . . . *Cauchy* is the only one occupied with pure mathematics. *Poisson, Fourier, Ampère,* etc., busy themselves exclusively with magnetism and other physical subjects. Mr. *Laplace* writes nothing now, I believe. His last work was a supplement to his Theory of Probabilities. I have often seen him at the Institut. He is a very jolly little chap. *Poisson* is a little fellow; he knows how to behave with a great deal of dignity; Mr. *Fourier* the same. *Lacroix* is quite old. Mr. *Hachette* is going to present me to several of these men.

"The French are much more reserved with strangers than the Germans. It is extremely difficult to gain their intimacy, and I do not dare to urge my pretensions as far as that; finally every beginner has a great deal of difficulty in getting noticed here. I have just finished an extensive treatise on a certain class of transcendental functions [his masterpiece] to present it to the Institut [Academy of Sciences], which will be done next Monday. I showed it to Mr. *Cauchy,* but he scarcely deigned to glance at it. And I dare to say, without bragging, that it is a good piece of work. I am curious to hear the opinion of the Institut on it. I shall not fail to share it with you. . . ."

He then tells what he is doing and continues with a rather disturbed forecast of his prospects. "I regret having set two years for my travels, a year and a half would have sufficed."

He has got all there is to be got out of Continental Europe and is anxious to be able to devote his time to working up what he has invented. "So many things remain for me to do, but so long as I am abroad, all that goes badly enough. If I had my professorship as Mr. *Kielhau* has his! My position is not assured, it is true, but I am not uneasy about it; if fortune deserts me in one quarter perhaps she will smile on me in another."

From a letter of earlier date to the astronomer Hansteen we take two extracts, the first relating to Abel's great project of re-establishing mathematical analysis as it existed in his day on a firm foundation, the second showing something of his human side. (Both are free translations.)

"In the higher analysis too few propositions are proved with conclusive rigor. Everywhere we find the unfortunate procedure of reasoning from the special to the general, and the miracle is that after such a process it is only seldom that we find what are called paradoxes. It is indeed exceedingly interesting to seek the reason for this. This reason, in my opinion, resides in the fact that the functions which have hitherto occurred in analysis can be expressed for the most part as powers.... When we proceed by a general method, it is not too difficult [to avoid pitfalls]; but I have had to be very circumspect, because propositions without rigorous proof (i.e. without any proof) have taken root in me to such an extent that I constantly run the risk of using them without further examination. These trifles will appear in the journal published by Mr. *Crelle*."

Immediately following this he expresses his gratitude for his treatment in Berlin. "It is true that few persons are interested in me, but these few are infinitely dear to me, because they have shown me so much kindness. Perhaps I can respond in some way to their hopes of me, for it must be hard for a benefactor to see his trouble lost."

He tells then how Crelle has been begging him to take up his residence permanently in Berlin. Crelle was already using all his human engineering skill to hoist the Norwegian Abel into a professorship in the University of Berlin. Such was the Germany of 1826. Abel of course was already great, and the sure promise of what he had in him indicated him as the likeliest mathematical successor to Gauss. That he was a foreigner made no difference; Berlin in 1826 wanted the best in mathematics. A century later the best in mathematical physics was not good enough, and Berlin quite forcibly got rid of Einstein. Thus do we progress. But to return to the sanguine Abel.

"At first I counted on going directly from Berlin to Paris, happy in the promise that Mr. *Crelle* would accompany me. But Mr. *Crelle* was prevented, and I shall have to travel alone. Now I am so constituted that I cannot endure solitude. Alone, I am depressed, I get cantankerous, and I have little inclination for work. So I said to myself it would be much better to go with Mr. *Boeck* to Vienna, and this trip seems to me to be justified by the fact that at Vienna there are men like *Littrow*, *Burg*, and still others, all indeed excellent mathematicians; add to this that I shall make but this one voyage in my life. Could one find anything but reasonableness in this wish of mine to see something of the life of the South? I could work assiduously enough while travelling. Once in Vienna and leaving there for Paris, it is almost a bee-line via Switzerland. Why shouldn't I see a little of it too? My God! I, even I, have some taste for the beauties of nature, like everybody else. This whole trip would bring me to Paris two months later, that's all. I could quickly catch up the time lost. Don't you think such a trip would do me good?"

So Abel went South, leaving his masterpiece in Cauchy's care to be presented to the Institut. The prolific Cauchy was so busy laying eggs of his own and cackling about them that he had no time to examine the veritable roc's egg which the modest Abel had deposited in the nest. Hachette, a mere pot-washer of a mathematician, presented Abel's *Memoir on a general property of a very extensive class of transcendental functions* to the Paris Academy of Sciences on the tenth of October, 1826. This is the work which Legendre later described in the words of Horace as *"monumentum aere perennius,"* and the five hundred years' work which Hermite said Abel had laid out for future generations of mathematicians. It is one of the crowning achievements of modern mathematics.

What happened to it? Legendre and Cauchy were appointed as referees. Legendre was seventy four, Cauchy thirty nine. The veteran was losing his edge, the captain was in his self-centred prime. Legendre complained (letter to Jacobi, 8 April, 1829) that "we perceived that the memoir was barely legible; it was written in ink almost white, the letters badly formed; it was agreed between us that the author should be asked for a neater copy to be read." What an alibi! Cauchy took the memoir home, mislaid it, and forgot all about it.

To match this phenomenal feat of forgetfulness we have to imagine an Egyptologist mislaying the Rosetta Stone. Only by a sort of miracle was the memoir unearthed after Abel's death. Jacobi heard of it from Legendre, with whom Abel corresponded after returning to Norway, and in a letter dated 14 March, 1829, Jacobi exclaims, "What a discovery is this of Mr. Abel's! . . . Did anyone ever see the like? But how comes it that this discovery, perhaps the most important mathematical discovery that has been made in our Century, having been communicated to your Academy two years ago, has escaped the attention of your colleagues?" The enquiry reached Norway. To make a long story short, the Norwegian consul at Paris raised a diplomatic row about the missing manuscript and Cauchy dug it up in 1830. Finally it was printed, but not till 1841, in the *Mémoires présentés par divers savants à l'Académie royale des sciences de l'Institut de France*, vol. 7, pp. 176–264. To crown this epic *in parvo* of crass incompetence, the editor, or the printers, or both between them, succeeded in losing the manuscript before the proof-sheets were read.* The Academy (in 1830) made amends to Abel by awarding him the Grand Prize in Mathematics jointly with Jacobi. Abel, however, was dead.

The opening paragraphs of the memoir indicate its scope.

"The transcendental functions hitherto considered by mathematicians are very few in number. Practically the entire theory of transcendental functions is reduced to that of logarithmic functions, circular and exponential functions, functions which, at bottom, form but a single species. It is only recently that some other functions have begun to be considered. Among the latter, the elliptic transcendents, several of whose remarkable and elegant properties have been developed by Mr. Legendre, hold the first place. The author [Abel] has considered, in the memoir which he has the honor to present to the Academy, a very extended class of functions, namely: all those whose derivatives are expressible by means of algebraic

---

*Libri, a *soi-disant* mathematician, who saw the work through the press adds, "by permission of the Academy," a smug footnote acknowledging the genius of the lamented Abel. This is the last straw; the Academy might have come out with all the facts or have held its official tongue. But at all costs the honor and dignity of a stuffed shirt must be upheld. Finally it may be recalled that valuable manuscripts and books had an unaccountable trick of vanishing when Libri was round.

equations whose coefficients are rational functions of one variable, and he has proved for these functions properties analogous to those of logarithmic and elliptic functions . . . and he has arrived at the following theorem:

"If we have several functions whose derivatives can be roots of *one and the same algebraic equation*, all of whose coefficients are *rational* functions of one variable, we can always express the sum of any number of such functions by an *algebraic* and *logarithmic* function, provided that we establish a certain number of *algebraic* relations between the variables of the functions in question.

"The number of these relations does not depend at all upon the number of functions, but only upon the nature of the particular functions considered. . . ."

The theorem which Abel thus briefly describes is today known as Abel's Theorem. His proof of it has been described as nothing more than "a marvellous exercise in the integral calculus." As in his algebra, so in his analysis, Abel attained his proof with a superb parsimony. The proof, it may be said without exaggeration, is well within the purview of any seventeen-year-old who has been through a good *first* course in the calculus. There is nothing high-falutin' about the classic simplicity of Abel's own proof. The like cannot be said for some of the nineteenth century expansions and geometrical reworkings of the original proof. Abel's proof is like a statue by Phidias; some of the others resemble a Gothic cathedral smothered in Irish lace, Italian confetti, and French pastry.

There is ground for a possible misunderstanding in Abel's opening paragraph. Abel no doubt was merely being kindly courteous to an old man who had patronized him—in the bad sense—on first acquaintance, but who, nevertheless, had spent most of his long working life on an important problem without seeing what it was all about. It is not true that Legendre had discussed the elliptic *functions*, as Abel's words might imply; what Legendre spent most of his life over was elliptic *integrals*, which are as different from elliptic *functions* as a horse is from the cart it pulls, and therein precisely is the crux and the germ of one of Abel's greatest contributions to mathematics. The matter is quite simple to anyone who has had a school course in trigonometry; to obviate tedious explanations of elementary matters this much will be assumed in what follows presently.

For those who have forgotten all about trigonometry, however, the essence, the *methodology*, of Abel's epochal advance can be analogized thus. We alluded to the cart and the horse. The frowsy proverb about putting the cart before the horse describes what Legendre did; Abel saw that if the cart was to move forward the horse should precede it. To take another instance: Francis Galton, in his statistical studies of the relation between poverty and chronic drunkenness, was led, by his impartial mind, to a reconsideration of all the self-righteous platitudes by which indignant moralists and economic crusaders with an axe to grind evaluate such social phenomena. Instead of assuming that people are depraved *because* they drink to excess, Galton *inverted* this hypothesis and assumed temporarily that people drink to excess *because* they have inherited no moral guts from their ancestors, in short, *because* they *are* depraved. Brushing aside all the vaporous moralizings of the reformers, Galton took a firm grip on a *scientific*, unemotional, *workable* hypothesis to which he could apply the impartial machinery of mathematics. His work has not yet registered socially. For the moment we need note only that Galton, like Abel, *inverted* his problem—turned it upside-down and inside-out, back-end-to and foremost-end-backward. Like Hiawatha and his fabulous mittens, Galton put the skinside inside and the inside outside.

All this is far from being obvious or a triviality. It is one of the most powerful methods of mathematical discovery (or invention) ever devised, and Abel was the first human being to use it consciously as an engine of research. "You must always invert," as Jacobi said when asked the secret of his mathematical discoveries. He was recalling what Abel and he had done. If the solution of a problem becomes hopelessly involved, try turning the problem backwards, put the quaesita for the data and vice versa. Thus if we find Cardan's character incomprehensible when we think of him as *a* son of his father, shift the emphasis, *invert* it, and see what we get when we analyse Cardan's father as *the* begetter and endower of his son. Instead of studying "inheritance" concentrate on "endowing." To return to those who remember some trigonometry.

Suppose mathematicians had been so blind as not to see that $\sin x$, $\cos x$ and the other *direct* trigonometric functions are simpler to use, in the addition formulas and elsewhere, than the *inverse* functions $\sin^{-1} x$, $\cos^{-1} x$. Recall the formula $\sin(x + y)$ in terms of sines and cosines of x and y, and contrast it with the formula for $\sin^{-1}(x + y)$

in terms of x and y. Is not the former incomparably simpler, more elegant, more "natural" than the latter? Now, in the integral calculus, the *inverse* trigonometric functions present themselves naturally as definite integrals of simple algebraic irrationalities (second degree); such integrals appear when we seek to find the length of an arc of a circle by means of the integral calculus. Suppose the *inverse* trigonometric functions had *first* presented themselves this way. Would it not have been "more natural" to consider the *inverses* of these functions, that is, the familiar trigonometric functions themselves as the *given* functions to be studied and analyzed? Undoubtedly; but in shoals of more advanced problems, the simplest of which is that of finding the length of the arc of an *ellipse* by the integral calculus, the awkward *inverse* "elliptic" (not "circular," as for the arc of a circle) functions presented themselves *first*. It took Abel to see that *these* functions should be "inverted" and studied, precisely as in the case of $\sin x$, $\cos x$ instead of $\sin^{-1} x$, $\cos^{-1} x$. Simple, was it not? Yet Legendre, a great mathematician, spent more than *forty years* over his "elliptic integrals" (the awkward "inverse functions" of his problem) without ever once suspecting that he should *invert*.* This extremely simple, uncommonsensical way of looking at an apparently simple but profoundly recondite problem was one of the greatest mathematical advances of the nineteenth century.

All this however was but the beginning, although a sufficiently tremendous beginning—like Kipling's dawn coming up like thunder—of what Abel did in his magnificent theorem and in his work on elliptic functions. The trigonometric or circular functions have a single real period, thus $\sin (x + 2\pi) = \sin x$, etc. Abel discovered that his new functions provided by the inversion of an elliptic integral have precisely *two* periods, whose ratio is imaginary. After that, Abel's followers in this direction—Jacobi, Rosenhain, Weierstrass, Riemann, and many more—mined deeply into Abel's great theorem and by carrying on and extending his ideas discovered functions of n variables having $2n$ periods. Abel himself carried the exploitation of his discoveries far. His successors have applied all this work to geometry, mechanics, parts of mathematical physics, and other tracts of mathe-

---

*In ascribing priority to Abel, rather than "joint discovery" to Abel and Jacobi, in this matter, I have followed Mittag-Leffler. From a thorough acquaintance with all the published evidence, I am convinced that Abel's claim is indisputable, although Jacobi's compatriots argue otherwise.

matics, solving important problems which, without this work initiated by Abel, would have been unsolvable.

While in Paris Abel consulted good physicians for what he thought was merely a persistent cold. He was told that he had tuberculosis of the lungs. He refused to believe it, wiped the mud of Paris off his boots, and returned to Berlin for a short visit. His funds were running low; about seven dollars was the extent of his fortune. An urgent letter brought a loan from Holmboë after some delay. It must not be supposed that Abel was a chronic borrower on no prospects. He had good reason for believing that he should have a paying job when he got home. Moreover, money was still owed to him. On Holmboë's loan of about sixty dollars Abel existed and researched from March till May, 1827. Then, all his resources exhausted, he turned homeward and arrived in Kristiania completely destitute.

But all was soon to be rosy, he hoped. Surely the University job would be forthcoming now. His genius had begun to be recognized. There was a vacancy. Abel did not get it. Holmboë reluctantly took the vacant chair which he had intended Abel to fill only after the governing board threatened to import a foreigner if Holmboë did not take it. Holmboë was in no way to blame. It was assumed that Holmboë would be a better teacher than Abel, although Abel had amply demonstrated his ability to teach. Anyone familiar with the current American pedagogical theory, fostered by professional Schools of Education, that the less a man knows about what he is to teach the better he will teach it, will understand the situation perfectly.

Nevertheless things did brighten up. The University paid Abel the balance of what it owed on his travel money and Holmboë sent pupils his way. The professor of astronomy took a leave of absence and suggested that Abel be employed to carry part of his work. A well-to-do couple, the Schjeldrups, took him in and treated him as if he were their own son. But with all this he could not free himself of the burden of his dependents. To the last they clung to him, leaving him practically nothing for himself, and to the last he never uttered an impatient word.

By the middle of January, 1829, Abel knew that he had not long to live. The evidence of a hemorrhage is not to be denied. "I will fight for my life!" he shouted in his delirium. But in more tranquil moments, exhausted and trying to work, he drooped "like a sick eagle looking at the sun," knowing that his weeks were numbered.

Abel spent his last days at Froland, in the home of an English family where his fiancée (Crelly Kemp) was governess. His last thoughts were for her future, and he wrote to his friend Kielhau, "She is not beautiful; she has red hair and freckles, but she is an admirable woman." It was Abel's wish that Crelly and Kielhau should marry after his death; and although the two had never met, they did as Abel had half-jokingly proposed. Toward the last Crelly insisted on taking care of Abel without help, "to possess these last moments alone." Early in the morning of the sixth of April, 1829, he died, aged twenty six years, eight months.

Two days after Abel's death Crelle wrote to say that his negotiations had at last proved successful and that Abel would be appointed to the professorship of mathematics in the University of Berlin.

CHAPTER EIGHTEEN

The Great Algorist

JACOBI

It is the increasingly pronounced tendency of modern analysis to substitute ideas for calculation; nevertheless there are certain branches of mathematics where calculation conserves its rights.—P. G. LEJEUNE DIRICHLET

THE NAME JACOBI appears frequently in the sciences, not always meaning the same man. In the 1840's one very notorious Jacobi—M. H.—had a comparatively obscure brother, C. G. J., whose reputation then was but a tithe of M. H.'s. Today the situation is reversed: C. G. J. is immortal—or seemingly so, while M. H. is rapidly receding into the obscurity of limbo. M. H. achieved fame as the founder of the fashionable quackery of galvanoplastics; C. G. J.'s much narrower but also much higher reputation is based on mathematics. During his lifetime the mathematician was always being confused with his more famous brother, or worse, being congratulated for his involuntary kinship to the sincerely deluded quack. At last C. G. J. could stand it no longer. "Pardon me, beautiful lady," he retorted to an enthusiastic admirer of M. H. who had complimented him on having so distinguished a brother, "but *I* am my brother." On other occasions C. G. J. would blurt out, "I am not *his* brother, he is *mine.*" There is where fame has left the relationship today.

Carl Gustav Jacob Jacobi, born at Potsdam, Prussia, Germany, on December 10, 1804, was the second son of a prosperous banker, Simon Jacobi, and his wife (family name Lehmann). There were in all four children, three boys, Moritz, Carl, and Eduard, and a girl, Therese. Carl's first teacher was one of his maternal uncles, who taught the boy classics and mathematics, preparing him to enter the Potsdam Gymnasium in 1816 in his twelfth year. From the first Jacobi gave evidence of the "universal mind" which the rector of the Gymnasium declared him to be on his leaving the school in 1821 to enter the University of Berlin. Like Gauss, Jacobi could easily have made a high reputation in philology had not mathematics attracted him more strongly. Having seen that the boy had mathematical genius, the teacher (Heinrich Bauer) let Jacobi work by himself—after a prolonged tussle in which Jacobi rebelled at learning mathematics by rote and by rule.

Young Jacobi's mathematical development was in some respects curiously parallel to that of his greater rival Abel. Jacobi also went to the masters; the works of Euler and Lagrange taught him algebra and the calculus, and introduced him to the theory of numbers. This earliest self-instruction was to give Jacobi's first outstanding work—in elliptic functions—its definite direction, for Euler, the master of ingenious devices, found in Jacobi his brilliant successor. For sheer manipulative ability in tangled algebra Euler and Jacobi have had no rival, unless it be the Indian mathematical genius, Srinivasa Ramanujan, in our own century. Abel also could handle formulas like a master when he wished, but his genius was more philosophical, less formal than Jacobi's. Abel is closer to Gauss in his insistence upon rigor than Jacobi was by nature—not that Jacobi's work lacked rigor, for it did not, but its inspiration appears to have been formalistic rather than rigoristic.

Abel was two years older than Jacobi. Unaware that Abel had attacked the general quintic in 1820, Jacobi in the same year attempted a solution, reducing the general quintic to the form $x^5 - 10q^2x = p$ and showing that the solution of this equation would follow from that of a certain equation of the tenth degree. Although the attempt was abortive it taught Jacobi a great deal of algebra and he ascribed considerable importance to it as a step in his mathematical education. But it does not seem to have occurred to him, as it did to Abel, that the general quintic might be unsolvable algebraically. This oversight, or lack of imagination, or whatever we wish to call it, on Jacobi's part is typical of the difference between him and Abel. Jacobi, who had a magnificently objective mind and not a particle of envy or jealousy in his generous nature, himself said of one of Abel's masterpieces, "It is above my praises as it is above my own works."

Jacobi's student days at Berlin lasted from April, 1821, to May, 1825. During the first two years he spent his time about equally between philosophy, philology, and mathematics. In the philological seminar Jacobi attracted the favorable attention of P. A. Boeckh, a renowned classical scholar who brought out (among other works) a fine edition of Pindar. Boeckh, luckily for mathematics, failed to convert his most promising pupil to classical studies as a life interest. In mathematics not much was offered for an ambitious student and Jacobi continued his private study of the masters. The university lectures in mathematics he characterized briefly and sufficiently as twaddle. Jacobi was usually blunt and to the point, although he knew how to be as subservient as any courtier when trying to insinuate some deserving mathematical friend into a worthy position.

While Jacobi was diligently making a mathematician of himself Abel was already well started on the very road which was to lead Jacobi to fame. Abel had written to Holmboë on August 4, 1823, that he was busy with elliptic functions: "This little work, you will recall, deals with the inverses of the elliptic transcendents, and I proved something [that seemed] impossible; I begged Degen to read it as soon as he could from one end to the other, but he could find no false conclusion, nor understand where the mistake was; God knows how I shall get myself out of it." By a curious coincidence Jacobi at last made up his mind to put his all on mathematics almost exactly when Abel wrote this. Two years' difference in the ages of young men around twenty (Abel was twenty one, Jacobi nineteen) count for more than two decades of maturity. Abel got a tremendous start but Jacobi, unaware that he had a competitor in the race, soon caught up. Jacobi's first great work was in Abel's field of elliptic functions. Before considering this we shall outline his busy life.

Having decided to go into mathematics for all he was worth, Jacobi wrote to his uncle Lehmann his estimate of the labor he had undertaken. "The huge colossus which the works of Euler, Lagrange, and Laplace have raised demands the most prodigious force and exertion of thought if one is to penetrate into its inner nature and not merely rummage about on its surface. To dominate this colossus and not to fear being crushed by it demands a strain which permits neither rest nor peace till one stands on top of it and surveys the work in its entirety. Then only, when one has comprehended its spirit, is it possible to work justly and in peace at the completion of its details."

With this declaration of willing servitude Jacobi forthwith became one of the most terrific workers in the history of mathematics. To a timid friend who complained that scientific research is exacting and likely to impair bodily health, Jacobi retorted:

"Of course! Certainly I have sometimes endangered my health by overwork, but what of it? Only cabbages have no nerves, no worries. And what do they get out of their perfect wellbeing?"

In August, 1825, Jacobi received his Ph.D. degree for a dissertation on partial fractions and allied topics. There is no need to explain the nature of this—it is not of any great interest and is now a detail in the second course of algebra or the integral calculus. Although Jacobi handled the general case of his problem and showed considerable ingenuity in manipulating formulas, it cannot be said that the dissertation exhibited any marked originality or gave any definite hint of the author's superb talent. Concurrently with his examination for the Ph.D. degree, Jacobi rounded off his training for the teaching profession.

After his degree Jacobi lectured at the University of Berlin on the applications of the calculus to curved surfaces and twisted curves (roughly, curves determined by the intersections of surfaces). From the very first lectures it was evident that Jacobi was a born teacher. Later, when he began developing his own ideas at an amazing speed, he became the most inspiring mathematical teacher of his time.

Jacobi seems to have been the first regular mathematical instructor in a university to train students in research by lecturing on his own latest discoveries and letting the students see the creation of a new subject taking place before them. He believed in pitching young men into the icy water to learn to swim or drown by themselves. Many students put off attempting anything on their own account till they have mastered everything relating to their problem that has been done by others. The result is that but few ever acquire the knack of independent work. Jacobi combated this dilatory erudition. To drive home the point to a gifted but diffident young man who was always putting off doing anything until he had learned something more, Jacobi delivered himself of the following parable. "Your father would never have married, and you wouldn't be here now, if he had insisted on knowing *all* the girls in the world before marrying *one*."

Jacobi's entire life was spent in teaching and research except for one ghastly interlude, to be related, and occasional trips to attend scientific meetings in England and on the Continent, or forced vacations to recuperate after too intensive work. The chronology of his life is not very exciting—a professional scientist's seldom is except to himself.

Jacobi's talents as a teacher secured him the position of lecturer at the University of Königsberg in 1826 after only half a year in a similar position at Berlin. A year later some results which Jacobi had published in the theory of numbers (relating to cubic reciprocity; see chapter on Gauss) excited Gauss' admiration. As Gauss was not an easy man to stir up, the Ministry of Education took prompt notice and promoted Jacobi over the heads of his colleagues to an assistant professorship—quite a step for a young man of twenty three. Naturally the men he had stepped over resented the promotion; but two years later (1829) when Jacobi published his first masterpiece, *Fundamenta Nova Theoriae Functionum Ellipticarum* (New Foundations of the Theory of Elliptic Functions) they were the first to say that no more than justice had been done and to congratulate their brilliant young colleague.

In 1832 Jacobi's father died. Up till this he need not have worked for a living. His prosperity continued about eight years longer, when the family fortune went to smash in 1840. Jacobi was cleaned out himself at the age of thirty six and in addition had to provide for his mother, also ruined.

Gauss all this time had been watching Jacobi's phenomenal activity with more than a mere scientific interest, as many of Jacobi's discoveries overlapped some of those of his own youth which he had never published. He had also (it is said) met the young man personally: Jacobi called on Gauss (no account of the visit has survived) in September, 1839, on his return trip to Königsberg after a vacation in Marienbad to recuperate from overwork. Gauss appears to have feared that Jacobi's financial collapse would have a disastrous effect on his mathematics, but Bessel reassured him: "Fortunately such a talent cannot be destroyed, but I should have liked him to have the sense of freedom which money assures."

The loss of his fortune had no effect whatever on Jacobi's mathematics. He never alluded to his reverses but kept on working as assiduously as ever. In 1842 Jacobi and Bessel attended the meeting of the British Association at Manchester, where the German Jacobi and the Irish Hamilton met in the flesh. It was to be one of Jacobi's greatest glories to continue the work of Hamilton in dynamics and, in a sense, to complete what the Irishman had abandoned in favor of a will-o-the-wisp (which will be followed when we come to it).

At this point in his career Jacobi suddenly attempted to blossom out into something showier than a mere mathematician. Not to interrupt the story of his scientific life when we take it up, we shall dispose here of the illustrious mathematician's singular misadventures in politics.

The year following his return from the trip of 1842, Jacobi had a complete breakdown from overwork. The advancement of science in the 1840's in Germany was in the hands of the benevolent princes and kings of the petty states which were later to coalesce into the German Empire. Jacobi's good angel was the King of Prussia, who seems to have appreciated fully the honor which Jacobi's researches conferred on the Kingdom. Accordingly, when Jacobi fell ill, the benevolent King urged him to take as long a vacation as he liked in the mild climate of Italy. After five months at Rome and Naples with Borchardt (whom we shall meet later in the company of Weierstrass) and Dirichlet, Jacobi returned to Berlin in June, 1844. He was now permitted to stay on in Berlin until his health should be completely restored but, owing to jealousies, was not given a professorship in the University, although as a member of the Academy he was permitted to lecture on anything he chose. Further, out of his own pocket, practically, the King granted Jacobi a substantial allowance.

After all this generosity on the part of the King one might think that Jacobi would have stuck to his mathematics. But on the utterly imbecilic advice of his physician he began meddling in politics "to benefit his nervous system." If ever a more idiotic prescription was handed out by a doctor to a patient whose complaint he could not diagnose it has yet to be exhumed. Jacobi swallowed the dose. When the democratic upheaval of 1848 began to erupt Jacobi was ripe for office. On the advice of a friend—who, by the way, happened to be one of the men over whose head Jacobi had been promoted some twenty years before—the guileless mathematician stepped into the arena of politics with all the innocence of an enticingly plump missionary setting foot on a cannibal island. They got him.

The mildly liberal club to which his slick friend had introduced him ran Jacobi as their candidate for the May election of 1848. But he never saw the inside of parliament. His eloquence before the club convinced the wiser members that Jacobi was no candidate for them. Quite properly, it would seem, they pointed out that Jacobi, the King's pensioner, might possibly be the liberal he now professed to be, but that it was more probable he was a trimmer, a turncoat, and a stoolpigeon for the royalists. Jacobi refuted these base insinuations in a magnificent speech packed with irrefutable logic—oblivious of the axiom that logic is the last thing on earth for which a practical politician has any use. They let him hang himself in his own noose. He was not elected. Nor was his nervous system benefited by the uproar over his candidacy which rocked the beer halls of Berlin to their cellars.

Worse was to come. Who can blame the Minister of Education for enquiring the following May whether Jacobi's health had recovered sufficiently for him to return safely to Königsberg? Or who can wonder that his allowance from the King was stopped a few days later? After all even a King may be permitted a show of petulance when the mouth he tries to feed bites him. Nevertheless Jacobi's desperate plight was enough to excite anybody's sympathy. Married and practically penniless he had seven small children to support in addition to his wife. A friend in Gotha took in the wife and children, while Jacobi retired to a dingy hotel room to continue his researches.

He was now (1849) in his forty fifth year and, except for Gauss, the most famous mathematician in Europe. Hearing of his plight, the University of Vienna began angling for him. As an item of interest here, Littrow, Abel's Viennese friend, took a leading part in the negotiations. At last, when a definite and generous offer was tendered, Alexander von Humboldt talked the sulky King round; the allowance was restored, and Jacobi was not permitted to rob Germany of her second greatest man. He remained in Berlin, once more in favor but definitely out of politics.

The subject, elliptic functions, in which Jacobi did his first great work, has already been given what may seem like its share of space; for after all it is today more or less of a detail in the vaster theory of functions of a complex variable which, in its turn, is fading from the ever changing scene as a thing of living interest. As the theory of elliptic functions will be mentioned several times in succeeding chapters we shall attempt a brief justification of its apparently unmerited prominence.

No mathematician would dispute the claim of the theory of functions of a complex variable to have been one of the major fields of nineteenth century mathematics. One of the reasons why this theory was of such importance may be repeated here. Gauss had shown that *complex* numbers are both necessary and sufficient to provide every algebraic equation with a root. Are any further, more general, kinds of "numbers" possible? How might such "numbers" arise?

Instead of regarding *complex* numbers as having first presented themselves in the attempt to solve certain simple equations, say $x^2 + 1 = 0$, we may also see their origin in another problem of elementary algebra, that of *factorization*. To resolve $x^2 - y^2$ into factors of the *first* degree we need nothing more mysterious than the positive and negative integers: $(x^2 - y^2) = (x+y)(x-y)$. But the same problem for $x^2 + y^2$ demands "imaginaries": $x^2 + y^2 = (x + y\sqrt{-1})(x - y\sqrt{-1})$. Carrying this up a step in one of many possible ways open, we might seek to resolve $x^2 + y^2 + z^2$ into two factors of the *first* degree. Are the positives, negatives, and imaginaries sufficient? Or must some new kind of "number" be invented to solve the problem? The latter is the case. It was found that for the new "numbers" necessary the rules of common algebra break down in one important particular: it is no longer true that the *order* in which "numbers" are *multiplied* together is indifferent; that is, for the new numbers it is not true that $a \times b$ is equal to $b \times a$. More will be said on this when we come to Hamilton. For the moment we note that the elementary algebraic problem of factorization quickly leads us into regions where complex numbers are inadequate.

How far can we go, what are the *most general numbers possible*, if we insist that for these numbers *all* the familiar laws of common algebra are to hold? It was proved in the latter part of the nineteenth century that the complex numbers $x + iy$, where x,y are real numbers and $i = \sqrt{-1}$, are the most general for which common algebra is true. The real numbers, we recall, correspond to the distances measured along a fixed straight line in either direction (positive, negative) from a fixed point, and the graph of a function $f(x)$, plotted as $y = f(x)$, in Cartesian geometry, gives us a picture of a function y of a *real* variable x. The mathematicians of the seventeenth and eighteenth centuries imagined their functions as being of this kind. But if the common algebra and its extensions into the calculus which they applied to their functions are equally applicable to complex numbers, which include the real numbers as a very degenerate case, it was but natural that many of the things the early analysts found were less than half the whole story possible. In particular the integral calculus presented many inexplicable anomalies which were cleared up only when the field of operations was enlarged to its fullest possible extent and functions of *complex* variables were introduced by Gauss and Cauchy.

The importance of elliptic functions in all this vast and fundamental development cannot be overestimated. Gauss, Abel, and Jacobi, by their extensive and detailed elaboration of the theory of elliptic functions, in which complex numbers appear inevitably, provided a testing ground for the discovery and improvement of general theorems in the theory of functions of a complex variable. The two theories seemed to have been designed by fate to complement and supplement one another—there is a reason for this, also for the deep connection of elliptic functions with the Gaussian theory of quadratic forms, which considerations of space force us to forego. Without the innumerable clues for a general theory provided by the special instances of more inclusive theorems occurring in elliptic functions, the theory of functions of a complex variable would have developed much more slowly than it did—Liouville's theorem, the entire subject of multiple periodicity with its impact on the theory of algebraic functions and their integrals, may be recalled to mathematical readers. If some of these great monuments of nineteenth century mathematics are already receding into the mists of yesterday, we need only remind ourselves that Picard's theorem on exceptional values in the neighborhood of an essential singularity, one of the most suggestive in current analysis, was first proved by devices originating in the theory of elliptic functions. With this partial summary of the reason why elliptic functions were important in the mathematics of the nineteenth century we may pass on to Jacobi's cardinal part in the development of the theory.

The history of elliptic functions is quite involved, and although of considerable interest to specialists, is not likely to appeal to the general reader. Accordingly we shall omit the evidence (letters of Gauss, Abel, Jacobi, Legendre, and others) on which the following bare summary is based.

First, it is established that Gauss anticipated both Abel and Jacobi by as much as twenty seven years in some of their most striking work. Indeed Gauss says that "Abel has followed exactly the same road that I did in 1798." That this claim is just will be admitted by anyone who will study the evidence published only after Gauss' death. Second, it seems to be agreed that Abel anticipated Jacobi in certain important details, but that Jacobi made his great start in entire ignorance of his rival's work.

A capital property of the elliptic functions is their *double periodicity* (discovered in 1825 by Abel): if $E(x)$ is an elliptic function, then there are two distinct numbers, say p_1, p_2, such that

$$E(x + p_1) = E(x), \text{ and } E(x + p_2) = E(x)$$

for all values of the variable x.

Finally, on the historical side, is the somewhat tragic part played by Legendre. For forty years he had slaved over elliptic *integrals* (*not* elliptic *functions*) without noticing what both Abel and Jacobi saw almost at once, namely that by *inverting* his point of view the whole subject would become infinitely simpler. Elliptic integrals first present themselves in the problem of finding the length of an arc of an ellipse. To what was said about inversion in connection with Abel the following statement in symbols may be added. This will bring out more clearly the point which Legendre missed.

If $R(t)$ denotes a polynomial in t, an integral of the type

$$\int_0^x \frac{1}{\sqrt{R(t)}} dt$$

is called an *elliptic integral* if $R(t)$ is of either the third or the fourth degree; if $R(t)$ is of degree higher than the fourth, the integral is called *Abelian* (after Abel, some of whose greatest work concerned such integrals). If $R(t)$ is of only the second degree, the integral can be calculated out in terms of elementary functions. In particular

$$\int_0^x \frac{1}{\sqrt{1-t^2}} dt = \sin^{-1} x,$$

($\sin^{-1} x$ is read, "an angle whose sine is x"). That is, if

$$y = \int_0^x \frac{1}{\sqrt{1-t^2}} dt$$

we consider the *upper limit*, x, of the integral, as a function of the integral itself, namely of y. This *inversion* of the problem removed most of the difficulties which Legendre had grappled with for forty years. The true theory of these important integrals rushed forth almost of itself after this obstruction had been removed—like a log-jam going down the river after the king log has been snaked out.

When Legendre grasped what Abel and Jacobi had done he encouraged them most cordially, although he realized that their simpler approach (that of inversion) nullified what was to have been his own masterpiece of forty years' labor. For Abel, alas, Legendre's praise came too late, but for Jacobi it was an inspiration to surpass himself. In one of the finest correspondences in the whole of scientific literature the young man in his early twenties and the veteran in his late seventies strive to outdo one another in sincere praise and gratitude. The only jarring note is Legendre's outspoken disparagement of Gauss, whom Jacobi vigorously defends. But as Gauss never condescended to publish his researches—he had planned a major work on elliptic functions when Abel and Jacobi anticipated him in publication—Legendre can hardly be blamed for holding a totally mistaken opinion. For lack of space we must omit extracts from this beautiful correspondence (the letters are given in full in vol. 1 of Jacobi's *Werke*—in French).

The joint creation with Abel of the theory of elliptic functions was only a small if highly important part of Jacobi's huge output. Only to enumerate all the fields he enriched in his brief working life of less than a quarter of a century would take more space than can be devoted to one man in an account like the present, so we shall merely mention a few of the other great things he did.

Jacobi was the first to apply elliptic functions to the theory of numbers. This was to become a favorite diversion with some of the greatest mathematicians who followed Jacobi. It is a curiously recondite subject, where arabesques of ingenious algebra unexpectedly reveal hitherto unsuspected relations between the common whole numbers. It was by this means that Jacobi proved the famous assertion of Fermat that every integer 1,2,3, . . . is a sum of four integer squares (zero being counted as an integer) and, moreover, his beautiful analysis told him *in how many ways* any given integer may be expressed as such a sum.*

For those whose tastes are more practical we may cite Jacobi's work in dynamics. In this subject, of fundamental importance in both applied science and mathematical physics, Jacobi made the first significant ad-

---

*If n is odd, the number of ways is 8 times the sum of all the divisors of n (1 and n included); if n is even, the number of ways is 24 times the sum of all the odd divisors of n.

vance beyond Lagrange and Hamilton. Readers acquainted with quantum mechanics will recall the important part played in some presentations of that revolutionary theory by the Hamilton-Jacobi equation. His work in differential equations began a new era.

In algebra, to mention only one thing of many, Jacobi cast the theory of determinants into the simple form now familiar to every student in a second course of school algebra.

To the Newton-Laplace-Lagrange theory of attraction Jacobi made substantial contributions by his beautiful investigations on the functions which recur repeatedly in that theory and by applications of elliptic and Abelian functions to the attraction of ellipsoids.

Of a far higher order of originality is his great discovery in Abelian functions. Such functions arise in the inversion of an Abelian integral, in the same way that the elliptic functions arise from the inversion of an elliptic integral. (The technical terms were noted earlier in this chapter.) Here he had nothing to guide him, and for long he wandered lost in a maze that had no clue. The appropriate inverse functions in the simplest case are functions of *two* variables having *four* periods; in the general case the functions have n variables and $2n$ periods; the elliptic functions correspond to $n = 1$. This discovery was to nineteenth century analysis what Columbus' discovery of America was to fifteenth century geography.

Jacobi did not suffer an early death from overwork, as his lazier friends predicted that he should, but from smallpox (February 18, 1851) in his forty seventh year. In taking leave of this large-minded man we may quote his retort to the great French mathematical physicist Fourier, who had reproached both Abel and Jacobi for "wasting" their time on elliptic functions while there were still problems in heat-conduction to be solved.

"It is true," Jacobi says, "that M. Fourier had the opinion that the principal aim of mathematics was public utility and the explanation of natural phenomena; but a philosopher like him should have known that the sole end of science is the honor of the human mind, and that under this title a question about numbers is worth as much as a question about the system of the world."

If Fourier could revisit the glimpses of the moon he might be disgusted at what has happened to the analysis he invented for "public utility and the explanation of natural phenomena." So far as mathematical physics is concerned Fourier analysis today is but a detail in the infinitely vaster theory of boundary-value problems, and it is in the purest of pure mathematics that the analysis which Fourier invented finds its interest and its justification. Whether "the human mind" is honored by these modern researches may be put up to the experts—provided the behaviorists have left anything of the human mind to be honored.

CHAPTER NINETEEN

An Irish Tragedy

HAMILTON

*In mathematics he was greater
Than Tycho Brahe or Erra Pater;
For he by geometric scale
Could take the size of pots of ale.*
— SAMUEL BUTLER

WILLIAM ROWAN HAMILTON is by long odds the greatest man of science that Ireland has produced. His nationality is emphasized because one of the driving impulses behind Hamilton's incessant activity was his avowed desire to put his superb genius to such uses as would bring glory to his native land. Some have claimed that he was of Scotch descent. Hamilton himself insisted that he was Irish, and it is certainly difficult for a Scot to see anything Scotch in Ireland's greatest and most eloquent mathematician.

Hamilton's father was a solicitor in Dublin, Ireland, where William, the youngest of three brothers and one sister, was born on August 3, 1805.* The father was a first-rate business man with an "exuberant eloquence," a religious zealot, and last, but unfortunately not least, a very convivial man, all of which traits he passed on to his gifted son. Hamilton's extraordinary intellectual brilliance was probably inherited from his mother, Sarah Hutton, who came of a family well known for its brains.

However, on the father's side, the swirling clouds of eloquence, "both of lips and pen," which made the jolly toper the life of every party he graced with his reeling presence, condensed into something less gaseous in William's uncle, the Reverend James Hamilton, curate of the village of Trim (about twenty miles from Dublin). Uncle

---

*The date on his tombstone is August 4, 1805. Actually he was born at midnight; hence the confusion in dates. Hamilton, who had a passion for accuracy in such trifles, chose August 3rd until in later life he shifted to August 4th for sentimental reasons.

James was in fact an inhumanly accomplished linguist—Greek, Latin, Hebrew, Sanskrit, Chaldee, Pali, and heaven knows what other heathen dialects, came to the tip of his tongue as readily as the more civilized languages of Continental Europe and Ireland. This polyglot fluency played no inconsiderable part in the early and extremely extensive miseducation of the hapless but eager William, for at the age of three, having already given signs of genius, he was relieved of his doting mother's affection and packed off by his somewhat stupid father to glut himself with languages under the expert tutelage of the supervoluble Uncle James.

Hamilton's parents had very little to do with his upbringing; his mother died when he was twelve, his father two years later. To James Hamilton belongs whatever credit there may be for having wasted young William's abilities in the acquisition of utterly useless languages and turning him out, at the age of thirteen, as one of the most shocking examples of a linguistic monstrosity in history. That Hamilton did not become an insufferable prig under his misguided parson-uncle's instruction testifies to the essential soundness of his Irish common sense. The education he suffered might well have made a permanent ass of even a humorous boy, and Hamilton had no humor.

The tale of Hamilton's infantile accomplishments reads like a bad romance, but it is true: at three he was a superior reader of English and was considerably advanced in arithmetic; at four he was a good geographer; at five he read and translated Latin, Greek, and Hebrew, and loved to recite yards of Dryden, Collins, Milton, and Homer—the last in Greek; at eight he added a mastery of Italian and French to his collection and extemporized fluently in Latin, expressing his unaffected delight at the beauty of the Irish scene in Latin hexameters when plain English prose offered too plebeian a vent for his nobly exalted sentiments; and finally, before he was ten he had laid a firm foundation for his extraordinary scholarship in oriental languages by beginning Arabic and Sanskrit.

The tally of Hamilton's languages is not yet complete. When William was three months under ten years old his uncle reports that "His thirst for the Oriental languages is unabated. He is now master of most, indeed of all except the minor and comparatively provincial ones. The Hebrew, Persian, and Arabic are about to be confirmed by the superior and intimate acquaintance with the Sanskrit, in which he is already a proficient. The Chaldee and Syriac he is grounded in, also the Hindoostanee, Malay, Mahratta, Bengali, and others. He is about to commence the Chinese, but the difficulty of procuring books is very great. It cost me a large sum to supply him from London, but I hope the money was well expended." To which we can only throw up our hands and ejaculate Good God! What was the sense of it all?

By thirteen William was able to brag that he had mastered one language for each year he had lived. At fourteen he composed a flowery welcome in Persian to the Persian Ambassador, then visiting Dublin, and had it transmitted to the astonished potentate. Wishing to follow up his advantage and slay the already slain, young Hamilton called on the Ambassador, but that wily oriental, forewarned by his faithful secretary, "much regretted that on account of a bad headache he was unable to receive me [Hamilton] personally." Perhaps the Ambassador had not yet recovered from the official banquet, or he may have read the letter. In translation at least it is pretty awful—just the sort of thing a boy of fourteen, taking himself with devastating seriousness and acquainted with all the stickiest and most bombastic passages of the Persian poets, might imagine a sophisticated oriental out on a wild Irish spree would relish as a pick-me-up the morning after. Had young Hamilton really wished to view the Ambassador he should have sent in a salt herring, not a Persian poem.

Except for his amazing ability, the maturity of his conversation and his poetical love of nature in all her moods, Hamilton was like any other healthy boy. He delighted in swimming and had none of the grind's interesting if somewhat repulsive pallor. His disposition was genial and his temper—rather unusually so for a sturdy Irish boy—invariably even. In later life however Hamilton showed his Irish by challenging a detractor—who had called him a liar—to mortal combat. But the affair was amicably arranged by Hamilton's second, and Sir William cannot be legitimately counted as one of the great mathematical duellists. In other respects young Hamilton was not a normal boy. The infliction of pain or suffering on beast or man he would not tolerate. All his life Hamilton loved animals and, what is regrettably rarer, respected them as equals.

Hamilton's redemption from senseless devotion to useless languages began when he was twelve and was completed before he was fourteen. The humble instrument selected by Providence to turn Hamilton from the path of error was the American calculating boy, Zerah Colburn (1804–1839), who at the time had been attending

Westminster School in London. Colburn and Hamilton were brought together in the expectation that the young Irish genius would be able to penetrate the secret of the American's methods, which Colburn himself did not fully understand (as was seen in the chapter on Fermat). Colburn was entirely frank in exposing his tricks to Hamilton, who in his turn improved upon what he had been shown. There was but little abstruse or remarkable about Colburn's methods. His feats were largely a matter of memory. Hamilton's acknowledgment of Colburn's influence occurs in a letter written when he was seventeen (August, 1822) to his cousin Arthur.

By the age of seventeen Hamilton had mastered mathematics through the integral calculus and had acquired enough mathematical astronomy to be able to calculate eclipses. He read Newton and Lagrange. All this was his recreation; the classics were still his serious study, although only a second love. What is more important, he had already made "some curious *discoveries*," as he wrote to his sister Eliza.

The discoveries to which Hamilton refers are probably the germs of his first great work, that on systems of rays in optics. Thus in his seventeenth year Hamilton had already begun his career of fundamental discovery. Before this he had brought himself to the attention of Dr. Brinkley, Professor of Astronomy at Dublin, by the detection of an error in Laplace's attempted proof of the parallelogram of forces.

Hamilton never attended any school before going to the University but received all his preliminary training from his uncle and by private study. His forced devotion to the classics in preparation for the entrance examinations to Trinity College, Dublin, did not absorb all of his time, for on May 31, 1823, he writes to his cousin Arthur, "In Optics I have made a very curious discovery—at least it seems so to me. . . ."

If, as has been supposed, this refers to the "characteristic function," which Hamilton will presently describe for us, the discovery marks its author as the equal of any mathematician in history for genuine precocity. On July 7, 1823, young Hamilton passed, easily first out of one hundred candidates, into Trinity College. His fame had preceded him, and as was only to be expected, he quickly became a celebrity; indeed his classical and mathematical prowess, while he was yet an undergraduate, excited the curiosity of academic circles in England and Scotland as well as in Ireland, and it was even declared by some that a second Newton had arrived. The tale of his undergraduate triumphs can be imagined—he carried off practically all the available prizes and obtained the highest honors in both classics and mathematics. But more important than all these triumphs, he completed the first draft of Part I of his epoch-making memoir on systems of rays. "This young man," Dr. Brinkley remarked, when Hamilton presented his memoir to the Royal Irish Academy, "I do not say *will* be, but *is*, the first mathematician of his age."

Even his laborious drudgeries to sustain his brilliant academic record and the hours spent more profitably on research did not absorb all of young Hamilton's superabundant energies. At nineteen he experienced the first of his three serious love affairs. Being conscious of his own "unworthiness"—especially as concerned his material prospects—William contented himself with writing poems to the young lady, with the usual result: a solider, more prosaic man married the girl. Early in May, 1825, Hamilton learned from his sweetheart's mother that his love had married his rival. Some idea of the shock he experienced can be inferred from the fact that Hamilton, a deeply religious man to whom suicide was a deadly sin, was tempted to drown himself. Fortunately for science he solaced himself with another poem. All his life Hamilton was a prolific versifier. But his true poetry, as he told his friend and ardent admirer, William Wordsworth, was his mathematics. From this no mathematician will dissent.

We may dispose here of Hamilton's lifelong friendships with some of the shining literary lights of his day—the poets Wordsworth, Southey, and Coleridge, of the so-called Lake School, Aubrey de Vere, and the didactic novelist Maria Edgeworth—a litteratrice after Hamilton's own pious heart. Wordsworth and Hamilton first met on the latter's trip of September, 1827, to the English Lake District. Having "waited on Wordsworth at tea," Hamilton oscillated back and forth with the poet all night, each desperately trying to see the other home. The following day Hamilton sent Wordsworth a poem of ninety iron lines which the poet himself might have warbled in one of his heavier flights. Naturally Wordsworth did not relish the eager young mathematician's unconscious plagiarism, and after damning it with faint praise, proceeded to tell the hopeful author—at great length—that "the workmanship (what else could be expected from so young a writer?) is not what it ought to be." Two years later, when Hamilton was already installed as astronomer at the Dunsink Observatory, Wordsworth returned the visit. Hamilton's sister Eliza, on being introduced to the poet, felt herself "involuntarily parodying the first lines of his own poem *Yarrow Visited*—

> *And this is* Wordsworth! *this the man*
> *Of whom my fancy cherished*
> *So faithfully a waking dream,*
> *An image that hath perished!"*

One great benefit accrued from Wordsworth's visit: Hamilton realized at last that "his path must be the path of Science, and not that of Poetry; that he must renounce the hope of habitually cultivating both, and that, therefore, he must brace himself up to bid a painful farewell to Poetry." In short, Hamilton grasped the obvious truth that there was not a spark of poetry in him, in the *literary* sense. Nevertheless he continued to versify all his life. Wordsworth's opinion of Hamilton's intellect was high. In fact he graciously said (in effect) that only two men he had ever known gave him a feeling of inferiority, Coleridge and Hamilton.

Hamilton did not meet Coleridge till 1832, when the poet had practically ceased to be anything but a spurious copy of a mediocre German metaphysician. Nevertheless each formed a high estimate of the other's capacity, as Hamilton had for long been a devoted student of Kant in the original. Indeed philosophical speculation always fascinated Hamilton, and at one time he declared himself a wholehearted believer—intellectually, but not intestinally—in Berkeley's devitalized idealism. Another bond between the two was their preoccupation with the theological side of philosophy (if there is such a side), and Coleridge favored Hamilton with his half-digested ruminations on the Holy Trinity, by which the devout mathematician set considerable store.

The close of Hamilton's undergraduate career at Trinity College was even more spectacular than its beginning; in fact it was unique in university annals. Dr. Brinkley resigned his professorship of astronomy to become Bishop of Cloyne. According to the usual British custom the vacancy was advertised, and several distinguished astronomers, including George Biddell Airy (1801-1892), later Astronomer Royal of England, sent in their credentials. After some discussion the

Governing Board passed over all the applicants and unanimously elected Hamilton, then (1827) an undergraduate of twenty two, to the professorship. Hamilton had not applied. "Straight for him was the path of gold" now, and Hamilton resolved not to disappoint the hopes of his enthusiastic electors. Since the age of fourteen he had had a passion for astronomy, and once as a boy he had pointed out the Observatory on its hill at Dunsink, commanding a beautiful view, as the place of all others where he would like to live were he free to choose. He now, at the age of twenty two, had his ambition by the bit; all he had to do was to ride straight ahead.

He started brilliantly. Although Hamilton was no practical astronomer, and although his assistant observer was incompetent, these drawbacks were not serious. From its situation the Dunsink Observatory could never have cut any important figure in modern astronomy, and Hamilton did wisely in putting his major efforts on his mathematics. At the age of twenty three he published the completion of the "curious discoveries" he had made as a boy of seventeen, Part I of *A Theory of Systems of Rays*, the great classic which does for optics what Lagrange's *Mécanique analytique* does for mechanics and which, in Hamilton's own hands, was to be extended to dynamics, putting that fundamental science in what is perhaps its ultimate, perfect form.

The techniques which Hamilton introduced into applied mathematics in this, his first masterpiece, are today indispensable in mathematical physics, and it is the aim of many workers in particular branches of theoretical physics to sum up the whole of a theory in a Hamiltonian principle. This magnificent work is that which caused Jacobi, fourteen years later at the British Association meeting at Manchester in 1842, to assert that "Hamilton is the Lagrange of your country"—(meaning of the English-speaking race). As Hamilton himself took great pains to describe the essence of his new methods in terms comprehensible to non-specialists, we shall quote from his own abstract presented to the Royal Irish Academy on April 23, 1827.

"A Ray, in Optics, is to be considered here as a straight or bent or curved line, along which light is propagated; and a *System of Rays* as a collection or aggregate of such lines, connected by some common bond, some similarity of origin or production, in short some optical unity. Thus the rays which diverge from a luminous point compose one optical system, and, after they have been reflected at a mirror, they compose another. To investigate the geometrical relations of the rays of a system of which we know (as in these simple cases) the optical origin and history, to inquire how they are disposed among themselves, how they diverge or converge, or are parallel, what surfaces or curves they touch or cut, and at what angles of section, how they can be combined in partial pencils, and how each ray in particular can be determined and distinguished from every other, is to study that System of Rays. And to generalize this study of one system so as to become able to pass, without change of plan, to the study of other systems, to assign general rules and a general method whereby these separate optical arrangements may be connected and harmonised together, is to form a *Theory of Systems of Rays*. Finally, to do this in such a manner as to make available the powers of the modern mathesis, replacing figures by functions and diagrams by formulas, is to construct an Algebraic Theory of such Systems, or an *Application of Algebra to Optics*.

"Towards constructing such an application it is natural, or rather necessary, to employ the method introduced by Descartes for the application of Algebra to Geometry. That great and philosophical mathematician conceived the possibility, and employed the plan, of representing or expressing algebraically the position of any point in space by three co-ordinate numbers which answer respectively how far the point is in three rectangular directions (such as north, east, and west), from some fixed point or origin selected or assumed for the purpose; the three dimensions of space thus receiving their three algebraical equivalents, their appropriate conceptions and symbols in the general science of progression [order]. A plane or curved surface became thus algebraically defined by assigning as *its equation* the relation connecting the three co-ordinates of any point upon it, and common to all those points: and a line, straight or curved, was expressed according to the same method, by the assigning two such relations, correspondent to two surfaces of which the line might be regarded as the intersection. In this manner it became possible to conduct general investigations respecting surfaces and curves, and to discover properties common to all, through the medium of general investigations respecting equations between three variable numbers: every geometrical problem could be at least algebraically expressed, if not at once resolved, and every improvement or discovery in Algebra became susceptible of application or interpretation in Geometry. The sciences of Space and Time (to adopt here a view of Algebra which I have elsewhere ventured to propose) became intimately intertwined and indissolubly connected with each other. Henceforth it was almost impossible to improve either science without improving the other also. The problem of drawing tangents to curves led to the discovery of Fluxions or Differentials: those of rectification and quadrature to the inversion of Fluents or Integrals: the investigation of curvatures of surfaces required the Calculus of Partial Differentials: the isoperimetrical problems resulted in the formation of the Calculus of Variations. And reciprocally, all these great steps in Algebraic Science had immediately their applications to Geometry, and led to the discovery of new relations between points or lines or surfaces. But even if the applications of the method had not been so manifold and important, there would still have been derivable a high intellectual pleasure from the contemplation of it *as* a method.

"The first important application of this algebraical method of co-ordinates to the study of optical systems was made by Malus, a French officer of engineers in Napoleon's army in Egypt, and who has acquired celebrity in the history of Physical Optics as the discoverer of polarization of light by reflexion. Malus presented to the Institute of France, in 1807, a profound mathematical work which is of the kind above alluded to, and is entitled *Traité d'Optique*. The method employed in that treatise may be thus described:—The direction of a straight ray of any final optical system being considered as dependent on the position of some assigned point on the ray, according to some law which characterizes the particular system and distinguishes it from others; this law may be algebraically expressed by assigning three expressions for the three co-ordinates of some other point of the ray, as *functions* of the three co-ordinates of the point proposed. Malus accordingly introduces general symbols denoting three such functions (or at least three functions equivalent to these), and proceeds to draw several important general conclusions, by very complicated yet symmetric calculations; many of which conclusions, along with many others, were also obtained afterwards by myself, when, by a method nearly similar, without knowing what Malus had done, I began my own attempt to apply Algebra to Optics. But my researches soon conducted me to substitute, for this method of Malus, a very different, and (as I conceive that I have proved) a much more *appropriate* one, for the study of optical systems; by which, instead of employing the *three* functions above mentioned, or at least their *two* ratios, it becomes sufficient to employ *one function*, which I call *characteristic* or *principal*. And thus, whereas he made his deductions by setting out with the *two equations of a ray*, I on the other hand establish and employ the *one equation of a system*.

"The function which I have introduced for this purpose, and made the basis of my method of *deduction* in mathematical Optics, had, in another connexion, presented itself to former writers as expressing the result of a very high and extensive *induction* in that science. This known result is usually called the *law of least action*, but sometimes also the principle of *least time* [see chapter on Fermat], and includes all that has hitherto been discovered respecting the rules which determine the forms and positions of the lines along which light is propagated, and the changes of direction of those lines produced by reflexion or refraction, ordinary or extraordinary [the latter as in a doubly refracting crystal, say Iceland spar, in which a single ray is split into two, both refracted, on entering the crystal]. A certain quantity which in one physical theory is the *action*, and in another the *time*, expended by light in going from any first to any second point, is found to be less than if the light had gone in any other than its actual path, or at least to have what is technically called its variation null, the extremities of the path being unvaried. The mathematical novelty of my method consists in considering this quantity as a *function* of the co-ordinates of these extremities, which varies when they vary, according to a law which I have called the *law of varying action;* and in *reducing all researches respecting optical systems of rays to the study of this single function:* a reduction which presents mathematical Optics under an entirely novel view, and one analogous (as it appears to me) to the aspect under which Descartes presented the application of Algebra to Geometry."

Nothing need be added to this account of Hamilton's, except possibly the remark that no science, no matter how ably expounded, is understood as readily as any novel, no matter how badly written. The whole extract will repay a second reading.

In this great work on systems of rays Hamilton had built better than even he knew. Almost exactly one hundred years after the above abstract was written the methods which Hamilton introduced into optics were found to be just what was required in the wave mechanics associated with the modern quantum theory and the theory of atomic structure. It may be recalled that Newton had favored an emission, or corpuscular, theory of light, while Huygens and his successors up to almost our own time sought to explain the phenomena of light wholly by means of a wave theory. Both points of view were united and, in a purely mathematical sense, reconciled in the modern quantum theory, which came into being in 1925-6. In 1834, when he was twenty eight, Hamilton realized his ambition of extending the principles which he had introduced into optics to the whole of dynamics.

Hamilton's theory of rays, shortly after its publication when its author was but twenty seven, had one of the promptest and most spectacular successes of any of the classics of mathematics. The theory purported to deal with phenomena of the actual physical universe as it is observed in everyday life and in scientific laboratories. Unless any such mathematical theory is capable of predictions which experiments later verify, it is no better than a concise dictionary of the subject it systematizes, and it is almost certain to be superseded shortly by a more imaginative picture which does not reveal its whole meaning at the first glance. Of the famous predictions which have certified the value of truly mathematical theories in physical science, we may recall three: the mathematical discovery by John Couch Adams (1819-1892) and Urbain-Jean-Joseph Leverrier (1811-1877) of the planet Neptune, independently and almost simultaneously in 1845, from an analysis of the perturbations of the planet Uranus according to the Newtonian theory of gravitation; the mathematical prediction of wireless waves by James Clerk Maxwell (1831-1879) in 1864, as a consequence of his own electromagnetic theory of light; and finally, Einstein's prediction in 1915-16, from his theory of general relativity, of the deflection of a ray of light in a gravitational field, first confirmed by observations of the solar eclipse on the historic May 29, 1919, and his prediction, also from his theory, that the spectral lines in light issuing from a massive body would be shifted by an amount, which Einstein stated, toward the red end of the spectrum—also confirmed. The last two of these instances—Maxwell's and Einstein's—are of a different order from the first: in both, *totally unknown and unforeseen phenomena* were predicted mathematically; that is, these predictions were *qualitative*. Both Maxwell and Einstein amplified their qualitative foresight by precise *quantitative* predictions which precluded any charge of mere guessing when their prophecies were finally verified experimentally.

Hamilton's prediction of what is called conical refraction in optics was of this same qualitative plus quantitative order. From his theory of systems of rays he predicted mathematically that a wholly unexpected phenomenon would be found in connection with the refraction of light in biaxal crystals. While polishing the Third Supplement to his memoir on rays he surprised himself by a discovery which he thus describes:

"The law of the reflexion of light at ordinary mirrors appears to have been known to Euclid; that of ordinary refraction at a surface of water, glass, or other uncrystallized medium, was discovered at a much later date by Snellius; Huygens discovered, and Malus confirmed, the law of extraordinary refraction produced by uniaxal crystals, such as Iceland spar; and finally the law of the extraordinary double refraction at the faces of biaxal crystals, such as topaz or arragonite, was found in our own time by Fresnel. But even in these cases of extraordinary or crystalline refraction, no more than *two* refracted rays had ever been observed or even suspected to exist, if we except a theory of Cauchy, that there might possibly be a *third* ray, though probably imperceptible to our senses. Professor Hamilton, however, in investigating by his general method the consequences of the law of Fresnel, was led to conclude that there ought to be in certain cases, which he assigned, not merely two, nor three, nor any finite number, but an *infinite* number, or a *cone* of refracted rays *within* a biaxal crystal, corresponding to and resulting from a *single* incident ray; and that in certain other cases, a single ray within such a crystal should give rise to an infinite number of emergent rays, arranged in a certain other cone. He was led, therefore, to anticipate from theory two new laws of light, to which he gave the names of *Internal and External Conical Refraction*."

The prediction and its experimental verification by Humphrey Lloyd evoked unbounded admiration for young Hamilton from those who could appreciate what he had done. Airy, his former rival for the professorship of astronomy, estimated Hamilton's achievement thus: "Perhaps the most remarkable prediction that has ever been made is that lately made by Professor Hamilton." Hamilton himself considered this, like any similar prediction, "a subordinate and secondary result" compared to the grand object which he had in view, "to introduce harmony and unity into the contemplations and reasonings of optics, regarded as a branch of pure science."

According to some this spectacular success was the high-water mark in Hamilton's career; after the great work on optics and dynamics his tide ebbed. Others, particularly members of what has been styled the High Church of Quaternions, hold that Hamilton's greatest work was still to come—the creation of what Hamilton himself considered his masterpiece and his title to immortality, his theory of quaternions. Leaving quaternions out of the indictment for the moment, we may simply state that, from his twenty seventh year till his death at sixty, two disasters raised havoc with Hamilton's scientific career, marriage and alcohol. The second was partly, but not wholly, a consequence of the unfortunate first.

After a second unhappy love affair, which ended with a thoughtless remark that meant nothing but which the hypersensitive suitor took to heart, Hamilton married his third fancy, Helen Maria Bayley, in the spring of 1833. He was then in his twenty eighth year. The bride was the daughter of a country parson's widow. Helen was "of pleasing ladylike appearance, and early made a favourable impression upon him [Hamilton] by her truthful nature and by the religious principles which he knew her to possess, although to these recommendations was not added any striking beauty of face or force of intellect." Now, any fool can tell the truth, and if truthfulness is all a fool has to recommend her, whoever commits matrimony with her will get the short end of the indiscretion. In the summer of 1832 Miss Bayley "passed through a dangerous illness, . . . , and this event doubtless drew his [the lovelorn Hamilton's] thoughts especially toward her, in the form of anxiety for her recovery, and, coming at a time [when he had just broken with the girl he really wanted] when he felt obliged to suppress his former passion, prepared the way for tenderer and warmer feelings." Hamilton in short was properly hooked by an ailing female who was to become a semi-invalid for the rest of her life and who, either through incompetence or ill-health, let her husband's slovenly servants run his house as they chose, which at least in some quarters—especially his study—came to resemble a pigsty. Hamilton needed a sympathetic woman with backbone to keep him and his domestic affairs in some semblance of order; instead he got a weakling.

Ten years after his marriage Hamilton tried to pull himself up short on the slippery trail he realized with a brutal shock he was treading. As a young man, fêted and toasted at dinners, he had rather let himself go, especially as his great gifts for eloquence and conviviality were naturally enough heightened by a drink or two. After his marriage, irregular meals or no meals at all, and his habit of working twelve or fourteen hours at a stretch, were compensated for by taking nourishment from a bottle.

It is a moot question whether mathematical *inventiveness* is accelerated or retarded by moderate indulgence in alcohol, and until an exhaustive set of *controlled* experiments is carried out to settle the matter, the doubt must remain a doubt, precisely as in any other biological research. If, as some maintain, poetic and mathematical inventiveness are akin, it is by no means obvious that reasonable alcoholic indulgence (if there is such a thing) is destructive of mathematical inventiveness; in fact numerous well-attested instances would seem to indicate the contrary. In the case of poets, of course, "wine and song" have often gone together, and in at least one instance—Swinburne—without the first the second dried up almost completely. Mathematicians have frequently remarked on the terrific strain induced by prolonged concentration on a difficulty, and some have found the let-down occasioned by a drink a decided relief. But poor Hamilton quickly passed beyond this stage and became careless, not only in the untidy privacy of his study, but also in the glaring publicity of a banquet hall. He got drunk at a scientific dinner. Realizing what had overtaken him, he resolved never to touch alcohol again, and for two years he kept his resolution. Then, during a scientific meeting at the estate of Lord Rosse (owner of the largest and most useless telescope then in existence), his old rival, Airy, jeered at him for drinking nothing but water. Hamilton gave in, and thereafter took all he wanted—which was more than enough. Still, even this handicap could not put him out of the race, although without it he probably would have gone farther and have reached a greater height than he did. However, he got high enough, and moralizing may be left to moralists.

Before considering what Hamilton regarded as his masterpiece, we may briefly summarize the principal honors which came his way. At thirty he held an influential office in the British Association for the Advancement of Science at its Dublin meeting, and at the same time the Lord Lieutenant bade him to "Kneel down, Professor Hamilton," and then, having dubbed him on both shoulders with the sword of State, to "Rise up, Sir William Rowan Hamilton." This was one of the few occasions in his life on which Hamilton had nothing whatever to say. At thirty two he became President of the Royal Irish Academy, and at thirty eight was awarded a Civil List life pension of two hundred pounds a year from the British Government, Sir Robert Peel, Ireland's reluctant friend, being then Premier. Shortly before this Hamilton had made his capital invention—quaternions.

An honor which pleased him more than any he had ever received was the last, as he lay on his deathbed: he was elected the first foreign member of the National Academy of Sciences of the United States, which was founded during the Civil War. This honor was in recognition of his work in quaternions, principally, which for some unfathomable reason stirred American mathematicians of the time (there were only one or two in existence, Benjamin Peirce of Harvard being the chief) more profoundly than had any other British mathematics since Newton's *Principia*. The early popularity of quaternions in the United States is somewhat of a mystery. Possibly the turgid eloquence of the *Lectures on Quaternions* captivated the taste of a young and vigorous nation which had yet to outgrow its morbid addiction to senatorial oratory and Fourth of July verbal fireworks.

Quaternions has too long a history for the whole story to be told here. Even Gauss with his anticipation of 1817 was not the first in the field; Euler preceded him with an isolated result which is most simply interpreted in terms of quaternions. The origin of quaternions may go back even farther than this, for Augustus de Morgan once half-jokingly offered to trace their history for Hamilton from the ancient Hindus to Queen Victoria. However, we need glance here only at the lion's share in the invention and consider briefly what inspired Hamilton.

The British school of algebraists, as will be seen in the chapter on Boole, put common algebra on its own feet during the first half of the nineteenth century. Anticipating the currently accepted procedure in developing any branch of mathematics carefully and rigorously they founded algebra *postulationally*. Before this, the various kinds of "numbers"—fractions, negatives, irrationals—which enter mathematics when it is *assumed* that all algebraic equations have roots, had been allowed to function on precisely the same footing as the common positive integers which were so staled by custom that all mathematicians believed them to be "natural" and in some vague sense completely understood—they are not, even today, as will be seen when the work of Georg Cantor is discussed. This naïve faith in the self-consistency of a system founded on the blind, formal juggling of mathematical symbols may have been sublime but it was also slightly idiotic. The climax of this credulity was reached in the notorious *principle of permanence of form*, which stated in effect that a set of rules which yield consistent results for one kind of numbers—say the positive integers—will continue to yield consistency when applied to any other kind—say the imaginaries—even when no interpretation of the results is evident. It does not seem surprising that this faith in the integrity of meaningless symbols frequently led to absurdity.

The British school changed all this, although they were unable to take the final step and *prove* that their postulates for common algebra will never lead to a contradiction. That step was taken only in our own generation by the German workers in the foundations of mathematics. In this connection it must be kept in mind that algebra deals only with *finite* processes; when *infinite* processes enter, as for example in summing an infinite series, we are thrust out of algebra into another domain. This is emphasized because the usual elementary text labelled "Algebra" contains a great deal—infinite geometric progressions, for instance—that is *not* algebra in the modern meaning of the word.

The nature of what Hamilton did in his creation of quaternions will show up more clearly against the background of a set of postulates (taken from L. E. Dickson's *Algebras and Their Arithmetics*, Chicago, 1923) for common algebra or, as it is technically called, a *field* (English writers sometimes use *corpus* as the equivalent of the German *Körper* or French *corps*).

"A field F is a system consisting of a set S of elements $a,b,c, \ldots$ and two operations, called addition and multiplication which may be performed upon any two (equal or distinct) elements a and b of S, taken in that order, to produce uniquely determined elements $a \oplus b$ and $a \odot b$ of S, such that postulates I–V are satisfied. For simplicity we shall write $a + b$ for $a \oplus b$, and ab for $a \odot b$, and call them the *sum* and *product*, respectively, of a and b. Moreover, elements of S will be called elements of F.

"I. If a and b are any two elements of F, $a + b$ and ab are uniquely determined elements of F, and

$$b + a = a + b, \quad ba = ab.$$

"II. If a,b,c are any three elements of F,

$$(a + b) + c = a + (b + c), \quad (ab)c = a(bc), \quad a(b + c) = ab + ac.$$

"III. There exist in F two distinct elements, denoted by 0, 1, such that if a is any element of F, $a + 0 = a$, $a1 = a$ (whence $0 + a = a$, $1a = a$, by I).

"IV. Whatever be the element a of F, there exists in F an element x such that $a + x = 0$ (whence $x + a = 0$ by I).

"V. Whatever be the element a (distinct from 0) of F, there exists in F an element y such that $ay = 1$ (whence $ya = 1$, by I)."

From these simple postulates the whole of common algebra follows. A word or two about some of the statements may be helpful to those who have not seen algebra for years. In II, the statement $(a + b) + c = a + (b + c)$, called the *associative law of addition*, says that if a and b are added, and to this sum is added c, the result is the same as if a and the sum of b and c are added. Similarly, with respect to multiplication, for the second statement in II. The third statement in II is called the *distributive law*. In III a "zero" and "unity" are postulated; in IV, the postulated x gives the negative of a; and the first parenthetical remark in V forbids "division by zero." The demands in Postulate I are called the *commutative laws of addition and multiplication* respectively.

Such a set of postulates may be regarded as a distillation of experience. Centuries of working with numbers and getting useful results according to the rules of arithmetic—empirically arrived at—suggested most of the rules embodied in these precise postulates, but once the suggestions of experience are understood, the *interpretation* (here common arithmetic) furnished by experience is deliberately suppressed or forgotten, and the *system* defined by the postulates is developed *abstractly*, on its own merits, by common logic plus mathematical tact.

Notice in particular IV, which *postulates the existence* of negatives. We do not attempt to *deduce* the existence of negatives from the behavior of positives. When negative numbers first appeared in experience, as in debits instead of credits, they, *as numbers*, were held in the same abhorrence as "unnatural" monstrosities as were later the "imaginary" numbers $\sqrt{-1}$, $\sqrt{-2}$, etc., arising from the *formal* solution of equations such as $x^2 + 1 = 0$, $x^2 + 2 = 0$, etc. If the reader will glance back at what Gauss did for complex numbers he will appreciate more fully the complete simplicity of the following partial statement of Hamilton's original way of stripping "imaginaries" of their silly, purely imaginary mystery. This simple thing was one of the steps which led Hamilton to his quaternions, although strictly it has nothing to do with them. It is the *method* and the *point of view* behind this ingenious recasting of the algebra of complex numbers which are of importance for the sequel.

If as usual i denotes $\sqrt{-1}$, a "complex number" is a number of the type $a + bi$, where a,b are "real numbers" or, if preferred, and more generally, elements of the field F defined by the above postulates. Instead of regarding $a + bi$ as one "number," Hamilton conceived it as an *ordered couple* of "numbers," and he designated this couple by writing it (a,b). He then proceeded to impose definitions of *sum* and *product* on these couples, as suggested by the *formal* rules of combination sublimated from the experience of algebraists in manipulating complex numbers *as if* the laws of common algebra did in fact hold for them. One advantage of this new way of approaching complex numbers was this: the definitions for sum and product of couples were seen to be *instances* of the general, abstract definitions of sum and product as in a field. Hence, if the consistency of the system defined by the postulates for a field is proved, the like follows, without further proof, for complex numbers and the usual rules by which they are combined. It will be sufficient to state the definitions of sum and product in Hamilton's theory of complex numbers considered as couples (a,b) (c,d), etc.

The *sum* of (a,b) and (c,d) is $(a + b, c + d)$; their *product* is $(ac - bd, ad + bc)$. In the last, the minus sign is as in a field; namely, the element x postulated in IV is denoted by $- a$. To the 0, 1 of a field correspond here the couples $(0,0)$, $(1,0)$. With these definitions it is easily verified that Hamilton's couples satisfy all the stated postulates for a field. But they also accord with the *formal* rules for manipulating complex numbers. Thus, to (a,b), (c,d) correspond respectively $a + bi$, $c + di$, and the formal "sum" of these two is $(a + c) + i(b + d)$, to which corresponds the couple $(a + c, b + d)$. Again, formal multiplication of $a + bi$, $c + id$ gives $(ac - bd) + i(ad + bc)$, to which corresponds the couple $(ac - bd, ad + bc)$. If this sort of thing is new to any reader, it will repay a second inspection, as it is an example of the way in which modern mathematics eliminates mystery. So long as there is a shred of mystery attached to any concept that concept is not mathematical.

Having disposed of complex numbers by *couples*, Hamilton sought to extend his device to ordered *triples* and *quadruples*. Without some idea of what is sought to be accomplished such an undertaking is of course so vague as to be meaningless. Hamilton's object was to invent an algebra which would do for rotations in space of *three* dimensions what complex numbers, or his couples, do for rotations in space of *two* dimensions, both spaces being Euclidean as in elementary geometry. Now, a complex number $a + bi$ can be thought of as representing a *vector*, that is, a line segment having both *length and direction*, as is evident from the diagram, in which the directed segment (indicated by the arrow) represents the vector *OP*.

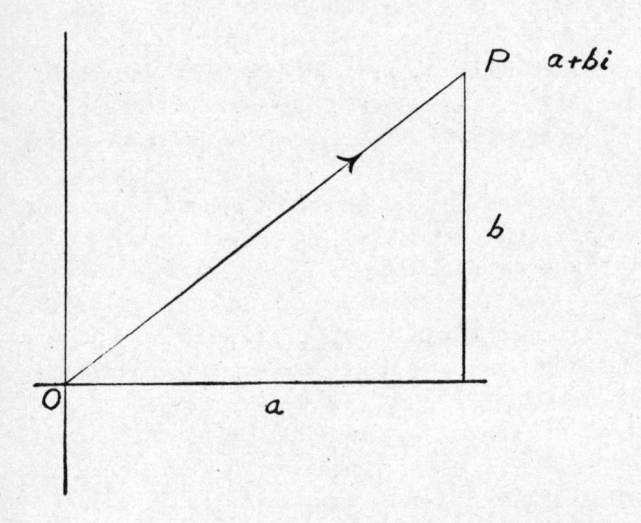

But on attempting to symbolize the behavior of vectors in three dimensional space so as to preserve those properties of vectors which are of use in physics, particularly in the combination of rotations, Hamilton was held up for years by an unforeseen difficulty whose very nature he for long did not even suspect. We may glance in passing at one of the clues he followed. That this led him anywhere—as he insisted it did—is all the more remarkable as it is now almost universally regarded as an absurdity, or at best a metaphysical speculation without foundation in history or in mathematical experience.

Objecting to the purely abstract, postulational formulation of algebra advocated by his British contemporaries, Hamilton sought to found algebra on something "more real," and for this strictly meaningless enterprise he drew on his knowledge of Kant's mistaken notions—exploded by the creation of non-Euclidean geometry—of space as "a pure form of sensuous intuition." Indeed Hamilton, who seems to have been unacquainted with non-Euclidean geometry, followed Kant in believing that "Time and space are two sources of knowledge from which various *a priori* synthetical cognitions can be derived. Of this, pure mathematics gives a splendid example in the case of our cognition of space and its various relations. As they are both pure forms of sensuous intuition, they render synthetic propositions *a priori* possible." Of course any not utterly illiterate mathematician today knows that Kant was mistaken in this conception of mathematics, but in the 1840's, when Hamilton was on his way to quaternions, the Kantian philosophy of mathematics still made sense to those—and they were nearly all—who had never heard of Lobatchewsky. By what looks like a bad mathematical pun, Hamilton applied the Kantian doctrine to algebra and drew the remarkable conclusion that, since geometry is the science of space, and since time and space are "pure sensuous forms of intuition," therefore the rest of mathematics must belong to time, and he wasted much of his own time in elaborating the bizarre doctrine that *algebra is the science of pure time*.

This queer crotchet has attracted many philosophers, and quite recently it has been exhumed and solemnly dissected by owlish metaphysicians seeking the philosopher's stone in the gall bladder of mathematics. Just because "algebra as the science of pure time" is of no earthly mathematical significance, it will continue to be discussed with animation till time itself ends. The opinion of a great mathematician on the "pure time" aspect of algebra may be of interest. "I cannot myself recognize the connection of algebra with the notion of time," Cayley confessed; "granting that the notion of continuous progression presents itself and is of importance, I do not see that it is in anywise the fundamental notion of the science."

Hamilton's difficulties in trying to construct an algebra of vectors and rotations for three-dimensional space were rooted in his subconscious conviction that the most important laws of common algebra must persist in the algebra he was seeking. How were vectors in three-dimensional space to be multiplied together?

To sense the difficulty of the problem it is essential to bear in mind (see Chapter on Gauss) that *ordinary complex numbers $a + bi$ ($i = \sqrt{-1}$)* had been given a simple interpretation in terms of *rotations in a plane*, and further that *complex numbers obey all the rules of common algebra*, in particular the *commutative law of multiplication*: if A, B are any complex numbers, then $A \times B = B \times A$, whether A, B are interpreted *algebraically*, or *in terms of rotations in a plane*. It was but human then to anticipate that *the same commutative law* would hold for the *generalizations of complex numbers* which represent *rotations in space of three dimensions*.

Hamilton's great discovery—or invention—was an algebra, one of the "natural" algebras of rotations in space of three dimensions, in which the commutative law of multiplication does not hold. In this Hamiltonian algebra of *quaternions* (as he called his invention), a multiplication appears in which $A \times B$ is *not* equal to $B \times A$ but to minus $B \times A$, that is, $A \times B = -B \times A$.

That a consistent, practically useful system of algebra could be constructed in defiance of the commutative law of multiplication was a discovery of the first order, comparable, perhaps, to the conception of non-Euclidean geometry. Hamilton himself was so impressed by the magnitude of what suddenly dawned on his mind (after fifteen years of fruitless thought) one day (October 16, 1843) when he was out walking with his wife that he carved the fundamental formulas of the new algebra in the stone of the bridge on which he found himself at the moment. His great invention showed algebraists the way to other algebras until today, following Hamilton's lead, mathematicians manufacture algebras practically at will by negating one or more of the postulates for a field and developing the consequences. Some of these "algebras" are extremely useful; the general theories embracing swarms of them include Hamilton's great invention as a mere detail, although a highly important one.

In line with Hamilton's quaternions the numerous brands of *vector analysis* favored by physicists of the past two generations sprang into being. Today all of these, including quaternions, *so far as physical applications are concerned*, are being swept aside by the incomparably simpler and more general *tensor analysis* which came into vogue with general relativity in 1915. Something will be said about this later.

In the meantime it is sufficient to remark that Hamilton's deepest tragedy was neither alcohol nor marriage but his obstinate belief that quaternions held the key to the mathematics of the physical universe. History has shown that Hamilton tragically deceived himself when he insisted ". . . I still must assert that this discovery appears to me to be as important for the middle of the nineteenth century as the discovery of fluxions [the calculus] was for the close of the seventeenth." Never was a great mathematician so hopelessly wrong.

The last twenty two years of Hamilton's life were devoted almost exclusively to the elaboration of quaternions, including their application to dynamics, astronomy, and the wave theory of light, and his voluminous correspondence. The style of the overdeveloped *Elements of Quaternions,* published the year after Hamilton's death, shows plainly the effects of the author's mode of life. After his death from gout on September 2, 1865 in the sixty first year of his age, it was found that Hamilton had left behind a mass of papers in indescribable confusion and about sixty huge manuscript books full of mathematics. An adequate edition of his works is now in progress. The state of his papers testified to the domestic difficulties under which the last third of his life had been lived: innumerable dinner plates with the remains of desiccated, unviolated chops were found buried in the mountainous piles of papers, and dishes enough to supply a large household were dug out from the confusion. During his last period Hamilton lived as a recluse, ignoring the meals shoved at him as he worked, obsessed by the dream that the last tremendous effort of his magnificent genius would immortalize both himself and his beloved Ireland, and stand forever unshaken as the greatest mathematical contribution to science since the *Principia* of Newton.

His early work, on which his imperishable glory rests, he came to regard as a thing of but little moment in the shadow of what he believed was his masterpiece. To the end he was humble and devout, and wholly without anxiety for his scientific reputation. "I have very long admired Ptolemy's description of his great astronomical master, Hipparchus, as ἀνὴρ φιλόπονος καὶ φιλαληθής; a labor-loving and truth-loving man. Be such my epitaph."

CHAPTER TWENTY

Genius and Stupidity

GALOIS

Against stupidity the gods themselves fight unvictorious.—SCHILLER

ABEL WAS DONE TO DEATH BY POVERTY, Galois by stupidity. In all the history of science there is no completer example of the triumph of crass stupidity over untamable genius than is afforded by the all too brief life of Évariste Galois. The record of his misfortunes might well stand as a sinister monument to all self-assured pedagogues, unscrupulous politicians, and conceited academicians. Galois was no "ineffectual angel," but even his magnificent powers were shattered before the massed stupidity aligned against him, and he beat his life out fighting one unconquerable fool after another.

The first eleven years of Galois' life were happy. His parents lived in the little village of Bourg-la-Reine, just outside Paris, where Évariste was born on October 25, 1811. Nicolas-Gabriel Galois, the father of Évariste, was a relic of the eighteenth century, cultivated, intellectual, saturated with philosophy, a passionate hater of royalty and an ardent lover of liberty. During the Hundred Days after Napoleon's escape from Elba, Galois was elected mayor of the village. After Waterloo he retained his office and served faithfully under the King, backing the villagers against the priest and delighting social gatherings with the old-fashioned rhymes which he composed himself. These harmless activities were later to prove the amiable man's undoing. From his father, Évariste acquired the trick of rhyming and a hatred of tyranny and baseness.

Until the age of twelve Galois had no teacher but his mother, Adélaïde-Marie Demante. Several of the traits of Galois' character were inherited from his mother, who came from a long line of distinguished jurists. Her father appears to have been somewhat of a Tartar. He gave his daughter a thorough classical and religious education, which she in turn passed on to her eldest son, not as she had received it, but fused into a virile stoicism in her own independent mind. She had not rejected Christianity, nor had she accepted it without question; she had merely contrasted its teachings with those of Seneca and Cicero, reducing all to their basic morality. Her friends remembered her as a woman of strong character with a mind of her own, generous, with a marked vein of originality, quizzical, and, at times, inclined to be paradoxical. She died in 1872 at the age of eighty four. To the last she retained the full vigor of her mind. She, like her husband, hated tyranny.

There is no record of mathematical talent on either side of Galois' family. His own mathematical genius came on him like an explosion, probably at early adolescence. As a child he was affectionate and rather serious, although he entered readily enough into the gaiety of the recurrent celebrations in his father's honor, even composing rhymes and dialogues to entertain the guests. All this changed under the first stings of petty persecution and stupid misunderstanding, not by his parents, but by his teachers.

In 1823, at the age of twelve, Galois entered the lycée of Louis-le-Grand in Paris. It was his first school. The place was a dismal horror. Barred and grilled, and dominated by a provisor who was more of a political gaoler than a teacher, the place looked like a prison, and it was. The France of 1823 still remembered the Revolution. It was a time of plots and counterplots, of riots and rumors of revolution. All this was echoed in the school. Suspecting the provisor of scheming to bring back the Jesuits, the students struck, refusing to chant in chapel. Without even notifying their parents the provisor expelled those whom he thought most guilty. They found themselves in the street. Galois was not among them, but it would have been better for him if he had been.

Till now tyranny had been a mere word to the boy of twelve. Now he saw it in action, and the experience warped one side of his character for life. He was shocked into unappeasable rage. His studies, owing to his mother's excellent instruction in the classics, went very well and he won prizes. But he had also gained something more lasting than any prize, the stubborn conviction, right or wrong, that neither fear nor the utmost severity of discipline can extinguish the sense of justice and fair dealing in young minds experiencing their first unselfish devotion. This his fellow students had taught him by their courage. Galois never forgot their example. He was too young not to be embittered.

The following year marked another crisis in the young boy's life. Docile interest in literature and the classics gave way to boredom; his mathematical genius was already stirring. His teachers advised that he be demoted. Évariste's father objected, and the boy continued with his interminable exercises in rhetoric, Latin, and Greek. His work was reported as mediocre, his conduct "dissipated," and the teachers had their way. Galois was demoted. He was forced to lick up the stale leavings which his genius had rejected. Bored and disgusted he gave his work perfunctory attention and passed it without effort or interest. Mathematics was taught more or less as an aside to the serious business of digesting the classics, and the pupils of various grades and assorted ages took the elementary mathematical course at the convenience of their other studies.

It was during this year of acute boredom that Galois began mathematics in the regular school course. The splendid geometry of Legendre came his way. It is said that two years was the usual time required by even the better mathematicians among the boys to master Legendre. Galois read the geometry from cover to cover as easily as other boys read a pirate yarn. The book aroused his enthusiasm; it was no textbook written by a hack, but a work of art composed by a creative mathematician. A single reading sufficed to reveal the whole structure of elementary geometry in crystal clarity to the fascinated boy. He had mastered it.

His reaction to algebra is illuminating. It disgusted him, and for a very good reason when we consider what sort of mind Galois had. Here was no master like Legendre to inspire him. The text in algebra was a schoolbook and nothing more. Galois contemptuously tossed it aside. It lacked, he said, the creator's touch that only a creative mathematician can give. Having made the acquaintance of one great mathematician through his work, Galois took matters into his own hands. Ignoring the meticulous pettifogging of his teacher, Galois went directly for his algebra to the greatest master of the age, Lagrange. Later he read Abel. The boy of fourteen or fifteen absorbed masterpieces of algebraical analysis addressed to mature professional mathematicians—the memoirs on the numerical solution of equations, the theory of analytical functions, and the calculus of functions. His class work in mathematics was mediocre: the traditional course was trivial to a mathematical genius and not necessary for the mastering of real mathematics.

Galois' peculiar gift of being able to carry on the most difficult mathematical investigations almost entirely in his head helped him with neither teachers nor examiners. Their insistence upon details which to him were obvious or trivial exasperated him beyond endurance, and he frequently lost his temper. Nevertheless he carried off the prize in the general examination. To the amazement of teachers and students alike Galois had taken his own kingdom by assault while their backs were turned.

With this first realization of his tremendous power, Galois' character underwent a profound change. Knowing his kinship to the great masters of algebraical analysis he felt an immense pride and longed to rush on to the front rank to match his strength with theirs. His family—even his unconventional mother—found him strange. At school he seems to have inspired a curious mixture of fear and anger in the minds of his teachers and fellow students. His teachers were good men and patient, but they were stupid, and to Galois stupidity was the unpardonable sin. At the beginning of the year they had reported him as "very gentle, full of innocence and good qualities, *but* —" And they went on to say that "there is something strange about him." No doubt there was. The boy had unusual brains. A little later they admit that he is not "wicked," but merely "original and queer," "argumentative," and they complain that he delights to tease his comrades. All very reprehensible, no doubt, but they might have used their eyes. The boy had discovered mathematics and he was already being driven by his daemon. By the end of the year of awakening we learn that "his queerness has alienated him from all his companions," and his teachers observe "something secret in his character." Worse, they accuse him of "*affecting* ambition and originality." But it is admitted by some that Galois is good in mathematics. His rhetoric teachers indulge in a little classical sarcasm: "His cleverness is now a legend that we cannot credit." They rail that there is only slovenliness and eccentricity in his assigned tasks—when he deigns to pay any attention to them—and that he goes out of his way to weary his teachers by incessant "dissipation." The last does not refer to vice, because Galois had no viciousness in him. It is merely a strong word to describe the heinous inability of a mathematical genius of the first rank to squander his intellect on the futilities of rhetoric as expounded by pedants.

One man, to the everlasting credit of his pedagogical insight, declared that Galois was as able in literary studies as he was in mathematics. Galois appears to have been touched by this man's kindness. He promised to give rhetoric a chance. But his mathematical devil was now fully aroused and raging to get out, and poor Galois fell from grace. In a short time the dissenting teacher joined the majority and made the vote unanimous. Galois, he sadly admitted, was beyond salvation, "conceited, with an insufferable affectation of originality." But the pedagogue redeemed himself by one excellent, exasperated suggestion. Had it been followed, Galois might have lived to eighty. "The mathematical madness dominates this boy. I think his parents had better let him take only mathematics. He is wasting his time here, and all he does is to torment his teachers and get into trouble."

At the age of sixteen Galois made a curious mistake. Unaware that Abel at the beginning of his career had convinced himself that he had done the impossible and had solved the general equation of the fifth degree, Galois repeated the error. For a time—a very short time, however—he believed that he had done what cannot be done. This is merely one of several extraordinary similarities in the careers of Abel and Galois.

While Galois at the age of sixteen was already well started on his career of fundamental discovery, his mathematical teacher—Vernier —kept fussing over him like a hen that has hatched an eaglet and does not know how to keep the unruly creature's feet on the good dirt of the barnyard. Vernier implored Galois to work systematically. The advice was ignored and Galois, without preparation, took the competitive examinations for entrance to the École Polytechnique. This great school, the mother of French mathematicians, founded during the French Revolution (some say by Monge), to give civil and military engineers the best scientific and mathematical education available anywhere in the world, made a double appeal to the ambitious Galois. At the Polytechnique his mathematical talent would be recognized and encouraged to the utmost. And his craving for liberty and freedom of utterance would be gratified; for were not the virile, audacious young Polytechnicians, among them the future leaders of the army, always a thorn in the side of reactionary schemers who would undo the glorious work of the Revolution and bring back the corrupt priesthood and the divine right of kings? The fearless Polytechnicians, at least in Galois' boyish eyes, were no race of puling rhetoricians like the browbeaten nonentities at Louis-le-Grand, but a consecrated band of young patriots. Events were presently to prove him at least partly right in his estimate.

Galois failed in the examinations. He was not alone in believing his failure the result of a stupid injustice. The comrades he had teased unmercifully were stunned. They believed that Galois had mathematical genius of the highest order and they suspected his examiners of incompetence in their office. Nearly a quarter of a century later Terquem, editor of the *Nouvelles Annales de Mathématiques*, the mathematical journal devoted to the interests of candidates for the Polytechnique and Normale schools, reminded his readers that the controversy was not yet dead. Commenting on the failure of Galois and on the inscrutable decrees of the examiners in another instance, Terquem remarks, "A candidate of superior intelligence is lost with an examiner of inferior intelligence. *Hic ego barbarus sum quia non intelligor illis* [Because *they* don't understand *me*, *I* am a barbarian.] . . . Examinations are mysteries before which I bow. Like the mysteries of theology, the reason must admit them with humility, without seeking to understand them." As for Galois, the failure was almost the finishing touch. It drove him in upon himself and embittered him for life.

In 1828 Galois was seventeen. It was his great year. For the first time he met a man who had the capacity to understand his genius, Louis-Paul-Émile Richard (1795–1849), teacher of advanced mathematics (*mathématiques spéciales*) at Louis-le-Grand. Richard was no conventional pedagogue, but a man of talent who followed the advanced lectures on geometry at the Sorbonne in his spare time and kept himself abreast of the progress of living mathematicians to pass it on to his pupils. Timid and unambitious on his own account, he threw all his talent on the side of his pupils. The man who would not go a step out of his way to advance his own interests counted no sacrifice too great where the future of one of his students was at stake. In his zeal to advance mathematics through the work of abler men he forgot himself completely, although his scientific friends urged him to write, and to his inspired teaching more than one outstanding French mathematician of the nineteenth century has paid grateful tribute: Leverrier, codiscoverer with Adams by pure mathematical analysis of the planet Neptune; Serret, a geometer of repute and author of a classic on higher algebra in which he gave the first systematic exposition of Galois' theory of equations; Hermite, master algebraist and arithmetician of the first rank; and last, Galois.

Richard recognized instantly what had fallen into his hands—"the Abel of France." The original solutions to difficult problems which Galois handed in were proudly explained to the class, with just praise for the young author, and Richard shouted from the housetops that this extraordinary pupil should be admitted to the Polytechnique without examination. He gave Galois the first prize and wrote in his term report, "This pupil has a marked superiority above all his fellow students; he works only at the most advanced parts of mathematics." All of which was the literal truth. Galois at seventeen was making discoveries of epochal significance in the theory of equations, discoveries whose consequences are not yet exhausted after more than a century. On the first of March, 1829, Galois published his first paper, on continued fractions. This contains no hint of the great things he had done, but it served to announce him to his fellow students as no mere scholar but an inventive mathematician.

The leading French mathematician of the time was Cauchy. In fertility of invention Cauchy has been equalled by but few; and as we have seen, the mass of his collected works is exceeded in bulk only by the outputs of Euler and Cayley,* the most prolific mathematicians of history. Whenever the Academy of Sciences wished an authoritative opinion on the merits of a mathematical work submitted for its consideration it called upon Cauchy. As a rule he was a prompt and just referee. But occasionally he lapsed. Unfortunately the occasions of his lapses were the most important of all. To Cauchy's carelessness mathematics is indebted for two of the major disasters in its history: the neglect of Galois and the shabby treatment of Abel. For the latter Cauchy was only partly to blame, but for the inexcusable laxity in Galois' case Cauchy alone is responsible.

Galois had saved the fundamental discoveries he had made up to the age of seventeen for a memoir to be submitted to the Academy. Cauchy promised to present this, but he forgot. To put the finishing touch to his ineptitude he lost the author's abstract. That was the last

---

*That is, so far as actually published work is concerned up to the present (1936). Euler undoubtedly will surpass Cayley in bulk when the full edition of his works is finally printed.

Galois ever heard of Cauchy's generous promise. This was only the first of a series of similar disasters which fanned the thwarted boy's sullen contempt of academies and academicians into a fierce hate against the whole of the stupid society in which he was condemned to live.

In spite of his demonstrated genius the harassed boy was not even now left to himself at school. The authorities gave him no peace to harvest the rich field of his discoveries, but pestered him to distraction with petty tasks and goaded him to open revolt by their everlasting preachings and punishments. Still they could find nothing in him but conceit and an iron determination to be a mathematician. He already was one, but they did not know it.

Two further disasters in his eighteenth year put the last touches to Galois' character. He presented himself a second time for the entrance examinations at the Polytechnique. Men who were not worthy to sharpen his pencils sat in judgment on him. The result was what might have been anticipated. Galois failed. It was his last chance; the doors of the Polytechnique were closed forever against him.

That examination has become a legend. Galois' habit of working almost entirely in his head put him at a serious disadvantage before a blackboard. Chalk and erasers embarrassed him—till he found a proper use for one of them. During the oral part of the examination one of the inquisitors ventured to argue a mathematical difficulty with Galois. The man was both wrong and obstinate. Seeing all his hopes and his whole life as a mathematician and polytechnic champion of democratic liberty slipping away from him, Galois lost all patience. He knew that he had officially failed. In a fit of rage and despair he hurled the eraser at his tormentor's face. It was a hit.

The final touch was the tragic death of Galois' father. As the mayor of Bourg-la-Reine the elder Galois was a target for the clerical intrigues of the times, especially as he had always championed the villagers against the priest. After the stormy elections of 1827 a resourceful young priest organized a scurrilous campaign against the mayor. Capitalizing the mayor's well-known gift for versifying, the ingenious priest composed a set of filthy and stupid verses against a member of the mayor's family, signed them with Mayor Galois' name, and circulated them freely among the citizens. The thoroughly decent mayor developed a persecution mania. During his wife's absence one day he slipped off to Paris and, in an apartment but a stone's throw from the school where his son sat at his studies, committed suicide. At the funeral serious disorder broke out. Stones were hurled by the enraged citizens; a priest was gashed on the forehead. Galois saw his father's coffin lowered into the grave in the midst of an unseemly riot. Thereafter, suspecting everywhere the injustice which he hated, he could see no good in anything.

After his second failure at the Polytechnique, Galois returned to school to prepare for a teaching career. The school now had a new director, a timeserving, somewhat cowardly stoolpigeon for the royalists and clerics. This man's shilly-shally temporizing in the political upheaval which was presently to shake France to its foundations had a tragic influence on Galois' last years.

Still persecuted and maliciously misunderstood by his preceptors, Galois prepared himself for the final examinations. The comments of his examiners are interesting. In mathematics and physics he got "very good." The final oral examination drew the following comments: "This pupil is sometimes obscure in expressing his ideas, but he is intelligent and shows a remarkable spirit of research. He has communicated to me some new results in applied analysis." In literature: "This is the only student who has answered me poorly; he knows absolutely nothing. I was told that this student has an extraordinary capacity for mathematics. This astonishes me greatly; for, after his examination, I believed him to have but little intelligence. He succeeded in hiding such as he had from me. If this pupil is really what he has seemed to me to be, I seriously doubt whether he will ever make a good teacher." To which Galois, remembering some of his own good teachers, might have replied, "God forbid."

In February, 1830, at the age of nineteen, Galois was definitely admitted to university standing. Again his sure knowledge of his own transcendent ability was reflected in a withering contempt for his plodding teachers and he continued to work in solitude on his own ideas. During this year he composed three papers in which he broke new ground. These papers contain some of his great work on the theory of algebraic equations. It was far in advance of anything that had been done, and Galois had hopefully submitted it all (with further results) in a memoir to the Academy of Sciences, in competition for the Grand Prize in Mathematics. This prize was still the blue ribbon in mathematical research; only the foremost mathematicians of the day could sensibly compete. Experts agree that Galois' memoir was more than worthy of the prize. It was work of the highest originality. As Galois said with perfect justice, "I have carried out researches which will halt many savants in theirs."

The manuscript reached the Secretary safely. The Secretary took it home with him for examination, but died before he had time to look at it. When his papers were searched after his death no trace of the manuscript was found, and that was the last Galois ever heard of it. He can scarcely be blamed for ascribing his misfortunes to something less uncertain than blind chance. After Cauchy's lapse a repetition of the same sort of thing looked too providential to be a mere accident. "Genius," he said, "is condemned by a malicious social organization to an eternal denial of justice in favor of fawning mediocrity." His hatred grew, and he flung himself into politics on the side of republicanism, then a forbidden radicalism.

The first shots of the revolution of 1830 filled Galois with joy. He tried to lead his fellow students into the fray, but they hung back, and the temporizing director put them on their honor not to quit the school. Galois refused to pledge his word, and the director begged him to stay in till the following day. In his speech the director displayed a singular lack of tact and a total absence of common sense. Enraged, Galois tried to escape during the night, but the wall was too high for him. Thereafter, all through "the glorious three days" while the heroic young Polytechnicians were out in the streets making history, the director prudently kept his charges under lock and key. Whichever way the cat should jump the director was prepared to jump with it. The revolt successfully accomplished, the astute director very generously placed his pupils at the disposal of the temporary government. This put the finishing touch to Galois' political creed. During the vacation he shocked his family and boyhood friends with his fierce championship of the rights of the masses.

The last months of 1830 were as turbulent as is usual after a thorough political stir-up. The dregs sank to the bottom, the scum rose to the top, and suspended between the two the moderate element of the population hung in indecision. Galois, back at college, contrasted the timeserving vacillations of the director and the wishy-washy loyalty of the students with their exact opposites at the Polytechnique. Unable to endure the humiliation of inaction longer he wrote a blistering letter to the *Gazette des Écoles* in which he let both students and director have what he thought was their due. The students could have saved him. But they lacked backbone, and Galois was expelled. Incensed, Galois wrote a second letter to the *Gazette*, addressed to the students. "I ask nothing of you for myself," he wrote; "but speak out for your honor and according to your conscience." The letter was unanswered, for the apparent reason that those to whom Galois appealed had neither honor nor conscience.

Footloose now, Galois announced a private class in higher algebra, to meet once a week. Here he was at nineteen, a creative mathematician of the very first rank, peddling lessons to no takers. The course was to have included "a new theory of imaginaries [what is now known as the theory of 'Galois Imaginaries,' of great importance in algebra and the theory of numbers]; the theory of the solution of equations by radicals, and the theory of numbers and elliptic functions treated by pure algebra"—all his own work.

Finding no students, Galois temporarily abandoned mathematics and joined the artillery of the National Guard, two of whose four battalions were composed almost wholly of the liberal group calling themselves "Friends of the People." He had not yet given up mathematics entirely. In one last desperate effort to gain recognition, encouraged by Poisson, he had sent a memoir on the general solution of equations—now called the "Galois theory"—to the Academy of Sciences. Poisson, whose name is remembered wherever the mathematical theories of gravitation, electricity, and magnetism are studied, was the referee. He submitted a perfunctory report. The memoir, he said was "incomprehensible," but he did not state how long it had taken him to reach his remarkable conclusion. This was the last straw. Galois devoted all his energies to revolutionary politics. "If a carcase is needed to stir up the people," he wrote, "I will donate mine."

The ninth of May, 1831, marked the beginning of the end. About two hundred young republicans held a banquet to protest against the royal order disbanding the artillery which Galois had joined. Toasts were drunk to the Revolutions of 1789 and 1793, to Robespierre, and to the Revolution of 1830. The whole atmosphere of the gathering was revolutionary and defiant. Galois rose to propose a toast, his glass in one hand, his open pocket knife in the other: "To Louis Philippe"—the King. His companions misunderstood the purpose of the toast and whistled him down. Then they saw the open knife. Interpreting this as a threat against the life of the King they howled their approval. A friend of Galois, seeing the great Alexander Dumas and other notables passing by the open windows, implored Galois to sit down, but the uproar continued. Galois was the hero of the moment, and the artillerists adjourned to the street to celebrate their exuberance by dancing all night. The following day Galois was arrested at his mother's house and thrown into the prison of Sainte-Pélagie.

A clever lawyer, with the help of Galois' loyal friends, devised an ingenious defence, to the effect that Galois had really said: "To Louis Philippe, *if he turns traitor*." The open knife was easily explained; Galois had been using it to cut his chicken. This was the fact. The saving clause in his toast, according to his friends who swore they had heard it, was drowned by the whistling, and only those close to the speaker caught what was said. Galois would not claim the saving clause.

During the trial Galois' demeanor was one of haughty contempt for the court and his accusers. Caring nothing for the outcome, he launched into an impassioned tirade against all the forces of political injustice. The judge was a human being with children of his own. He warned the accused that he was not helping his own case and sharply silenced him. The prosecution quibbled over the point whether the restaurant where the incident occurred was or was not a "public place" when used for a semiprivate banquet. On this nice point of law hung the liberty of Galois. But it was evident that both court and jury were moved by the youth of the accused. After only ten minutes' deliberation the jury returned a verdict of not guilty. Galois picked up his knife from the evidence table, closed it, slipped it in his pocket, and left the courtroom without a word.

He did not keep his freedom long. In less than a month, on July 14, 1831, he was arrested again, this time as a precautionary measure. The republicans were about to hold a celebration, and Galois, being a "dangerous radical" in the eyes of the authorities, was locked up *on no charge whatever*. The government papers of all France played up this brilliant coup of the police. They now had "the dangerous *republican*, Évariste Galois," where he could not possibly start a revolution. But they were hard put to it to find a legal accusation under which he could be brought to trial. True, he had been armed to the teeth when arrested, but he had not resisted arrest. Galois was no fool. Should they accuse him of plotting against the government? Too strong; it wouldn't go; no jury would convict. Ah! After two months of incessant thought they succeeded in trumping up a charge. When arrested Galois had been wearing his artillery uniform. But the artillery had been disbanded. Therefore Galois was guilty of illegally wearing a uniform. This time they convicted him. A friend, arrested with him, got three months; Galois got six. He was to be incarcerated in Sainte-Pélagie till April 29, 1832. His sister said he looked about fifty years old at the prospect of the sunless days ahead of him. Why not? "Let justice prevail though the heavens fall."

Discipline in the jail for political prisoners was light, and they were treated with reasonable humanity. The majority spent their waking hours promenading in the courtyard reserved for their use, or boozing in the canteen—the private graft of the governor of the prison. Soon Galois, with his somber visage, abstemious habits, and perpetual air of intense concentration, became the butt of the jovial swillers. He was concentrating on his mathematics, but he could not help hearing the taunts hurled at him.

"What! You drink only water? Quit the Republican Party and go back to your mathematics."—"Without wine and women you'll never be a man." Goaded beyond endurance Galois seized a bottle of brandy, not knowing or caring what it was, and drank it down. A decent fellow prisoner took care of him till he recovered. His humiliation when he realized what he had done devastated him.

At last he escaped from what one French writer of the time calls the foulest sewer in Paris. The cholera epidemic of 1832 caused the solicitous authorities to transfer Galois to a hospital on the sixteenth of March. The "important political prisoner" who had threatened the life of Louis Philippe was too precious to be exposed to the epidemic.

Galois was put on parole, so he had only too many occasions to see outsiders. Thus it happened that he experienced his one and only love affair. In this, as in everything else, he was unfortunate. Some worthless girl ("*quelque coquette de bas étage*") initiated him. Galois took it violently and was disgusted with love, with himself, and with his girl. To his devoted friend Auguste Chevalier he wrote, "Your letter, full of apostolic unction, has brought me a little peace. But how obliterate the mark of emotions as violent as those which I have experienced? . . . On re-reading your letter, I note a phrase in which you accuse me of being inebriated by the putrefied slime of a rotten world which has defiled my heart, my head, and my hands. . . . Inebriation! I am disillusioned of everything, even love and fame. How can a world which I detest defile me?" This is dated May 25, 1832. Four days later he was at liberty. He had planned to go into the country to rest and meditate.

What happened on May 29th is not definitely known. Extracts from two letters suggest what is usually accepted as the truth: Galois had run foul of political enemies immediately after his release. These "patriots" were always spoiling for a fight, and it fell to the unfortunate Galois' lot to accommodate them in an affair of "honor." In a "Letter to All Republicans," dated 29 May, 1832, Galois writes:

"I beg patriots and my friends not to reproach me for dying otherwise than for my country. I die the victim of an infamous coquette. It is in a miserable brawl that my life is extinguished. Oh! why die for so trivial a thing, die for something so despicable! . . . Pardon for those who have killed me, they are of good faith." In another letter to two unnamed friends: "I have been challenged by two patriots—it was impossible for me to refuse. I beg your pardon for having advised neither of you. But my opponents had put me on my honor not to warn any patriot. Your task is very simple: prove that I fought in spite of myself, that is to say after having exhausted every means of accommodation. . . . Preserve my memory since fate has not given me life enough for my country to know my name. I die your friend

E. GALOIS."

These were the last words he wrote. All night, before writing these letters, he had spent the fleeting hours feverishly dashing off his scientific last will and testament, writing against time to glean a few of the great things in his teeming mind before the death which he foresaw could overtake him. Time after time he broke off to scribble in the margin "I have not time; I have not time," and passed on to the next frantically scrawled outline. What he wrote in those desperate last hours before the dawn will keep generations of mathematicians busy for hundreds of years. He had found, once and for all, the true solution of a riddle which had tormented mathematicians for centuries: under what conditions can an equation be solved? But this was only one thing of many. In this great work, Galois used the theory of groups (see chapter on Cauchy) with brilliant success. Galois was indeed one of the great pioneers in this abstract theory, today of fundamental importance in all mathematics.

In addition to this distracted letter Galois entrusted his scientific executor with some of the manuscripts which had been intended for the Academy of Sciences. Fourteen years later, in 1846, Joseph Liouville edited some of the manuscripts for the *Journal de Mathématiques pures et appliquées*. Liouville, himself a distinguished and original mathematician, and editor of the great *Journal*, writes as follows in his introduction:

"The principal work of Évariste Galois has as its object the conditions of solvability of equations by radicals. The author lays the foundations of a general theory which he applies in detail to equations whose degree is a prime number. At the age of sixteen, and while a student at the college of Louis-le-Grand . . . Galois occupied himself with this difficult subject." Liouville then states that the referees at the Academy had rejected Galois' memoirs on account of their obscurity. He continues: "An exaggerated desire for conciseness was the cause of this defect which one should strive above all else to avoid when treating the abstract and mysterious matters of pure Algebra. Clarity is, indeed, all the more necessary when one essays to lead the reader farther from the beaten path and into wilder territory. As Descartes said, 'When transcendental questions are under discussion be transcendentally clear.' Too often Galois neglected this precept; and we can understand how illustrious mathematicians may have judged it proper to try, by the harshness of their sage advice, to turn a beginner, full of genius but inexperienced, back on the right road. The author they censured was before them, ardent, active; he could profit by their advice.

"But now everything is changed. Galois is no more! Let us not indulge in useless criticisms; let us leave the defects there and look at the merits." Continuing, Liouville tells how he studied the manuscripts, and singles out one perfect gem for special mention.

"My zeal was well rewarded, and I experienced an intense pleasure at the moment when, having filled in some slight gaps, I saw the complete correctness of the method by which Galois proves, in particular, this beautiful theorem: *In order that an irreducible equation of* *prime degree be solvable by radicals it is necessary and sufficient that all its roots be rational functions of any two of them.*"*

Galois addressed his will to his faithful friend Auguste Chevalier, to whom the world owes its preservation. "My dear friend," he began, "I have made some new discoveries in analysis." He then proceeds to outline such as he has time for. They were epoch-making. He concludes: "Ask Jacobi or Gauss publicly to give their opinion, not as to the truth, but as to the importance of these theorems. Later there will be, I hope, some people who will find it to their advantage to decipher all this mess. *Je t'embrasse avec effusion.* E. Galois."

Confiding Galois! Jacobi was generous; what would Gauss have said? What did he say of Abel? What did he omit to say of Cauchy, or of Lobatchewsky? For all his bitter experience Galois was still a hopeful boy.

At a very early hour on the thirtieth of May, 1832, Galois confronted his adversary on the "field of honor." The duel was with pistols at twenty five paces. Galois fell, shot through the intestines. No surgeon was present. He was left lying where he had fallen. At nine o'clock a passing peasant took him to the Cochin Hospital. Galois knew he was about to die. Before the inevitable peritonitis set in, and while still in the full possession of his faculties, he refused the offices of a priest. Perhaps he remembered his father. His young brother, the only one of his family who had been warned, arrived in tears. Galois tried to comfort him with a show of stoicism. "Don't cry," he said, "I need all my courage to die at twenty."

Early in the morning of May 31, 1832, Galois died, being then in the twenty first year of his age. He was buried in the common ditch of the South Cemetery, so that today there remains no trace of the grave of Évariste Galois. His enduring monument is his collected works. They fill sixty pages.

---

*The significance of this theorem will be clear if the reader will glance through the extracts from Abel in Chapter 17.

CHAPTER TWENTY ONE

Invariant Twins

CAYLEY AND SYLVESTER

The theory of Invariants sprang into existence under the strong hand of Cayley, but that it emerged finally a complete work of art, for the admiration of future generations of mathematicians, was largely owing to the flashes of inspiration with which Sylvester's intellect illuminated it.—P. A. MACMAHON

"IT IS DIFFICULT to give an idea of the vast extent of modern mathematics. The word 'extent' is not the right one: I mean extent crowded with beautiful detail—not an extent of mere uniformity such as an objectless plain, but of a tract of beautiful country seen at first in the distance, but which will bear to be rambled through and studied in every detail of hillside and valley, stream, rock, wood, and flower. But, as for every thing else, so for a mathematical theory—beauty can be perceived but not explained."

These words from Cayley's presidential address in 1883 to the British Association for the Advancement of Science might well be applied to his own colossal output. For prolific inventiveness Euler, Cauchy, and Cayley are in a class by themselves, with Poincaré (who died younger than any of the others) a far second. This applies only to the bulk of these men's work; its quality is another matter, to be judged partly by the frequency with which the ideas originated by these giants recur in mathematical research, partly by mere personal opinion, and partly by national prejudice.

Cayley's remarks about the vast extent of modern mathematics suggest that we confine our attention to some of those features of his own work which introduced distinctly new and far-reaching ideas. The work on which his greatest fame rests is in the theory of invariants and what grew naturally out of that vast theory of which he, brilliantly sustained by his friend Sylvester, was the originator and unsurpassed developer. The concept of invariance is of great importance for modern physics, particularly in the theory of relativity, but this is not its chief claim to attention. Physical theories are notoriously subject to revision and rejection; the theory of invariance as a permanent addition to pure mathematical thought appears to rest on firmer ground.

Another of the ideas originated by Cayley, that of the geometry of "higher space" (space of n dimensions) is likewise of present scientific significance but of incomparably greater importance as pure mathematics. Similarly for the theory of matrices, again an invention of Cayley's. In non-Euclidean geometry Cayley prepared the way for Klein's splendid discovery that the geometry of Euclid and the non-Euclidean geometries of Lobatchewsky and Riemann are, all three, merely different aspects of a more general kind of geometry which includes them as special cases. The nature of these contributions of Cayley's will be briefly indicated after we have sketched his life and that of his friend Sylvester.

The lives of Cayley and Sylvester should be written simultaneously, if that were possible. Each is a perfect foil to the other, and the life of each, in large measure, supplies what is lacking in that of the other. Cayley's life was serene; Sylvester, as he himself bitterly remarks, spent much of his spirit and energy "fighting the world." Sylvester's thought was at times as turbulent as a millrace; Cayley's was always strong, steady, and unruffled. Only rarely did Cayley permit himself the printed expression of anything less severe than a precise mathematical statement—the simile quoted at the beginning of this chapter is one of the rare exceptions; Sylvester could hardly talk about mathematics without at once becoming almost orientally poetic, and his unquenchable enthusiasm frequently caused him to go off half-cocked. Yet these two became close friends and inspired one another to some of the best work that either of them did, for example in the theories of invariants and matrices (described later).

With two such temperaments it is not surprising that the course of friendship did not always run smoothly. Sylvester was frequently on the point of exploding; Cayley sat serenely on the safety valve, confident that his excitable friend would presently cool down, when he would calmly resume whatever they had been discussing as if Sylvester had never blown off, while Sylvester for his part ignored his hot-headed indiscretion—till he got himself all steamed up for another. In many ways this strangely congenial pair were like a honeymoon couple, except that one party to the friendship never lost his temper.

Although Sylvester was Cayley's senior by seven years, we shall begin with Cayley. Sylvester's life breaks naturally into the calm stream of Cayley's like a jagged rock in the middle of a deep river.

Arthur Cayley was born on August 16, 1821, at Richmond, Surrey, the second son of his parents, then residing temporarily in England. On his father's side Cayley traced his descent back to the days of the Norman Conquest (1066) and even before, to a baronial estate in Normandy. The family was a talented one which, like the Darwin family, should provide much suggestive material for students of heredity. His mother was Maria Antonia Doughty, by some said to have been of Russian origin. Cayley's father was an English merchant engaged in the Russian trade; Arthur was born during one of the periodical visits of his parents to England.

In 1829, when Arthur was eight, the merchant retired, to live thenceforth in England. Arthur was sent to a private school at Blackheath and later, at the age of fourteen, to King's College School in London. His mathematical genius showed itself very early. The first manifestations of superior talent were like those of Gauss; young Cayley developed an amazing skill in long numerical calculations which he undertook for amusement. On beginning the formal study of mathematics he quickly outstripped the rest of the school. Presently he was in a class by himself, as he was later when he went up to the University, and his teachers agreed that the boy was a born mathematician who should make mathematics his career. In grateful contrast to Galois' teachers, Cayley's recognized his ability from the beginning and gave him every encouragement. At first the retired merchant objected strongly to his son's becoming a mathematician but finally, won over by the Principal of the school, gave his consent, his blessing, and his money. He decided to send his son to Cambridge.

Cayley began his university career at the age of seventeen at Trinity College, Cambridge. Among his fellow students he passed as "a mere mathematician" with a queer passion for novel-reading. Cayley was indeed a lifelong devotee of the somewhat stilted fiction, now considered classical, which charmed readers of the 1840's and '50's. Scott appears to have been his favorite, with Jane Austen a close second. Later he read Thackeray and disliked him; Dickens he could never bring himself to read. Byron's tales in verse excited his admiration, although his somewhat puritanical Victorian taste rebelled at the best of the lot and he never made the acquaintance of that diverting scapegrace Don Juan. Shakespeare's plays, especially the comedies were a perpetual delight to him. On the more solid—or stodgier—side he read and reread Grote's interminable *History of Greece* and Macaulay's rhetorical *History of England*. Classical Greek, acquired at school, remained a reading-language for him all his life; French he read and wrote as easily as English, and his knowledge of German and Italian gave him plenty to read after he had exhausted the Victorian classics (or they had exhausted him). The enjoyment of solid fiction was only one of his diversions; others will be noted as we go.

By the end of his third year at Cambridge Cayley was so far in front of the rest in mathematics that the head examiner drew a line under his name, putting the young man in a class by himself "above the first." In 1842, at the age of twenty one, Cayley was senior wrangler in the mathematical tripos, and in the same year he was placed first in the yet more difficult test for the Smith's prize.

Under an excellent plan Cayley was now in line for a fellowship which would enable him to do as he pleased for a few years. He was elected Fellow of Trinity and assistant tutor for a period of three years. His appointment might have been renewed had he cared to take holy orders, but although Cayley was an orthodox Church of England Christian he could not quite stomach the thought of becoming a parson to hang onto his job or to obtain a better one—as many did, without disturbing either their faith or their conscience.

His duties were light almost to the point of nonexistence. He took a few pupils, but not enough to hurt either himself or his work. Making the best possible use of his liberty he continued the mathematical researches which he had begun as an undergraduate. Like Abel, Galois, and many others who have risen high in mathematics, Cayley went to the masters for his inspiration. His first work, published in 1841 when he was an undergraduate of twenty, grew out of his study of Lagrange and Laplace.

With nothing to do but what he wanted to do after taking his degree Cayley published eight papers the first year, four the second, and thirteen the third. These early papers by the young man who was not yet twenty five when the last of them appeared map out much of the work that is to occupy him for the next fifty years. Already he has begun the study of geometry of n dimensions (which he originated), the theory of invariants, the enumerative geometry of plane curves, and his distinctive contributions to the theory of elliptic functions.

During this extremely fruitful period he was no mere grind. In 1843, when he was twenty two, and occasionally thereafter till he left Cambridge at the age of twenty five, he escaped to the Continent for delightful vacations of tramping, mountaineering, and water-color sketching. Although he was slight and frail in appearance he was tough and wiry, and often after a long night spent in tramping over hilly country, would turn up as fresh as the dew for breakfast and ready to put in a few hours at his mathematics. During his first trip he visited Switzerland and did a lot of mountaineering. Thus began another lifelong passion. His description of the "extent of modern mathematics" is no mere academic exercise by a professor who had never climbed a mountain or rambled lovingly over a tract of beautiful country, but the accurate simile of a man who had known nature intimately at first hand.

During the last four months of his first vacation abroad he became acquainted with northern Italy. There began two further interests which were to solace him for the rest of his life: an understanding appreciation of architecture and a love of good painting. He himself delighted in water-colors, in which he showed marked talent. With his love of good literature, travel, painting, and architecture, and with his deep understanding of natural beauty, he had plenty to keep him from degenerating into the "mere mathematician" of conventional literature—written, for the most part, by people who may indeed have known some pedantic college professor of mathematics, but who never in their lives saw a real mathematician in the flesh.

In 1846, when he was twenty five, Cayley left Cambridge. No position as a mathematician was open to him unless possibly he could square his conscience to the formality of "holy orders." As a mathematician Cayley felt no doubt that it would be easier to square the circle. Anyhow, he left. The law, which with the India Civil Service has absorbed much of England's most promising intellectual capital at one time or another, now attracted Cayley. It is somewhat astonishing to see how many of England's leading barristers and judges in the nineteenth century were high wranglers in the Cambridge tripos, but it does not follow, as some have claimed, that a mathematical training is a good preparation for the law. What seems less doubtful is that it may be a social imbecility to put a young man of Cayley's demonstrated mathematical genius to drawing up wills, transfers, and leases.

Following the usual custom of those looking toward an English legal career of the more gentlemanly grade (that is, above the trade of solicitor), Cayley entered Lincoln's Inn to prepare himself for the Bar. After three years as a pupil of a Mr. Christie, Cayley was called to the Bar in 1849. He was then twenty eight. On being admitted to the Bar, Cayley made a wise resolve not to let the law run off with his brains. Determined not to rot, he rejected more business than he accepted. For fourteen mortal years he stuck it, making an ample living and deliberately turning away the opportunity to smother himself in money and the somewhat blathery sort of renown that comes to prominent barristers, in order that he might earn enough, but no more than enough, to enable him to get on with his work.

His patience under the deadening routine of dreary legal business was exemplary, almost saintly, and his reputation in his branch of the profession (conveyancing) rose steadily. It is even recorded that his name has passed into one of the law books in connection with an exemplary piece of legal work he did. But it is extremely gratifying to record also that Cayley was no milk-and-water saint but a normal human being who could, when the occasion called for it, lose his temper. Once he and his friend Sylvester were animatedly discussing some point in the theory of invariants in Cayley's office when the boy entered and handed Cayley a large batch of legal papers for his perusal. A glance at what was in his hands brought him down to earth with a jolt. The prospect of spending days straightening out some petty muddle to save a few pounds to some comfortable client's already plethoric income was too much for the man with real brains in his head. With an exclamation of disgust and a contemptuous reference to the "wretched rubbish" in his hands, he hurled the stuff to the floor and went on talking mathematics. This, apparently, is the only instance on record when Cayley lost his temper. Cayley got out of the law at the first opportunity—after fourteen years of it. But during his period of servitude he had published between two and three hundred mathematical papers, many of which are now classic.

As Sylvester entered Cayley's life during the legal phase we shall introduce him here.

James Joseph—to give him first the name with which he was born

—was the youngest of several brothers and sisters, and was born of Jewish parents on September 3, 1814, in London. Very little is known of his childhood, as Sylvester appears to have been reticent about his early years. His eldest brother emigrated to the United States, where he took the name of Sylvester, an example followed by the rest of the family. But why an orthodox Jew should have decorated himself with a name favored by Christian popes hostile to Jews is a mystery. Possibly that eldest brother had a sense of humor; anyhow, plain James Joseph, son of Abraham Joseph, became henceforth and forevermore James Joseph Sylvester.

Like Cayley's, Sylvester's mathematical genius showed itself early. Between the ages of six and fourteen he attended private schools. The last five months of his fourteenth year were spent at the University of London, where he studied under De Morgan. In a paper written in 1840 with the somewhat mystical title *On the Derivation of Coexistence*, Sylvester says "I am indebted for this term [recurrents] to Professor De Morgan, whose pupil I may boast to have been."

In 1829, at the age of fifteen, Sylvester entered the Royal Institution at Liverpool, where he stayed less than two years. At the end of his first year he won the prize in mathematics. By this time he was so far ahead of his fellow students in mathematics that he was placed in a special class by himself. While at the Royal Institution he also won another prize. This is of particular interest as it establishes the first contact of Sylvester with the United States of America where some of the happiest—also some of the most wretched—days of his life were to be spent. The American brother, by profession an actuary, had suggested to the Directors of the Lotteries Contractors of the United States that they submit a difficult problem in arrangements to young Sylvester. The budding mathematician's solution was complete and practically most satisfying to the Directors, who gave Sylvester a prize of five hundred dollars for his efforts.

The years at Liverpool were far from happy. Always courageous and open, Sylvester made no bones about his Jewish faith, but proudly proclaimed it in the face of more than petty persecution at the hands of the sturdy young barbarians at the Institution who humorously called themselves Christians. But there is a limit to what one lone peacock can stand from a pack of dull jays, and Sylvester finally fled to Dublin with only a few shillings in his pocket. Luckily he was recognized on the street by a distant relative who took him in, straightened him out, and paid his way back to Liverpool.

Here we note another curious coincidence: Dublin, or at least one of its citizens, accorded the religious refugee from Liverpool decent human treatment on his first visit; on his second, some eleven years later, Trinity College, Dublin granted him the academic degrees (B.A. and M.A.) which his own alma mater, Cambridge University, had refused him because he could not, being a Jew, subscribe to that remarkable compost of nonsensical statements known as the Thirty-Nine Articles prescribed by the Church of England as the minimum of religious belief permissible to a rational mind. It may be added here however that when English higher education finally unclutched itself from the stranglehold of the dead hand of the Church in 1871 Sylvester was promptly given his degrees *honoris causa*. And it should be remarked that in this as in other difficulties Sylvester was no meek, long-suffering martyr. He was full of strength and courage, both physical and moral, and he knew how to put up a devil of a fight to get justice for himself—and frequently did. He was in fact a born fighter with the untamed courage of a lion.

In 1831, when he was just over seventeen, Sylvester entered St. John's College, Cambridge. Owing to severe illnesses his university career was interrupted, and he did not take the mathematical tripos till 1837. He was placed second. The man who beat him was never heard of again as a mathematician. Not being a Christian, Sylvester was ineligible to compete for the Smith's prizes.

In the breadth of his intellectual interests Sylvester resembles Cayley. Physically the two men were nothing alike. Cayley, though wiry and full of physical endurance as we have seen, was frail in appearance and shy and retiring in manner. Sylvester, short and stocky, with a magnificent head set firmly above broad shoulders, gave the impression of tremendous strength and vitality, and indeed he had both. One of his students said he might have posed for the portrait of Hereward the Wake in Charles Kingsley's novel of the same name. As to interests outside of mathematics, Sylvester was much less restricted and far more liberal than Cayley. His knowledge of the Greek and Latin classics in the originals was broad and exact, and he retained his love of them right up to his last illness. Many of his papers are enlivened by quotations from these classics. The quotations are always singularly apt and really do illuminate the matter in hand.

The same may be said for his allusions from other literatures. It might amuse some literary scholar to go through the four volumes of the collected *Mathematical Papers* and reconstruct Sylvester's wide range of reading from the credited quotations and the curious hints thrown out without explicit reference. In addition to the English and classical literatures he was well acquainted with the French, German, and Italian in the originals. His interest in language and literary form was keen and penetrating. To him is due most of the graphic terminology of the theory of invariants. Commenting on his extensive coinage of new mathematical terms from the mint of Greek and Latin, Sylvester referred to himself as the "mathematical Adam."

On the literary side it is quite possible that had he not been a very great mathematician he might have been something a little better than a merely passable poet. Verse, and the "laws" of its construction, fascinated him all his life. On his own account he left much verse (some of which has been published), a sheaf of it in the form of sonnets. The subject matter of his verse is sometimes rather apt to raise a smile, but he frequently showed that he understood what poetry is. Another interest on the artistic side was music, in which he was an accomplished amateur. It is said that he once took singing lessons from Gounod and that he used to entertain workingmen's gatherings with his songs. He was prouder of his "high C" than he was of his invariants.

One of the many marked differences between Cayley and Sylvester may be noted here: Cayley was an omnivorous reader of other mathematicians' work; Sylvester found it intolerably irksome to attempt to master what others had done. Once, in later life, he engaged a young man to teach him something about elliptic functions as he wished to apply them to the theory of numbers (in particular to the theory of partitions, which deals with the number of ways a given number can be made up by adding together numbers of a given kind, say all odd, or some odd and some even). After about the third lesson Sylvester had abandoned his attempt to learn and was lecturing to the young man on his own latest discoveries in algebra. But Cayley seemed to know everything, even about subjects in which he seldom worked, and his advice as a referee was sought by authors and editors from all over Europe. Cayley never forgot anything he had seen; Sylvester had difficulty in remembering his own inventions and once even disputed that a certain theorem of his own could possibly be true. Even comparatively trivial things that every working mathematician knows were sources of perpetual wonder and delight to Sylvester. As a consequence almost any field of mathematics offered an enchanting world for discovery to Sylvester, while Cayley glanced serenely over it all, saw what he wanted, took it, and went on to something fresh.

In 1838, at the age of twenty four, Sylvester got his first regular job, that of Professor of Natural Philosophy (science in general, physics in particular) at University College, London, where his old teacher De Morgan was one of his colleagues. Although he had studied chemistry at Cambridge, and retained a lifelong interest in it, Sylvester found the teaching of science thoroughly uncongenial and, after about two years, abandoned it. In the meantime he had been elected a Fellow of the Royal Society at the unusually early age of twenty five. Sylvester's mathematical merits were so conspicuous that they could not escape recognition, but they did not help him into a suitable position.

At this point in his career Sylvester set out on one of the most singular misadventures of his life. Depending upon how we look at it, this mishap is silly, ludicrous, or tragic. Sanguine and filled with his usual enthusiasm, he crossed the Atlantic to become Professor of Mathematics at the University of Virginia in 1841—the year in which Boole published his discovery of invariants.

Sylvester endured the University only about three months. The refusal of the University authorities to discipline a young gentleman who had insulted him caused the professor to resign. For over a year after this disastrous experience Sylvester tried vainly to secure a suitable position, soliciting—unsuccessfully—both Harvard and Columbia Universities. Failing, he returned to England.

Sylvester's experiences in America gave him his fill of teaching for the next ten years. On returning to London he became an energetic actuary for a life insurance company. Such work for a creative mathematician is poisonous drudgery, and Sylvester almost ceased to be a mathematician. However, he kept alive by taking a few private pupils, one of whom was to leave a name that is known and revered in every country of the world today. This was in the early 1850's, the "potatoes, prunes, and prisms" era of female propriety when young women were not supposed to think of much beyond dabbling in paints and piety. So it is rather surprising to find that Sylvester's most distinguished pupil was a young woman, Florence Nightingale, the first human being to get some decency and cleanliness into military hospitals—over the outraged protests of bull-headed military officialdom. Sylvester at the time was in his late thirties, Miss Nightingale six years younger than her teacher. Sylvester escaped from his makeshift ways of earning a living in the same year (1854) that Miss Nightingale went out to the Crimean War.

Before this however he had taken another false step that landed him nowhere. In 1846, at the age of thirty two, he entered the Inner Temple (where he coyly refers to himself as "a dove nestling among hawks") to prepare for a legal career, and in 1850 was called to the Bar. Thus he and Cayley came together at last.

Cayley was twenty nine, Sylvester thirty six at the time; both were out of the real jobs to which nature had called them. Lecturing at Oxford thirty five years later Sylvester paid grateful tribute to "Cayley, who, though younger than myself is my spiritual progenitor —who first opened my eyes and purged them of dross so that they could see and accept the higher mysteries of our common Mathematical faith." In 1852, shortly after their acquaintance began, Sylvester refers to "Mr. Cayley, who habitually discourses pearls and rubies." Mr. Cayley for his part frequently mentions Mr. Sylvester, but always in cold blood, as it were. Sylvester's earliest outburst of gratitude in print occurs in a paper of 1851 where he says, "The theorem above enunciated [it is his relation between the minor determinants of linearly equivalent quadratic forms] was in part suggested in the course of a conversation with Mr. Cayley (to whom I am indebted for my restoration to the enjoyment of mathematical life) . . . . "

Perhaps Sylvester overstated the case, but there was a lot in what he said. If he did not exactly rise from the dead he at least got a new pair of lungs: from the hour of his meeting with Cayley he breathed and lived mathematics to the end of his days. The two friends used to tramp round the Courts of Lincoln's Inn discussing the theory of invariants which both of them were creating and later, when Sylvester moved away, they continued their mathematical rambles, meeting about halfway between their respective lodgings. Both were bachelors at the time.

The theory of algebraic invariants from which the various extensions of the concept of invariance have grown naturally originated in an extremely simple observation. As will be noted in the chapter on Boole, the earliest instance of the idea appears in Lagrange, from whom it passed into the arithmetical works of Gauss. But neither of these men noticed that the simple but remarkable algebraical phenomenon before them was the germ of a vast theory. Nor does Boole seem to have fully realized what he had found when he carried on and greatly extended the work of Lagrange. Except for one slight tiff, Sylvester was always just and generous to Boole in the matter of priority, and Cayley, of course, was always fair.

The simple observation mentioned above can be understood by anyone who has ever seen a quadratic equation solved, and is merely this. A necessary and sufficient condition that the equation $ax^2 + 2bx + c = 0$ shall have two equal roots is that $b^2 - ac$ shall be zero. Let us replace the variable x by its value in terms of y obtained by the transformation $y = (px + q)/(rx + s)$. Thus x is to be replaced by the result of solving this for x, namely $x = (q - sy)/(ry - p)$. This transforms the given equation into another in y; say the new equation is $Ay^2 + 2By + C = 0$. Carrying out the algebra we find that the new coefficients A, B, C are expressed in terms of the old a, b, c as follows,

$$A = as^2 - 2bsr + cr^2,$$
$$B = -aqs + b(qr + sp) - cpr,$$
$$C = aq^2 - 2bpq + cp^2.$$

From these it is easy to show (by brute-force reductions, if necessary, although there is a simpler way of reasoning the result out, without actually calculating A, B, C) that

$$B^2 - AC = (ps - qr)^2(b^2 - ac).$$

Now $b^2 - ac$ is called the discriminant of the quadratic equation in x; hence the discriminant of the quadratic in y is $B^2 - AC$, and it has been shown that *the discriminant of the transformed equation is equal to the discriminant of the original equation, times the factor* $(ps - qr)^2$ *which depends only upon the coefficients p, q, r, s in the transformation* $y = (px + q)/(rx + s)$ *by means of which x was expressed in terms of y.*

Boole was the first (in 1841) to observe something worth looking at in this particular trifle. Every algebraic equation has a discriminant, that is, a certain expression (such as $b^2 - ac$ for the quadratic) which is equal to zero if, and only if, two or more roots of the equation are equal. Boole first asked, does the discriminant of every equation when its x is replaced by the related y (as was done for the quadratic) come back unchanged except for a factor depending only on the coefficients of the transformation? He found that this was true. Next he asked whether there might not be expressions other than discriminants constructed from the coefficients having this same property of *invariance* under *transformation*. He found two such for the general equation of the fourth degree. Then another man, the brilliant young German mathematician, F. M. G. Eisenstein (1823-1852) following up a result of Boole's, in 1844, discovered that certain expressions involving *both the coefficients and the x* of the original equations exhibit the same sort of invariance: the original coefficients and the original x pass into the transformed coefficients and y (as for the quadratic), and the expressions in question constructed from the originals differ from those constructed from the transforms only by a factor which depends solely on the coefficients of the transformation.

Neither Boole nor Eisenstein had any *general* method for finding such *invariant* expressions. At this point Cayley entered the field in 1845 with his pathbreaking memoir, *On the Theory of Linear Transformations*. At the time he was twenty four. He set himself the problem of finding uniform methods which would give him *all* the invariant expressions of the kind described. To avoid lengthy explanations the problem has been stated in terms of equations; actually it was attacked otherwise, but this is of no importance here.

As this question of invariance is fundamental in modern scientific thought we shall give three further illustrations of what it means, none of which involves any symbols or algebra. Imagine any figure consisting of intersecting straight lines and curves drawn on a sheet of paper. Crumple the paper in any way you please without tearing it, and try to think what is the most obvious property of the figure that is the same before and after crumpling. Do the same for any figure drawn on a sheet of rubber, stretching but not tearing the rubber in any complicated manner dictated by whim. In this case it is obvious that sizes of areas and angles, and lengths of lines, have *not* remained "invariant." By suitably stretching the rubber the straight lines may be distorted into curves of almost any tortuosity you like, and at the same time the original curves—or at least some of them—may be transformed into straight lines. Yet *something* about the whole figure has remained unchanged; its very simplicity and obviousness might well cause it to be overlooked. This is the order of the points on any one of the lines of the figure which mark the places where other lines intersect the given one. Thus, if moving the pencil along a given line from A to C, we had to pass over the point B on the line before the figure was distorted, we shall have to pass over B in going from A to C after distortion. The *order* (as described) is an *invariant* under the particular *transformations* which crumpled the sheet of paper into a crinkly ball, say, or which stretched the sheet of rubber.

This illustration may seem trivial, but anyone who has read a non-mathematical description of the intersections of "world-lines" in general relativity, and who recalls that an intersection of two such lines marks a physical *"point-event,"* will see that what we have been discussing is of the same stuff as one of our pictures of the physical universe. The mathematical machinery powerful enough to handle such complicated "transformations" and actually to produce the invariants was the creation of many workers, including Riemann, Christoffel, Ricci, Levi-Civita, Lie, and Einstein—all names well known to readers of popular accounts of relativity; the whole vast program was originated by the early workers in the theory of algebraic invariants, of which Cayley and Sylvester were the true founders.

As a second example, imagine a knot to be looped in a string whose ends are then tied together. Pulling at the knot, and running it along the string, we distort it into any number of "shapes." What remains "invariant," what is "conserved," under all these distortions which, in this case, are our transformations? Obviously neither the shape nor the size of the knot is invariant. But the "style" of the knot itself is invariant; in a sense that need not be elaborated, it is the *same sort* of a knot whatever we do to the string provided we do not untie its ends. Again, in the older physics, energy was "conserved"; the total amount of energy in the universe was assumed to be an invariant, the same under all transformations from one form, such as electrical energy, into others, such as heat and light.

Our third illustration of invariance need be little more than an allusion to physical science. An observer fixes his "position" in space and time with reference to three mutually perpendicular axes and a standard timepiece. Another observer, moving relatively to the first, wishes to describe the same physical event that the first describes. He also has his space-time reference system; his movement relatively to the first-observer can be expressed as a transformation of his own coordinates (or of the other observer's). The descriptions given by the two may or may not differ in mathematical form, according to the particular kind of transformation concerned. If their descriptions do differ, the difference is not, obviously, inherent in the physical event they are both observing, but in their reference systems and the transformation. The problem then arises to formulate only those mathematical expressions of natural phenomena which shall be independent, mathematically, of any *particular* reference system and therefore be expressed by all observers in the same form. This is equivalent to finding the invariants of the transformation which expresses the most general shift in "space-time" of one reference system with respect to any other. Thus the problem of finding the mathematical expressions for the intrinsic laws of nature is replaced by an attackable one in the theory of invariants. More will be said on this when we come to Riemann.

In 1863 Cambridge University established a new professorship of mathematics (the Sadlerian) and offered the post to Cayley, who promptly accepted. The same year, at the age of forty two, he married Susan Moline. Although he made less money as a professor of mathematics than he had at the law, Cayley did not regret the change. Some years later the affairs of the University were reorganized and Cayley's salary was raised. His duties also were increased from one course of lectures during one term to two. His life was now devoted almost entirely to mathematical research and university administration. In the latter his sound business training, even temper, impersonal judgment, and legal experience proved invaluable. He never had a great deal to say, but what he said was usually accepted as final, for he never gave an opinion without having reasoned the matter through. His marriage and home life were happy; he had two children, a son and a daughter. As he gradually aged his mind remained as vigorous as ever and his nature became, if anything, gentler. No harsh judgment uttered in his presence was allowed to pass without a quiet protest. To younger men and beginners in mathematical careers he was always generous with his help, encouragement, and sound advice.

During his professorship the higher education of women was a hotly contested issue. Cayley threw all his quiet, persuasive influence on the side of civilization and largely through his efforts women were at last admitted as students (in their own nunneries of course) to the monkish seclusion of medieval Cambridge.

While Cayley was serenely mathematicizing at Cambridge his friend Sylvester was still fighting the world. Sylvester never married. In 1854, at the age of forty, he applied for the professorship of mathematics at the Royal Military Academy, Woolwich. He did not get it. Nor did he get another position for which he applied at Gresham College, London. His trial lecture was too good for the governing board. However, the successful Woolwich candidate died the following year and Sylvester was appointed. Among his not too generous emoluments was the right of pasturage on the common. As Sylvester kept neither horse, cow, nor goat, and did not eat grass himself, it is difficult to see what particular benefit he got out of this inestimable boon.

Sylvester held the position at Woolwich for sixteen years, till he was forcibly retired as "superannuated" in 1870 at the age of fifty six. He was still full of vigor but could do nothing against the hidebound officialdom against him. Much of his great work was still in the future, but his superiors took it for granted that a man of his age must be through.

Another aspect of his forced retirement roused all his fighting instincts. To put the matter plainly, the authorities attempted to swindle Sylvester out of part of the pension which was legitimately his. Sylvester did not take it lying down. To their chagrin the would-be gyppers learned that they were not browbeating some meek old professor but a man who could give them a little better than he took. They came through with the full pension.

While all these disagreeable things were happening in his material affairs Sylvester had no cause to complain on the scientific side. Honors frequently came his way, among them one of those most highly prized by scientific men, foreign correspondent of the French Academy of Sciences. Sylvester was elected in 1863 to the vacancy in the section of geometry caused by the death of Steiner.

After his retirement from Woolwich Sylvester lived in London, versifying, reading the classics, playing chess, and enjoying himself generally, but not doing much mathematics. In 1870 he published his pamphlet, *The Laws of Verse*, by which he set great store. Then, in 1876, he suddenly came to mathematical life again at the age of sixty two. The "old" man was simple inextinguishable.

The Johns Hopkins University had been founded at Baltimore in

1875 under the brilliant leadership of President Gilman. Gilman had been advised to start off with an outstanding classicist and the best mathematician he could afford as the nucleus of his faculty. All the rest would follow, he was told, and it did. Sylvester at last got a job where he might do practically as he pleased and in which he could do himself justice. In 1876 he again crossed the Atlantic and took up his professorship at Johns Hopkins. His salary was generous for those days, five thousand dollars a year. In accepting the call Sylvester made one curious stipulation; his salary was "to be paid in gold." Perhaps he was thinking of Woolwich, which gave him the equivalent of $2750.00 (plus pasturage), and wished to be sure that this time he really got what was coming to him, pension or no pension.

The years from 1876 to 1883 spent at Johns Hopkins were probably the happiest and most tranquil Sylvester had thus far known. Although he did not have to "fight the world" any longer he did not recline on his honors and go to sleep. Forty years seemed to fall from his shoulders and he became a vigorous young man again, blazing with enthusiasm and scintillating with new ideas. He was deeply grateful for the opportunity Johns Hopkins gave him to begin his second mathematical career at the age of sixty three, and he was not backward in expressing his gratitude publicly, in his address at the Commemoration Day Exercises of 1877.

In this Address he outlined what he hoped to do (he did it) in his lectures and researches.

"There are things called Algebraical Forms. Professor Cayley calls them Quantics. [Examples: $ax^2 + 2bxy + cy^2$, $ax^3 + 3bx^2y + 3cxy^2 + dy^3$; the numerical coefficients 1,2,1 in the first, 1,3,3,1 in the second, are binomial coefficients, as in the third and fourth lines of Pascal's triangle (Chapter 5); the next in order would be $x^4 + 4x^3y + 6x^2y^2 + 4xy^3 + y^4$]. They are not, properly speaking, Geometrical Forms, although capable, to some extent, of being embodied in them, but rather schemes of process, or of operations for forming, for calling into existence, as it were, Algebraic quantities.

"To every such Quantic is associated an infinite variety of other forms that may be regarded as engendered from and floating, like an atmosphere, around it—but infinite as were these derived existences, these emanations from the parent form, it is found that they admit of being obtained by composition, by mixture, so to say, of a certain limited number of fundamental forms, standard rays, as they might be termed in the Algebraic Spectrum of the Quantic to which they belong. And, as it is a leading pursuit of the Physicists of the present day [1877, and even today] to ascertain the fixed lines in the spectrum of every chemical substance, so it is the aim and object of a great school of mathematicians to make out the fundamental derived forms, the *Covariants* [that kind of 'invariant' expression, already described, which involves *both* the variables *and* the coefficients of the form or quantic] and *Invariants*, as they are called, of these Quantics."

To mathematical readers it will be evident that Sylvester is here giving a very beautiful analogy for the fundamental system and the syzygies for a given form; the nonmathematical reader may be recommended to reread the passage to catch the spirit of the algebra Sylvester is talking about, as the analogy is really a close one and as fine an example of "popularized" mathematics as one is likely to find in a year's marching.

In a footnote Sylvester presently remarks "I have at present a class of from eight to ten students attending my lectures on the Modern Higher Algebra. One of them, a young engineer, engaged from eight in the morning to six at night in the duties of his office, with an interval of an hour and a half for his dinner or lectures, has furnished me with the best proof, and the best expressed, I have ever seen of what I call [a certain theorem]. . . . " Sylvester's enthusiasm—he was past sixty—was that of a prophet inspiring others to see the promised land which he had discovered or was about to discover. Here was teaching at its best, at the only level, in fact, which justifies advanced teaching at all.

He had complimentary things to say (in footnotes) about the country of his adoption: " . . . I believe there is no nation in the world where ability with character counts for so much, and the mere possession of wealth (in spite of all that we hear about the Almighty dollar), for so little as in America, . . . "

He also tells how his dormant mathematical instincts were again aroused to full creative power. "But for the persistence of a student of this University [Johns Hopkins] in urging upon me his desire to study with me the modern Algebra, I should never have been led into this investigation. . . . He stuck with perfect respectfulness, but with invincible pertinacity, to his point. He would have the New Algebra (Heaven knows where he had heard about it, for it is almost unknown on this continent), that or nothing. I was obliged to yield, and what was the consequence? In trying to throw light on an obscure explanation in our text-book, my brain took fire, I plunged with requickened zeal into a subject which I had for years abandoned, and found food for thoughts which have engaged my attention for a considerable time past, and will probably occupy all my powers of contemplation advantageously for several months to come."

Almost any public speech or longer paper of Sylvester's contains much that is quotable *about* mathematics in addition to technicalities. A refreshing anthology for beginners and even for seasoned mathematicians could be gathered from the pages of his collected works. Probably no other mathematician has so transparently revealed his personality through his writings as has Sylvester. He liked meeting people and infecting them with his own contagious enthusiasm for mathematics. Thus he says, truly in his own case, "So long as a man remains a gregarious and sociable being, he cannot cut himself off from the gratification of the instinct of imparting what he is learning, of propagating through others the ideas and impressions seething in his own brain, without stunting and atrophying his moral nature and drying up the surest sources of his future intellectual replenishment."

As a pendant to Cayley's description of the extent of modern mathematics, we may hang Sylvester's beside it. "I should be sorry to suppose that I was to be left for long in sole possession of so vast a field as is occupied by modern mathematics. Mathematics is not a book confined within a cover and bound between brazen clasps, whose contents it needs only patience to ransack; it is not a mine, whose treasures may take long to reduce into possession, but which fill only a limited number of veins and lodes; it is not a soil, whose fertility can be exhausted by the yield of successive harvests; it is not a continent or an ocean, whose area can be mapped out and its contour defined: it is limitless as that space which it finds too narrow for its aspirations; its possibilities are as infinite as the worlds which are forever crowding in and multiplying upon the astronomer's gaze; it is as incapable of being restricted within assigned boundaries or being reduced to definitions of permanent validity, as the consciousness, the life, which seems to slumber in each monad, in every atom of matter, in each leaf and bud and cell, and is forever ready to burst forth into new forms of vegetable and animal existence."

In 1878 the *American Journal of Mathematics* was founded by Sylvester and placed under his editorship by Johns Hopkins University.

The *Journal* gave mathematics in the United States a tremendous urge in the right direction—research. Today it is still flourishing mathematically but hard pressed financially.

Two years later occurred one of the classic incidents in Sylvester's career. We tell it in the words of Dr. Fabian Franklin, Sylvester's successor in the chair of mathematics at Johns Hopkins for a few years and later editor of the Baltimore *American*, who was an eye (and ear) witness.

"He [Sylvester] made some excellent translations from Horace and from German poets, besides writing a number of pieces of original verse. The tours de force in the way of rhyming, which he performed while in Baltimore, were designed to illustrate the theories of versification of which he gives illustrations in his little book called 'The Laws of Verse.' The reading of the Rosalind poem at the Peabody Institute was the occasion of an amusing exhibition of absence of mind. The poem consisted of no less than four hundred lines, all rhyming with the name Rosalind (the long and short sound of the i both being allowed). The audience quite filled the hall, and expected to find much interest or amusement in listening to this unique experiment in verse. But Professor Sylvester had found it necessary to write a large number of explanatory footnotes, and he announced that in order not to interrupt the poem he would read the footnotes in a body first. Nearly every footnote suggested some additional extempore remark, and the reader was so interested in each one that he was not in the least aware of the flight of time, or of the amusement of the audience. When he had dispatched the last of the notes, he looked up at the clock, and was horrified to find that he had kept the audience an hour and a half before beginning to read the poem they had come to hear. The astonishment on his face was answered by a burst of good-humored laughter from the audience; and then, after begging all his hearers to feel at perfect liberty to leave if they had engagements, he read the Rosalind poem."

Doctor Franklin's estimate of his teacher sums the man up admirably: "Sylvester was quick-tempered and impatient, but generous, charitable and tender-hearted. He was always extremely appreciative of the work of others and gave the warmest recognition to any talent or ability displayed by his pupils. He was capable of flying into a passion on slight provocation, but he did not harbor resentment, and was always glad to forget the cause of quarrel at the earliest opportunity."

Before taking up the thread of Cayley's life where it crossed Sylvester's again, we shall let the author of *Rosalind* describe how he made one of his most beautiful discoveries, that of what are called "canonical forms." [This means merely the reduction of a given "quantic" to a "standard" form. For example $ax^2 + 2bxy + cy^2$ can be expressed as the sum of two squares, say $X^2 + Y^2$; $ax^5 + 5bx^4y + 10cx^3y^2 + 10dx^2y^3 + 5exy^4 + fy^5$ can be expressed as a sum of three fifth powers, $X^5 + Y^5 + Z^5$.]

"I discovered and developed the whole theory of canonical binary forms for odd degrees, and, so far as yet made out, for even degrees* too, at one sitting, with a decanter of port wine to sustain nature's flagging energies, in a back office in Lincoln's Inn Fields. The work was done, and well done, but at the usual cost of racking thought—a brain on fire, and feet feeling, or feelingless, as if plunged in an ice-pail. *That night we slept no more.*" Experts agree that the symptoms are unmistakable. But it must have been ripe port, to judge by what Sylvester got out of the decanter.

Cayley and Sylvester came together again professionally when Cayley accepted an invitation to lecture at Johns Hopkins for half a year in 1881–82. He chose Abelian functions, in which he was researching at the time, as his topic, and the 67-year-old Sylvester faithfully attended every lecture of his famous friend. Sylvester had still several prolific years ahead of him, Cayley not quite so many.

We shall now briefly describe three of Cayley's outstanding contributions to mathematics in addition to his work on the theory of algebraic invariants. It has already been mentioned that he invented the theory of matrices, the geometry of space of n dimensions, and that one of his ideas in geometry threw a new light (in Klein's hands) on non-Euclidean geometry. We shall begin with the last because it is the hardest.

Desargues, Pascal, Poncelet, and others had created *projective* geometry (see chapters 5, 13) in which the object is to discover those properties of figures which are invariant under projection. Measurements—sizes of angles, lengths of lines—and theorems which depend upon measurement, as for example the Pythagorean proposition that the

---

*This part of the theory was developed many years later by E. K. Wakeford (1894–1916), who lost his life in the World War. "Now thanked be God who matched us with this hour." (Rupert Brooke.)

square on the longest side of a right triangle is equal to the sum of the squares on the other two sides, are not projective but *metrical*, and are not handled by *ordinary* projective geometry. It was one of Cayley's greatest achievements in geometry to transcend the barrier which, before he leapt it, had separated projective from metrical properties of figures. From his higher point of view metrical geometry also became projective, and the great power and flexibility of projective methods were shown to be applicable, by the introduction of "imaginary" elements (for instance points whose coordinates involve $\sqrt{-1}$) to metrical properties. Anyone who has done any analytic geometry will recall that two circles intersect in four points, two of which are always "imaginary." (There are cases of apparent exception, for example concentric circles, but this is close enough for our purpose.) The fundamental notions in metrical geometry are the distance between two points and the angle between two lines. Replacing the concept of distance by another, also involving "imaginary" elements, Cayley provided the means for unifying Euclidean geometry and the common non-Euclidean geometries into one comprehensive theory. Without the use of some algebra it is not feasible to give an intelligible account of how this may be done; it is sufficient for our purpose to have noted Cayley's main advance of uniting projective and metrical geometry with its cognate unification of the other geometries just mentioned.

The matter of n-dimensional geometry when Cayley first put it out was much more mysterious than it seems to us today, accustomed as we are to the special case of four dimensions (space-time) in relativity. It is still sometimes said that a four-dimensional geometry is inconceivable to human beings. This is a superstition which was exploded long ago by Plücker; it is easy to put four-dimensional figures on a flat sheet of paper, and so far as *geometry* is concerned the *whole* of a four-dimensional "space" can be easily imagined. Consider first a rather unconventional three-dimensional space: *all* the *circles* that may be drawn in a *plane*. This "all" is a three-dimensional "space" for the simple reason that it takes *precisely three numbers*, or *three coordinates*, to individualize any one of the swarm of circles, namely *two* to fix the position of the center with reference to any arbitrarily given pair of axes, and *one* to give the length of the radius.

If the reader now wishes to visualize a four-dimensional space he may think of *straight lines*, instead of *points*, as the *elements* out of which our common "solid" space is built. Instead of our familiar solid space looking like an agglomeration of infinitely fine birdshot it now resembles a cosmic haystack of infinitely thin, infinitely long straight straws. That it is indeed four-dimensional in *straight lines* can be seen easily if we convince ourselves (as we may do) that *precisely four numbers* are necessary and sufficient to individualize a particular straw in our haystack. The "dimensionality" of a "space" can be anything we choose to make it, provided we suitably select the elements (points, circles, lines, etc.) out of which we build it. Of course if we take *points* as the elements out of which our space is to be constructed, nobody outside of a lunatic asylum has yet succeeded in visualizing a space of more than three dimensions.

Modern physics is fast teaching some to shed their belief in a mysterious "absolute space" over and above the mathematical "spaces" —like Euclid's, for example—that were *constructed* by geometers to correlate their physical experiences. Geometry today is largely a matter of analysis, but the old terminology of "points," "lines," "distances," and so on, is helpful in suggesting interesting things to do with our sets of coordinates. But it does not follow that these particular things are the most useful that might be done in analysis; it may turn out some day that all of them are comparative trivialities by more significant things which we, hidebound in outworn traditions, continue to do merely because we lack imagination.

If there is any mysterious virtue in talking about situations which arise in analysis as if we were back with Archimedes drawing diagrams in the dust, it has yet to be revealed. Pictures after all may be suitable only for very young children; Lagrange dispensed entirely with such infantile aids when he composed his analytical mechanics. Our propensity to "geometrize" our analysis may only be evidence that we have not yet grown up. Newton himself, it is known, first got his marvellous results analytically and re-clothed them in the demonstrations of an Apollonius partly because he knew that the multitude —mathematicians less gifted than himself—would believe a theorem true only if it were accompanied by a pretty picture and a stilted Euclidean demonstration, partly because he himself still lingered by preference in the pre-Cartesian twilight of geometry.

The last of Cayley's great inventions which we have selected for mention is that of matrices and their algebra in its broad outline. The subject originated in a memoir of 1858 and grew directly out of simple observations on the way in which the transformations (linear) of the theory of algebraic invariants are combined. Glancing back at what was said on discriminants and their invariance we note the transformation (the arrow is here read "is replaced by") $y \to \dfrac{px + q}{rx + s}$. Suppose we have two such transformations,

$$y \to \frac{px + q}{rx + s}, \quad x \to \frac{Pz + Q}{Rz + S},$$

the second of which is to be applied to the x in the first. We get

$$y \to \frac{(pP + qR)z + (pQ + qS)}{(rP + sR)z + (rQ + sS)}.$$

Attending only to the coefficients in the three transformations we write them in square arrays, thus

$$\left\| \begin{matrix} p & q \\ r & s \end{matrix} \right\|, \quad \left\| \begin{matrix} P & Q \\ R & S \end{matrix} \right\|, \quad \left\| \begin{matrix} pP + qR & pQ + qS \\ rP + sR & rQ + sS \end{matrix} \right\|,$$

and see that the result of performing the first two transformations successively could have been written down by the following **rule of "multiplication,"**

$$\left\| \begin{matrix} p & q \\ r & s \end{matrix} \right\| \times \left\| \begin{matrix} P & Q \\ R & S \end{matrix} \right\| = \left\| \begin{matrix} pP + qR & pQ + qS \\ rP + sR & rQ + sS \end{matrix} \right\|,$$

where the *rows* of the array on the right are obtained, in an obvious way, by applying the *rows* of the first array on the left onto the columns of the second. Such arrays (of any number of rows and columns) are called *matrices*. Their algebra follows from a few simple postulates, of which we need cite only the following. The matrices $\left\| \begin{matrix} a & b \\ c & d \end{matrix} \right\|$ and $\left\| \begin{matrix} A & B \\ C & D \end{matrix} \right\|$ are *equal* (by definition) when, and only when, $a = A, b = B, c = C, d = D$. The *sum* of the two matrices just written is the matrix $\left\| \begin{matrix} a + A & b + B \\ c + C & d + D \end{matrix} \right\|$. The result of multiplying $\left\| \begin{matrix} a & b \\ c & d \end{matrix} \right\|$ by m (any *number*) is the matrix $\left\| \begin{matrix} ma & mb \\ mc & md \end{matrix} \right\|$. The rule for "multiplying," $\times$, (or "compounding") matrices is as exemplified for $\left\| \begin{matrix} p & q \\ r & s \end{matrix} \right\|, \left\| \begin{matrix} P & Q \\ R & S \end{matrix} \right\|$ above.

A distinctive feature of these rules is that multiplication is *not commutative*, except for *special* kinds of matrices. For example, by the rule we get

$$\left\| \begin{matrix} P & Q \\ R & S \end{matrix} \right\| \times \left\| \begin{matrix} p & q \\ r & s \end{matrix} \right\| = \left\| \begin{matrix} Pp + Qr & Pq + Qs \\ Rp + Sr & Rq + Ss \end{matrix} \right\|,$$

and the matrix on the right is not equal to that which arises from the multiplication

$$\left\| \begin{matrix} p & q \\ r & s \end{matrix} \right\| \times \left\| \begin{matrix} P & Q \\ R & S \end{matrix} \right\|.$$

All this detail, particularly the last, has been given to illustrate a phenomenon of frequent occurrence in the history of mathematics: the necessary mathematical tools for scientific applications have often been invented decades before the science to which the mathematics is the key was imagined. The bizarre rule of "multiplication" for matrices, by which we get different results according to the order in which we do the multiplication (unlike common algebra where $x \times y$ is always equal to $y \times x$), seems about as far from anything of scientific or practical use as anything could possibly be. Yet sixty seven years after Cayley invented it, Heisenberg in 1925 recognized in the algebra of matrices exactly the tool which he needed for his revolutionary work in quantum mechanics.

Cayley continued in creative activity up to the week of his death, which occurred after a long and painful illness, borne with resignation and unflinching courage, on January 26, 1895. To quote the closing sentences of Forsyth's biography: "But he was more than a mathematician. With a singleness of aim, which Wordsworth would have chosen for his 'Happy Warrior,' he persevered to the last in his nobly lived ideal. His life had a significant influence on those who knew him [Forsyth was a pupil of Cayley and became his successor at Cambridge]: they admired his character as much as they respected his genius: and they felt that, at his death, a great man had passed from the world."

Much of what Cayley did has passed into the main current of mathematics, and it is probable that much more in his massive *Collected Mathematical Papers* (thirteen large quarto volumes of about 600 pages each, comprising 966 papers) will suggest profitable forays to adventurous mathematicians for generations to come. At present the fashion is away from the fields of Cayley's greatest interest, and the same may be said for Sylvester; but mathematics has a habit of returning to its old problems to sweep them up into more inclusive syntheses.

In 1883 Henry John Stephen Smith, the brilliant Irish specialist in the theory of numbers and Savilian Professor of Geometry in Oxford University, died in his scientific prime at the age of fifty seven. Oxford invited the aged Sylvester, then in his seventieth year, to take the vacant chair. Sylvester accepted, much to the regret of his innumerable friends in America. But he felt homesick for his native land which had treated him none too generously; possibly also it gave him a certain satisfaction to feel that "the stone which the builders rejected, the same is become the head of the corner."

The amazing old man arrived in Oxford to take up his duties with a brand-new mathematical theory ("Reciprocants"—differential invariants) to spring on his advanced students. Any praise or just recognition always seemed to inspire Sylvester to outdo himself. Although he had been partly anticipated in his latest work by the French mathematician Georges Halphen, he stamped it with his peculiar genius and enlivened it with his ineffaceable individuality.

The inaugural lecture, delivered on December 12, 1885, at Oxford when Sylvester was seventy one, has all the fire and enthusiasm of his early years, perhaps more, because he now felt secure and knew that he was recognized at last by that snobbish world which had fought him. Two extracts will give some idea of the style of the whole.

"The theory I am about to expound, or whose birth I am about to announce, stands to this ["the great theory of Invariants"] in the relation not of a younger sister, but of a brother, who, though of later birth, on the principle that the masculine is more worthy than the feminine, or at all events, according to the regulations of the Salic law, is entitled to take precedence over his elder sister, and exercise supreme sway over their united realms."

Commenting on the unaccountable absence of a term in a certain algebraic expression he waxes lyric.

"Still, in the case before us, this unexpected absence of a member of the family, whose appearance might have been looked for, made an impression on my mind, and even went to the extent of acting on my emotions. I began to think of it as a sort of lost Pleiad in an Algebraical Constellation, and in the end, brooding over the subject, my feelings found vent, or sought relief, in a rhymed effusion, a *jeu de sottise*, which, not without some apprehension of appearing singular or extravagant, I will venture to rehearse. It will at least serve as an interlude, and give some relief to the strain upon your attention before I proceed to make my final remarks on the general theory.

TO A MISSING MEMBER
OF A FAMILY OF TERMS IN AN ALGEBRAICAL FORMULA.

Lone and discarded one! divorced by fate,
From thy wished-for fellows—whither art flown?
Where lingerest thou in thy bereaved estate,
Like some lost star or buried meteor stone?
Thou mindst me much of that presumptuous one
Who loth, aught less than greatest, to be great,
From Heaven's immensity fell headlong down
To live forlorn, self-centred, desolate:
Or who, new Heraklid, hard exile bore,
Now buoyed by hope, now stretched on rack of fear,
Till throned Astraea, wafting to his ear
Words of dim portent through the Atlantic roar,
Bade him 'the sanctuary of the Muse revere
And strew with flame the dust of Isis' shore.'

Having refreshed ourselves and bathed the tips of our fingers in the Pierian spring, let us turn back for a few brief moments to a light banquet of the reason, and entertain ourselves as a sort of after-course with some general reflections arising naturally out of the previous matter of my discourse."

If the Pierian spring was the old boy's finger bowl at this astonishing feast of reason, it is a safe bet that the faithful decanter of port was never very far from his elbow.

Sylvester's sense of the kinship of mathematics to the finer arts found frequent expression in his writings. Thus, in a paper on Newton's rule for the discovery of imaginary roots of algebraic equations, he asks in a footnote "May not Music be described as the Mathematic of sense, Mathematic as Music of the reason? Thus the musician *feels* Mathematic, the mathematician *thinks* Music—Music the dream, Mathematic the working life—each to receive its consummation from the other when the human intelligence, elevated to its perfect type, shall shine forth glorified in some future Mozart-Dirichlet or Beethoven-Gauss—a union already not indistinctly foreshadowed in the genius and labors of a Helmholtz!"

Sylvester loved life, even when he was forced to fight it, and if ever a man got the best that is in life out of it, he did. He gloried in the fact that the great mathematicians, except for what may be classed as avoidable or accidental deaths, have been long-lived and vigorous of mind to their dying days. In his presidential address to the British Association in 1869 he called the honor roll of some of the greatest mathematicians of the past and gave their ages at death to bear out his thesis that ". . . there is no study in the world which brings into more harmonious action all the faculties of the mind than [mathematics], . . . or, like this, seems to raise them, by successive steps of initiation, to higher and higher states of conscious intellectual being. . . . The mathematician lives long and lives young; the wings of the soul do not early drop off, nor do its pores become clogged with the earthy particles blown from the dusty highways of vulgar life."

Sylvester was a living example of his own philosophy. But even he at last began to bow to time. In 1893—he was then seventy nine—his eyesight began to fail, and he became sad and discouraged because he could no longer lecture with his old enthusiasm. The following year he asked to be relieved of the more onerous duties of his professorship, and retired to live, lonely and dejected, in London or at Tunbridge Wells. All his brothers and sisters had long since died, and he had outlived most of his dearest friends.

But even now he was not through. His mind was still vigorous, although he himself felt that the keen edge of his inventiveness was dulled forever. Late in 1896, in the eighty second year of his age, he found a new enthusiasm in a field which had always fascinated him, and he blazed up again over the theory of compound partitions and Goldbach's conjecture that every even number is the sum of two primes.

He had not much longer. While working at his mathematics in his London rooms early in March, 1897, he suffered a paralytic stroke which destroyed his power of speech. He died on March 15, 1897, at the age of eighty three. His life can be summed up in his own words, "I really love my subject."

CHAPTER TWENTY TWO

Master and Pupil

WEIERSTRASS AND SONJA KOWALEWSKI

The theory that has had the greatest development in recent times is without any doubt the theory of functions.—VITO VOLTERRA

YOUNG DOCTORS IN MATHEMATICS, anxiously seeking positions in which their training and talents may have some play, often ask whether it is possible for a man to do elementary teaching for long and keep alive mathematically. It is. The life of Boole is a partial answer; the career of Weierstrass, the prince of analysts, "the father of modern analysis," is conclusive.

Before considering Weierstrass in some detail, we place him chronologically with respect to those of his German contemporaries, each of whom, like him, gave at least one vast empire of mathematics a new outlook during the second half of the nineteenth century and the first three decades of the twentieth. The year 1855, which marks the death of Gauss and the breaking of the last link with the outstanding mathematicians of the preceding century, may be taken as a convenient point of reference. In 1855 Weierstrass (1815–1897) was forty; Kronecker (1823–1891), thirty two; Riemann (1826–1866), twenty nine; Dedekind (1831–1916), twenty four; while Cantor (1845–1918) was a small boy of ten. Thus German mathematics did not lack recruits to carry on the great tradition of Gauss. Weierstrass was just gaining recognition; Kronecker was well started; some of Riemann's greatest work was already behind him, and Dedekind was entering the field (the theory of numbers) in which he was to gain his greatest fame. Cantor, of course, had not yet been heard from.

We have juxtaposed these names and dates because four of the men mentioned, dissimilar and totally unrelated as much of their finest work was, came together on one of the central problems of all mathematics, that of irrational numbers: Weierstrass and Dedekind resumed the discussion of irrationals and continuity practically where

Eudoxus had left it in the fourth century B.C.; Kronecker, a modern echo of Zeno, made Weierstrass' last years miserable by skeptical criticism of the latter's revision of Eudoxus; while Cantor, striking out on a new road of his own, sought to compass the actual infinite itself which is implicit—according to some—in the very concept of continuity. Out of the work of Weierstrass and Dedekind developed the modern epoch of analysis, that of critical logical precision in analysis (the calculus, the theory of functions of a complex variable, and the theory of functions of real variables) in distinction to the looser intuitive methods of some of the older writers—invaluable as heuristic guides to discovery but quite worthless from the standpoint of the Pythagorean ideal of mathematical proof. As has already been noted, Gauss, Abel, and Cauchy inaugurated the first period of rigor; the movement started by Weierstrass and Dedekind was on a higher plane, suitable to the more exacting demands of analysis in the second half of the century, for which the earlier precautions were inadequate.

One discovery by Weierstrass in particular shocked the intuitive school of analysts into a decent regard for caution: he produced a continuous curve which has no tangent at any point. Gauss once called mathematics "the science of the eye"; it takes more than a good pair of eyes to "see" the curve which Weierstrass presented to the advocates of sensual intuition.

Since to every action there is an equal and opposite reaction it was but natural that all this revamped rigor should engender its own opposition. Kronecker attacked it vigorously, even viciously, and quite exasperatingly. He denied that it meant anything. Although he succeeded in hurting the venerable and kindly Weierstrass, he made but little impression on his conservative contemporaries and practically none on mathematical analysis. Kronecker was a generation ahead of his time. Not till the second decade of the twentieth century did his strictures on the currently accepted doctrines of continuity and irrational numbers receive serious consideration. Today it is true that not all mathematicians regard Kronecker's attack as merely the release of his pent-up envy of the more famous Weierstrass which some of his contemporaries imagined it to be, and it is admitted that there may be something—not much, perhaps—in his disturbing objections. Whether there is or not, Kronecker's attack was partly responsible for the *third* period of rigor in modern mathematical reasoning, that which we ourselves are attempting to enjoy. Weierstrass was not the only fellow-mathematician whom Kronecker harried; Cantor also suffered deeply under what he considered his influential colleague's malicious persecution. All these men will speak for themselves in the proper place; here we are only attempting to indicate that their lives and work were closely interwoven in at least one corner of the gorgeous pattern.

To complete the picture we must indicate other points of contact between Weierstrass, Kronecker, and Riemann on one side and Kronecker and Dedekind on the other. Abel, we recall, died in 1829, Galois in 1832, and Jacobi in 1851. In the epoch under discussion one of the outstanding problems in mathematical analysis was the completion of the work of Abel and Jacobi on multiply periodic functions—elliptic functions, Abelian functions (see chapters 17, 18). From totally different points of view Weierstrass and Riemann accomplished what was to be done—Weierstrass indeed considered himself in some degree a successor of Abel; Kronecker opened up new vistas in elliptic functions but he did not compete with the other two in the field of Abelian functions. Kronecker was primarily an arithmetician and an algebraist; some of his best work went into the elaboration and extension of the work of Galois in the theory of equations. Thus Galois found a worthy successor not too long after his death.

Apart from his forays into the domain of continuity and irrational numbers, Dedekind's most original work was in the higher arithmetic, which he revolutionized and renovated. In this Kronecker was his able and sagacious rival, but again their whole approaches were entirely different and characteristic of the two men: Dedekind overcame his difficulties in the theory of algebraic numbers by taking refuge in the infinite (in his theory of "ideals," as will be indicated in the proper place); Kronecker sought to solve his problems in the finite.

Karl Wilhelm Theodor Weierstrass, the eldest son of Wilhelm Weierstrass (1790–1869) and his wife Theodora Forst, was born on October 31, 1815, at Ostenfelde in the district of Münster, Germany. The father was then a customs officer in the pay of the French. It may be recalled that 1815 was the year of Waterloo; the French were still dominating Europe. That year also saw the birth of Bismarck, and it is interesting to observe that whereas the more famous statesman's life work was shot to pieces in the World War, if not earlier, the contributions of his comparatively obscure contemporary to science and the advancement of civilization in general are even more highly esteemed today than they were during his lifetime.

The Weierstrass family were devout liberal Catholics all their lives; the father had been converted from Protestantism, probably at the time of his marriage. Karl had a brother, Peter (died in 1904), and two sisters, Klara (1823–1896), and Elise (1826–1898) who looked after his comfort all their lives. The mother died in 1826, shortly after Elise's birth, and the father married again the following year. Little is known of Karl's mother, except that she appears to have regarded her husband with a restrained aversion and to have looked on her marriage with moderated disgust. The stepmother was a typical German housewife; her influence on the intellectual development of her stepchildren was probably nil. The father, on the other hand, was a practical idealist, and a man of culture who at one time had been a teacher. The last ten years of his life were spent in peaceful old age in the house of his famous son in Berlin, where the two daughters also lived. None of the children ever married, although poor Peter once showed an inclination toward matrimony which was promptly squelched by his father and sisters.

One possible discord in the natural sociability of the children was the father's uncompromising righteousness, domineering authority, and Prussian pigheadedness. He nearly wrecked Peter's life with his everlasting lecturing and came perilously close to doing the same by Karl, whom he attempted to force into an uncongenial career without ascertaining where his brilliant young son's abilities lay. Old Weierstrass had the audacity to preach at his younger son and meddle in his affairs till the "boy" was nearly forty. Luckily Karl was made of more resistant stuff. As we shall see his fight against his father—although he himself was probably quite unaware that he was fighting the tyrant—took the not unusual form of making a mess of the life his father had chosen for him. It was as neat a defense as he could possibly have devised, and the best of it was that neither he nor his father ever dreamed what was happening, although a letter of Karl's when he was sixty shows that he had at last realized the cause of his early difficulties. Karl at last got his way, but it was a long, roundabout way, beset with trials and errors. Only a shaggy man like himself, huge and rugged of body and mind, could have won through to the end.

Shortly after Karl's birth the family moved to Westernkotten, Westphalia, where the father became a customs officer at the salt works. Westernkotten, like other dismal holes in which Weierstrass spent the best years of his life, is known in Germany today only because Weierstrass once was condemned to rot there—only he did not rust; his first published work is dated as having been written in 1841 (he was then 26) at Westernkotten. There being no school in the village, Karl was sent to the adjacent town of Münster whence, at fourteen, he entered the Catholic Gymnasium at Paderborn. Like Descartes under somewhat similar conditions, Weierstrass thoroughly enjoyed his school and made friends of his expert, civilized instructors. He traversed the set course in considerably less than the standard time, making a uniformly brilliant record in all his studies. He left in 1834 at the age of nineteen. Prizes fell his way with unfailing regularity; one year he carried off seven; he was usually first in German and in two of the three, Latin, Greek, and mathematics. By a beautiful freak of irony he never won a prize for calligraphy, although he was destined to teach penmanship to little boys but recently emancipated from their mothers' apron strings.

As mathematicians often have a liking for music it is of interest to note here that Weierstrass, broad as he was, could not tolerate music in any form. It meant nothing to him and he did not pretend that it did. When he had become a success his solicitous sisters tried to get him to take music lessons to make him more conventional socially, but after a halfhearted lesson or two he abandoned the distasteful project. Concerts bored him and grand opera put him to sleep—when they could drag him out to either.

Like his good father, Karl was not only an idealist but was also extremely practical—for a time. In addition to capturing most of the prizes in purely impractical studies he secured a paying job, at the age of fifteen, as accountant for a prosperous female merchant in the ham and butter business.

All of these successes had a disastrous effect on Karl's future. Old Weierstrass, like many parents, drew the wrong conclusion from his son's triumphs. He "reasoned" as follows. Because the boy has won a cartload of prizes, therefore he must have a good mind—this much may be admitted; and because he has kept himself in pocket money by posting the honored female butter and ham merchant's books efficiently, therefore he will be a brilliant bookkeeper. Now what is the acme of all bookkeeping? Obviously a government nest—in the higher branches of course—in the Prussian civil service. But to prepare for this exalted position, a knowledge of the law is desirable in order to pluck effectively and to avoid being plucked.

As the grand conclusion of all this logic, paterfamilias Weierstrass shoved his gifted son, at the age of nineteen, headfirst into the University of Bonn to master the chicaneries of commerce and the quibblings of the law.

Karl had more sense than to attempt either. He devoted his great bodily strength, his lightning dexterity and his keen mind almost exclusively to fencing and the mellow sociability that is induced by nightly and liberal indulgence in honest German beer. What a shocking example for ant-eyed Ph.D.'s who shrink from a spell of schoolteaching lest their dim lights be dimmed forever! But to do what Weierstrass did, and get away with it, one must have at least a tenth of his constitution and not less than one tenth of one percent of his brains.

Bonn found Weierstrass unbeatable. His quick eye, his long reach, his devilish accuracy, and his lightning speed in fencing made him an opponent to admire but not to touch. As a matter of historical fact he never was touched: no jagger scar adorned his cheeks, and in all his bouts he never lost a drop of blood. Whether or not he was ever put under the table in the subsequent celebrations of his numerous victories is not known. His discreet biographers are somewhat reticent on this important point, but to anyone who has ever contemplated one of Weierstrass' mathematical masterpieces it is inconceivable that so strong a head as his could ever have nodded over a half-gallon stein. His four misspent years in the university were perhaps after all well spent.

His experiences at Bonn did three things of the greatest moment for Weierstrass: they cured him of his father fixation without in any way damaging his affection for his deluded parent; they made him a human being capable of entering fully into the pathetic hopes and aspirations of human beings less gifted than himself—his pupils—and thus contributed directly to his success as probably the greatest mathematical teacher of all time; and last, the humorous geniality of his boyhood became a fixed life-habit. So the "student years" were not the loss his disappointed father and his fluttering sisters—to say nothing of the panicky Peter—thought they were when Karl returned, after four "empty" years at Bonn, without a degree, to the bosom of his wailing family.

There was a terrific row. They lectured him—"sick of body and soul" as he was, possibly the result of not enough law, too little mathematics, and too much beer; they sat around and glowered at him and, worst of all, they began to discuss him as if he were dead: what was to be done with the corpse? Touching the law, Weierstrass had only one brief encounter with it at Bonn, but it sufficed: he astonished the Dean and his friends by his acute "opposition" of a candidate for the doctor degree in law. As for the mathematics at Bonn—it was inconsiderable. The one gifted man, Julius Plücker, who might have done Weierstrass some good was so busy with his manifold duties that he had no time to spare on individuals and Weierstrass got nothing out of him.

But like Abel and so many other mathematicians of the first rank, Weierstrass had gone to the masters in the interludes between his fencing and drinking: he had been absorbing the Celestial Mechanics of Laplace, thereby laying the foundations for his lifelong interest in dynamics and systems of simultaneous differential equations. Of course he could get none of this through the head of his cultured, petty-official father, and his obedient brother and his dismayed sisters knew not what the devil he was talking about. The fact alone was sufficient: brother Karl, the genius of the timorous little family, on whom such high hopes of bourgeois respectability had been placed, had come home, after four years of rigid economy on father's part, without a degree.

At last—after weeks—a sensible friend of the family who had sympathized with Karl as a boy, and who had an intelligent amateur's interest in mathematics, suggested a way out: let Karl prepare himself at the neighboring Academy of Münster for the state teachers' examination. Young Weierstrass would not get a Ph.D. out of it, but his job as a teacher would provide a certain amount of evening leisure in which he could keep alive mathematically provided he had the right stuff in him. Freely confessing his "sins" to the authorities, Weierstrass begged the opportunity of making a fresh start. His plea was granted, and Weierstrass matriculated on May 22, 1839 at Münster to prepare himself for a secondary school-teaching career. This was a most important stepping stone to his later mathematical eminence, although at the time it looked like a total rout.

What made all the difference to Weierstrass was the presence at Münster of Christof Gudermann (1798–1852) as Professor of Mathematics. Gudermann at the time (1839) was an enthusiast for elliptic functions. We recall that Jacobi had published his *Fundamenta nova* in 1829. Although few are now familiar with Gudermann's elaborate investigations (published at the instigation of Crelle in a series of articles in his *Journal*), he is not to be dismissed as contemptuously as it is sometimes fashionable to do merely because he is outmoded. For his time Gudermann had what appears to have been an original idea. The theory of elliptic functions can be developed in many different ways—too many for comfort. At one time some particular way seems the best; at another, a slightly different approach is highly advertised for a season and is generally regarded as being more chic.

Gudermann's idea was to base everything on the *power series* expansion of the functions. (This statement will have to do for the moment; its meaning will become clear when we describe one of the leading motivations of the work of Weierstrass.) This really was a good new idea, and Gudermann slaved over it with overwhelming German thoroughness for years without, perhaps, realizing what lay behind his inspiration, and himself never carried it through. The important thing to note here is that Weierstrass made the theory of power series—Gudermann's inspiration—the nerve of all his work in analysis. He got the idea from Gudermann, whose lectures he attended. In later life, contemplating the scope of the methods he had developed in analysis, Weierstrass was wont to exclaim, "There is nothing but power series!"

At the opening lecture of Gudermann's course on elliptic functions (he called them by a different name, but that is of no importance) there were thirteen auditors. Being in love with his subject the lecturer quickly left the earth and was presently soaring practically alone in the aether of pure thought. At the second lecture only one auditor appeared and Gudermann was happy. The solitary student was Weierstrass. Thereafter no incautious third party ventured to profane the holy communion between the lecturer and his unique disciple. Gudermann and Weierstrass were fellow Catholics; they got along splendidly together.

Weierstrass was duly grateful for the pains Gudermann lavished on him, and after he had become famous he seized every opportunity —the more public the better—to proclaim his gratitude for what Gudermann had done for him. The debt was not inconsiderable: it is not every professor who can drop a hint like the one—power series representation of functions as a point of attack—which inspired Weierstrass. In addition to the lectures on elliptic functions, Gudermann also gave Weierstrass private lessons on "analytical spherics" —whatever that may have been.

In 1841, at the age of twenty six, Weierstrass took his examinations for his teacher's certificate. The examination was in two sections, written and oral. For the first he was allowed six months in which to write out essays on three topics acceptable to the examiners. The third question inspired a fine dissertation on the Socratic method in secondary teaching, a method which Weierstrass followed with brilliant success when he became the foremost mathematical teacher of advanced students in the world.

A teacher—at least in higher mathematics—is judged by his students. If his students are enthusiastic about his "beautifully clear lectures," of which they take copious notes, but never do any original mathematics themselves after getting their advanced degrees, the teacher is a flat failure as a university instructor and his proper sphere —if anywhere—is in a secondary school or a small college where the aim is to produce tame gentlemen but not independent thinkers. Weierstrass' lectures were models of perfection. But if they had been nothing more than finished expositions they would have been pedagogically worthless. To perfection of form Weierstrass added that intangible something which is called inspiration. He did not rant about the sublimity of mathematics and he never orated; but somehow or another he made creative mathematicians out of a disproportionately large fraction of his students.

The examination which admitted Weierstrass after a year of probationary teaching to the profession of secondary school work is one of the most extraordinary of its kind on record. One of the essays which he submitted must be the most abstruse production ever accepted in a teachers' examination. At the candidate's request Gudermann had set Weierstrass a real mathematical problem: to find the power series developments of the elliptic functions. There was more than this, but the part mentioned was probably the most interesting.

Gudermann's report on the work might have changed the course of Weierstrass' life had it been listened to, but it made no practical impression where it might have done good. In a postscript to the official report Gudermann states that "This problem, which in general would be far too difficult for a young analyst, was set at the candidate's express request with the consent of the commission." After the acceptance of his written work and the successful conclusion of his oral examination, Weierstrass got a special certificate on his original contribution to mathematics. Having stated what the candidate had done, and having pointed out the originality of the attack and the novelty of some of the results attained, Gudermann declares that the work evinces a fine mathematical talent "which, provided it is not frittered away, will inevitably contribute to the advancement of science. For the author's sake and that of science it is to be desired that he shall not become a secondary teacher, but that favorable conditions will make it possible for him to function in academic instruction. . . . The candidate hereby enters by birthright into the ranks of the famous discoverers."

These remarks, in part underlined by Gudermann, were very properly stricken from the official report. Weierstrass got his certificate and that was all. At the age of twenty six he entered his trade of secondary teaching which was to absorb nearly fifteen years of his life, including the decade from thirty to forty which is usually rated as the most fertile in a scientific man's career.

His work was excessive. Only a man with iron determination and a rugged physique could have done what Weierstrass did. The nights were his own and he lived a double life. Not that he became a dull drudge; far from it. Nor did he pose as the village scholar absorbed in mysterious meditations beyond the comprehension of ordinary mortals. With quiet satisfaction in his later years he loved to dwell on the way he had fooled them all; the gay government officials and the young officers found the amiable school teacher a thoroughly good fellow and a lively tavern companion.

But in addition to these boon companions of an occasional night out, Weierstrass had another, unknown to his happy-go-lucky fellows —Abel, with whom he kept many a long vigil. He himself said that Abel's works were never very far from his elbow. When he became the leading analyst in the world and the greatest mathematical teacher in Europe his first and last advice to his numerous students was "Read Abel!" For the great Norwegian he had an unbounded admiration undimmed by any shadow of envy. "Abel, the lucky fellow!" he would exclaim: "He has done something everlasting! His ideas will always exercise a fertilizing influence on our science."

The same might be said for Weierstrass, and the creative ideas with which he fertilized mathematics were for the most part thought out while he was an obscure schoolteacher in dismal villages where advanced books were unobtainable, and at a time of economic stress when the postage on a letter absorbed a prohibitive part of the teacher's meagre weekly wage. Being unable to afford postage, Weierstrass was barred from scientific correspondence. Perhaps it is as well that he was: his originality developed unhampered by the fashionable ideas of the time. The independence of outlook thus acquired characterized his work in later years. In his lectures he aimed to develop everything from the ground up in his own way and made almost no reference to the work of others. This occasionally mystified his auditors as to what was the master's and what another's.

It will be of interest to mathematical readers to note one or two stages in Weierstrass' scientific career. After his probationary year as a teacher at the Gymnasium at Münster, Weierstrass wrote a memoir on analytic functions in which, among other things, he arrived independently at Cauchy's integral theorem—the so-called fundamental theorem of analysis. In 1842 he heard of Cauchy's work but claimed no priority (as a matter of fact Gauss had anticipated them both away back in 1811, but as usual had laid his work aside to ripen). In 1842, at the age of twenty seven, Weierstrass applied the methods he had developed to systems of differential equations—such as those occurring in the Newtonian problem of three bodies, for example; the treatment was mature and rigorous. These works were undertaken without thought of publication merely to prepare the ground on which Weierstrass' life work (on Abelian functions) was to be built.

In 1842 Weierstrass was assistant teacher of mathematics and physics at the Pro-Gymnasium in Deutsch-Krone, West Prussia. Presently he was promoted to the dignity of ordinary teacher. In addition to the subjects mentioned the leading analyst in Europe also taught German, geography, and writing to the little boys under his charge; gymnastics was added in 1845.

In 1848, at the age of thirty three, Weierstrass was transferred as ordinary teacher to the Gymnasium at Braunsberg. This was something of a promotion, but not much. The head of the school was an excellent man who did what he could to make things agreeable for Weierstrass although he had only a remote conception of the intellectual eminence of his colleague. The school boasted a very small library of carefully selected books on mathematics and science.

It was in this year that Weierstrass turned aside for a few weeks from his absorbing mathematics to indulge in a little delicious mischief. The times were somewhat troubled politically; the virus of liberty had infected the patient German people and at least a few of the bolder souls were out on the warpath for democracy. The royalist party in power clamped a strict censorship on all spoken or printed sentiments not sufficiently laudatory to their regime. Fugitive hymns to liberty began appearing in the papers. The authorities of course could tolerate nothing so subversive of law and order as this, and when Braunsberg suddenly blossomed out with a lush crop of democratic poets all singing the praises of liberty in the local paper, as yet uncensored, the flustered government hastily appointed a local civil servant as censor and went to sleep, believing that all would be well.

Unfortunately the newly appointed censor had a violent aversion to all forms of literature, poetry especially. He simply could not bring himself to read the stuff. Confining his supervision to blue-pencilling the dull political prose, he turned over all the literary effusions to schoolteacher Weierstrass for censoring. Weierstrass was delighted. Knowing that the official censor would never glance at any poem, Weierstrass saw to it that the most inflammatory ones were printed in full right under the censor's nose. This went merrily on to the great delight of the populace till a higher official stepped in and put an end to the farce. As the censor was the officially responsible offender, Weierstrass escaped scot-free.

The obscure hamlet of Deutsch-Krone has the honor of being the place where Weierstrass (in 1842–43) first broke into print. German schools publish occasional "programs" containing papers by members of the staff. Weierstrass contributed *Remarks on Analytical Factorials*. It is not necessary to explain what these are; the point of interest here is that the subject of factorials was one which had caused the older analysts many a profitless headache. Until Weierstrass attacked the problems connected with factorials the nub of the matter had been missed.

Crelle, we recall, wrote extensively on factorials, and we have seen how interested he was when Abel somewhat rashly informed him that his work contained serious oversights. Crelle now enters once more, and again in the same fine spirit he showed Abel.

Weierstrass' work was not published till 1856, fourteen years after it had been written, when Crelle printed it in his *Journal*. Weierstrass was then famous. Admitting that the rigorous treatment by Weierstrass clearly exposes the errors of his own work, Crelle continues as follows: "I have never taken the personal point of view in my work, nor have I striven for fame and praise, but only for the advancement of truth to the best of my ability; and it is all one to me whoever it may be that comes nearer to the truth—whether it is I or someone else, provided only a closer approximation to the truth is attained." There was nothing neurotic about Crelle. Nor was there about Weierstrass.

Whether or not the tiny village of Deutsch-Krone is conspicuous on the map of politics and commerce it stands out like the capital of an empire in the history of mathematics, for it was there that Weierstrass, without even an apology for a library and with no scientific connections whatever, laid the foundations of his life work—"to complete the life work of Abel and Jacobi growing out of Abel's Theorem and Jacobi's discovery of multiply periodic functions of several variables."

Abel, he observes, cut down in the flower of his youth, had no opportunity to follow out the consequences of his tremendous discovery, and Jacobi had failed to see clearly that the true meaning of his own work was to be sought in Abel's Theorem. "The consolidation and extension of these gains—the task of actually exhibiting the functions and working out their properties—is one of the major problems of mathematics." Weierstrass thus declares his intention of devoting his energies to this problem as soon as he shall have understood it deeply and have developed the necessary tools. Later he tells how slowly he progressed: "The fabrication of methods and other difficult problems occupied my time. Thus years slipped away before I could get at the main problem itself, hampered as I was by an unfavorable environment."

The whole of Weierstrass' work in analysis can be regarded as a grand attack on his main problem. Isolated results, special developments and even extensive theories—for example that of irrational numbers as developed by him—all originated in some phase or another of the central problem. He early became convinced that for a clear understanding of what he was attempting to do a radical revision of the fundamental concepts of mathematical analysis was necessary, and from this conviction he passed to another, of more significance today perhaps than the central problem itself: analysis must be founded on the common whole numbers 1,2,3, ... The irrationals which give us the concepts of limits and continuity, from which analysis springs, must be referred back by irrefrangible reasoning to the integers; shoddy proofs must be discarded or reworked, gaps must be filled up, and obscure "axioms" must be dragged out into the light of critical inquiry till all are understood and all are stated in comprehensible language in terms of the integers. This in a sense is the Pythagorean dream of basing all mathematics on the integers, but Weierstrass gave the program constructive definiteness and made it work.

Thus originated the nineteenth century movement known as *the arithmetization of analysis*—something quite different from Kronecker's arithmetical program, at which we shall glance in a later chapter; indeed the two approaches were mutually antagonistic.

In passing it may be pointed out that Weierstrass' plan for his life work and his magnificent accomplishment of most of what he set himself as a young man to do, is a good illustration of the value of the advice Felix Klein once gave a perplexed student who had asked him the secret of mathematical discovery. "You must have a problem," Klein replied. "Choose one definite objective and drive ahead toward it. You may never reach your goal, but you will find something of interest on the way."

From Deutsch-Krone Weierstrass moved to Braunsberg, where he taught in the Royal Catholic Gymnasium for six years, beginning in 1848. The school "program" for 1848–49 contains a paper by Weierstrass which must have astonished the natives: *Contributions to the Theory of Abelian Integrals*. If this work had chanced to fall under the eyes of any of the professional mathematicians of Germany, Weierstrass would have been made. But, as his Swedish biographer, Mittag-Leffler, dryly remarks, one does not look for epochal papers on pure mathematics in secondary-school programs. Weierstrass might as well have used his paper to light his pipe.

His next effort fared better. The summer vacation of 1853 (Weierstrass was then 38) was passed in his father's house at Westernkotten. Weierstrass spent the vacation writing up a memoir on Abelian functions. When it was completed he sent it to Crelle's great *Journal*. It was accepted and appeared in volume 47 (1854).

This may have been the paper whose composition was responsible for an amusing incident in Weierstrass' career as a schoolteacher at

Braunsberg. Early one morning the director of the school was startled by a terrific uproar proceeding from the classroom where Weierstrass was supposed to be holding forth. On investigation he discovered that Weierstrass had not shown up. He hurried over to Weierstrass' dwelling, and on knocking was bidden to enter. There sat Weierstrass pondering by the glimmering light of a lamp, the curtains of the room still drawn. He had worked the whole night through and had not noticed the approach of dawn. The director called his attention to the fact that it was broad daylight and told him of the uproar in his classroom. Weierstrass replied that he was on the trail of an important discovery which would rouse great interest in the scientific world and he could not possibly interrupt his work.

The memoir on Abelian functions published in Crelle's *Journal* in 1854 created a sensation. Here was a masterpiece from the pen of an unknown schoolmaster in an obscure village nobody in Berlin had ever heard of. This in itself was sufficiently astonishing. But what surprised those who could appreciate the magnitude of the work even more was the almost unprecedented fact that the solitary worker had published no preliminary bulletins announcing his progress from time to time, but with admirable restraint had held back everything till the work was completed.

Writing to a friend some ten years later, Weierstrass gives his modest version of his scientific reticence: " . . . the infinite emptiness and boredom of those years [as a schoolteacher] would have been unendurable without the hard work that made me a recluse—even if I was rated rather a good fellow by the circle of my friends among the junkers, lawyers, and young officers of the community. . . . The present offered nothing worth mentioning, and it was not my custom to speak of the future."

Recognition was immediate. At the University of Königsberg, where Jacobi had made his great discoveries in the field which Weierstrass had now entered with a masterpiece of surpassing excellence, Richelot, himself a worthy successor of Jacobi in the theory of multiply periodic functions, was Professor of Mathematics. His expert eyes saw at once what Weierstrass had done. He forthwith persuaded his university to confer the degree of doctor, *honoris causa*, on Weierstrass and himself journeyed to Braunsberg to present the diploma.

At the dinner organized by the director of the Gymnasium in

Weierstrass' honor Richelot asserted that "we have all found our master in Mr. Weierstrass." The Ministry of Education immediately promoted him and granted him a year's leave to prosecute his scientific work. Borchardt, the editor of Crelle's *Journal* at the time, hurried to Braunsberg to congratulate the greatest analyst in the world, thus starting a warm friendship which lasted till Borchardt's death a quarter of a century later.

None of this went to Weierstrass' head. Although he was deeply moved and profoundly grateful for all the generous recognition so promptly accorded him, he could not refrain from casting a backward glance over his career. Years later, thinking of the happiness of the occasion and of what that occasion had opened up for him when he was forty years of age, he remarked sadly that "everything in life comes too late."

Weierstrass did not return to Braunsberg. No really suitable position being open at the time, the leading German mathematicians did what they could to tide over the emergency and got Weierstrass appointed Professor of Mathematics at the Royal Polytechnic School in Berlin. This appointment dated from July 1, 1856; in the autumn of the same year he was made Assistant Professor (in addition to the other post) at the University of Berlin and was elected to the Berlin Academy.

The excitement of novel working conditions and the strain of too much lecturing presently brought on a nervous breakdown. Weierstrass had also been overworking at his researches. In the summer of 1859 he was forced to abandon his course and take a rest cure. Returning in the fall he continued his work, apparently refreshed, but in the following March was suddenly attacked by spells of vertigo, and he collapsed in the middle of a lecture.

All the rest of his life he was bothered with the same trouble off and on, and after resuming his work—as full professor, with a considerably lightened load—never trusted himself to write his own formulas on the board. His custom was to sit where he could see the class and the blackboard, and dictate to some student delegated from the class what was to be written. One of these "mouthpieces" of the master developed a rash propensity to try to improve on what he had been told to write. Weierstrass would reach up and rub out the amateur's efforts and make him write what he had been told. Occasionally the battle between the professor and the obstinate student would go to several rounds, but in the end Weierstrass always won. He had seen little boys misbehaving before.

As the fame of his work spread over Europe (and later to America), Weierstrass' classes began to grow rather unwieldy and he would sometimes regret that the quality of his auditors lagged far behind their rapidly mounting quantity. Nevertheless he gathered about him an extremely able band of young mathematicians who were absolutely devoted to him and who did much to propagate his ideas, for Weierstrass was always slow about publication, and without the broadcasting of his lectures which his disciples took upon themselves his influence on the mathematical thought of the nineteenth century would have been considerably retarded.

Weierstrass was always accessible to his students and sincerely interested in their problems, whether mathematical or human. There was nothing of the "great man" complex about him, and he would as gladly walk home with any of the students—and there were many—who cared to join him as with the most famous of his colleagues, perhaps more gladly when the colleague happened to be Kronecker. He was happiest when, sitting at a table over a glass of wine with a few of his devoted disciples, he became a jolly student again himself and insisted on paying the bill for the crowd.

An anecdote (about Mittag-Leffler) may suggest that the Europe of the present century has partly lost something it had in the 1870's. The Franco-Prussian war (1870–71) had left France pretty sore at Germany. But it had not befogged the minds of mathematicians regarding one another's merits irrespective of their nationalities. The like holds for the Napoleonic wars and the mutual esteem of the French and British mathematicians. In 1873 Mittag-Leffler arrived in Paris from Stockholm all set and full of enthusiasm to study analysis under Hermite. "You have made a mistake, sir," Hermite told him: "you should follow Weierstrass' course at Berlin. He is the master of all of us."

Mittag-Leffler took the sound advice of the magnanimous Frenchman and not so long afterward made a capital discovery of his own which is to be found today in all books on the theory of functions. "Hermite was a Frenchman and a patriot," Mittag-Leffler remarks;

"I learned at the same time in what degree he was also a mathematician."

The years (1864-97) of Weierstrass' career at Berlin as Professor of Mathematics were full of scientific and human interests for the man who was acknowledged as the leading analyst in the world. One phase of these interests demands more than the passing reference that might suffice in a purely scientific biography of Weierstrass: his friendship with his favorite pupil, Sonja (or Sophie) Kowalewski.

Madame Kowalewski's maiden name was Sonja Corvin-Kroukowsky; she was born at Moscow, Russia, on January 15, 1850, and died at Stockholm, Sweden, on February 10, 1891, six years before the death of Weierstrass.

At fifteen Sonja began the study of mathematics. By eighteen she had made such rapid progress that she was ready for advanced work and was enamored of the subject. As she came of an aristocratic and prosperous family, she was enabled to gratify her ambition for foreign study and matriculated at the University of Heidelberg.

This highly gifted girl became not only the leading woman mathematician of modern times, but also made a reputation as a leader in the movement for the emancipation of women, particularly as regarded their age-old disabilities in the field of higher education.

In addition to all this she was a brilliant writer. As a young girl she hesitated long between mathematics and literature as a career. After the composition of her most important mathematical work (the prize memoir noted later), she turned to literature as a relaxation and wrote the reminiscences of her childhood in Russia in the form of a novel (published first in Swedish and Danish). Of this work it is reported that "the literary critics of Russia and Scandinavia were unanimous in declaring that Sonja Kowalewski had equalled the best writers of Russian literature in style and thought." Unfortunately this promising start was blocked by her premature death, and only fragments of other literary works survive. Her one novel was translated into many languages.

Although Weierstrass never married he was no panicky bachelor who took to his heels every time he saw a pretty woman coming. Sonja, according to competent judges who knew her, was extremely good-looking. We must first tell how she and Weierstrass met.

Weierstrass used to enjoy his summer vacations in a thoroughly human manner. The Franco-Prussian war caused him to forego his usual summer trip in 1870, and he stayed in Berlin, lecturing on elliptic functions. Owing to the war his class had dwindled to only twenty instead of the fifty who heard the lectures two years before. Since the autumn of 1869 Sonja Kowalewski, then a dazzling young woman of nineteen, had been studying elliptic functions under Leo Königsberger (born 1837) at the University of Heidelberg, where she had also followed the lectures on physics by Kirchhoff and Helmholtz and had met Bunsen the famous chemist under rather amusing circumstances—to be related presently. Königsberger, one of Weierstrass' first pupils, was a first-rate publicity agent for his master. Sonja caught her teacher's enthusiasm and resolved to go directly to the master himself for inspiration and enlightenment.

The status of unmarried women students in the 1870's was somewhat anomalous. To forestall gossip, Sonja at the age of eighteen contracted what was to have been a nominal marriage, left her husband in Russia, and set out for Germany. Her one indiscretion in her dealings with Weierstrass was her neglect to inform him at the beginning that she was married.

Having decided to learn from the master himself, Sonja took her courage in her hands and called on Weierstrass in Berlin. She was twenty, very earnest, very eager, and very determined; he was fifty five, vividly grateful for the lift Gudermann had given him toward becoming a mathematician by taking him on as a pupil, and sympathetically understanding of the ambitions of young people. To hide her trepidation Sonja wore a large and floppy hat, "so that Weierstrass saw nothing of those marvelous eyes whose eloquence, when she wished it, none could resist."

Some two or three years later, on a visit to Heidelberg, Weierstrass learned from Bunsen—a crabbed bachelor—that Sonja was "a dangerous woman." Weierstrass enjoyed his friend's terror hugely, as Bunsen at the time was unaware that Sonja had been receiving frequent private lessons from Weierstrass for over two years.

Poor Bunsen based his estimate of Sonja on bitter personal experience. He had proclaimed for years that no woman, and especially no Russian woman, would ever be permitted to profane the masculine sanctity of his laboratory. One of Sonja's Russian girl friends, desiring ardently to study chemistry in Bunsen's laboratory, and having been thrown out herself, prevailed upon Sonja to try her powers of persuasion on the crusty chemist. Leaving her hat at home, Sonja interviewed Bunsen. He was only too charmed to accept Sonja's friend as a student in his laboratory. After she left he woke up to what she had done to him. "And now *that woman* has made me eat my own words," he lamented to Weierstrass.

Sonja's evident earnestness on her first visit impressed Weierstrass favorably and he wrote to Königsberger inquiring about her mathematical aptitudes. He asked also whether "the lady's personality offers the necessary guarantees." On receiving an enthusiastic reply, Weierstrass tried to get the university senate to admit Sonja to his mathematical lectures. Being brusquely refused he took care of her himself on his own time. Every Sunday afternoon was devoted to teaching Sonja at his house, and once a week Weierstrass returned her visit. After the first few lessons Sonja lost her hat. The lessons began in the autumn of 1870 and continued with slight interruptions due to vacations or illnesses till the autumn of 1874. When for any reason the friends were unable to meet they corresponded. After Sonja's death in 1891 Weierstrass burnt all her letters to him, together with much of his other correspondence and probably more than one mathematical paper.

The correspondence between Weierstrass and his charming young friend is warmly human, even when most of a letter is given over to mathematics. Much of the correspondence was undoubtedly of considerable scientific importance, but unfortunately Sonja was a very untidy woman when it came to papers, and most of what she left behind was fragmentary or in hopeless confusion.

Weierstrass himself was no paragon in this respect. Without keeping records he loaned his unpublished manuscripts right and left to students who did not always return what they borrowed. Some even brazenly rehashed parts of their teacher's work, spoiled it, and published the results as their own. Although Weierstrass complains about this outrageous practice in letters to Sonja his chagrin is not over the petty pilfering of his ideas but of their bungling in incompetent hands and the consequent damage to mathematics. Sonja of course never descended to anything of this sort, but in another respect she was not entirely blameless. Weierstrass sent her one of his unpublished works by which he set great store, and that was the last he ever saw of it. Apparently she lost it, for she discreetly avoids the topic—to judge from his letters—whenever he brings it up.

To compensate for this lapse Sonja tried her best to get Weierstrass to exercise a little reasonable caution in regard to the rest of his unpublished work. It was his custom to carry about with him on his frequent travels a large white wooden box in which he kept all his working notes and the various versions of papers which he had not yet perfected. His habit was to rework a theory many times until he found the best, the "natural" way in which it should be developed. Consequently he published slowly and put out a work under his own name only when he had exhausted the topic from some coherent point of view. Several of his rough-hewn projects are said to have been confided to the mysterious box. In 1880, while Weierstrass was on a vacation trip, the box was lost in the baggage. It has never been heard of since.

After taking her degree *in absentia* from Göttingen in 1874, Sonja returned to Russia for a rest as she was worn out by excitement and overwork. Her fame had preceded her and she "rested" by plunging into the hectic futilities of a crowded social season in St. Petersburg while Weierstrass, back in Berlin, pulled wires all over Europe trying to get his favorite pupil a position worthy of her talents. His fruitless efforts disgusted him with the narrowness of the orthodox academic mind.

In October 1875, Weierstrass received from Sonja the news that her father had died. She apparently never replied to his tender condolences, and for nearly three years she dropped completely out of his life. In August, 1878, he writes to ask whether she ever received a letter he had written her so long before that he has forgotten its date. "Didn't you get my letter? Or what can be preventing you from confiding freely in me, your best friend as you so often called me, as you used to do? This is a riddle whose solution only you can give me. . . ."

In the same letter Weierstrass rather pathetically begs her to contradict the rumor that she has abandoned mathematics: Tchebycheff, a Russian mathematician, had called on Weierstrass when he was out, but had told Borchardt that Sonja had "gone social," as indeed she had. "Send your letter to Berlin at the old address," he concludes; "it will certainly be forwarded to me."

Man's ingratitude to man is a familiar enough theme; Sonja now demonstrated what a woman can do in that line when she puts her mind to it. She did not answer her old friend's letter for two years although she knew he had been unhappy and in poor health.

The answer when it did come was rather a letdown. Sonja's sex had got the better of her ambitions and she had been living happily with her husband. Her misfortune at the time was to be the focus for the flattery and unintelligent, sideshow wonder of a superficially brilliant mob of artists, journalists, and dilettant litterateurs who gabbled incessantly about her unsurpassable genius. The shallow praise warmed and excited her. Had she frequented the society of her intellectual peers she might still have lived a normal life and have kept her enthusiasm. And she would not have been tempted to treat the man who had formed her mind as shabbily as she did.

In October, 1878, Sonja's daughter "Foufie" was born.

The forced quiet after Foufie's arrival roused the mother's dormant mathematical interests once more, and she wrote to Weierstrass for technical advice. He replied that he must look up the relevant literature before venturing an opinion. Although she had neglected him, he was still ready with his ungrudging encouragement. His only regret (in a letter of October, 1880) is that her long silence has deprived him of the opportunity of helping her. "But I don't like to dwell so much on the past—so let us keep the future before our eyes."

Material tribulations aroused Sonja to the truth. She was a born mathematician and could no more keep away from mathematics than a duck can from water. So in October, 1880 (she was then thirty), she wrote begging Weierstrass to advise her again. Not waiting for his reply she packed up and left Moscow for Berlin. His reply, had she received it, might have caused her to stay where she was. Nevertheless when the distracted Sonja arrived unexpectedly he devoted a whole day to going over her difficulties with her. He must have given her some pretty straight talk, for when she returned to Moscow three months later she went after her mathematics with such fury that her gay friends and silly parasites no longer recognized her. At Weierstrass' suggestion she attacked the problem of the propagation of light in a crystalline medium.

In 1882 the correspondence takes two new turns, one of which is of mathematical interest. The other is Weierstrass' outspoken opinion that Sonja and her husband are unsuited to one another, especially as the latter has no true appreciation of her intellectual merits. The mathematical point refers to Poincaré, then at the beginning of his career. With his sure instinct for recognizing young talent, Weierstrass hails Poincaré as a coming man and hopes that he will outgrow his propensity to publish too rapidly and let his researches ripen without scattering them over too wide a field. "To publish an article of real merit every week—that is impossible," he remarks, referring to Poincaré's deluge of papers.

Sonja's domestic difficulties presently resolved themselves through the sudden death of her husband in March 1883. She was in Paris at the time, he in Moscow. The shock prostrated her. For four days she shut herself up alone, refused food, lost consciousness the fifth day, and on the sixth recovered, asked for paper and pencil, and covered the paper with mathematical formulas. By autumn she was herself again, attending a scientific congress at Odessa.

Thanks to Mittag-Leffler, Madame Kowalewski at last obtained a position where she could do herself justice; in the autumn of 1884 she was lecturing at the University of Stockholm, where she was to be appointed (in 1889) as professor for life. A little later she suffered a rather embarrassing setback when the Italian mathematician Vito Volterra pointed out a serious mistake in her work on the refraction of light in crystalline media. This oversight had escaped Weierstrass, who at the time was so overwhelmed with official duties that outside of them he had "time only for eating, drinking, and sleeping. . . . In short," he says, "I am what the doctors call brain-weary." He was now nearly seventy. But as his bodily ills increased his intellect remained as powerful as ever.

The master's seventieth birthday was made the occasion for public honors and a gathering of his disciples and former pupils from all over Europe. Thereafter he lectured publicly less and less often, and for ten years received a few of his students at his own house. When they saw that he was tired out they avoided mathematics and talked of other things, or listened eagerly while the companionable old man reminisced of his student pranks and the dreary years of his isolation from all scientific friends. His eightieth birthday was celebrated by an even more impressive jubilee than his seventieth and he became in some degree a national hero of the German people.

One of the greatest joys Weierstrass experienced in his declining years was the recognition won at last by his favorite pupil. On Christmas Eve, 1888, Sonja received in person the Bordin Prize of the French

Academy of Sciences for her memoir *On the rotation of a solid body about a fixed point.*

As is the rule in competition for such prizes, the memoir had been submitted anonymously (the author's name being in a sealed envelope bearing on the outside the same motto as that inscribed on the memoir, the envelope to be opened only if the competing work won the prize), so there was no opportunity for jealous rivals to hint at undue influence. In the opinion of the judges the memoir was of such exceptional merit that they raised the value of the prize from the previously announced 3000 francs to 5000. The monetary value, however, was the least part of the prize.

Weierstrass was overjoyed. "I do not need to tell you," he writes, "how much your success has gladdened the hearts of myself and my sisters, also of your friends here. I particularly experienced a true satisfaction; competent judges have now delivered their verdict that my 'faithful pupil,' my 'weakness' is indeed not a 'frivolous humbug.'"

We may leave the friends in their moment of triumph. Two years later (February 10, 1891) Sonja died in Stockholm at the age of forty one after a brief attack of influenza which at the time was epidemic. Weierstrass outlived her six years, dying peacefully in his eighty second year on February 19, 1897, at his home in Berlin after a long illness followed by influenza. His last wish was that the priest say nothing in his praise at the funeral but restrict the services to the customary prayers.

Sonja is buried in Stockholm, Weierstrass with his two sisters in a Catholic cemetery in Berlin. Sonja also was of the Catholic faith, belonging to the Greek Church.

We shall now give some intimation of two of the basic ideas on which Weierstrass founded his work in analysis. Details or an exact description are out of the question here, but may be found in the earlier chapters of any competently written book on the theory of functions.

A *power series* is an expression of the form

$$a_0 + a_1 z + a_2 z^2 + \ldots + a_n z^n + \ldots,$$

in which the coefficients $a_0, a_1, a_2, \ldots, a_n, \ldots$ are constant numbers and z is a variable number; the numbers concerned may be real or complex.

The sums of 1,2,3, ... terms of the series, namely a_0, $a_0 + a_1z$, $a_0 + a_1z + a_2z^2$, ... are called the *partial sums*. If for some particular value of z these partial sums give a sequence of numbers which converge to a definite limit, the power series is said to converge to the same limit for that value of z.

All the values of z for which the power series converges to a limit constitute the *domain of convergence* of the series; for any value of the variable z in this domain the series *converges*; for other values of z it *diverges*.

If the series converges for some value of z, its value can be calculated to any desired degree of approximation, for that value, by taking a sufficiently large number of terms.

Now, in the majority of mathematical problems which have applications to science, the "answer" is indicated as the solution in series of a differential equation (or system of such equations), and this solution is only rarely obtainable as a finite expression in terms of mathematical functions which have been tabulated (for instance logarithms, trigonometric functions, elliptic functions, etc.). In such problems it then becomes necessary to do two things: prove that the series converges, if it does; calculate its numerical value to the required accuracy.

If the series does not converge it is usually a sign that the problem has been either incorrectly stated or wrongly solved. The multitude of functions which present themselves in pure mathematics are treated in the same way, whether they are ever likely to have scientific applications or not, and finally a general theory of convergence has been elaborated to cover vast tracts of all this, so that the individual examination of a particular series is often referred to more inclusive investigations already carried out.

Finally, all this (both pure and applied) is extended to power series in 2, 3, 4, ... variables instead of the single variable z above; for example, in two variables,

$$a + b_0z + b_1w + c_0z^2 + c_1zw + c_2w^2 + \dots$$

It may be said that without the theory of power series most of mathematical physics (including much of astronomy and astro-physics) as we know it would not exist.

Difficulties arising with the concepts of limits, continuity, and convergence drove Weierstrass to the creation of his theory of irrational numbers.

Suppose we extract the square root of 2 as we did in school, carrying the computation to a large number of decimal places. We get as successive approximations to the required square root the *sequence* of numbers 1, 1.4, 1.41, 1.412, ... With sufficient labor, proceeding by well-defined steps according to the usual rule, we could if necessary exhibit the first thousand, or the first million, of the *rational* numbers 1, 1.4, ... constituting this sequence of approximations. Examining this sequence we see that when we have gone far enough we have determined a perfectly definite rational number containing as many decimal places as we please (say 1000), and that *this* rational number differs from any of the *succeeding* rational numbers in the sequence by a number (decimal), such as .000 ....000 ..., in which a correspondingly large number of zeros occur *before* another digit (1,2, ... or 9) appears.

This illustrates what is meant by a *convergent sequence* of numbers: the *rationals* 1, 1.4, ... constituting the sequence give us ever closer approximations to the "irrational number" which we *call* the square root of 2, and which we conceive of as having been *defined* by the *convergent sequence of rationals*, this definition being in the sense that a method has been indicated (the usual school one) of calculating *any particular member of the sequence in a finite number of steps.*

Although it is impossible actually to exhibit the whole sequence, as it does not stop at any finite number of terms, nevertheless we regard the *process* for constructing *any* member of the sequence as a sufficiently clear conception of the whole sequence as a single definite object which we can reason about. Doing so, we have a workable method for *using* the square root of 2 and similarly for any irrational number, in mathematical analysis.

As has been indicated it is impossible to make this precise in an account like the present, but even a careful statement might disclose some of the logical objections glaringly apparent in the above description—objections which inspired Kronecker and others to attack Weierstrass' "sequential" definition of irrationals.

Nevertheless, right or wrong, Weierstrass and his school made the theory *work*. The most useful results they obtained have not yet been questioned, at least on the ground of their great utility in mathematical analysis and its applications, by any competent judge in his right mind. This does not mean that objections cannot be well taken: it merely calls attention to the fact that in mathematics, as in everything else, this earth is not yet to be confused with the Kingdom of Heaven, that perfection is a chimaera, and that, in the words of Crelle, we can only hope for closer and closer approximations to mathematical truth—whatever that may be, if anything—precisely as in the Weierstrassian theory of convergent sequences of rationals defining irrationals.

After all, why should mathematicians, who are human beings like the rest of us, always be so pedantically exact and so inhumanly perfect? As Weierstrass said, "It is true that a mathematician who is not also something of a poet will never be a perfect mathematician." That is the answer: a perfect mathematician, by the very fact of his poetic perfection, would be a mathematical impossibility.

CHAPTER TWENTY THREE

Complete Independence

BOOLE

Pure Mathematics was discovered by Boole in a work which he called The Laws of Thought.—BERTRAND RUSSELL

"OH, WE NEVER READ ANYTHING the English mathematicians do." This characteristically continental remark was the reply of a distinguished European mathematician when he was asked whether he had seen some recent work of one of the leading English mathematicians. The "we" of his frank superiority included Continental mathematicians in general.

This is not the sort of story that mathematicians like to tell on themselves, but as it illustrates admirably that characteristic of British mathematicians—insular originality—which has been the chief claim to distinction of the British school, it is an ideal introduction to the life and work of one of the most insularly original mathematicians England has produced, George Boole. The fact is that British mathematicians have often serenely gone their own way, doing the things that interested them personally as if they were playing cricket for their own amusement only, with a self-satisfied disregard for what others, shouting at the top of their scientific lungs, have assured the world is of supreme importance. Sometimes, as in the prolonged idolatry of Newton's methods, indifference to the leading fashions of the moment has cost the British school dearly, but in the long run the take-it-or-leave-it attitude of this school has added more new fields to mathematics than a slavish imitation of the continental masters could ever have done. The theory of invariance is a case in point; Maxwell's electrodynamic field theory is another.

Although the British school has had its share of powerful developers of work started elsewhere, its greater contribution to the progress of mathematics has been in the direction of originality. Boole's work is a striking illustration of this. When first put out it was ignored *as mathematics*, except by a few, chiefly Boole's own more unorthodox countrymen, who recognized that here was the germ of something of supreme interest for all mathematics. Today the natural development of what Boole started is rapidly becoming one of the major divisions of pure mathematics, with scores of workers in practically all countries extending it to all fields of mathematics where attempts are being made to consolidate our gains on firmer foundations. As Bertrand Russell remarked some years ago, pure mathematics was *discovered* by George Boole in his work *The Laws of Thought* published in 1854. This may be an exaggeration, but it gives a measure of the importance in which mathematical logic and its ramifications are held today. Others before Boole, notably Leibniz and De Morgan, had dreamed of adding logic itself to the domain of algebra; Boole did it.

George Boole was not, like some of the other originators in mathematics, born into the lowest economic stratum of society. His fate was much harder. He was born on November 2, 1815, at Lincoln, England, and was the son of a petty shopkeeper. If we can credit the picture drawn by English writers themselves of those hearty old days—1815 was the year of Waterloo—to be the son of a small tradesman at that time was to be damned by foreordination.

The whole class to which Boole's father belonged was treated with a contempt a trifle more contemptuous than that reserved for enslaved scullery maids and despised second footmen. The "lower classes," into whose ranks Boole had been born, simply did not exist in the eyes of the "upper classes"—including the more prosperous wine merchants and moneylenders. It was taken for granted that a child in Boole's station should dutifully and gratefully master the shorter catechism and so live as never to transgress the strict limits of obedience imposed by that remarkable testimonial to human conceit and class-conscious snobbery.

To say that Boole's early struggles to educate himself into a station above that to which "it had pleased God to call him" were a fair imitation of purgatory is putting it mildly. By an act of divine providence Boole's great spirit had been assigned to the meanest class; let it stay there then and stew in its own ambitious juice. Americans may like to recall that Abraham Lincoln, only six years older than Boole, had his struggle about the same time. Lincoln was not sneered at but encouraged.

The schools where young gentlemen were taught to knock one another about in training for their future parts as leaders in the sweatshop and coal mine systems then coming into vogue were not for the likes of George Boole. No; his "National School" was designed chiefly with the end in view of keeping the poor in their proper, unwashable place.

A wretched smattering of Latin, with perhaps a slight exposure to Greek, was one of the mystical stigmata of a gentleman in those incomprehensible days of the sooty industrial revolution. Although few of the boys ever mastered Latin enough to enable them to read it without a crib, an assumed knowledge of its grammar was one of the hall marks of gentility, and its syntax, memorized by rote was, oddly enough, esteemed as mental discipline of the highest usefulness in preparation for the ownership and conservation of property.

Of course no Latin was taught in the school that Boole was permitted to attend. Making a pathetically mistaken diagnosis of the abilities which enabled the propertied class to govern those beneath them in the scale of wealth, Boole decided that he must learn Latin and Greek if he was ever to get his feet out of the mire. This was Boole's mistake. Latin and Greek had nothing to do with the cause of his difficulties. He did teach himself Latin with his poor struggling father's sympathetic encouragement. Although the poverty-stricken tradesman knew that he himself should never escape he did what he could to open the door for his son. He knew no Latin. The struggling boy appealed to another tradesman, a small bookseller and friend of his father. This good man could only give the boy a start in the elementary grammar. Thereafter Boole had to go it alone. Anyone who has watched even a good teacher trying to get a normal child of eight through Caesar will realize what the untutored Boole was up against. By the age of twelve he had mastered enough Latin to translate an ode of Horace into English verse. His father, hopefully proud but understanding nothing of the technical merits of the translation, had it printed in the local paper. This precipitated a scholarly row, partly flattering to Boole, partly humiliating.

A classical master denied that a boy of twelve could have produced such a translation. Little boys of twelve often know more about some things than their forgetful elders give them credit for. On the technical side grave defects showed up. Boole was humiliated and resolved to supply the deficiencies of his self-instruction. He had also taught himself Greek. Determined now to do a good job or none he spent the next two years slaving over Latin and Greek, again without help. The effect of all this drudgery is plainly apparent in the dignity and marked Latinity of much of Boole's prose.

Boole got his early mathematical instruction from his father, who had gone considerably beyond his own meager schooling by private study. The father had also tried to interest his son in another hobby, that of making optical instruments, but Boole, bent on his own ambition, stuck to it that the classics were the key to dominant living. After finishing his common schooling he took a commercial course. This time his diagnosis was better, but it did not help him greatly. By the age of sixteen he saw that he must contribute at once to the support of his wretched parents. Schoolteaching offered the most immediate opportunity of earning steady wages—in Boole's day "ushers," as assistant teachers were called, were not paid salaries but wages. There is more than a monetary difference between the two. It may have been about this time that the immortal Squeers, in Dickens' *Nicholas Nickleby*, was making his great but unappreciated contribution to modern pedagogy at Dotheboys Hall with his brilliant anticipation of the "project" method. Young Boole may even have been one of Squeers' ushers; he taught at two schools.

Boole spent four more or less happy years teaching in these elementary schools. The chilly nights, at least, long after the pupils were safely and mercifully asleep, were his own. He still was on the wrong track. A third diagnosis of his social unworthiness was similar to his second but a considerable advance over both his first and second. Lacking anything in the way of capital—practically every penny the young man earned went to the support of his parents and the barest necessities of his own meager existence—Boole now cast an appraising eye over the gentlemanly professions. The Army at that time was out of his reach as he could not afford to purchase a commission. The Bar made obvious financial and educational demands which he had no prospect of satisfying. Teaching, of the grade in which he was then engaged, was not even a reputable trade, let alone a profession. What remained? Only the Church. Boole resolved to become a clergyman.

In spite of all that has been said for and against God, it must be admitted even by his severest critics that he has a sense of humor. Seeing the ridiculousness of George Boole's ever becoming a clergyman, he skilfully turned the young man's eager ambition into less preposterous channels. An unforeseen affliction of greater poverty than any they had yet enjoyed compelled Boole's parents to urge their son to forego all thoughts of ecclesiastical eminence. But his four years of private preparation (and rigid privation) for the career he had planned were not wholly wasted; he had acquired a mastery of French, German, and Italian, all destined to be of indispensable service to him on his true road.

At last he found himself. His father's early instruction now bore fruit. In his twentieth year Boole opened up a civilized school of his own. To prepare his pupils properly he had to teach them some mathematics as it should be taught. His interest was aroused. Soon the ordinary and execrable textbooks of the day awoke his wonder, then his contempt. Was this stuff mathematics? Incredible. What did the great masters of mathematics say? Like Abel and Galois, Boole went directly to great headquarters for his marching orders. It must be remembered that he had had no mathematical training beyond the rudiments. To get some idea of his mental capacity we can imagine the lonely student of twenty mastering, by his own unaided efforts, the *Mécanique céleste* of Laplace, one of the toughest masterpieces ever written for a conscientious student to assimilate, for the mathematical reasoning in it is full of gaps and enigmatical declarations that "it is easy to see," and then we must think of him making a thorough, understanding study of the excessively abstract *Mécanique analytique* of Lagrange, in which there is not a single diagram to illuminate the analysis from beginning to end. Yet Boole, self-taught, found his way and saw what he was doing. He even got his first contribution to mathematics out of his unguided efforts. This was a paper on the calculus of variations.

Another gain that Boole got out of all this lonely study deserves a separate paragraph to itself. He discovered invariants. The significance of this great discovery which Cayley and Sylvester were to develop in grand fashion has been sufficiently explained; here we repeat that without the mathematical theory of invariance (which grew out of the early algebraic work) the theory of relativity would have been impossible. Thus at the very threshold of his scientific career Boole noticed something lying at his feet which Lagrange himself might easily have seen, picked it up, and found that he had a gem of the first water. That Boole saw what others had overlooked was due no doubt to his strong feeling for the symmetry and beauty of algebraic relations—when of course they happen to be both symmetrical and beautiful; they are not always. Others might have thought his find merely pretty. Boole recognized that it belonged to a higher order.

Opportunities for mathematical publication in Boole's day were inadequate unless an author happened to be a member of some learned society with a journal or transactions of its own. Luckily for Boole, *The Cambridge Mathematical Journal*, under the able editorship of the Scotch mathematician, D. F. Gregory, was founded in 1837. Boole submitted some of his work. Its originality and style impressed Gregory favorably, and a cordial mathematical correspondence began a friendship which lasted out Boole's life.

It would take us too far afield to discuss here the great contribution which the British school was making at the time to the understanding of algebra as *algebra*, that is, as the abstract development of the consequences of a set of postulates without necessarily any interpretation or application to "numbers" or anything else, but it may be mentioned that the modern conception of algebra began with the British "reformers," Peacock, Herschel, De Morgan, Babbage, Gregory, and Boole. What was a somewhat heretical novelty when Peacock published his *Treatise on Algebra* in 1830 is today a commonplace in any competently written schoolbook. Once and for all Peacock broke away from the superstition that the $x, y, z, \ldots$ in such relations as $x + y = y + x$, $xy = yx$, $x(y + z) = xy + xz$, and so on, as we find them in elementary algebra, necessarily "represent numbers"; they do not, and that is one of the most important things about algebra and the source of its power in applications. The $x, y, z, \ldots$ are merely arbitrary marks, combined according to certain operations, one of which is symbolized as $+$, another by $\times$ (or simply as xy instead of $x \times y$), in accordance with postulates laid down at the beginning, like the specimens $x + y = y + x$, etc., above.

Without this realization that algebra is of itself nothing more than an abstract system, algebra might still have been stuck fast in the arithmetical mud of the eighteenth century, unable to move forward to its modern and extremely useful variants under the direction of Hamilton. We need only note here that this renovation of algebra gave Boole his first opportunity to do fine work appreciated by his contemporaries. Striking out on his own initiative he separated the *symbols* of mathematical *operations* from the things upon which they operate and proceeded to investigate these operations on their own account. How did they combine? Were they too subject to some sort of symbolic algebra? He found that they were. His work in this direction is extremely interesting, but it is overshadowed by the contribution which is peculiarly his own, the creation of a simple, workable system of symbolic or mathematical logic.

To introduce Boole's splendid invention properly we must digress slightly and recall a famous row of the first half of the nineteenth century, which raised a devil of a din in its own day but which is now almost forgotten except by historians of pathological philosophy. We mentioned Hamilton a moment ago. There were two Hamiltons of public fame at this time, one the Irish mathematician Sir William Rowan Hamilton (1805–1865), the other the Scotch philosopher Sir William Hamilton (1788–1856). Mathematicians usually refer to the philosopher as the *other* Hamilton. After a somewhat unsuccessful career as a Scotch barrister and candidate for official university positions the eloquent philosopher finally became Professor of Logic and Metaphysics in the University of Edinburgh. The mathematical Hamilton, as we have seen, was one of the outstanding original mathematicians of the nineteenth century. This is perhaps unfortunate for the *other* Hamilton, as the latter had no earthly use for mathematics, and hasty readers sometimes confuse the two famous Sir Williams. This causes the other one to turn and shiver in his grave.

Now, if there is anything more obtuse mathematically than a thickheaded Scotch metaphysician it is probably a mathematically thickerheaded German metaphysician. To surpass the ludicrous absurdity of some of the things the Scotch Hamilton said about mathematics we have to turn to what Hegel said about astronomy or Lotze about non-Euclidean geometry. Any depraved reader who wishes to fuddle himself can easily run down all he needs. It was the metaphysician Hamilton's misfortune to have been too dense or too lazy to get more than the most trivial smattering of elementary mathematics at school, but "omniscience was his foible," and when he began lecturing and writing on philosophy, he felt constrained to tell the world exactly how worthless mathematics is.

Hamilton's attack on mathematics is probably the most famous of all the many savage assaults mathematics has survived, undented. Less than ten years ago lengthy extracts from Hamilton's diatribe were vigorously applauded when a pedagogical enthusiast retailed them at a largely attended meeting of our own National Educational Association. Instead of applauding, the auditors might have got more out of the exhibition if they had paused to swallow some of Hamilton's philosophy as a sort of compulsory sauce for the proper enjoyment of his mathematical herring. To be fair to him we shall pass on a few of his hottest shots and let the reader make what use of them he pleases.

"Mathematics [Hamilton always used "mathematics" as a plural, not a singular, as customary today] freeze and parch the mind"; "an excessive study of mathematics absolutely incapacitates the mind for those intellectual energies which philosophy and life require"; "mathematics can not conduce to logical habits at all"; "in mathematics dullness is thus elevated into talent, and talent degraded into incapacity"; "mathematics may distort, but can never rectify, the mind."

This is only a handful of the birdshot; we have not room for the cannon balls. The whole attack is most impressive—for a man who knew far less mathematics than any intelligent child of ten knows. One last shot deserves special mention, as it introduces the figure of mathematical importance in the whole wordy war, De Morgan (1806–1871), one of the most expert controversialists who ever lived, a mathematician of vigorous independence, a great logician who prepared the way for Boole, the remorselessly good-humored enemy of all cranks, charlatans, and humbugs, and finally father of the famous novelist (*Alice for Short*, etc.). Hamilton remarks, "This [a perfectly nonsensical reason that need not be repeated] is why Mr. De Morgan among other mathematicians so often argues right. Still, had Mr. De Morgan been less of a Mathematician, he might have been more of a Philosopher; and be it remembered, that mathematics and dram-drinking tell especially, in the long run." Although the esoteric punctuation is obscure the meaning is clear enough. But it was not De Morgan who was given to tippling.

De Morgan, having gained some fame from his pioneering studies in logic, allowed himself in an absent-minded moment to be trapped into a controversy with Hamilton over the latter's famous principle of "the quantification of the predicate." There is no need to explain what this mystery is (or was); it is as dead as a coffin nail. De Morgan had made a real contribution to the syllogism; Hamilton thought he detected De Morgan's diamond in his own blue mud; the irate Scottish lawyer-philosopher publicly accused De Morgan of plagiarism—

an insanely unphilosophical thing to do—and the fight was on. On De Morgan's side, at least, the row was a hilarious frolic. De Morgan never lost his temper; Hamilton had never learned to keep his.

If this were merely one of the innumerable squabbles over priority which disfigure scientific history it would not be worth a passing mention. Its historical importance is that Boole by now (1848) was a firm friend and warm admirer of De Morgan. Boole was still teaching school, but he knew many of the leading British mathematicians personally or by correspondence. He now came to the aid of his friend—not that the witty De Morgan needed any mortal's aid, but because he knew that De Morgan was right and Hamilton wrong. So, in 1848, Boole published a slim volume, *The Mathematical Analysis of Logic*, his first public contribution to the vast subject which his work inaugurated and in which he was to win enduring fame for the boldness and perspicacity of his vision. The pamphlet—it was hardly more than that—excited De Morgan's warm admiration. Here was the master, and De Morgan hastened to recognize him. The booklet was only the promise of greater things to come six years later, but Boole had definitely broken new, stubborn ground.

In the meantime, reluctantly turning down his mathematical friends' advice that he proceed to Cambridge and take the orthodox mathematical training there, Boole went on with the drudgery of elementary teaching, without a complaint, because his parents were now wholly dependent upon his support. At last he got an opportunity where his conspicuous abilities as an investigator and a lecturer could have some play. He was appointed Professor of Mathematics at the recently opened Queen's College at what was then called the city of Cork, Ireland. This was in 1849.

Needless to say, the brilliant man who had known only poverty and hard work all his life made excellent use of his comparative freedom from financial worry and everlasting grind. His duties would now be considered onerous; Boole found them light by contrast with the dreary round of elementary teaching to which he had been accustomed. He produced much notable miscellaneous mathematical work, but his main effort went on licking his masterpeice into shape. In 1854 he published it: *An Investigation of the Laws of Thought, on which are founded the Mathematical Theories of Logic and Probabilities*. Boole was thirty nine when this appeared. It is somewhat unusual for a mathematician as old as that to produce work of such profound originality, but the phenomenon is accounted for when we remember the long, devious path Boole was compelled to follow before he could set his face fairly toward his goal. (Compare the careers of Boole and Weierstrass.)

A few extracts will give some idea of Boole's style and the scope of his work.

"The design of the following treatise is to investigate the fundamental laws of those operations of the mind by which reasoning is performed; to give expression to them in the language of a Calculus, and upon this foundation to establish the science of Logic and construct its method; to make that method itself the basis of a general method for the application of the mathematical doctrine of probabilities; and, finally, to collect from the various elements of truth brought to view in the course of these inquiries some probable intimations concerning the nature and constitution of the human mind. . . . "

"Shall we then err in regarding that as the true science of Logic which, laying down certain elementary laws, confirmed by the very testimony of the mind, permits us thence to deduce, by uniform processes, the entire chain of its secondary consequences, and furnishes, for its practical applications, methods of perfect generality? . . . "

"There exist, indeed, certain general principles founded in the very nature of language, by which the use of symbols, which are but the elements of scientific language, is determined. To a certain extent these elements are arbitrary. Their interpretation is purely conventional: we are permitted to employ them in whatever sense we please. But this permission is limited by two indispensable conditions, —first, that from the sense once conventionally established we never, in the same process of reasoning, depart; secondly, that the laws by which the process is conducted be founded exclusively upon the above fixed sense or meaning of the symbols employed. In accordance with these principles, any agreement which may be established between the laws of the symbols of Logic and those of Algebra can but issue in an agreement of processes. The two provinces of interpretation remain apart and independent, each subject to its own laws and conditions.

"Now the actual investigations of the following pages exhibit Logic, in its practical aspect, as a system of processes carried on by the aid of symbols having a definite interpretation, and subject to laws founded upon that interpretation alone. But at the same time they exhibit those laws as identical in form with the laws of the general symbols of Algebra, with this single addition, viz., that the symbols of Logic are further subject to a special law [$x^2 = x$ in the algebra of logic, which can be interpreted, among other ways, as "the class of all those things common to a class x and itself is merely the class x"], to which the symbols of quantity, as such, are not subject." (That is, in common algebra, it is not true that *every* x is equal to its square, whereas in the Boolean algebra of logic, this *is* true.)

This program is carried out in detail in the book. Boole reduced logic to an extremely easy and simple type of algebra. "Reasoning" upon appropriate material becomes in this algebra a matter of elementary manipulations of formulas far simpler than most of those handled in a second year of school algebra. Thus logic itself was brought under the sway of mathematics.

Since Boole's pioneering work his great invention has been modified, improved, generalized, and extended in many directions. Today symbolic or mathematical logic is indispensable in any serious attempt to understand the nature of mathematics and the state of its foundations on which the whole colossal superstructure rests. The intricacy and delicacy of the difficulties explored by the *symbolic* reasoning would, it is safe to say, defy human reason if only the old, pre-Boole methods of *verbal* logical arguments were at our disposal. The daring originality of Boole's whole project needs no signpost. It is a landmark in itself.

Since 1899, when Hilbert published his classic on the foundations of geometry, much attention has been given to the postulational formulation of the several branches of mathematics. This movement goes back as far as Euclid, but for some strange reason—possibly because the techniques invented by Descartes, Newton, Leibniz, Euler, Gauss, and others gave mathematicians plenty to do in developing their subject freely and somewhat uncritically—the Euclidean method was for long neglected in everything but geometry. We have already seen that the British school applied the method to algebra in the first half of the nineteenth century. Their successes seem to have made no very great impression on the work of their contemporaries and immediate successors, and it was only with the work of Hilbert that the postulational method came to be recognized as the clearest and most rigorous approach to any mathematical discipline.

Today this tendency to abstraction, in which the symbols and rules of operation in a particular subject are emptied of all meaning and discussed from a purely formal point of view, is all the rage, rather to the neglect of applications (practical or mathematical) which some say are the ultimate human justification for any scientific activity. Nevertheless the abstract method does give insights which looser attacks do not, and in particular the true simplicity of Boole's algebra of logic is most easily seen thus.

Accordingly we shall state the postulates for Boolean algebra (the algebra of logic) and, having done so, see that they can indeed be given an interpretation consistent with classical logic. The following set of postulates is taken from a paper by E. V. Huntington, in the *Transactions of the American Mathematical Society*, (vol. 35, 1933, pp. 274-304). The whole paper is easily understandable by anyone who has had a week of algebra, and may be found in most large public libraries. As Huntington points out, this first set of his which we transcribe is not as elegant as some of his others. But as its interpretation in terms of class inclusion as in formal logic is more immediate than the like for the others, it is to be preferred here.

The set of postulates is expressed in terms of K, $+$, $\times$, where K is a class of undefined (wholly arbitrary, without any assigned meaning or properties beyond those given in the postulates) elements a, b, c, . . . , and $a + b$ and $a \times b$ (written also simply as ab) are the results of two undefined binary operations, $+$, $\times$ ("binary," because each of $+$, $\times$ operates on *two* elements of K). There are ten postulates, I a–VI:

"I a. *If a and b are in the class K, then $a + b$ is in the class K.*

"I b. *If a and b are in the class K, then ab is in the class K.*

"II a. *There is an element Z such that $a + Z = a$ for every element a.*

"II b. *There is an element U such that $aU = a$ for every element a.*

"III a. $a + b = b + a$.

"III b. $ab = ba$.

"IV a. $a + bc = (a + b)(a + c)$.

"IV b. $a(b + c) = ab + ac$.

"V. *For every element a there is an element a' such that $a + a' = U$ and $aa' = Z$.*

"VI. *There are at least two distinct elements in the class K.*"

It will be readily seen that these postulates are satisfied by the following interpretation: a, b, c, . . . are *classes*; $a + b$ is the class of all those things that are in *at least one* of the classes a, b; ab is the class of all those things that are in *both* of the classes a, b; Z is the "null class" —the class that has no members; U is the "universal class"—the class that contains *all* the things in *all* the classes under discussion. Postulate V then states that given any class a, there is a class a' consisting of all those things which are not in a. Note that VI implies that U, Z are not the same class.

From such a simple and obvious set of statements it seems rather remarkable that the whole of classical logic can be built up symbolically by means of the easy algebra generated by the postulates. From these postulates a theory of what may be called "logical equations" is developed: problems in logic are translated into such equations, which are then "solved" by the devices of the algebra; the solution is then reinterpreted in terms of the logical data, giving the solution of the original problem. We shall close this description with the symbolic equivalent of "inclusion"—also interpretable, when *propositions* rather than *classes* are the elements of K, as "implication."

"The relation $a < b$ [read, a is included in b] is defined by any one of the following equations

$$a + b = b,\ ab = a,\ a' + b = U,\ ab' = Z."$$

To see that these are reasonable, consider for example the second, $ab = a$. This states that if a is included in b, then everything that is in *both* a and b is the whole of a.

From the stated postulates the following theorems on inclusion (with thousands of more complicated ones, if desired) can be *proved*. The specimens selected all agree with our intuitive conception of what "inclusion" means.

(1) $a < a$.

(2) *If* $a < b$ *and* $b < c$, *then* $a < c$.

(3) *If* $a < b$ *and* $b < a$, *then* $a = b$.

(4) $Z < a$ (*where Z is the element in* II a—it is proved to be the *only* element satisfying II a).

(5) $a < U$ (*where U is the element in* II b—likewise unique).

(6) $a < a + b$; *and if* $a < y$ *and* $b < y$, *then* $a + b < y$.

(7) $ab < a$; *and if* $x < a$ *and* $x < b$, *then* $x < ab$.

(8) *If* $x < a$ *and* $x < a'$, *then* $x = Z$; *and if* $a < y$ *and* $a' < y$, *then* $y = U$.

(9) *If* $a < b'$ *is false, then there is at least one element x, distinct from Z, such that* $x < a$ *and* $x < b$.

It may be of interest to observe that "<" in arithmetic and analysis is the symbol for "less than." Note that if $a, b, c, \ldots$ are real numbers, and Z denotes zero, then (2) is satisfied for this interpretation of "<," and similarly for (4), provided a is positive; but that (1) is not satisfied, nor is the second part of (6)—as we see from $5 < 10$, $7 < 10$, but $5 + 7 < 10$ is false.

The tremendous power and fluent ease of the method can be readily appreciated by seeing what it does in any work on symbolic logic. But, as already emphasized, the importance of this "symbolic reasoning" is in its applicability to subtle questions regarding the foundations of all mathematics which, were it not for this precise method of fixing meanings of "words" or other "symbols" once for all, would probably be unapproachable by ordinary mortals.

Like nearly all novelties, symbolic logic was neglected for many years after its invention. As late as 1910 we find eminent mathematicians scorning it as a "philosophical" curiosity without mathematical significance. The work of Whitehead and Russell in *Principia Mathematica* (1910–1913) was the first to convince any considerable body of professional mathematicians that symbolic logic might be worth their serious attention. One staunch hater of symbolic logic may be mentioned—Cantor, whose work on the infinite will be noticed in the concluding chapter. By one of those little ironies which make mathematical history such amusing reading for the open-minded, symbolic logic was to play an important part in the drastic criticism of Cantor's work that caused its author to lose faith in himself and his theory.

Boole did not long survive the production of his masterpiece. The year after its publication, still subconsciously striving for the social respectability that he once thought a knowledge of Greek could confer, he married Mary Everest, niece of the Professor of Greek in Queen's College. His wife became his devoted disciple. After her husband's death, Mary Boole applied some of the ideas which she had acquired from him to rationalizing and humanizing the education of young children. In her pamphlet, *Boole's Psychology*, Mary Boole records an interesting speculation of Boole's which readers of *The Laws of Thought* will recognize as in keeping with the unexpressed but implied personal philosophy in certain sections. Boole told his wife that in 1832, when he was about seventeen, it "flashed upon" him as he was walking across a field that besides the knowledge gained from direct observation, man derives knowledge from some source undefinable and invisible—which Mary Boole calls "the unconscious." It will be interesting (in a later chapter) to hear Poincaré expressing a similar opinion regarding the genesis of mathematical "inspirations" in the "subconscious mind." Anyhow, Boole was inspired, if ever a mortal was, when he wrote *The Laws of Thought*.

Boole died, honored and with a fast-growing fame, on December 8, 1864, in the fiftieth year of his age. His premature death was due to pneumonia contracted after faithfully keeping a lecture engagement when he was soaked to the skin. He fully realized that he had done great work.

CHAPTER TWENTY FOUR

The Man, Not the Method

HERMITE

Talk with M. Hermite: he never evokes a concrete image; yet you soon perceive that the most abstract entities are for him like living creatures.—HENRI POINCARÉ

OUTSTANDING UNSOLVED PROBLEMS demand new methods for their solution, while powerful new methods beget new problems to be solved. But, as Poincaré observed, it is the man, not the method, that solves a problem.

Of old problems responsible for new methods in mathematics that of motion and all it implies for mechanics, terrestrial and celestial, may be recalled as one of the principal instigators of the calculus and present attempts to put reasoning about the infinite on a firm basis. An example of new problems suggested by powerful new methods is the swarm which the tensor calculus, popularized to geometers by its successes in relativity, let loose in geometry. And finally, as an illustration of Poincaré's remark, it was Einstein, and not the method of tensors, that solved the problem of giving a coherent mathematical account of gravitation. All three theses are sustained in the life of Charles Hermite, the leading French mathematician of the second half of the nineteenth century—if we except Hermite's pupil Poincaré, who belonged partly to our own century.

Charles Hermite, born at Dieuze, Lorraine, France, on December 24, 1822, could hardly have chosen a more propitious era for his birth than the third decade of the nineteenth century. His was just the rare combination of creative genius and the ability to master the best in the work of other men which was demanded in the middle of the century to coordinate the arithmetical creations of Gauss with the discoveries of Abel and Jacobi in elliptic functions, the striking advances of Jacobi in Abelian functions, and the vast theory of algebraic invariants in process of rapid development by the English mathematicians Boole, Cayley, and Sylvester.

Hermite almost lost his life in the French Revolution—although the last head had fallen nearly a quarter of a century before he was born. His paternal grandfather was ruined by the Commune and died in prison; his grandfather's brother went to the guillotine. Hermite's father escaped owing to his youth.

If Hermite's mathematical ability was inherited, it probably came from the side of the father, who had studied engineering. Finding engineering uncongenial, Hermite senior gave it up, and after an equally distasteful start in the salt industry, finally settled down in business as a cloth merchant. This resting place was no doubt chosen by the rolling stone because he had married his employer's daughter, Madeleine Lallemand, a domineering woman who wore the breeches in her family and ran everything from the business to her husband. She succeeded in building both up to a state of solid bourgeois prosperity. Charles was the sixth of seven children—five sons and two daughters. He was born with a deformity of the right leg which rendered him lame for life—possibly a disguised blessing, as it effectively barred him from any career even remotely connected with the army—and he had to get about with a cane. His deformity never affected the uniform sweetness of his disposition.

Hermite's earliest education was received from his parents. As the business continued to prosper, the family moved from Dieuze to Nancy when Hermite was six. Presently the growing demands of the business absorbed all the time of the parents and Hermite was sent as a boarder to the *lycée* at Nancy. This school proving unsatisfactory the prosperous parents decided to give Charles the best and packed him off to Paris. There he studied for a short time at the Lycée Henri IV, moving on at the age of eighteen (1840) to the more famous (or infamous) Louis-le-Grand the "Alma" Mater of the wretched Galois—to prepare for the Polytechnique.

For a while it looked as if Hermite was to repeat the disaster of his untamable predecessor at Louis-le-Grand. He had the same dislike for rhetoric and the same indifference to the elementary mathematics of the classroom. But the competent lectures on physics fascinated him and won his cordial cooperation in the bilateral process of acquiring an education. Later on, unpestered by pedants, Hermite became a good classicist and the master of a beautifully clear prose.

Those who hate examinations will love Hermite. There is something in the careers of these two most famous alumni of Louis-le-

Grand, Galois and Hermite, which might well cause the advocates of examinations as a reliable yardstick for arranging human beings in order of intellectual merit to ask themselves whether they have used their heads or their feet in arriving at their conclusions. It was only by the grace of God and the diplomatic persistence of the devoted and intelligent Professor Richard, who had done his unavailing best fifteen years before to save Galois for science, that Hermite was not tossed out by stupid examiners to rot on the rubbish heap of failure. While still a student at the *lycée*, Hermite, following in the steps of Galois, supplemented and neglected his elementary lessons by private reading at the library of Sainte-Geneviève, where he found and mastered the memoir of Lagrange on the solution of numerical equations. Saving up his pennies, he bought the French translation of the *Disquisitiones Arithmeticae* of Gauss and, what is more, mastered it as few before or since have mastered it. By the time he had followed what Gauss had done Hermite was ready to *go on*. "It was in these two books," he loved to say in later life, "that I learned Algebra." Euler and Laplace also instructed him through their works. And yet Hermite's performance in examinations was, to say the most flattering thing possible of it, mediocre. Mathematical nonentities beat him out of sight.

Mindful of the tragic end of Galois, Richard tried his best to steer Hermite away from original investigation to the less exciting though muddier waters of the competitive examinations for entrance to the École Polytechnique—the filthy ditch in which Galois had drowned himself. Nevertheless the good Richard could not refrain from telling Hermite's father that Charles was "a young Lagrange."

The *Nouvelles Annales de Mathématiques*, a journal devoted to the interests of students in the higher schools, was founded in 1842. The first volume contains two papers composed by Hermite while he was still a student at Louis-le-Grand. The first is a simple exercise in the analytic geometry of conic sections and betrays no originality. The second, which fills only six and a half pages in Hermite's collected works, is a horse of quite a different color. Its unassuming title is *Considerations on the algebraic solution of the equation of the fifth degree* (translation).

"It is known," the modest mathematician of twenty begins, "that Lagrange made the algebraic solution of the general equation of the fifth degree depend on the determination of a root of a *particular* equation of the sixth degree, which he calls a *reduced equation* [today, a 'resolvent']. .... So that, if this resolvent were decomposable into rational factors of the second or third degrees, we should have the solution of the equation of the fifth degree. I shall try to show that such a decomposition is impossible." Hermite not only succeeded in his attempt—by a beautifully simple argument—but showed also in doing so that he was an algebraist. With but a few slight changes this short paper will do all that is required.

It may seem strange that a young man capable of genuine mathematical reasoning of the caliber shown by Hermite in his paper on the general quintic should find elementary mathematics difficult. But it is not necessary to understand—or even to have heard of—much of classical mathematics as it has evolved in the course of its long history in order to be able to follow or work creatively in the mathematics that has been developed since 1800 and is still of living interest to mathematicians. The geometrical treatment (synthetic) of conic sections of the Greeks, for instance, need not be mastered today by anyone who wishes to follow modern geometry; nor need any geometry at all be learned by one whose tastes are algebraic or arithmetical. To a lesser degree the same is true for analysis, where such geometrical language as is used is of the simplest and is neither necessary nor desirable if up-to-date proofs are the object. As a last example, descriptive geometry, of great use to designing engineers, is of practically no use whatever to a working mathematician. Some quite difficult subjects that are still mathematically alive require only a school education in algebra and a clear head for their comprehension. Such are the theory of finite groups, the mathematical theory of the infinite, and parts of the theory of probabilities and the higher arithmetic. So it is not astonishing that large tracts of what a candidate is required to know for entrance to a technical or scientific school, or even for graduation from the same, are less than worthless for a mathematical career. This accounts for Hermite's spectacular success as a budding mathematician and his narrow escape from complete disaster as an examinee.

Late in 1842, at the age of twenty, Hermite sat for the entrance examinations to the École Polytechnique. He passed, but only as sixty eighth in order of merit. Already he was a vastly better mathematician than some of the men who examined him were, or were ever to become. The humiliating outcome of this test made an impression on the young master which all the triumphs of his manhood never effaced.

Hermite stayed only one year at the Polytechnique. It was not his head that disqualified him but his lame foot which, according to a ruling of the authorities, unfitted him for any of the positions open to successful students of the school. Perhaps it is as well that Hermite was thrown out; he was an ardent patriot and might easily have been embroiled in one or other of the political or military rows so precious to the effervescent French temperament. However, the year was by no means wasted. Instead of slaving over descriptive geometry, which he hated, Hermite spent his time on Abelian functions, then (1842) perhaps the topic of outstanding interest and importance to the great mathematicians of Europe. He had also made the acquaintance of Joseph Liouville (1809-1882), a first-rate mathematician and editor of the *Journal des Mathématiques*.

Liouville recognized genius when he saw it. In passing it may be amusing to recall that Liouville inspired William Thomson, Lord Kelvin, the famous Scotch physicist, to one of the most satisfying definitions of a mathematician that has ever been given. "Do you know what a mathematician is?" Kelvin once asked a class. He stepped to the board and wrote

$$\int_{-\infty}^{+\infty} e^{-x^2} dx = \sqrt{\pi}.$$

Putting his finger on what he had written, he turned to the class. "A mathematician is one to whom *that* is as obvious as that twice two makes four is to you. Liouville was a mathematician." Young Hermite's pioneering work in Abelian functions, well begun before he was twenty one, was as far beyond Kelvin's example in unobviousness as the example is beyond "twice two makes four." Remembering the cordial welcome the aged Legendre had accorded the revolutionary work of the young and unknown Jacobi, Liouville guessed that Jacobi would show a similar generosity to the beginning Hermite. He was not mistaken.

The first of Hermite's astonishing letters to Jacobi is dated from Paris, January, 1843. "The study of your [Jacobi's] memoir on quadruply periodic functions arising in the theory of Abelian functions has led me to a theorem, for the division of the arguments [variables] of these functions, analogous to that which you gave . . . to obtain the simplest expression for the roots of the equations treated by Abel. M. Liouville induced me to write to you, to submit this work to you;

dare I hope, Sir, that you will be pleased to welcome it with all the indulgence it needs?" With that he plunges at once into the mathematics.

To recall briefly the bare nature of the problem in question: the trigonometric functions are functions of *one* variable with *one* period, thus $\sin(x + 2\pi) = \sin x$, where x is the variable and 2π is the period; Abel and Jacobi, by "inverting" the elliptic integrals, had discovered functions of *one* variable and *two* periods, say $f(x + p + q) = f(x)$, where p, q are the periods (see Chapters 12, 18); Jacobi had discovered functions of *two* variables and *four* periods, say

$$F(x + a + b, y + c + d) = F(x,y),$$

where a,b,c,d are the periods. A problem early encountered in trigonometry is to express $\sin\left(\dfrac{x}{2}\right)$, or $\sin\left(\dfrac{x}{3}\right)$, or generally $\sin\left(\dfrac{x}{n}\right)$, where n is any given integer, in terms of $\sin x$ (and possibly other trigonometric functions of x). The corresponding problem for the functions of two variables and four periods was that which Hermite attacked. In the trigonometric problem we are finally led to quite simple equations; in Hermite's incomparably more difficult problem the upshot is again an equation (of degree n^4), and the unexpected thing about this equation is that it can be solved algebraically, that is, by radicals.

Barred from the Polytechnique by his lameness, Hermite now cast longing eyes on the teaching profession as a haven where he might earn his living while advancing his beloved mathematics. The career should have been flung wide open to him, degree or no degree, but the inexorable rules and regulations made no exceptions. Red tape always hangs the wrong man, and it nearly strangled Hermite.

Unable to break himself of his "pernicious originality," Hermite continued his researches to the last possible moment when, at the age of twenty four, he abandoned the fundamental discoveries he was making to master the trivialities required for his first degrees (bachelor of letters and science). Two harder ordeals would normally have followed the first before the young mathematical genius could be certified as fit to teach, but fortunately Hermite escaped the last and worst when influential friends got him appointed to a position where he could mock the examiners. He passed his examinations (in 1847–48) very badly. But for the friendliness of two of the inquisitors—Sturm and Bertrand, both fine mathematicians who recognized a fellow craftsman when they saw one—Hermite would probably not have passed at all. (Hermite married Bertrand's sister Louise in 1848.)

By an ironic twist of fate Hermite's first academic success was his appointment in 1848 as an examiner for admissions to the very Polytechnique which had almost failed to admit him. A few months later he was appointed quiz master (*répétiteur*) at the same institution. He was now securely established in a niche where no examiner could get at him. But to reach this "bad eminence" he had sacrificed nearly five years of what almost certainly was his most inventive period to propitiate the stupidities of the official system.

Having finally satisfied or evaded his rapacious examiners, Hermite settled down to become a great mathematician. His life was peaceful and uneventful. In 1848 to 1850 he substituted for Libri at the Collège de France. Six years later, at the early age of thirty four, he was elected to the Institut (as a member of the Academy of Sciences). In spite of his world-wide reputation as a creative mathematician Hermite was forty seven before he obtained a suitable position: he was appointed professor only in 1869 at the École Normale and finally, in 1870, he became professor at the Sorbonne, a position which he held till his retirement twenty seven years later. During his tenure of this influential position he trained a whole generation of distinguished French mathematicians, among whom Émile Picard, Gaston Darboux, Paul Appell, Émile Borel, Paul Painlevé and Henri Poincaré, may be mentioned. But his influence extended far beyond France, and his classic works helped to educate his contemporaries in all lands.

A distinguishing feature of Hermite's beautiful work is closely allied to his repugnance to take advantage of his authoritative position to re-create all his pupils in his own image: this is the unstinted generosity which he invariably displays to his fellow mathematicians. Probably no other mathematician of modern times has carried on such a voluminous scientific correspondence with workers all over Europe as Hermite, and the tone of his letters is always kindly, encouraging, and appreciative. Many a mathematician of the second half of the nineteenth century owed his recognition to the publicity which Hermite gave his first efforts. In this, as in other respects, there is no finer character than Hermite in the whole history of mathematics. Jacobi was as generous—with the one exception of his early treatment of

Eisenstein—but he had a tendency to sarcasm (often highly amusing, except possibly to the unhappy victim) which was wholly absent from Hermite's genial wit. Such a man deserved the generous reply of Jacobi when the unknown young mathematician ventured to approach him with his first great work on Abelian functions. "Do not be put out, Sir," Jacobi wrote, "if some of your discoveries coincide with old work of my own. As you must begin where I end, there is necessarily a small sphere of contact. In future, if you honor me with your communications, I shall have only to learn."

Encouraged by Jacobi, Hermite shared with him not only the discoveries in Abelian functions, but also sent him four tremendous letters on the theory of numbers, the first early in 1847. These letters, the first of which was composed when Hermite was only twenty four, break new ground (in what respect we shall indicate presently) and are sufficient alone to establish Hermite as a creative mathematician of the first rank. The generality of the problems he attacked and the bold originality of the methods he devised for their solution assure Hermite's remembrance as one of the born arithmeticians of history.

The first letter opens with an apology. "Nearly two years have elapsed without my answering the letter full of goodwill which you did me the honor to write to me. Today I shall beg you to pardon my long negligence and express to you all the joy I felt in seeing myself given a place in the repertory of your works. [Jacobi has published parts of Hermite's letter, with all due acknowledgment, in some work of his own.] Having been for long away from the work, I was greatly touched by such an attestation of your kindness; allow me, Sir, to believe that it will not desert me." Hermite then says that another research of Jacobi's has inspired him to his present efforts.

If the reader will glance at what was said about *uniform* functions of a single variable in the chapter on Gauss (a uniform function takes *only one* value for each value of the variable), the following statement of what Jacobi had proved should be intelligible: a *uniform* function of only *one* variable with *three* distinct periods is impossible. That uniform functions of *one* variable exist having either *one* period or *two* periods is proved by exhibiting the trigonometric functions and the elliptic functions. This theorem of Jacobi's, Hermite declares, gave him his own idea for the novel methods which he introduced into the higher arithmetic. Although these methods are too technical for description here, the spirit of one of them can be briefly indicated.

Arithmetic in the sense of Gauss deals with properties of the rational integers 1,2,3, ... ; irrationals (like the square root of 2) are excluded. In particular Gauss investigated the integer solutions of large classes of indeterminate equations in two or three unknowns, for example as in $ax^2 + 2bxy + cy^2 = m$, where a,b,c,m are any given integers and it is required to discuss all integer solutions x, y of the equation. The point to be noted here is that the problem is stated and is to be solved entirely in the domain of the rational integers, that is, in the realm of *discrete* number. To fit *analysis*, which is adapted to the investigation of *continuous* number, to such a *discrete* problem would seem to be an impossibility, yet this is what Hermite did. Starting with a *discrete* formulation, he applied *analysis* to the problem, and in the end came out with results in the discrete domain from which he had started. As analysis is far more highly developed than any of the discrete techniques invented for algebra and arithmetic, Hermite's advance was comparable to the introduction of modern machinery into a medieval handicraft.

Hermite had at his disposal much more powerful machinery, both algebraic and analytic, than any available to Gauss when he wrote the *Disquisitiones Arithmeticae*. With Hermite's own great invention these more modern tools enabled him to attack problems which would have baffled Gauss in 1800. At one stride Hermite caught up with *general* problems of the type which Gauss and Eisenstein had discussed, and he at least began the arithmetical study of quadratic forms in any number of unknowns. The general nature of the arithmetical "theory of forms" can be seen from the statement of a special problem. Instead of the Gaussian equation $ax^2 + 2bxy + cy^2 = m$ of degree *two* in *two* unknowns (x,y), it is required to discuss the integer solutions of similar equations of degree n in s unknowns, where n, s are *any* integers, and the degree of each term on the left of the equation is n (not 2 as in Gauss' equation). After stating how he had seen after much thought that Jacobi's researches on the periodicity of uniform functions depend upon deeper questions in the theory of quadratic forms, Hermite outlines his programs.

"But, having once arrived at this point of view, the problems— vast enough—which I had thought to propose to myself, seemed inconsiderable beside the great questions of the general theory of forms. In this boundless expanse of researches which Monsieur Gauss [Gauss was still living when Hermite wrote this, hence the polite "Monsieur"] has opened up to us, Algebra and the Theory of Numbers seem necessarily to be merged in the same order of analytical concepts, of which our present knowledge does not yet permit us to form an accurate idea."

He then makes a remark which, although not very clear, can be interpreted as meaning that the key to the subtle connections between algebra, the higher arithmetic, and certain parts of the theory of functions will be found in a thorough understanding of *what sort* of "numbers" are both necessary and sufficient for the explicit solution of all types of algebraic equations. Thus, for $x^3 - 1 = 0$, it is necessary and sufficient to understand $\sqrt[3]{1}$; for $x^5 + ax + b = 0$, where a,b are any given numbers, what sort of a "number" x must be invented in order that x may be expressed *explicitly* in terms of a,b? Gauss of course gave one kind of answer: any root x is a complex number. But this is only a beginning. Abel proved that if only a *finite* number of rational operations and extractions of roots are permitted, then there is *no* explicit formula giving x in terms of a,b. We shall return to this question later; Hermite even at this early date (1848; he was then twenty six) seems to have had one of his greatest discoveries somewhere at the back of his head.

In his attitude toward numbers Hermite was somewhat of a mystic in the tradition of Pythagoras and Descartes—the latter's mathematical creed, as will appear in a moment, was essentially Pythagorean. In other matters, too, the gentle Hermite exhibited a marked leaning toward mysticism. Up to the age of forty three he was a tolerant agnostic, like so many French men of science of his time. Then, in 1856, he fell suddenly and dangerously ill. In this debilitated condition he was no match for even the least persistent evangelist, and the ardent Cauchy, who had always deplored his brilliant young friend's open-mindedness on religious matters, pounced on the prostrate Hermite and converted him to Roman Catholicism. Thenceforth Hermite was a devout Catholic, and the practice of his religion gave him much satisfaction.

Hermite's number-mysticism is harmless enough and it is one of those personal things on which argument is futile. Briefly, Hermite believed that numbers have an existence of their own above all control by human beings. Mathematicians, he thought, are permitted now and then to catch glimpses of the superhuman harmonies regulating this ethereal realm of numerical existence, just as the great geniuses of ethics and morals have sometimes claimed to have visioned the celestial perfections of the Kingdom of Heaven.

It is probably right to say that no reputable mathematician today who has paid any attention to what has been done in the past fifty years (especially the last twenty five) in attempting to understand the nature of mathematics and the processes of mathematical reasoning would agree with the mystical Hermite. Whether this modern skepticism regarding the other-worldliness of mathematics is a gain or a loss over Hermite's creed must be left to the taste of the reader. What is now almost universally held by competent judges to be the wrong view of "mathematical existence" was so admirably expressed by Descartes in his theory of the eternal triangle that it may be quoted here as an epitome of Hermite's mystical beliefs.

"I imagine a triangle, although perhaps such a figure does not exist and never has existed anywhere in the world outside my thought. Nevertheless this figure has a certain nature, or form, or determinate essence which is immutable or eternal, which I have not invented and which in no way depends on my mind. This is evident from the fact that I can demonstrate various properties of this triangle, for example that the sum of its three interior angles is equal to two right angles, that the greatest angle is opposite the greatest side, and so forth. Whether I desire to or not, I recognize very clearly and convincingly that these properties are in the triangle although I have never thought about them before, and even if this is the first time I have imagined a triangle. Nevertheless no one can say that I have invented or imagined them." Transposed to such simple "eternal verities" as $1 + 2 = 3$, $2 + 2 = 4$, Descartes' everlasting geometry becomes Hermite's superhuman arithmetic.

One arithmetical investigation of Hermite's, although rather technical, may be mentioned here as an example of the prophetic aspect of pure mathematics. Gauss, we recall, introduced *complex integers* (numbers of the form $a + bi$, where a, b are rational integers and i denotes $\sqrt{-1}$) into the higher arithmetic in order to give the law of biquadratic reciprocity its simplest expression. Dirichlet and other followers of Gauss then discussed quadratic forms in which the rational integers appearing as variables and coefficients are replaced by Gaussian complex integers. Hermite passed to the general case of this situation and investigated the representation of integers in what are today called *Hermitian forms*. An example of such a form (for the special case of two complex variables x_1, x_2 and their "conjugates" $\bar{x}_1$, $\bar{x}_2$ instead of n variables) is

$$a_{11}x_1\bar{x}_1 + a_{12}x_1\bar{x}_2 + a_{21}x_2\bar{x}_1 + a_{22}x_2\bar{x}_2,$$

in which the bar over a letter denoting a complex number indicates the *conjugate* of that number; namely, if $x + iy$ is the complex number, its "conjugate" is $x - iy$; and the coefficients a_{11}, a_{12}, a_{21}, a_{22} are such that $a_{ij} = \bar{a}_{ji}$, for $(i,j) = (1,1), (1,2), (2,1), (2,2)$, so that a_{12} and a_{21} are conjugates, and each of a_{11}, a_{22} is its own conjugate (so that a_{11}, a_{22} are real numbers). It is easily seen that the entire form is real (free of i) if all products are multiplied out, but it is most "naturally" discussed in the shape given.

When Hermite invented such forms he was interested in finding what numbers are represented by the forms. Over seventy years later it was found that the algebra of Hermitian forms is indispensable in mathematical physics, particularly in the modern quantum theory. Hermite had no idea that his pure mathematics would prove valuable in science long after his death—indeed, like Archimedes, he never seemed to care much for the scientific applications of mathematics. But the fact that Hermite's work has given physics a useful tool is perhaps another argument favoring the side that believes mathematicians best justify their abstract existence when left to their own inscrutable devices.

Leaving aside Hermite's splendid discoveries in the theory of algebraic invariants as too technical for discussion here, we shall pass on in a moment to two of his most spectacular achievements in other fields. The high esteem in which Hermite's work in invariants was held by his contemporaries may however be indicated by Sylvester's characteristic remark that "Cayley, Hermite, and I constitute an Invariantive Trinity." Who was who in this astounding trinity Sylvester omitted to state; but perhaps this oversight is immaterial, as each member of such a trefoil would be capable of transforming himself into himself or into either of his coinvariantive beings.

The two fields in which Hermite found what are perhaps the most striking individual results in all his beautiful work are those of the general equation of the fifth degree and transcendental numbers. The nature of what he found in the first is clearly indicated in the introduction to his short note *Sur la résolution de l'équation du cinquième degré*

(On the Solution of the [general] Equation of the Fifth Degree; published in the *Comptes rendus de l'Académie des Sciences* for 1858, when Hermite was thirty six).

"It is known that the general equation of the fifth degree can be reduced, by a substitution [on the unknown x] whose coefficients are determined without using any irrationalities other than square roots or cube roots, to the form

$$x^5 - x - a = 0.$$

[That is, *if* we can solve *this* equation for x, *then* we can solve the general equation of the fifth degree.]

"This remarkable result, due to the English mathematician Jerrard, is the most important step that has been taken in the algebraic theory of equations of the fifth degree since Abel proved that a solution by radicals is impossible. This impossibility shows in fact the necessity for introducing some new analytic element [some new kind of function] in seeking the solution, and, on this account, it seems natural to take as an auxiliary the roots of the very simple equation we have just mentioned. Nevertheless, in order to legitimize its use rigorously as an essential element in the solution of the general equation, it remains to see if this simplicity of form actually permits us to arrive at some idea of the nature of its roots, to grasp what is peculiar and essential in the mode of existence of these quantities, of which nothing is known beyond the fact that they are not expressible by radicals.

"Now it is very remarkable that Jerrard's equation lends itself with the greatest ease to this research, and is, in the sense which we shall explain, susceptible of an actual analytic solution. For we may indeed conceive the question of the algebraic solution of equations from a point of view different from that which for long has been indicated by the solution of equations of the first four degrees, and to which we are especially committed.

"Instead of expressing the closely interconnected system of roots, considered as functions of the coefficients, by a formula involving many-valued radicals,* we may seek to obtain the roots expressed

---

* For example, as in the simple quadratic $x^2 - a = 0$: the roots are $x = + \sqrt{a}$, and $x = - \sqrt{a}$; the "many-valuedness" of the radical involved, here a square root, or irrationality of the *second* degree, appears in the double sign, $\pm$, when we say briefly that the *two* roots are $\sqrt{a}$. The formula giving the *three* roots of cubic equations involves the three-valued irrationality $\sqrt[3]{1}$, which has the *three* values $1, \frac{1}{2}(-1 + \sqrt{-3}), \frac{1}{2}(-1 - \sqrt{-3})$.

separately by as many distinct uniform [one-valued] functions of auxiliary variables, as in the case of the third degree. In this case, where the equation

$$x^3 - 3x + 2a = 0$$

is under discussion, it suffices, as we know, to represent the coefficient a by the sine of an angle, say A, in order that the roots be isolated as the following well-determined functions

$$2 \sin \frac{A}{3}, \; 2 \sin \frac{a + 2\pi}{3}, \; 2 \sin \frac{A + 4\pi}{3}.$$

[Hermite is here recalling the familiar "trigonometric solution" of the cubic usually discussed in the second course of school algebra. The "auxiliary variable" is A; the "uniform functions" are here sines.]

"Now it is an entirely similar fact which we have to exhibit concerning the equation

$$x^5 - x - a = 0.$$

Only, instead of sines or cosines, it is the elliptic functions which it is necessary to introduce. . . ."

In short order Hermite then proceeds to solve *the general equation of the fifth degree*, using for the purpose elliptic functions (strictly, elliptic modular functions, but the distinction is of no importance here). It is almost impossible to convey to a nonmathematician the spectacular brilliance of such a feat; to give a very inadequate simile, Hermite found the famous "lost chord" when no mortal had the slightest suspicion that such an elusive thing existed anywhere in time and space. Needless to say his totally unforeseen success created a sensation in the mathematical world. Better, it inaugurated a new department of algebra and analysis in which the grand problem is to discover and investigate those functions in terms of which the general equation of the nth degree can be solved explicitly in finite form. The best result so far obtained is that of Hermite's pupil, Poincaré (in the 1880's), who created the functions giving the required solution. These turned out to be a "natural" generalization of the elliptic functions. The characteristic of those functions that was generalized was periodicity. Further details would take us too far afield here, but if there is space we shall recur to this point when we reach Poincaré.

Hermite's other sensational isolated result was that which established the *transcendence* (explained in a moment) of the number denoted in mathematical analysis by the letter e, namely

$$1 + \frac{1}{1!} + \frac{1}{2!} + \frac{1}{3!} + \frac{1}{4!} + \ldots,$$

where 1! means 1, 2! = 1×2, 3! = $1 \times 2 \times 3$, 4! = $1 \times 2 \times 3 \times 4$, and so on; this number is the "base" of the so-called "natural" system of logarithms, and is approximately 2.718281828. . . . It has been said that it is impossible to conceive of a universe in which e and π (the ratio of the circumference of a circle to its diameter) are lacking. However that may be (as a matter of fact it is false), it is a fact that e turns up everywhere in current mathematics, pure and applied. Why this should be so, at least so far as applied mathematics is concerned, may be inferred from the following fact: e^x, considered as a function of x, is the *only* function of x whose rate of change with respect to x is equal to the function itself—that is, e^x is the only function which is equal to its derivative.*

The concept of "transcendence" is extremely simple, also extremely important. Any root of an algebraic equation whose coefficients are rational integers $(0, \pm1, \pm2, \ldots)$ is called an *algebraic number*. Thus $\sqrt{-1}$, 2.78 are algebraic numbers, because they are roots of the respective algebraic equations $x^2 + 1 = 0$, $50x - 139 = 0$, in which the coefficients (1, 1 for the first, 50, —139 for the second) are rational integers. A "number" which is *not* algebraic is called transcendental. Otherwise expressed, a transcendental number is one which satisfies *no* algebraic equation with rational integer coefficients.

Now, given any "number" constructed according to some definite law, it is a meaningful question to ask whether it is algebraic or transcendental. Consider, for example, the following simply defined number,

$$\frac{1}{10} + \frac{1}{10^2} + \frac{1}{10^6} + \frac{1}{10^{24}} + \frac{1}{10^{120}} + \ldots,$$

in which the exponents 2, 6, 24, 120, . . . are the successive "factorials," namely $2 = 1 \times 2$, $6 = 1 \times 2 \times 3$, $24 = 1 \times 2 \times 3 \times 4$, $120 = 1 \times 2 \times 3 \times 4 \times 5$, . . . , and the indicated series continues "to infinity" according to the same law as that for the terms given.

---

* Strictly, ae^x, where a does not depend upon x, is the most general, but the "multiplicative constant" a is trivial here.

The next term is $\frac{1}{10^{720}}$; the sum of the first three terms is $.1 + .01 + .000001$, or $.110001$, and it can be proved that the series does actually define some definite number which is less than $.12$. Is this number a root of *any* algebraic equation with rational integer coefficients? The answer is no, although to prove this without having been shown how to go about it is a severe test of high mathematical ability. On the other hand, the number defined by the infinite series

$$\frac{1}{10^5} + \frac{1}{10^8} + \frac{1}{10^{11}} + \frac{1}{10^{14}} + \cdots$$

is algebraic; it is the root of $99900\ x - 1 = 0$ (as may be verified by the reader who remembers how to sum an infinite convergent geometrical progression).

The first to prove that certain numbers are transcendental was Joseph Liouville (the same man who encouraged Hermite to write to Jacobi) who, in 1844, discovered a very extensive class of transcendental numbers, of which all those of the form

$$\frac{1}{n} + \frac{1}{n^2} + \frac{1}{n^6} + \frac{1}{n^{24}} + \frac{1}{n^{120}} + \cdots,$$

where n is a real number greater than 1 (the example given above corresponds to $n = 10$), are among the simplest. But it is probably a much more difficult problem to prove that *a particular* suspect, like e or π, is or is not transcendental than it is to invent a whole infinite class of transcendentals: the inventive mathematician dictates—to a certain extent—the working conditions, while the suspected number is entire master of the situation, and it is the mathematician in this case, not the suspect, who takes orders which he only dimly understands. So when Hermite proved in 1873 that e (defined a short way back) is transcendental, the mathematical world was not only delighted but astonished at the marvellous ingenuity of the proof.

Since Hermite's time many numbers (and classes of numbers) have been proved transcendental. What is likely to remain a high-water mark on the shores of this dark sea for some time may be noted in passing. In 1934 the young Russian mathematician Alexis Gelfond proved that *all* numbers of the type a^b, where a is neither 0 nor 1 and b is *any irrational algebraic number*, are transcendental. This disposes of the seventh of David Hilbert's list of twenty three outstanding mathematical problems which he called to the attention of mathematicians at the Paris International Congress in 1900. Note that "irrational" is *necessary* in the statement of Gelfond's theorem (if $b = n/m$, where n, m are rational integers, then a^b, where a is any algebraic number, is a root of $x^m - a^n = 0$, and it can be shown that this equation is equivalent to one in which all the coefficients are rational integers.

Hermite's unexpected victory over the obstinate e inspired mathematicians to hope that π would presently be subdued in a similar manner. For himself, however, Hermite had had enough of a good thing. "I shall risk nothing," he wrote to Borchardt, "on an attempt to prove the transcendence of the number π. If others undertake this enterprise, no one will be happier than I at their success, but believe me, my dear friend, it will not fail to cost them some efforts." Nine years later (in 1882) Ferdinand Lindemann of the University of Munich, using methods very similar to those which had sufficed Hermite to dispose of e, proved that π is transcendental, thus settling forever the problem of "squaring the circle." From what Lindemann proved it follows that it is impossible with straightedge and compass alone to construct a square whose area is equal to that of any given circle—a problem which had tormented generations of mathematicians since before the time of Euclid.

As cranks are still tormented by the problem, it may be in order to state concisely how Lindemann's proof settles the matter. He proved that π is *not* an *algebraic* number. But any *geometrical* problem that *is* solvable by the aid of straightedge and compass alone, when *restated* in its equivalent *algebraic* form, leads to one or more algebraic equations with rational integer coefficients which can be solved by successive extractions of *square roots*. As π satisfies no such equation, the circle cannot be "squared" with the implements named. If other mechanical apparatus is permitted, it is easy to square the circle. To all but mild lunatics the problem has been completely dead for over half a century. Nor is there any merit at the present time in computing π to a large number of decimal places—more accuracy in this respect is already available than is ever likely to be of use to the human race if it survives for a billion to the billionth power years. Instead of trying to do the impossible, mystics may like to contemplate the following useful relation between e, π, -1 and $\sqrt{-1}$ till it becomes as plain to them as Buddha's navel is to a blind Hindu swami,

$$e^{\pi\sqrt{-1}} = -1.$$

Anyone who can perceive this mystery intuitively will not need to square the circle.

Since Lindemann settled π the one outstanding unsolved problem that attracts amateurs is Fermat's "Last Theorem." Here an amateur with real genius undoubtedly has a chance. Lest this be taken as an invitation to all and sundry to swamp the editors of mathematical journals with attempted proofs, we recall what happened to Lindemann when he boldly tackled the famous theorem. If this does not suggest that more than ordinary talent will be required to settle Fermat, nothing can. In 1901 Lindemann published a memoir of seventeen pages purporting to contain the long-sought proof. The vitiating error being pointed out, Lindemann, undaunted, spent the best part of the next seven years in attempting to patch the unpatchable, and in 1907 published sixty three pages of alleged proof which were rendered nonsensical by a slip in reasoning near the very beginning.

Great as were Hermite's contributions to the technical side of mathematics, his steadfast adherence to the ideal that science is beyond nations and above the power of creeds to dominate or to stultify was perhaps an even more significant gift to civilization in the long view of things as they now appear to a harassed humanity. We can only look back on his serene beauty of spirit with a poignant regret that its like is nowhere to be found in the world of science today. Even when the arrogant Prussians were humiliating Paris in the Franco-Prussian war, Hermite, patriot though he was, kept his head, and he saw clearly that the mathematics of "the enemy" was mathematics and nothing else. Today, even when a man of science does take the civilized point of view, he is not impersonal about his supposed broad-mindedness, but aggressive, as befits a man on the defensive. To Hermite it was so obvious that knowledge and wisdom are not the prerogatives of any sect, any creed, or any nation that he never bothered to put his instinctive sanity into words. In respect of what Hermite knew by instinct our generation is two centuries behind him. He died, loved the world over, on January 14, 1901.

CHAPTER TWENTY FIVE

The Doubter

KRONECKER

All results of the profoundest mathematical investigation must ultimately be expressible in the simple form of properties of the integers.—LEOPOLD KRONECKER

PROFESSIONAL MATHEMATICIANS who could properly be called business men are extremely rare. The one who most closely approximates to this ideal is Kronecker (1823–1891), who did so well for himself by the time he was thirty that thereafter he was enabled to devote his superb talents to mathematics in considerably greater comfort than most mathematicians can afford.

The obverse of Kronecker's career is to be found—according to a tradition familiar to American mathematicians—in the exploits of John Pierpont Morgan, founder of the banking house of Morgan and Company. If there is anything in this tradition, Morgan as a student in Germany showed such extraordinary mathematical ability that his professors tried to induce him to follow mathematics as his life work and even offered him a university position in Germany which would have sent him off to a flying start. Morgan declined and dedicated his gifts to finance, with results familiar to all. Speculators (in academic studies, not Wall Street) may amuse themselves by reconstructing world history on the hypothesis that Morgan had stuck to mathematics.

What might have happened to Germany had Kronecker not abandoned finance for mathematics also offers a wide field for speculation. His business abilities were of a high order; he was an ardent patriot with an uncanny insight into European diplomacy and a shrewd cynicism—his admirers called it realism—regarding the unexpressed sentiments cherished by the great Powers for one another.

At first a liberal like so many intellectual young Jews, Kronecker quickly became a rock-ribbed conservative when he saw which side his own abundant bread was buttered on—after his financial exploits, and proclaimed himself a loyal supporter of that callous old truth-doctor Bismarck. The famous episode of the Ems telegram which, according to some, was the electric spark that touched off the Franco-Prussian war in 1870, had Kronecker's warm approval, and his grasp of the situation was so firm that *before* the battle of Weissenburg, when even the military geniuses of Germany were doubtful as to the outcome of their bold challenging of France, Kronecker confidently predicted the success of the entire campaign and was proved right in detail. At the time, and indeed all his life, he was on cordial terms with the leading French mathematicians, and he was clear-headed enough not to let his political opinions cloud his just perception of his scientific rivals' merits. It is perhaps as well that so realistic a man as Kronecker cast his lot with mathematics.

Leopold Kronecker's life was easy from the day of his birth. The son of prosperous Jewish parents, he was born on December 7, 1823, at Liegnitz, Prussia. By an unaccountable oversight Kronecker's official biographers (Heinrich Weber and Adolf Kneser) omit all mention of Leopold's mother, although he probably had one, and concentrate on the father, who owned a flourishing mercantile business. The father was a well educated man with an unquenchable thirst for philosophy which he passed on to Leopold. There was another son, Hugo, seventeen years younger than Leopold, who became a distinguished physiologist and professor at Berne. Leopold's early education under a private tutor was supervised by the father; Hugo's upbringing later became the loving duty of Leopold.

In the second stage of his education at the preparatory school for the Gymnasium Leopold was strongly influenced by the co-rector Werner, a man with philosophical and theological leanings, who later taught Kronecker when he entered the Gymnasium. Among other things Kronecker imbibed from Werner was a liberal draught of Christian theology, for which he acquired a lifelong enthusiasm. With what looks like his usual caution, Kronecker did not embrace the Christian faith till practically on his deathbed when, having seen that it did his six children no noticeable mischief, he permitted himself to be converted from Judaism to evangelical Christianity in his sixty eighth year.

Another of Kronecker's teachers at the Gymnasium also influenced him profoundly and became his lifelong friend, Ernst Eduard Kummer (1810–1893), subsequently professor at the University of Berlin and one of the most original mathematicians Germany has produced, of whom more will be said in connection with Dedekind. These three, Kronecker senior, Werner, and Kummer, capitalized Leopold's immense native abilities, formed his mind, and charted the future course of his life so cunningly that he could not have departed from it if he had wished.

Already in this early stage of his education we note an outstanding feature of Kronecker's genial character, his ability to get along with people and his instinct for forming lasting friendships with men who had risen in the world or were to rise, and who would be useful to him either in business or mathematics. This genius for friendships of the right sort, which is one of the successful business man's distinguishing traits, was one of Kronecker's more valuable assets and he never mislaid it. He was not consciously mercenary, nor was he a snob; he was merely one of those lucky mortals who is more at ease with the successful than with the unsuccessful.

Kronecker's performance at school was uniformly brilliant and many-sided. In addition to the Greek and Latin classics which he mastered with ease and for which he retained a lifelong liking, he shone in Hebrew, philosophy, and mathematics. His mathematical talent appeared early under the expert guidance of Kummer, from whom he received special instruction. Young Kronecker however did not concentrate to any great extent on mathematics, although it was obvious that his greatest talent lay in that field, but set himself to acquiring a broad liberal education commensurate with his manifold abilities. In addition to his formal studies he took music lessons and became an accomplished pianist and vocalist. Music, he declared when he was an old man, is the finest of all the fine arts, with the possible exception of mathematics, which he likened to poetry. These many interests he retained throughout his life. In none of them was he a mere dabbler: his love of the classics of antiquity bore tangible fruit in his affiliation with Graeca, a society dedicated to the translation and popularization of the Greek classics; his keen appreciation of art made him an acute critic of painting and sculpture, and his beautiful house in Berlin became a rendezvous for musicians, among them Felix Mendelssohn.

Entering the University of Berlin in the spring of 1841, Kronecker continued his broad education but began to concentrate on mathematics. Berlin at that time boasted Dirichlet (1805–1859), Jacobi (1804–1851) and Steiner (1796–1863) on its mathematical faculty; Eisenstein (1823-1852), the same age as Kronecker, also was about, and the two became friends.

The influence of Dirichlet on Kronecker's mathematical tastes (particularly in the application of analysis to the theory of numbers) is clear all through his mature writings. Steiner seems to have made no impression on him; Kronecker had no feeling for geometry. Jacobi gave him a taste for elliptic functions which he was to cultivate with striking originality and brilliant success, chiefly in novel applications of magical beauty to the theory of numbers.

Kronecker's university career was a repetition on a larger scale of his years at school: he attended lectures on the classics and the sciences and indulged his bent for philosophy by profounder studies than any he had as yet undertaken, particularly in the system of Hegel. The last is emphasized because some curious and competent reader may be moved to seek the origin of Kronecker's mathematical heresies in the abstrusities of Hegel's dialectic—a quest wholly beyond the powers of the present writer. Nevertheless there is a strange similarity between some of the weird unorthodoxies of recent doubts concerning the self-consistency of mathematics—doubts for which Kronecker's "revolution" was partly responsible—and the subtleties of Hegel's system. The ideal candidate for such an undertaking would be a Marxian communist with a sound training in Polish many-valued logic, though in what incense tree this rare bird is to be sought God only knows.

Following the usual custom of German students, Kronecker did not spend all his time at Berlin but moved about. Part of his course was pursued at the University of Bonn, where his old teacher and friend Kummer had taken the chair of mathematics. During Kronecker's residence at Bonn the University authorities were in the midst of a futile war to suppress the student societies whose chief object was the fostering of drinking, duelling, and brawling in general. With his customary astuteness, Kronecker allied himself secretly with the students and thereby made many friends who were later to prove useful.

Kronecker's dissertation, accepted by Berlin for his Ph.D. in 1845, was inspired by Kummer's work in the theory of numbers and dealt with the units in certain algebraic number fields. Although the problem is one of extreme difficulty when it comes to actually exhibiting the units, its nature can be understood from the following rough description of the *general* problem of units (for *any* algebraic number field, not merely for the *special* fields which interested Kummer and Kronecker). This sketch may also serve to make more intelligible some of the allusions in the present and subsequent chapters to the work of Kummer, Kronecker, and Dedekind in the higher arithmetic. The matter is quite simple but requires several preliminary definitions.

The common whole numbers 1,2,3, . . . are called the (positive) rational integers. If m is any rational integer, it is the root of an algebraic equation of the *first* degree, whose coefficients are *rational integers*, namely $x - m = 0$. This, among other properties of the rational integers, suggested the *generalization* of the concept of integers to the "numbers" defined as roots of algebraic equations. Thus if r is a root of the equation

$$x^n + a_1 x^{n-1} + \ldots + a_{n-1} x + a_n = 0,$$

where the a's are rational integers (positive or negative), and if further r satisfies no equation of degree less than n, all of whose coefficients are rational integers and whose leading coefficient is 1 (as it is in the above equation, namely the coefficient of the highest power, x^n, of x in the equation is 1), then r is called an *algebraic integer* of *degree n*. For example, $1 + \sqrt{-5}$ is an algebraic integer of degree 2, because it is a root of $x^2 - 2x + 6 = 0$, and is not a root of any equation of degree less than 2 with coefficients of the prescribed kind; in fact $1 + \sqrt{-5}$ is the root of $x - (1 + \sqrt{-5}) = 0$, and the last coefficient, $-(1 + \sqrt{-5})$, is not a rational integer.

If in the above definition of an algebraic *integer* of degree n we suppress the requirement that the leading coefficient be 1, and say that it can be any rational integer (other than zero, which is considered an integer), a root of the equation is then called an *algebraic number* of *degree n*. Thus $\frac{1}{2}(1 + \sqrt{-5})$ is an algebraic *number* of degree 2, but is not an algebraic *integer*; it is a root of $2x^2 - 2x + 3 = 0$.

Another concept, that of an *algebraic number field* of *degree n* is now introduced: if r is an algebraic number of degree n, the totality of all expressions that can be constructed from r by repeated additions, subtractions, multiplications, and divisions (division by zero is not defined and hence is not attempted or permitted), is called *the algebraic number field generated by r*, and may be denoted by $F[r]$. For example, from

r we get $r + r$, or $2r$; from this and r we get $2r/r$ or 2, $2r - r$ or r, $2r \times r$ or $2r^2$, etc. The *degree* of this $F[r]$ is n.

It can be proved that every member of $F[r]$ is of the form $c_0 r^{n-1} + c_1 r^{n-2} + \ldots + c_{n-1}$, where the c's are rational numbers, and further every member of $F[r]$ is an algebraic number of degree not greater than n (in fact the degree is some divisor of n). *Some*, but not all, algebraic numbers in $F[r]$ will be algebraic *integers*.

The central problem of the theory of algebraic numbers is to investigate the laws of arithmetical divisibility of algebraic integers in an algebraic number field of degree n. To make this problem definite it is necessary to lay down exactly what is meant by "arithmetical divisibility," and for this we must understand the like for the *rational* integers.

We say that one rational integer, m, is divisible by another, d, if we can find a rational integer, q, such that $m = q \times d$; d (also q) is called a *divisor* of m. For example 6 is a divisor of 12, because $12 = 2 \times 6$; 5 is not a divisor of 12 because there does not exist a rational integer q such that $12 = q \times 5$.

A (positive) rational *prime* is a rational integer greater than 1 whose only positive divisors are 1 and the integer itself. When we try to extend this definition to algebraic integers we soon see that we have not found the root of the matter, and we must seek some property of rational primes which can be carried over to algebraic integers. This property is the following: if a rational prime p divides the product $a \times b$ of two rational integers, then (it can be proved that) p divides at least one of the factors a, b of the product.

Considering the unit, 1, of rational arithmetic, we notice that 1 has the peculiar property that it divides *every* rational integer; -1 also has the same property, and 1, -1 are the *only* rational integers having this property.

These and other clues suggest something simple that will work, and we lay down the following definitions as the basis for a theory of arithmetical divisibility for algebraic integers. We shall suppose that all the integers considered lie in an algebraic number field of degree n.

If r, s, t are algebraic integers such that $r = s \times t$, each of s, t is called a *divisor* of r.

If j is an algebraic integer which divides *every* algebraic integer in the field, j is called a *unit* (in that field). A given field may contain an infinity of units, in distinction to the pair 1, -1 for the rational field, and this is one of the things that breeds difficulties.

The next introduces a radical and disturbing distinction between rational integers and algebraic integers of degree greater than 1.

An algebraic integer other than a unit whose only divisors are units and the integer itself, is called *irreducible*. An irreducible algebraic integer which has the property that *if* it divides the product of two algebraic integers, *then* it divides at least one of the factors, is called a *prime* algebraic integer. All primes are irreducibles, but not all irreducibles are primes in *some* algebraic number fields, for example in $F[\sqrt{-5}]$, as will be seen in a moment. In the common arithmetic of $1,2,3 \ldots$ the irreducibles and the primes are the same.

In the chapter on Fermat the fundamental theorem of (rational) arithmetic was mentioned: a rational integer is the product of (rational) primes *in only one way*. From this theorem springs all the intricate theory of divisibility for rational integers. Unfortunately the fundamental theorem does *not* hold in *all* algebraic number fields of degree greater than one, and the result is chaos.

To give an instance (it is the stock example usually exhibited in text books on the subject), in the field $F[\sqrt{-5}]$ we have

$$6 = 2 \times 3 = (1 + \sqrt{-5}) \times (1 - \sqrt{-5});$$

each of $2, 3, 1 + \sqrt{-5}, 1 - \sqrt{-5}$ is a prime in this field (as may be verified with some ingenuity), so that 6, in this field, is *not uniquely* decomposable into a product of primes.

It may be stated here that Kronecker overcame this difficulty by a beautiful method which is too detailed to be explained untechnically, and that Dedekind did likewise by a totally different method which is much easier to grasp, and which will be noted when we consider his life. Dedekind's method is the one in widest use today, but this does not imply that Kronecker's is less powerful, nor that it will not come into favor when more arithmeticians become familiar with it.

In his dissertation of 1845 Kronecker attacked the theory of the units in certain special fields—those defined by the equations arising from the algebraic formulation of Gauss' problem to divide the circumference of a circle into n equal parts or, what is the same, to construct a regular polygon of n sides.

We can now close up one part of the account opened by Fermat.

In struggling to prove Fermat's "Last Theorem" that $x^n + y^n = z^n$ is impossible in rational integers x, y, z (none zero) if n is an integer greater than 2, arithmeticians took what looks like a natural step and resolved the left-hand side, $x^n + y^n$, into its n factors of the first degree (as is done in the usual second course of school algebra). This led to the exhaustive investigation of the algebraic number field mentioned above in connection with Gauss' problem—after serious but readily understandable mistakes had been made.

The problem at first was studded with pitfalls, into which many a competent mathematician and at least one great one—Cauchy—tumbled headlong. Cauchy assumed as a matter of course that in the algebraic number field concerned the fundamental theorem of arithmetic must hold. After several exciting but premature communications to the French Academy of Sciences, he admitted his error. Being restlessly interested in a large number of other problems at the time, Cauchy turned aside and failed to make the great discovery which was well within the capabilities of his prolific genius and left the field to Kummer. The central difficulty was serious: here was a species of "integers"—those of the field concerned—which defied the fundamental theorem of arithmetic; how reduce them to law and order?

The solution of this problem by the invention of a totally new kind of "number" appropriate to the situation, which (in terms of these "numbers") automatically restored the fundamental theorem of arithmetic, ranks with the creation of non-Euclidean geometry as one of the outstanding scientific achievements of the nineteenth century, and it is well up in the high mathematical achievements of all history. The creation of the new "numbers"—so-called "ideal numbers"—was the invention of Kummer in 1845. These new "numbers" were not constructed for all algebraic number fields but only for those fields arising from the division of the circle.

Kummer too had fallen afoul of the net which snared Cauchy, and for a time he believed that he had proved Fermat's "Last Theorem." Then Dirichlet, to whom the supposed proof was submitted for criticism, pointed out by means of an example that the fundamental theorem of arithmetic, contrary to Kummer's tacit assumption, does *not* hold in the field concerned. This failure of Kummer's was one of the most fortunate things that ever happened in mathematics. Like Abel's initial mistake in the matter of the general quintic, Kummer's turned him into the right track, and he invented his "ideal numbers."

Kummer, Kronecker, and Dedekind in their invention of the modern theory of algebraic numbers, by enlarging the scope of arithmetic *ad infinitum* and bringing algebraic equations within the purview of number, did for the higher arithmetic and the theory of algebraic equations what Gauss, Lobatchewsky, Johann Bolyai, and Riemann did for geometry in emancipating it from slavery in Euclid's too narrow economy. And just as the inventors of non-Euclidean geometry revealed vast and hitherto unsuspected horizons to geometry and physical science, so the creators of the theory of algebraic numbers uncovered an entirely new light, illuminating the whole of arithmetic and throwing the theories of equations, of systems of algebraic curves and surfaces, and the very nature of number itself, into sharp relief against a firm background of shiningly simple postulates.

The creation of "ideals"—Dedekind's inspiration from Kummer's vision of "ideal numbers"—renovated not only arithmetic but the whole of the algebra which springs from the theory of algebraic equations and systems of such equations, and it proved also a reliable clue to the inner significance of the "enumerative geometry"* of Plücker, Cayley and others, which absorbed so large a fraction of the energies of the geometers of the nineteenth century who busied themselves with the intersections of nets of curves and surfaces. And last, if Kronecker's heresy against Weierstrassian analysis (noted later) is some day to become a stale orthodoxy, as all not utterly insane heresies sooner or later do, these renovations of our familiar integers, 1,2,3, . . . , on which all analysis strives to base itself, may ultimately indicate extensions of analysis, and the Pythagorean speculation may envisage generative properties of "number" that Pythagoras never dreamed of in all his wild philosophy.

Kronecker entered this beautifully difficult field of algebraic numbers in 1845 at the age of twenty two with his famous dissertation *De Unitatibus Complexis* (*On Complex Units*). The particular units he discussed were those in algebraic number fields arising from the Gaus-

---

*One problem in this subject: an algebraic curve may have loops on it, or places where the curve crosses its tangents; given the degree of the curve, how many such points are there? Or if we cannot answer that, what equations connecting the number of these and other exceptional points must hold? Similarly for surfaces.

sian problem of the division of the circumference of a circle into n equal arcs. For this work he got his Ph.D.

The German universities used to have—and may still have—a laudable custom in connection with the taking of a Ph.D.: the successful candidate was in honor bound to fling a party—usually a prolonged beer bust with all the trimmings—for his examiners. At such festivities a mock examination consisting of ridiculous questions and more ridiculous answers was sometimes part of the fun. Kronecker invited practically the whole faculty, including the Dean, and the memory of that undignified feast in celebration of his degree was, he declared in later years, the happiest of his life.

In at least one respect Kronecker and his scientific enemy Weierstrass were much alike: they were both very great gentlemen, as even those who did not particularly care for either admitted. But in nearly everything else they were almost comically different. The climax of Kronecker's career was his prolonged mathematical war against Weierstrass, in which quarter was neither given nor asked. One was a born algebraist, the other almost made a religion of analysis. Weierstrass was large and rambling, Kronecker a compact, diminutive man, not over five feet tall, but perfectly proportioned and sturdy. After his student days Weierstrass gave up his fencing; Kronecker was always an expert gymnast and swimmer and in later life a good mountaineer.

Eyewitnesses of the battles between this curiously mismatched pair tell how the big fellow, annoyed by the persistence of the little fellow, would stand shaking himself like a good-natured St. Bernard dog trying to rid himself of a determined fly, only to excite his persecutor to more ingenious attacks, till Weierstrass, giving up in despair, would amble off, Kronecker at his heels still talking maddeningly. But for all their scientific differences the two were good friends, and both were great mathematicians without a particle of the "great man" complex that too often inflates the shirts of the would-be mighty.

Kronecker was blessed with a rich uncle in the banking business. The uncle also controlled extensive farming enterprises. All this fell into young Kronecker's hands for administration on the death of the uncle, shortly after the budding mathematician had taken his degree at the age of twenty two. The eight years from 1845 to 1853 were spent in managing the estate and running the business, which Kronecker did with great thoroughness and financial success. To manage the landed property efficiently he even mastered the principles of agriculture.

In 1848, at the age of twenty five, the energetic young business man very prudently fell in love with his cousin, Fanny Prausnitzer, daughter of the defunct wealthy uncle, married her, and settled down to raise a family. They had six children, four of whom survived their parents. Kronecker's married life was ideally happy, and he and his wife—a gifted, pleasant woman—brought up their children with the greatest devotion. The death of Kronecker's wife a few months before his own last illness was the blow which broke him.

During his eight years in business Kronecker produced no mathematics. But that he did not stagnate mathematically is shown by his publication in 1853 of a fundamental memoir on the algebraic solution of equations. All through his activity as a man of affairs Kronecker had maintained a lively scientific correspondence with his former master, Kummer, and on escaping from business in 1853 he visited Paris, where he made the acquaintance of Hermite and other leading French mathematicians. Thus he did not sever communications with the scientific world when circumstances forced him into business, but kept his soul alive by making mathematics rather than whist, pinochle, or checkers his hobby.

In 1853, when Kronecker's memoir on the algebraic solvability of equations (the nature of the problem was discussed in the chapters on Abel and Galois) was published, the Galois theory of equations was understood by very few. Kronecker's attack was characteristic of much of his finest work. Kronecker had mastered the Galois theory, indeed he was probably the only mathematician of the time (the late 1840's) who had penetrated deeply into Galois' ideas; Liouville had contented himself with a sufficient insight into the theory to enable him to edit some of Galois' remains intelligently.

A distinguishing feature of Kronecker's attack was its comprehensive thoroughness. In this, as in other investigations in algebra and the theory of numbers, Kronecker took the refined gold of his predecessors, toiled over it like an inspired jeweler, added gems of his own, and made from the precious raw material a flawless work of art with the unmistakable impress of his artistic individuality upon it. He delighted in perfect things; a few of his pages will often exhibit a complete development of one isolated result with all its implications immanent but not loading the unique theme with expressed detail. Consequently even the shortest of his papers has suggested important developments to his successors, and his longer works are inexhaustible mines of beautiful things.

Kronecker was what is called an "algorist" in most of his works. He aimed to make concise, expressive formulas tell the story and automatically reveal the action from one step to the next so that, when the climax was reached, it was possible to glance back over the whole development and see the apparent inevitability of the conclusion from the premises. Details and accessory aids were ruthlessly pruned away until only the main trunk of the argument stood forth in naked strength and simplicity. In short, Kronecker was an artist who used mathematical formulas as his medium.

After Kronecker's works on the Galois theory the subject passed from the private ownership of a few into the common property of all algebraists, and Kronecker had wrought so artistically that the next phase of the theory of equations—the current postulational formulation of the theory and its extensions—can be traced back to him. His aim in algebra, like that of Weierstrass in analysis, was to find the "natural" way—a matter of intuition and taste rather than scientific definition—to the heart of his problems.

The same artistry and tendency to unification appeared in another of his most celebrated papers, which occupies only a couple of pages in his collected works, *On the Solution of the General Equation of the Fifth Degree*, first published in 1858. Hermite, we recall, had given the first solution, by means of elliptic (modular) functions in the same year. Kronecker attains Hermite's solution—or what is practically the same—by applying the ideas of Galois to the problem, thereby making the miracle appear more "natural." In another paper, also short, over which he has spent most of his time for five years, he returns to the subject in 1861, and seeks the reason *why* the general equation of the fifth degree is solvable in the manner in which it is, thus taking a step beyond Abel who settled the question of solvability "by radicals."

Much of Kronecker's work has a distinct arithmetical tinge, either of rational arithmetic or of the broader arithmetic of algebraic numbers. Indeed, if his mathematical activity had any guiding clue, it may be said to have been his desire, perhaps subconscious, to *arithmetize* all mathematics, from algebra to analysis. "God made the integers," he said, "all the rest is the work of man." Kronecker's demand that analysis be replaced by finite arithmetic was the root of his disagreement with Weierstrass. Universal arithmetization may be too narrow an ideal for the luxuriance of modern mathematics, but at least it has the merit of greater clarity than is to be found in some others.

Geometry never seriously attracted Kronecker. The period of specialization was already well advanced when Kronecker did most of his work, and it would probably have been impossible for any man to have done the profoundly perfect sort of work that Kronecker did as an algebraist and in his own peculiar type of analysis and at the same time have accomplished anything of significance in other fields. Specialization is frequently damned, but it has its virtues.

A distinguishing feature of many of Kronecker's technical discoveries was the intimate way in which he wove together the three strands of his greatest interests—the theory of numbers, the theory of equations, and elliptic functions—into one beautiful pattern in which unforeseen symmetries were revealed as the design developed and many details were unexpectedly imaged in others far away. Each of the tools with which he worked seemed to have been designed by fate for the more efficient functioning of the others. Not content to accept this mysterious unity as a mere mystery, Kronecker sought and found its underlying structure in Gauss' theory of binary quadratic forms, in which the main problem is to investigate the solutions in integers of indeterminate equations of the second degree in two unknowns.

Kronecker's great work in the theory of algebraic numbers was not part of this pattern. In another direction he also departed occasionally from his principal interests when, according to the fashion of his times, he occupied himself with the purely mathematical aspects of certain problems (in the theory of attraction as in Newton's gravitation) of mathematical physics. His contributions in this field were of mathematical rather than physical interest.

Up till the last decade of his life Kronecker was a free man with obligations to no employer. Nevertheless he voluntarily assumed scientific duties, for which he received no remuneration, when he availed himself of his privilege as a member of the Berlin Academy to lecture at the University of Berlin. From 1861 to 1883 he conducted regular courses at the university, principally on his personal researches, after the necessary introductions. In 1883 Kummer, then at Berlin, retired, and Kronecker succeeded his old master as ordinary professor. At this period of his life he travelled extensively and was a frequent and welcome participant in scientific meetings in Great Britain, France, and Scandinavia.

Throughout his career as a mathematical lecturer Kronecker competed with Weierstrass and other celebrities whose subjects were more popular than his own. Algebra and the theory of numbers have never appealed to so wide an audience as have geometry and analysis, possibly because the connections of the latter with physical science are more apparent.

Kronecker took his aristocratic isolation good-naturedly and even with a certain satisfaction. His beautifully clear introductions deluded his auditors into a belief that the subsequent course of lectures would be easy to follow. This belief evaporated rapidly as the course progressed, until after three sessions all but a faithful and obstinate few had silently stolen away—many of them to listen to Weierstrass. Kronecker rejoiced. A curtain could now be drawn across the room behind the first few rows of chairs, he joked, to bring lecturer and auditors into cosier intimacy. The few disciples he retained followed him devotedly, walking home with him to continue the discussions of the lecture room and frequently affording the crowded sidewalks of Berlin the diverting spectacle of an excited little man talking with his whole body—especially his hands—to a spellbound group of students blocking the traffic. His house was always open to his pupils, for Kronecker really liked people, and his generous hospitality was one of the greatest satisfactions of his life. Several of his students became eminent mathematicians, but his "school" was the whole world and he made no effort to acquire an artifically large following.

The last is characteristic of Kronecker's own most startlingly independent work. In an atmosphere of confident belief in the soundness of analysis Kronecker assumed the unpopular rôle of the philosophical doubter. Not many of the great mathematicians have taken philosophy seriously; in fact the majority seem to have regarded philosophical speculations with repugnance, and any epistemological doubt affecting the soundness of their work has usually been ignored or impatiently brushed aside.

With Kronecker it was different. The most original part of his work, in which he was a true pioneer, was a natural outgrowth of his philosophical inclinations. His father, Werner, Kummer, and his own wide reading in philosophical literature had influenced him in the direction of a critical outlook on all human knowledge, and when he contemplated mathematics from this questioning point of view he did not spare it because it happened to be the field of his own particular interest, but infused it with an acid, beneficial skepticism. Although but little of this found its way into print it annoyed some of his contemporaries intensely and it has survived. The doubter did not address himself to the living but, as he said, "to those who shall come after me." Today these followers have arrived, and due to their united efforts—although they often succeed only in contradicting one another—we are beginning to get a clearer insight into the nature and meaning of mathematics.

Weierstrass (Chapter 22) would have constructed mathematical analysis on his conception of irrationals as defined by infinite sequences of rationals. Kronecker not only disputes Weierstrass; he would nullify Eudoxus. For him as for Pythagoras only the God-given integers $1, 2, 3, \ldots,$ "exist"; all the rest is a futile attempt of mankind to improve on the creator. Weierstrass on the other hand believed that he had at last made the square root of 2 as comprehensible and as safe to handle as 2 itself; Kronecker denied that the square root of 2 "exists," and he asserted that it is impossible to reason consistently with or about the Weierstrassian construction for this root or for any other irrational. Neither his older colleagues nor the young to whom Kronecker addressed himself gave his revolutionary idea a very enthusiastic welcome.

Weierstrass himself seems to have felt uneasy; certainly he was hurt. His strong emotion is released mostly in one tremendous German sentence* like a fugue, which it is almost impossible to preserve in English. "But the worst of it is," he complains, "that Kronecker uses his authority to proclaim that *all* those who up to now have labored to establish the theory of functions are sinners before the Lord. When a whimsical eccentric like Christoffel [the man whose somewhat neglected work was to become, years after his death, an important tool in differential geometry as it is cultivated today in the mathematics of relativity] says that in twenty or thirty years the present theory of functions will be buried and that the whole of analysis will be referred to the theory of forms, we reply with a shrug. But when Kronecker delivers himself of the following verdict which I repeat *word for word*: 'If time and strength are granted me, I myself

---

*In a letter to Sonja Kowalewski, 1885.

will show the mathematical world that not only geometry, but also arithmetic can point the way to analysis, and certainly a more rigorous way. If I cannot do it myself those who come after me will . . . and they will recognize the incorrectness of *all* those conclusions with which *so-called* analysis works at present'—such a verdict from a man whose eminent talent and distinguished performance in mathematical research I admire as sincerely and with as much pleasure as all his colleagues, is not only humiliating for those whom he adjures to acknowledge as an error and to forswear the substance of what has constituted the object of their thought and unremitting labor, but it is a direct appeal to the younger generation to desert their present leaders and rally around him as the disciple of a new system which *must* be founded. Truly it is sad, and it fills me with a bitter grief, to see a man, whose glory is without flaw, let himself be driven by the well justified feeling of his own worth to utterances whose injurious effect upon others he seems not to perceive.

"But enough of these things, on which I have touched only to explain to you the reason why I can no longer take the same joy that I used to take in my teaching, even if my health were to permit me to continue it a few years longer. But you must not speak of it; I should not like others, who do not know me as well as you, to see in what I say the expression of a sentiment which is in fact foreign to me."

Weierstrass was seventy and in poor health when he wrote this. Could he have lived till today he would have seen his own great system still flourishing like the proverbial green bay tree. Kronecker's doubts have done much to instigate a critical re-examination of the foundations of all mathematics, but they have not yet destroyed analysis. They go deeper, and if anything of far-reaching significance is to be replaced by something firmer but as yet unknown, it seems likely that a good part of Kronecker's own work will go too, for the critical attack which he foresaw has uncovered weaknesses where he suspected nothing. Time makes fools of us all. Our only comfort is that greater shall come after us.

Kronecker's "revolution," as his contemporaries called his subversive assault on analysis, would banish all but the positive integers from mathematics. Geometry since Descartes has been largely an affair of analysis applied to ordered pairs, triples, . . . of real numbers (the "numbers" which correspond to the distances measured on a given straight line from a fixed point on the line); hence it too would come under the sway of Kronecker's program. So familiar a concept as that of a negative integer, -2 for instance, would not appear in the mathematics Kronecker prophesied, nor would common fractions.

Irrationals, as Weierstrass points out, roused Kronecker's special displeasure. To speak of $x^2 - 2 = 0$ having a root would be meaningless. All of these dislikes and objections are of course themselves meaningless unless they can be backed by a definite program to replace what is rejected.

Kronecker actually did this, at least in outline, and indicated how the whole of algebra and the theory of numbers, including algebraic numbers, can be reconstructed in accordance with his demand. To get rid of $\sqrt{-1}$, for example, we need only put a letter for it temporarily, say i, and consider polynomials containing i and other letters, say $x,y,z,\ldots$ Then we manipulate these polynomials as in elementary algebra, treating i like any of the other letters, till the last step, when every polynomial containing i is divided by $i^2 + 1$ and everything but the remainder obtained from this division is discarded. Anyone who remembers a little elementary algebra may readily convince himself that this leads to all the familiar properties of the mysteriously misnamed "imaginary" numbers of the text books. In a similar manner negatives and fractions and *all* algebraic numbers (other than the positive rational integers) are eliminated from mathematics—if desired —and only the blessed positive integers remain. The inspiration about discarding $\sqrt{-1}$ goes back to Cauchy in 1847. This was the germ of Kronecker's program.

Those who dislike Kronecker's "revolution" call it a *Putsch*, which is more like a drunken brawl than an orderly revolution. Nevertheless it has led in recent years to two constructively critical movements in the whole of mathematics: the demand that a construction in a finite number of steps be given or proved to be possible for any "number" or other mathematical "entity" whose "existence" is indicated, and the banishment from mathematics of all definitions that cannot be stated explicitly in a finite number of words. Insistence upon these demands has already done much to clarify our conception of the nature of mathematics, but a vast amount remains to be done. As this work is still in progress we shall defer further consideration of it until we come to Cantor, when it will be possible to exhibit examples.

Kronecker's disagreement with Weierstrass should not leave an unpleasant impression, as it may do if we ignore the rest of Kronecker's generous life. Kronecker had no intention of wounding his kindly old senior; he merely let his tongue run away with him in the heat of a purely mathematical argument, and Weierstrass, when he was in good spirits, laughed the whole attack off, as he should have done, knowing well that just as he had improved on Eudoxus, so his successors would probably improve upon him. Possibly if Kronecker had been six or seven inches taller than he was he would not have felt constrained to overemphasize his objections to analysis so vociferously. Much of the whole wordy dispute sounds suspiciously like the overcorrection of an unjustified inferiority complex.

The reaction of many mathematicians to Kronecker's "revolution" was summed up by Poincaré when he said that Kronecker had been enabled to do so much fine mathematics because he frequently forgot his own mathematical philosophy. Like not a few epigrams this one is just untrue enough to be witty.

Kronecker died of a bronchial illness in Berlin on December 29, 1891, in his sixty ninth year.

CHAPTER TWENTY SIX

Anima Candida

RIEMANN

A geometer like Riemann might almost have foreseen the more important features of the actual world.—A. S. EDDINGTON

IT HAS BEEN SAID OF COLERIDGE that he wrote but little poetry of the highest order of excellence, but that that little should be bound in gold. The like has been said of Bernhard Riemann, the mathematical fruits of whose all too brief summer fill only one octavo volume. It may also be truly said of Riemann that he touched nothing that he did not in some measure revolutionize. One of the most original mathematicians of modern times, Riemann unfortunately inherited a poor constitution, and he died before he had reaped a tithe of the golden harvests in his fertile mind. Had he been born a century later than he was, medical science could probably have leased him twenty or thirty more years of life, and mathematics would not now be waiting for his successor.

Georg Friedrich Bernhard Riemann, the son of a Lutheran pastor, and the second of six children (two boys, four girls), was born in the little village of Breselenz, in Hanover, Germany, on September 17, 1826. His father had fought in the Napoleonic wars, and on settling down to a less barbarous mode of living had married Charlotte Ebell, daughter of a court councillor. Hanover in 1826 was not exactly prosperous, and the circumstances of an obscure country parson with a wife and six children to feed and clothe were far from affluent. It is claimed by some biographers, apparently with justice, that the frail health and early deaths of most of the Riemann children were the result of undernourishment in their youth and were not due to poor stamina. The mother also died before her children were grown.

In spite of poverty the home life was happy, and Riemann always retained the warmest affection—and homesickness, when he was absent—for all his lovable family. From his earliest years he was a timid, diffident soul with a horror of speaking in public or attracting attention to himself. In later life this chronic shyness proved a very serious handicap and occasioned him much agonized misery till he overcame it by diligent preparation for every public utterance he was likely to make. The engaging bashfulness of Riemann's boyhood and early manhood, which endeared him to all who met him, was in strange contrast to the ruthless boldness of his matured scientific thought. Supreme in the world of his own creation, he realized his transcendent powers and shrank from nobody, real or imaginary.

While Riemann was still an infant his father was transferred to the pastorate of Quickborn. There young Riemann received his first instruction, from his father, who appears to have been an excellent teacher. From the very first lessons Bernhard showed an unquenchable thirst for learning. His earliest interests were historical, particularly in the romantic and tragic history of Poland. As a boy of five Bernhard gave his father no peace about unhappy Poland, but demanded to be told over and over again the legend of that heroic country's gallant (and at times slightly fatuous) struggles for liberty and, in the late Woodrow Wilson's rich, fruity phrase, "self-determination."

Arithmetic, begun at about six, offered something less harrowing for the sensitive young boy to dwell on. His inborn mathematical genius now asserted itself. Bernhard not only solved all the problems shoved at him, but invented more difficult teasers to exasperate his brother and sisters. Already the creative impulse in mathematics dominated the boy's mind. At the age of ten he received instruction in more advanced arithmetic and geometry from a professional teacher, one Schulz, a fairly good pedagogue. Schulz soon found himself following his pupil, who often had better solutions than he.

At fourteen Riemann went to stay with his grandmother at Hanover, where he entered his first Gymnasium, in the upper third class. Here he endured his first overwhelming loneliness. His shyness made him the butt of his schoolfellows and drove him in upon his own resources. After a temporary setback his schoolwork was uniformly excellent, but it gave him no comfort, and his only solace was the joy of buying such inconsiderable presents as his pocket money would permit, to send home to his parents and brother and sisters on their birthdays. One present for his parents he invented and made himself, an original perpetual calendar, much to the astonishment of his incredulous schoolfellows. On the death of his grandmother two years later, Riemann was transferred to the Gymnasium at Lüneburg, where he studied till he was prepared, at the age of nineteen, to enter the University of Göttingen. At Lüneburg Riemann was within walking distance of home. He took full advantage of his opportunities to escape to the warmth of his own fireside. These years of his secondary education, while his health was still fair, were the happiest of his life. The tramps back and forth between the Gymnasium and Quickborn taxed his strength, but in spite of his mother's anxiety that he might wear himself out, Riemann continued to over-exert himself in order that he might be with his family as often as possible.

While still at the Gymnasium Riemann suffered from the itch for finality and perfection which was later to slow up his scientific publication. This defect—if such it was—caused him great difficulty in his written language exercises and at first made it doubtful whether he would "pass." But this same trait was responsible later for the finished form of two of his masterpieces, one of which even Gauss declared to be perfect. Things improved when Seyffer, the teacher of Hebrew, took young Riemann into his own house as a boarder and ironed him out.

The two studied Hebrew together, Riemann frequently giving more than he took, as the future mathematician at that time was all set to gratify his father's wishes and become a great preacher—as if Riemann, with his tongue-tied bashfulness, could ever have thumped hell and damnation or redemption and paradise out of any pulpit. Riemann himself was enamored of the pious prospect, and although he never got as far as a probationary sermon, he did employ his mathematical talents in an attempted demonstration, in the manner of Spinoza, of the truth of Genesis. Undaunted by his failure, young Riemann persevered in his faith and remained a sincere Christian all his life. As his biographer (Dedekind) states, "He reverently avoided disturbing the faith of others; for him the main thing in religion was daily self-examination." By the end of his Gymnasium course it was plain even to Riemann that Great Headquarters could have but little use for him as a router of the devil, but might be able to employ him profitably in the conquest of nature. Thus once again, as in the cases of Boole and Kummer, a brand was plucked from the burning, *ad majoram Dei gloriam.*

The director of the Gymnasium, Schmalfuss, having observed Riemann's talent for mathematics, had given the boy the run of his private library and had excused him from attending mathematical classes. In this way Riemann discovered his inborn aptitude for mathematics, but his failure to realize immediately the extent of his ability is so characteristic of his almost pathological modesty as to be ludicrous.

Schmalfuss had suggested that Riemann borrow some mathematical book for private study. Riemann said that would be nice, provided the book was not too easy, and at the suggestion of Schmalfuss carried off Legendre's *Théorie des Nombres* (Theory of Numbers). This is a mere trifle of 859 large quarto pages, many of them crabbed with very close reasoning indeed. Six days later Riemann returned the book. "How far did you read?" Schmalfuss asked. Without replying directly, Riemann expressed his appreciation of Legendre's classic. "That is certainly a wonderful book. I have mastered it." And in fact he had. Some time later when he was examined he answered perfectly, although he had not seen the book for months.

No doubt this is the origin of Riemann's interest in the riddle of prime numbers. Legendre has an empirical formula estimating the approximate number of primes less than any preassigned number; one of Riemann's profoundest and most suggestive works (only eight pages long) was to be in the same general field. In fact "Riemann's hypothesis," originating in his attempt to improve on Legendre, is today one of the outstanding challenges, if not *the* outstanding challenge, to pure mathematicians.

To anticipate slightly, we may state here what this hypothesis is. It occurs in the famous memoir *Ueber die Anzahl der Primzahlen unter einer gegebenen Grösse* (On the number of prime numbers under a given magnitude), printed in the monthly notices of the Berlin Academy for November, 1859, when Riemann was thirty three. The problem concerned is to give a formula which will state how many primes there are less than any given number n. In attempting to solve this Riemann was driven to an investigation of the infinite series

$$1 + \frac{1}{2^s} + \frac{1}{3^s} + \frac{1}{4^s} + \frac{1}{5^s} + \cdots,$$

in which s is a complex number, say $s = u + iv$ ($i = \sqrt{-1}$), where u and v are real numbers, so chosen that the series converges. With this proviso the infinite series is a definite function of s, say $\zeta(s)$ (the Greek zeta, ζ, is always used to denote this function, which is called

"Riemann's zeta function"); and as s varies, $\zeta(s)$ continuously takes on different values. *For what values of s will $\zeta(s)$ be zero?* Riemann conjectured that *all* such values of s for which u lies between 0 and 1 are of the form $\frac{1}{2} + iv$, namely, *all have their real part equal to $\frac{1}{2}$.*

This is the famous hypothesis. Whoever proves or disproves it will cover himself with glory and incidentally dispose of many extremely difficult questions in the theory of prime numbers, other parts of the higher arithmetic, and in some fields of analysis. Expert opinion favors the truth of the hypothesis. In 1914 the English mathematician G. H. Hardy proved that *an infinity* of values of s satisfy the hypothesis, but an infinity is not necessarily all. A decision one way or the other disposing of Riemann's conjecture would probably be of greater interest to mathematicians than a proof or disproof of Fermat's Last Theorem. Riemann's hypothesis is not the sort of problem that can be attacked by elementary methods. It has already give rise to an extensive and thorny literature.

Legendre was not the only great mathematician whose works Riemann absorbed by himself—always with amazing speed—at the Gymnasium; he became familiar with the calculus and its ramifications through the study of Euler. It is rather surprising that from such an antiquated start in analysis (Euler's approach was out of date by the middle 1840's owing to the work of Gauss, Abel, and Cauchy), Riemann later became the acute analyst that he did. But from Euler he may have picked up something which also has its place in creative mathematical work, an appreciation of symmetrical formulas and manipulative ingenuity. Although Riemann depended chiefly on what may be called deep philosophical ideas—those which get at the heart of a theory—for his greater inspirations, his work nevertheless is not wholly lacking in the "mere ingenuity" of which Euler was the peerless master and which it is now quite the fashion to despise. The pursuit of pretty formulas and neat theorems can no doubt quickly degenerate into a silly vice, but so also can the quest for austere generalities which are so very general indeed that they are incapable of application to any particular. Riemann's instinctive mathematical tact preserved him from the bad taste of either extreme.

In 1846, at the age of nineteen, Riemann matriculated as a student of philology and theology at the University of Göttingen. His desire to please his father and possibly help financially by securing a paying position as quickly as possible dictated the choice of theology. But he could not keep away from the mathematical lectures of Stern on the theory of equations and on definite integrals, those of Gauss on the method of least squares, and Goldschmidt's on terrestrial magnetism. Confessing all to his indulgent father, Riemann prayed for permission to alter his course. His father's ungrudging consent that Bernhard follow mathematics as a career made the young man supremely happy—also profoundly grateful.

After a year at Göttingen, where the instruction was decidedly antiquated, Riemann migrated to Berlin to receive from Jacobi, Dirichlet, Steiner, and Eisenstein his initiation into new and vital mathematics. From all of these masters he learned much—advanced mechanics and higher algebra from Jacobi, the theory of numbers and analysis from Dirichlet, modern geometry from Steiner, while from Eisenstein, three years older than himself, he learned not only elliptic functions but self-confidence, for he and the young master had a radical and most energizing difference of opinion as to how the theory should be developed. Eisenstein insisted on beautiful formulas, somewhat in the manner of a modernized Euler; Riemann wanted to introduce the complex variable and derive the entire theory, with a minimum of calculation, from a few simple, general principles. Thus, no doubt, originated at least the germs of one of Riemann's greatest contributions to pure mathematics. As the origin of Riemann's work in the theory of functions of a complex variable is of considerable importance in his own history and in that of modern mathematics, we shall glance at what is known about it.

Briefly, nothing definite. The definition of an analytic function of a complex variable, discussed in connection with Gauss' anticipation of Cauchy's fundamental theorem, was essentially that of Riemann. When expressed analytically instead of geometrically that definition leads to the pair of partial differential equations* which Riemann took as his point of departure for a theory of functions of a complex variable. According to Dedekind, "Riemann recognized in these partial

---

* If $z = x + iy$, and $w = u + iv$, is an analytic function of z, Riemann's equations are

$$\frac{\partial u}{\partial x} = \frac{\partial v}{\partial y}, \quad \frac{\partial u}{\partial y} = -\frac{\partial v}{\partial x}.$$

These equations had been given much earlier by Cauchy, and even Cauchy was not the first, as D'Alembert had stated the equations in the eighteenth century.

differential equations the essential definition of an [analytic] function of a complex variable. Probably these ideas, of the highest importance for his future career, were worked out by him in the fall vacation of 1847 [Riemann was then twenty one] for the first time."

Another version of the origin of Riemann's inspiration is due to Sylvester, who tells the following story, which is interesting even if possibly untrue. In 1896, the year before his death, Sylvester recalls staying at "a hotel on the river at Nuremberg, where I conversed outside with a Berlin bookseller, bound, like myself, for Prague. . . . He told me he was formerly a fellow pupil of Riemann, at the University, and that, one day, after receipt of some numbers of the *Comptes rendus* from Paris, the latter shut himself up for some weeks, and when he returned to the society of his friends, said (referring to the newly published papers of Cauchy), 'This is a new mathematic.'"

Riemann spent two years at the University of Berlin. During the political upheaval of 1848 he served with the loyal student corps and had one weary spell of sixteen hours' guard duty protecting the jittery if sacred person of the king in the royal palace. In 1849 he returned to Göttingen to complete his mathematical training for the doctorate. His interests were unusually broad for the pure mathematician he is commonly rated to be, and in fact he devoted as much of his time to physical science as he did to mathematics.

From this distance it seems as though Riemann's real interest was in mathematical physics, and it is quite possible that had he been granted twenty or thirty more years of life he would have become the Newton or Einstein of the nineteenth century. His physical ideas were bold in the extreme for his time. Not till Einstein realized Riemann's dream of a geometrized (macroscopic) physics did the physics which Riemann foreshadowed—somewhat obscurely, it may be—appear reasonable to physicists. In this direction his only understanding follower till our own century was the English mathematician William Kingdon Clifford (1845–1879), who also died long before his time.

During his last three semesters at Göttingen Riemann attended lectures on philosophy and followed the course of Wilhelm Weber in experimental physics with the greatest interest. The philosophical and psychological fragments left by Riemann at his death show that as a philosophical thinker he was as original as he was in mathematics and science. Weber recognized Riemann's scientific genius and became his warm friend and helpful counsellor. To a far higher degree than the majority of great mathematicians who have written on physical science, Riemann had a feeling for what is important—or likely to be so—in physics, and this feeling is no doubt due to his work in the laboratory and his contact with men who were primarily physicists and not mathematicians. The contributions of even great pure mathematicians to physical science have usually been characterized by a singular irrelevance so far as the universe observed by scientists is concerned. Riemann, as a physical mathematician, was in the same class as Newton, Gauss, and Einstein in his instinct for what is likely to be of scientific use in mathematics.

As a sequel to his philosophical studies with Johann Friedrich Herbart (1776–1841), Riemann came to the conclusion in 1850 (he was then twenty four) that "a complete, well-rounded mathematical theory can be established, which progresses from the elementary laws for individual points to the processes given to us in the plenum ('continuously filled space') of reality, without distinction between gravitation, electricity, magnetism, or thermostatics." This is probably to be interpreted as Riemann's rejection of all "action at a distance" theories in physical science in favor of field theories. In the latter the physical properties of the "space" surrounding a "charged particle," say, are the object of mathematical investigation. Riemann at this stage of his career seems to have believed in a space-filling "ether," a conception now abandoned. But as will appear from his epochal work on the foundations of geometry, he later sought the description and correlation of physical phenomena in the *geometry* of the "space" of human experience. This is in the current fashion, which rejects an existent, unobservable ether as a cumbersome superfluity.

Fascinated by his work in physics, Riemann let his pure mathematics slide for a while and in the fall of 1850 joined the seminar in mathematical physics which had just been founded by Weber, Ulrich, Stern, and Listing. Physical experiments in this seminar consumed the time that scholarly prudence would have reserved for the doctoral dissertation in mathematics, which Riemann did not submit till he was twenty five.

One of the leaders in the seminar, Johann Benedict Listing (1808–1882), may be noted in passing, as he probably influenced Riemann's thought in what was to be (1857) one of his greatest achievements, the introduction of topological methods into the theory of functions of a complex variable.

It will be recalled that Gauss had prophesied that analysis situs would become one of the most important fields of mathematics, and Riemann, by his inventions in the theory of functions, was to give a partial fulfillment of this prophecy. Although topology (now called analysis situs) as first developed bore but little resemblance to the elaborate theory which today absorbs all the energies of a prolific school, it may be of interest to state the trivial puzzle which apparently started the whole vast and intricate theory. In Euler's time seven bridges crossed the river Pregel in Königsberg, as in the diagram,

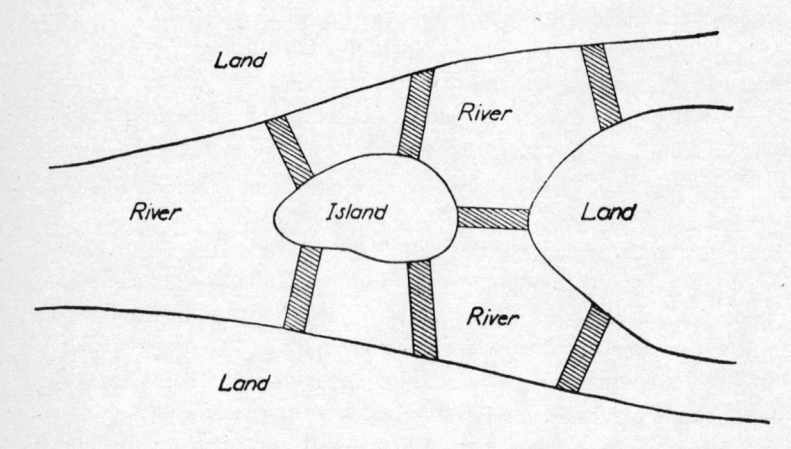

the shaded bars representing the bridges. Euler proposed the problem of crossing all seven bridges without passing twice over any one. The problem is impossible.

The nature of Riemann's use of topological methods in the theory of functions may be disposed of here, although an adequate description is out of the question in untechnical language. For the meaning of "uniformity" with respect to a function of a complex variable we must refer to what was said in the chapter on Gauss. Now, in the theory of Abelian functions, *multiform* functions present themselves inevitably; an n-valued function of z is a function which, except for certain values of z, takes precisely n distinct values for each value assigned to z. Illustrating *multiformity*, or *many-valuedness*, for functions of a real variable, we note that y, considered as a function of x, defined by the equation $y^2 = x$, is two-valued. Thus, if $x = 4$, we get $y^2 = 4$, and hence $y = 2$ or -2; if x is any real number except zero or "infinity,"

y has the two distinct values of $\sqrt{x}$ and $-\sqrt{x}$. In this simplest possible example y and x are connected by an algebraic equation, namely $y^2 - x = 0$. Passing at once to the general situation of which this is a very special case, we might discuss the n-valued function y which is defined, as a function of x, by the equation

$$P_0(x)y^n + P_1(x)y^{n-1} + \ldots + P_{n-1}(x)y + P_n(x) = 0,$$

in which the P's are polynomials in x. This equation defines y as an n-valued function of x. As in the case of $y^2 - x = 0$, there will be certain values of x for which two or more of these n values of y are equal. These values of x are the so-called *branch points* of the n-valued function defined by the equation.

All this is now extended to functions of complex variables, and the function w (also its integral) as defined by

$$P_0(z)w^n + P_1(z)w^{n-1} + \ldots + P_{n-1}(z)w + P_n(z) = 0,$$

in which z denotes the complex variable $s + it$, where s, t are real variables and $i = \sqrt{-1}$. The n values of w are called the *branches* of the function w. Here we must refer (chapter on Gauss) to what was said about the representation of *uniform* functions of z. Let the variable $z (= s + it)$ trace out any path in its plane, and let the *uniform* function $f(z)$ be expressed in the form $U + iV$, where U, V are functions of s,t. Then, to every value of z will correspond one, and only one, value for each of U, V, and, as z traces out its path in the s, t-plane, $f(z)$ will trace out a corresponding path in the U, V-plane: the path of $f(z)$ will be *uniquely* determined by that of z. But if w is a *multiform* (many-valued) function of z, such that precisely n distinct values of w are determined by each value of z (except at branch points, where several values of w may be equal), then it is obvious that *one* w-plane no longer suffices (if n is greater than 1) to represent the path, the "march" of the function w. In the case of a *two*-valued function w, such as that determined by $w^2 = z$, two w-planes would be required and, quite generally, for an n-valued function (n finite or infinite), precisely n such w-planes would be required.

The advantages of considering *uniform* (*one*-valued) functions instead of n-valued functions (n greater than 1) should be obvious even to a non-mathematician. What Riemann did was this: instead of the n distinct w-planes, he introduced an n-sheeted surface, of the sort roughly described in what follows, on which the *multiform* function is *uniform*, that is, on which, to each "place" on the surface corresponds one, and only one, value of the function represented.

Riemann *united*, as it were, all the n planes into a *single* plane, and he did this by what may at first look like an inversion of the representation of the n branches of the n-valued function on n distinct planes; but a moment's consideration will show that, in effect, he *restored uniformity*. For he superimposed n z-planes on one another; each of these planes, or *sheets*, is associated with a particular branch of the function so that, as long as z moves in a particular sheet, the corresponding branch of the function is traversed by w (the n-valued function of z under discussion), and as z passes from one sheet to another, the branches are changed, one into another, until, on the variable z having traversed all the sheets and having returned to its initial position, the original branch is restored. The passage of the variable z from one sheet to another is effected by means of *cuts* (which may be thought of as straight-line bridges) joining branch points; along a given cut providing passage from one sheet to another, one "lip" of the upper sheet is imagined as pasted or joined to the opposite lip of the under sheet, and similarly for the other lip of the upper sheet. Diagrammatically, in cross-section,

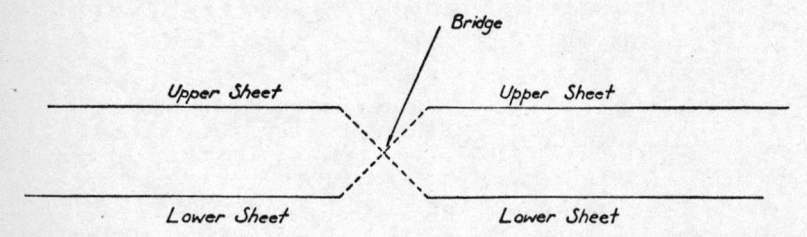

Bridge

Upper Sheet   *Upper Sheet*

Lower Sheet   *Lower Sheet*

The sheets are not joined along cuts (which may be drawn in many ways for given branch points) at random, but are so joined that, as z traverses its n-sheeted surface, passing from one sheet to another as a bridge or cut is reached, the *analytical* behavior of the function of z is pictured consistently, particularly as concerns the interchange of branches consequent on the variable z, if represented on a plane, having gone completely round a branch point. To this circuiting of a branch point on the *single* z-plane corresponds, on the n-sheeted Riemann surface, the passage from one sheet to another and the resultant interchange of the branches of the function.

There are many ways in which the variable may wander about the

n-sheeted *Riemann surface*, passing from one sheet to another. To each of these corresponds a particular interchange of the branches of the function, which may be symbolized by writing, one after another, letters denoting the several branches interchanged. In this way we get the symbols of certain *substitutions* (as in chapter 15) on n letters; all of these substitutions generate a group which, in some respects, pictures the nature of the function considered.

Riemann surfaces are not easy to represent pictorially, and those who use them content themselves with diagrammatical representations of the connection of the sheets, in much the same way that an organic chemist writes a "graphical" formula for a complicated carbon compound which recalls in a schematic manner the chemical behavior of the compound but which does not, and is not meant to, depict the actual spatial arrangement of the atoms in the compound. Riemann made wonderful advances, particularly in the theory of Abelian functions, by means of his surfaces and their topology— how shall the cuts be made so as to render the n-sheeted surface equivalent to a plane, being one question in this direction. But mathematicians are like other mortals in their ability to visualize complicated spatial relationships, namely, a high degree of spatial "intuition" is excessively rare.

Early in November, 1851, Riemann submitted his doctoral dissertation, *Grundlagen für eine allegemeine Theorie der Functionen einer veränderlichen complexen Grösse* (Foundations for a general theory of functions of a complex variable), for Gauss' consideration. This work by the young master of twenty five was one of the few modern contributions to mathematics that roused the enthusiasm of Gauss, then an almost legendary figure within four years of his death. When Riemann called on Gauss, after the latter had read the dissertation, Gauss told him that he himself had planned for years to write a treatise on the same topic. Gauss' official report to the Philosophical Faculty of the University of Göttingen is noteworthy as one of the rare formal pronouncements in which Gauss let himself go.

"The dissertation submitted by Herr Riemann offers convincing evidence of the author's thorough and penetrating investigations in those parts of the subject treated in the dissertation, of a creative, active, truly mathematical mind, and of a gloriously fertile originality. The presentation is perspicuous and concise and, in places, beautiful. The majority of readers would have preferred a greater clarity of arrangement. The whole is a substantial, valuable work, which not only satisfies the standards demanded for doctoral dissertations, but far exceeds them."

A month later Riemann passed his final examination, including the formality of a public "defense" of his dissertation. All went off successfully, and Riemann began to hope for a position in keeping with his talents. "I believe I have improved my prospects with my dissertation," he wrote to his father; "I hope also to learn to write more quickly and more fluently in time, especially if I mingle in society and if I get a chance to give lectures; therefore am I of good courage." He also apologizes to his father for not having gone after a vacant assistantship at the Göttingen Observatory more energetically, but as he hopes to be "habilitated" as a *Privatdozent* the outlook is not as dark as it might be.

For his *Habilitationsschrift* (probationary essay) Riemann had planned to submit a memoir on trigonometric series (Fourier series). But two and a half years were to pass before he might hang out his shingle as an unpaid university instructor picking up what he could in the way of fees from students not bound to attend his lectures. During the autumn of 1852 Riemann profited by Dirichlet's presence in Göttingen on a vacation and sought his advice on the embryonic memoir. Riemann's friends saw to it that the young man met the famous mathematician from Berlin—second only to Gauss—socially.

Dirichlet was captivated by Riemann's modesty and genius. "Next morning [after a dinner party] Dirichlet was with me for two hours," Riemann wrote his father. "He gave me the notes I needed for my probationary essay; otherwise I should have had to spend many hours in the library in laborious research. He also read over my dissertation with me and was very friendly—which I could hardly have expected, considering the great distance in rank between us. I hope he will remember me later on." During this visit of Dirichlet's there were excursions with Weber and others, and Riemann reported to his father that these human escapes from mathematics did him more good scientifically than if he had sat all day over his books.

From 1853 (Riemann was then twenty seven) onward he thought intensively about mathematical physics. By the end of the year he had completed the probationary essay, after many delays due to his growing passion for physical science.

There was still a trial lecture ahead of him before he could be appointed to the coveted—but unpaid—lectureship. For this ordeal he had submitted three titles for the faculty to choose from, hoping and expecting that one of the first two, on which he had prepared himself, would be selected. But he had incautiously included as his third offering a topic on which Gauss had pondered for sixty years or more—the foundations of geometry—and this he had not prepared. Gauss no doubt was curious to see what a Riemann's "gloriously fertile originality" would make of such a profound subject. To Riemann's consternation Gauss designated the third topic as the one on which Riemann should prove his mettle as a lecturer before the critical faculty. "So I am again in a quandary," the rash young man confided to his father, "since I have to work out this one. I have resumed my investigation of the connection between electricity, magnetism, light, and gravitation, and I have progressed so far that I can publish it without a qualm. I have become more and more convinced that Gauss has worked on this subject for years, and has talked to some friends (Weber among others) about it. I tell you this in confidence, lest I be thought arrogant—I hope it is not yet too late for me and that I shall gain recognition as an independent investigator."

The strain of carrying on two extremely difficult investigations simultaneously, while acting as Weber's assistant in the seminar in mathematical physics, combined with the usual handicaps of poverty, brought on a temporary breakdown. "I became so absorbed in my investigation of the unity of all physical laws that when the subject of the trial lecture was given me, I could not tear myself away from my research. Then, partly as a result of brooding on it, partly from staying indoors too much in this vile weather, I fell ill; my old trouble recurred with great pertinacity and I could not get on with my work. Only several weeks later, when the weather improved and I got more social stimulation, I began feeling better. For the summer I have rented a house in a garden, and since doing so my health has not bothered me. Having finished two weeks after Easter a piece of work I could not get out of, I began at once working on my trial lecture and finished it around Pentecost [that is, in about seven weeks]. I had some difficulty in getting a date for my lecture right away and almost had to return to Quickborn without having reached my goal. For Gauss is seriously ill and the physicians fear that his death is imminent. Being too weak to examine me, he asked me to wait till August, hoping that he might improve, especially as I would not lecture anyhow till fall. Then he decided anyway on the Friday after Pentecost to set the lecture for the next day at eleven thirty. On Saturday I was happily through with everything."

This is Riemann's own account of the historic lecture which was to revolutionize differential geometry and prepare the way for the geometrized physics of our own generation. In the same letter he tells how the work he had been doing around Easter turned out. Weber and some of his collaborators "had made very exact measurements of a phenomenon which up till then had never been investigated, the residual charge in a Leyden jar [after discharge it is found that the jar is not *completely* discharged] . . . I sent him [one of Weber's collaborators, Kohlrausch] my theory of this phenomenon, having worked it out specially for his purposes. I had found the explanation of the phenomenon through my general investigations of the connection between electricity, light, and magnetism. . . . This matter was important to me, because it was the first time I could apply my work to a phenomenon still unknown, and I hope that the publication [of it] will contribute to a favorable reception of my larger work."

The reception of Riemann's probationary lecture (June 10, 1854) was as cordial as even he could have wished in the scared secrecy of his modest heart. The lecture had made him sweat blood to prepare because he had determined to make it intelligible even to those members of the faculty who had but little knowledge of mathematics. In addition to being one of the great masterpieces of all mathematics, Riemann's essay *Ueber die Hypothesen, welche der Geometrie zu Grunde liegen* (On the hypotheses which lie at the foundations of geometry), is also a classic of presentation. Gauss was enthusiastic. "Against all tradition he had selected the third of the three topics submitted by the candidate, wishing to see how such a difficult subject would be handled by so young a man. He was surprised beyond all his expectations, and on returning from the faculty meeting expressed to Wilhelm Weber his highest appreciation of the ideas presented by Riemann, speaking with an enthusiasm that, for Gauss, was rare." What little can be said here about this masterpiece will be reserved for the conclusion of the present chapter.

After a rest at home with his family in Quickborn, Riemann returned in September to Göttingen, where he delivered a hastily prepared lecture (sitting up most of the night to get it ready on short notice) to a convention of scientists. His topic was the propagation of electricity in non-conductors. During the year he continued his researches in the mathematical theory of electricity and prepared a paper on Nobili's color rings because, as he wrote his sister Ida: "This subject is important, for very exact measurements can be made in connection with it, and the laws according to which electricity moves can be tested."

In the same letter (October 9, 1854) he expresses his unbounded joy at the success of his first academic lecture and his great satisfaction at the unexpectedly large number of auditors. Eight students had come to hear him! He had anticipated at the most two or three. Encouraged by this unhoped-for popularity, Riemann tells his father, "I have been able to hold my classes regularly. My first diffidence and constraint have subsided more and more, and I get accustomed to think more of the auditors than of myself, and to read in their expressions whether I should go on or explain the matter further."

When Dirichlet succeeded Gauss in 1855, Riemann's friends urged the authorities to appoint Riemann to the security of an assistant professorship, but the finances of the University could not be stretched so far. Nevertheless he was granted the equivalent of two hundred dollars a year, which was better than the uncertainty of half a dozen voluntary students' fees. His future worried him, and when presently he lost both his father and his sister Clara, making it impossible for him to escape for vacations to Quickborn, Riemann felt poor and miserable indeed. His three remaining sisters went to live with the other brother, a postal clerk in Bremen whose salary was princely beside that of the "economically valueless" mathematician.

The following year (1856; Riemann was then thirty) the outlook brightened a little. It was impossible for a creative genius like Riemann to be downed by despondency so long as he had the wherewithal to keep body and soul together in order that he might work. To this period belong part of his characteristically original work on Abelian functions, his classic on the hypergeometric series (see chapter on Gauss) and the differential equations—of great importance in mathematical physics—suggested by this series. In both of these works Riemann struck out on new directions of his own. The generality, the *intuitiveness*, of his approach was peculiarly his own. His work absorbed all his energies and made him happy in spite of material worries; possibly, too, the fatal optimism of the consumptive was already at work in him.

Riemann's development of the theory of Abelian functions is as unlike that of Weierstrass as moonlight is unlike sunlight. Weierstrass' attack was methodical, exact in all its details, like the advance of a perfectly disciplined army under a generalship that foresees everything and provides for all contingencies. Riemann, for his part, looked over the whole field, seeing everything but the details, which he left to take care of themselves, and was content to have grasped the key positions of the general topography in his imagination. The method of Weierstrass was arithmetical, that of Riemann geometrical and intuitive. To say that one is "better" than the other is meaningless; both cannot be seen from a common point of view.

Overwork and lack of reasonable comforts brought on a nervous breakdown early in his thirty first year, and Riemann was forced to spend a few weeks with a friend in the Hartz mountain country, where he was joined by Dedekind. The three took long tramps together into the mountains and Riemann soon recovered. Relieved of the strain of having to keep up academic appearances, Riemann indulged his sense of humor and kept his companions amused with his spontaneous wit. They also talked shop together—most mathematicians do when they get together, just as lawyers or doctors or business men do, provided they do not have to talk drivel to maintain the social conventions. One evening after a strenuous hike Riemann dipped into Brewster's life of Newton and discovered the letter to Bentley in which Newton himself asserts the impossibility of action at a distance without intervening media. This delighted Riemann and inspired him to an impromptu lecture. Today the "medium" which Riemann extolled is not the luminiferous ether, but his own "curved space," or its reflection in the space-time of relativity.

At last, in 1857, at the age of thirty one, Riemann got his assistant professorship. His salary was the equivalent of about three hundred dollars a year, but as he had had little all his life he missed less. However, a real disaster presently descended on him: his brother died and the care of three sisters fell to his lot. It figured out at exactly seventy five dollars a year for each of them. Love on nothing a year in a cottage may be paradise; existence on next to nothing in a university community is just plain hell. It was but little different in Riemann's day. No wonder he contracted consumption. However, the Lord, who had so generously given, shortly relieved Riemann of his youngest sister, Marie, so the individual budgets skyrocketed to one hundred dollars a year. If rations had to be watched, affection was free, and Riemann was more than repaid for his sacrifices by the self-confidence inspired in him by his sisters' devotion and encouragement. The Lord may have known that if ever a struggling mortal needed encouragement, poor Riemann did; still, it seems rather an odd way of providing what was required.

In 1858 Riemann produced his paper on electrodynamics, of which he told his sister Ida, "My discovery concerning the close connection between electricity and light I have dedicated to the Royal Society [of Göttingen]. From what I have heard, Gauss had devised another theory regarding this close connection, different from mine, and communicated it to his intimate friends. However, I am fully convinced that my theory is the correct one, and that in a few years it will be recognized as such. As is known, Gauss soon withdrew his memoir and did not publish it; probably he himself was not satisfied with it." Riemann would seem here to have been overoptimistic; Clerk Maxwell's electromagnetic theory is the one which today holds the field—in macroscopic phenomena. The present status of theories of light and the electromagnetic field is too complicated to be described here; it is sufficient to note that Riemann's theory has not survived.

Dirichlet died on May 5, 1859. He had always appreciated Riemann and had done his best to help the struggling young man along. This interest of Dirichlet's and Riemann's rapidly mounting reputation caused the government to promote Riemann to succeed Dirichlet. At thirty three Riemann thus became the second successor of Gauss. To ease his domestic difficulties the authorities let him reside at the Observatory, as Gauss had done. Recognition of the sincerest kind—praise from mathematicians who, although older than himself, were in some degree his rivals—now came in abundance. On a visit to Berlin he was fêted by Borchardt, Kummer, Kronecker, and Weierstrass. Learned societies, including the Royal Society of London and the French Academy of Sciences, honored him with membership, and in short he got the usual highest distinctions that can come to a man of science. A visit to Paris in 1860 acquainted him with the leading French mathematicians, particularly Hermite, whose admiration for Riemann was unbounded. This year, 1860, is memorable in the history of mathematical physics as that in which Riemann began intensive work on his memoir *Über eine Frage der Wärmeleitung* (On a Question in the Conduction of Heat), in which he develops the whole apparatus of quadratic differential forms (to be noticed in connection with Riemann's work in the foundations of geometry), which is today basic in the theory of relativity.

His material affairs having improved considerably with his appointment to the full professorship, Riemann was in a position to marry at the age of thirty six. His wife, Elise Koch, was a friend of his sisters. Barely a month after his marriage, Riemann fell ill in July 1862 with pleurisy. An incomplete recovery ended in consumption. Influential friends induced the Government to grant Riemann the funds for convalescence in the mild climate of Italy, where he spent the winter. The following spring on his return trip to Germany he took great delight in the art treasures of the many Italian cities he visited. This was the brief summer of his life.

Full of hope he left his beloved Italy, only to fall more seriously ill on reaching Göttingen. On the return journey he had grown careless, and while walking through deep snow in the Splügen Pass, had taken a severe chill. The following August (1863) he returned to Italy, stopping first at Pisa, where his daughter Ida (named after his older sister) was born. The winter was exceptionally harsh, the river Arno being frozen over. In May he moved to a small villa in the suburbs of Pisa. There his younger sister Helene died. His own illness, complicated by jaundice, grew steadily graver. To his great regret he was obliged to refuse a professorship offered to him at the University of Pisa. Göttingen generously extended his leave of absence to enable him to spend the following winter in Pisa, surrounded by his Italian mathematical friends. But further complications made him long for home, and after vainly seeking health in Leghorn and Genoa, he returned in October to Göttingen, where he spent a tolerable winter.

All this time he worked when he had the strength. At Göttingen he often expressed the desire to speak with Dedekind of the works he had not completed, but never felt quite strong enough to stand a visit. One of his last projects was a work on the mechanics of the ear, which he left incomplete. He had hoped to finish this, also some other things which he considered of great importance, and in a final attempt to regain his strength returned to Italy. His last days were spent in a villa at Selasca, Lago Maggiore.

Dedekind tells how his friend died. "But his strength declined rapidly; he felt himself that his end was near. The day before his death he worked under a fig tree, his soul filled with joy at the glorious landscape around him. . . . His life ebbed gently away, without strife or death agony; it seemed as though he followed with interest the separation of the soul from the body; his wife had to give him bread and wine . . . he said to her, 'Kiss our child.' She repeated the Lord's prayer with him; he could no longer speak; at the words 'Forgive us our trespasses' he looked up devoutly; she felt his hand grow colder in hers, and with a few last sighs his pure, noble heart had ceased to beat. The gentle mind which had been implanted in him in his father's house remained with him all his life, and he served his God faithfully, as his father had, but in a different way."

Thus Riemann died, in the full glory of his matured genius, on July 20, 1866, aged thirty nine. The inscription on his tombstone, erected by his Italian friends, closes with the words *"Denen die Gott lieben müssen alle Dinge zum Besten dienen,"* or as it is usually put in English, "All things work together for good to them that love the Lord."

Riemann's greatness as a mathematician resides in the powerful generality and unbounded scope of the methods and new points of view which he revealed to both pure and applied mathematics. Details never oppressed him; he saw the whole of a vast problem as a coherent unity. Even the fragmentary notes on uncompleted projects usually hint at some haunting novelty and sharpen our regret that Riemann died so long before his time. Only one of his great masterpieces can be described here, the memoir of 1854 on the foundations of geometry, and although it may not be quite fair to Clifford to use him merely to introduce another, we shall quote in its entirety his daring paper of 1870, *On the space-theory of matter*, as a singularly prophetic introduction to the body and spirit of Riemann's geometry. Clifford was no servile copyist but a man with a brilliantly original mind of his own, of whom it may be said, as Newton said of Cotes, "If he had lived we might have known something." The reader who is acquainted with any of the better available popular accounts of relativistic physics and the wave theory of electrons will recognize several curious adumbrations of current theories in Clifford's brief prophecy.

"Riemann has shown that as there are different kinds of lines and surfaces, so there are different kinds of space of three dimensions; and that we can only find out by experience to which of these kinds the space in which we live belongs. In particular, the axioms of plane geometry are true within the limits of experiment on the surface of a sheet of paper, and yet we know that the sheet is really covered with a number of small ridges and furrows, upon which (the total curvature being not zero) these axioms are not true. Similarly, he says, although the axioms of solid geometry are true within the limits of experiment for finite portions of our space, yet we have no reason to conclude that they are true for very small portions; and if any help can be got thereby for the explanation of physical phenomena, we may have reason to conclude that they are not true for very small portions of space.

"I wish here to indicate a manner in which these speculations may be applied to the investigation of physical phenomena. I hold in fact

(1) That small portions of space *are* in fact of a nature analogous to little hills on a surface which is on the average flat; namely, that the ordinary laws of geometry are not valid in them.

(2) That this property of being curved or distorted is continually being passed on from one portion of space to another after the manner of a wave.

(3) That this variation of the curvature of space is what really happens in that phenomenon which we call the *motion of matter*, whether ponderable or ethereal.

(4) That in the physical world nothing else takes place but this variation, subject (possibly) to the law of continuity.

"I am endeavoring in a general way to explain the laws of double refraction on this hypothesis, but have not yet arrived at any results sufficiently decisive to be communicated."

Riemann also believed that his new geometry would prove of scientific importance, as is shown by the conclusion of his memoir (Clifford's translation):

"Either therefore the reality which underlies space must form a discrete manifold, or we must seek the ground of its metric relations outside it, in binding forces which act upon it.

"The answer to these questions can only be got by starting from the conception of phenomena which has hitherto been justified by experience, and which Newton assumed as a foundation, and by making in this conception the successive changes required by facts which it cannot explain." And he goes on to say that researches like his own, starting from general notions, "can be useful in preventing this work from becoming hampered by too narrow views, and progress of knowledge of the interdependence of things from being checked by traditional prejudices.

"This leads us into the domain of another science, that of physics, into which the object of this work does not allow us to go today."

Riemann's work of 1854 put geometry in a new light. The geometry he visions is non-Euclidean, not in the sense of Lobatchewsky and Johann Bolyai, nor in that of Riemann's own elaboration of the hypothesis of the obtuse angle (as explained in chapter 16), but in a more comprehensive sense depending on the conception of *measurement*. To isolate *measure-relations* as the nerve of Riemann's theory is to do it an injustice; the theory contains much more than a workable philosophy of metrics, but this is one of its main features. No paraphrase of Riemann's concise memoir can bring out all that is in it; nevertheless, we shall attempt to describe some of his basic ideas, and we shall select three: the concept of a *manifold*, the definition of *distance*, and the notion of *curvature* of a manifold.

A manifold is a *class* of objects (at least in common mathematics) which is such that any member of the class can be completely specified by assigning to it certain numbers, in a definite order, corresponding to "numberable" properties of the elements, the assignment in the given order corresponding to a preassigned ordering of the "numberable" properties. Granted that this may be even less comprehensible than Riemann's definition, it is nevertheless a working basis from which to start, and all that it amounts to in plain mathematics is this: a manifold is a set of ordered "n-tuples" of numbers $(x_1, x_2, \ldots, x_n)$, where the parentheses, (), indicate that the numbers $x_1, x_2, \ldots, x_n$ are to be written in the order given. Two such n-tuples, $(x_1, x_2, \ldots, x_n)$ and $(y_1, y_2, \ldots, y_n)$ are *equal* when, and only when, corresponding numbers in them are respectively equal, namely, when, and only when, $x_1 = y_1, x_2 = y_2, \ldots, x_n = y_n$.

If precisely n numbers occur in each of these ordered n-tuples in the manifold, the manifold is said to be of n *dimensions*. Thus we are back again talking coordinates with Descartes. If each of the numbers in $(x_1, x_2, \ldots, x_n)$ is a positive, zero, or negative integer, or if it is an element of any countable set (a set whose elements may be counted off 1,2,3, ...), and if the like holds for every n-tuple in the set, the manifold is said to be *discrete*. If the numbers $x_1, x_2, \ldots, x_n$, may take on values *continuously* (as in the motion of a point along a line), the manifold is *continuous*.

This working definition has ignored—deliberately—the question of whether the set of ordered n-tuples is "the manifold" or whether something "represented by" these is "the manifold." Thus, when we say (x,y) are the coordinates of a point in a plane, we do not ask what "a point in a plane" is, but proceed to work with these *ordered couples of numbers* (x,y) where x,y run through all real numbers independently. On the other hand it may sometimes be advantageous to fix our attention on what such a symbol as (x,y) *represents*. Thus if x is the age in seconds of a man and y his height in centimeters, we may be interested in the *man* (or the class of all men) rather than in his *coordinates, with which alone the mathematics of our enquiry is concerned*. In this same order of ideas, geometry is no longer concerned with what "space" "is"—whether "is" means anything or not in relation to "space." Space, for a modern mathematician, is merely a number-manifold of the kind described above, and this conception of space grew out of Riemann's "manifolds."

Passing on to measurement, Riemann states that "Measurement consists in a superposition of the magnitudes to be compared. If this is lacking, magnitudes can be compared only when one is part of another, and then only the more or less, but not the how much, can be decided." It may be said in passing that a consistent and useful theory of measurement is at present an urgent desideratum in theoretical physics, particularly in all questions where quanta and relativity are of importance.

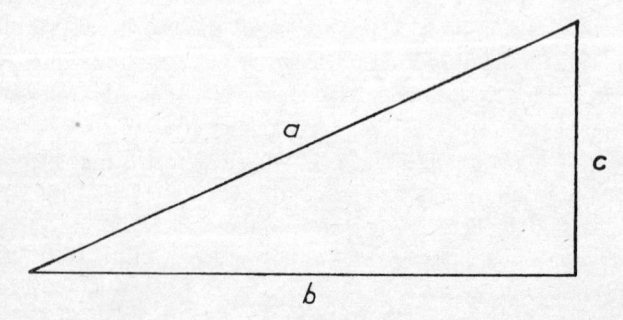

Descending once more from philosophical generalities to less mystical mathematics, Riemann proceeded to lay down a definition of *distance*, extracted from his concept of measurement, which has proved to be extremely fruitful in both physics and mathematics. The Pythagorean proposition that $a^2 = b^2 + c^2$ or $a = \sqrt{b^2 + c^2}$, where a is the length of the longest side of a right-angled triangle and b,c are the lengths of the other two sides, is the fundamental formula for the measurement of distances in a *plane*. How shall this be extended to a *curved surface*? To straight lines on the plane correspond geodesics (see chapter 14) on the surface; but on a sphere, for example, the Pythagorean proposition is not true for a right-angled triangle formed by geodesics. Riemann generalized the Pythagorean formula to any manifold as follows:

Let $(x_1, x_2, \ldots, x_n)$, $(x_1 + x_1', x_2 + x_2', \ldots, x_n + x_n')$ be the coordinates of two "points" in the manifold which are "infinitesimally near" one another. For our present purpose the meaning of "infinitesimally near" is that powers higher than the second of $x_1', x_2', \ldots, x_n'$, which measure the "separation" of the two points in the manifold, can be neglected. For simplicity we shall state the definition when $n = 4$—giving the distance between two neighboring points in a space of four dimensions: the distance is the square root of

$$g_{11}x_1'^2 + g_{22}x_2'^2 + g_{33}x_3'^2 + g_{44}x_4'^2$$
$$+ g_{12}x_1'x_2' + g_{13}x_1'x_3' + g_{14}x_1'x_4'$$
$$+ g_{23}x_2'x_3' + g_{24}x_2'x_4'$$
$$+ g_{34}x_3'x_4',$$

in which the ten coefficients $g_{11}, \ldots, g_{34}$ are functions of x_1, x_2, x_3, x_4. For a particular choice of the g's, one "space" is defined. Thus we might have $g_{11} = 1$, $g_{22} = 1$, $g_{33} = 1$, $g_{44} = -1$, and all the other g's zero; or we might consider a space in which all the g's except g_{44} and g_{34} were zero, and so on. A space considered in relativity is of this general kind in which all the g's except $g_{11}, g_{22}, g_{33}, g_{44}$ are zero, and these are certain simple expressions involving x_1, x_2, x_3, x_4.

In the case of an n-dimensional space the distance between *neighboring* points is defined in a similar manner; the general expression contains $\frac{1}{2}n(n + 1)$ terms. The generalized Pythagorean formula for the distance between neighboring points being given, it is a solvable problem in the integral calculus to find the distance between *any* two points of the space. A space whose *metric* (system of measurement) is defined by a formula of the type described is called *Riemannian*.

Curvature, as conceived by Riemann (and before him by Gauss; see chapter on the latter) is another generalization from common experience. A straight line has zero curvature; the "measure" of the amount by which a curved line departs from straightness may be the same for every point of the curve (as it is for a circle), or it may vary from point to point of the curve, when it becomes necessary again to express the "amount of curvature" through the use of infinitesimals. For curved surfaces, the curvature is measured similarly by the amount of departure from a plane, which has zero curvature. This may be generalized and made a little more precise as follows. For simplicity we state first the situation for a two-dimensional space, namely for a surface as we ordinarily imagine surfaces. It is possible from the formula

$$g_{11}x_1'^2 + g_{12}x_1'x_2' + g_{22}x_2'^2,$$

expressing (as before) the square of the distance between neighboring points on a given surface (determined when the functions g_{11}, g_{12}, g_{22} are given), to calculate the measure of curvature of any point of the surface *wholly in terms of the given functions* g_{11}, g_{12}, g_{22}. Now, in ordinary language, to speak of the "curvature" of a space of more than *two* dimensions is to make a meaningless noise. Nevertheless Riemann, generalizing Gauss, proceeded in the same *mathematical* way to build up an expression involving *all* the g's in the general case of an *n*-dimensional space, which is of *the same kind mathematically* as the Gaussian expression for the curvature of a *surface*, and this generalized expression is what he called the *measure of curvature* of the space. It is possible to exhibit visual representations of a curved space of more than two dimensions, but such aids to perception are about as useful as a pair of broken crutches to a man with no feet, for they add nothing to the understanding and they are mathematically useless.

Why did Riemann do all this and what has come out of it? Not attempting to answer the first, except to suggest that Riemann did what he did because his daemon drove him, we may briefly enumerate some of the gains that have accrued from Riemann's revolution in geometrical thought. First, it put the creation of "spaces" and "geometries" in unlimited number for specific purposes—use in dynamics, or in pure geometry, or in physical science—within the capabilities of professional geometers, and it baled together huge masses of important geometrical theorems into compact bundles that could be handled easily as wholes. Second, it clarified our conception of space, at least so far as mathematicians deal in "space," and stripped that mystic nonentity Space of its last shred of mystery. Riemann's achievement has taught mathematicians to disbelieve in *any* geometry, or in *any* space, as a *necessary* mode of human perception. It was the last nail in the coffin of absolute space, and the first in that of the "absolutes" of nineteenth century physics.

Finally, the curvature which Riemann defined, the processes which he devised for the investigation of quadratic differential forms (those giving the formula for the square of the distance between neighboring points in a space of any number of dimensions), and his recognition of the fact that the curvature is an invariant (in the technical sense explained in previous chapters), all found their physical interpretations in the theory of relativity. Whether the latter is in its final form or not is beside the point; since relativity our outlook on physical science is not what it was before. Without the work of Riemann this revolution in scientific thought would have been impossible—unless some later man had created the concepts and the mathematical methods that Riemann created.

CHAPTER TWENTY SEVEN

Arithmetic the Second

KUMMER AND DEDEKIND

We see therefore that ideal prime factors reveal the essence of complex numbers, make them transparent, as it were, and disclose their inner crystalline structure.
—E. E. KUMMER

The majority of my readers will be greatly disappointed to learn that by this commonplace observation the secret of continuity is to be revealed.—R. DEDEKIND

IT IS A CURIOUS FACT that although arithmetic—the theory of numbers—has been the fertile mother of more profound problems and powerful methods than any other discipline of mathematics, it is usually regarded as standing rather to one side of the main progress as a more or less cold-blooded spectator of the flashier achievements of geometry and analysis, particularly in their services to physical science, and comparatively few of the great mathematicians of the past two thousand years have expended their more serious efforts on the advancement of the science of "pure number."

Many causes have determined this strange neglect of what, after all, is mathematics par excellence. Among these we need note only the following: arithmetic at present is on a higher plane of intrinsic difficulty than the other great fields of mathematics; the immediate applications of the theory of numbers to science are few and not readily perceptible to the ordinary run of creative mathematicians, although some of the greatest have felt that the proper mathematics of nature will be found ultimately in the behavior of the common whole numbers; and, finally, it is only human for mathematicians—at least for some, even the great—to court reputation and popularity in their own generation by reaping the easier harvests of a spectacular success in analysis, geometry, or applied mathematics. Even Gauss succumbed, to his keen regret in middle life.

Modern arithmetic—after Gauss—began with Kummer. The origin of Kummer's theory in his attempt to prove Fermat's Last Theorem has already been noted (Chapter 25). Something of the man's long life may be told before we pass to Dedekind. Kummer was a typical German of the old school with all the blunt simplicity, good nature, and racy humor, which characterized that fast-vanishing species at its best. Museum specimens, aged in the wood, could be found behind the bar in any San Francisco German beer garden a generation ago.

Although Ernst Eduard Kummer (January 29, 1810–May 14, 1893) was born only five years before the deflation of Napoleon, the glorious Emperor of the French played an important if unwitting part in Kummer's life. The son of a physician of Sorau (then in the principality of Brandenburg), Germany, Kummer at the age of three lost his father: the lousy remnant of Napoleon's Grand Army, filtering back through Germany to France, brought with it the characteristically Russian gift of typhus, which it shared freely with the well-washed Germans. The overworked physician caught the disease, died of it, and left Ernst and an elder brother to the care of his widow. Young Kummer grew up in cramping poverty, but his struggling mother contrived somehow or another to see her sons through the local Gymnasium. The arrogance and exactions of the Napoleonic French, no less than the memory of his father, which the mother kept alive, made young Kummer an extremely practical patriot, and it was with real gusto that he devoted much of his superb scientific talent in later life to training German army officers in ballistics at the war college of Berlin. Many of his students gave good accounts of themselves in the Franco-Prussian War.

At the age of eighteen (in 1828) Kummer was sent by his mother to the University of Halle to study theology and otherwise fit himself for a career in the church. Owing to his poverty Kummer did not reside at the University, but tramped back and forth every day from Sorau to Halle with his food and books in a knapsack on his back. Regarding his theological studies Kummer makes the interesting observation that it is more or less a matter of accident or environment whether a mind with a gift for abstract speculation turns to philosophy or to mathematics. The accident in his own case was the presence at Halle of Heinrich Ferdinand Scherk (1798–1885) as professor of mathematics. Scherk was rather old fashioned, but he had an enthusiasm for algebra and the theory of numbers which he imparted to young Kummer. Under Scherk's guidance Kummer soon abandoned his moral and theological studies in favor of mathematics. Echoing

Descartes, Kummer said he preferred mathematics to philosophy because "mere errors and false views cannot enter mathematics." Had Kummer lived till today he might have modified his statement, for he was a broadminded man, and the present philosophical tendencies in mathematics are sometimes curiously reminiscent of medieval theology. In his third year at the University Kummer solved a prize problem in mathematics and was awarded his Ph.D. degree (September 10, 1831) at the age of twenty one. No university position being open at the time, Kummer began his career as a teacher in his old Gymnasium.

In 1832 he moved to Liegnitz, where he taught for ten years in the Gymnasium. It was there that he started Kronecker off on his revolutionary career. Fortunately Kummer was not so hard up as Weierstrass under similar circumstances and was able to afford postage for scientific correspondence. The eminent mathematicians (including Jacobi) with whom Kummer shared his mathematical discoveries saw to it that the young genius of a schoolteacher was lifted into a more suitable position at the earliest opportunity, and in 1842 Kummer was appointed Professor of Mathematics at the University of Breslau. He taught there till 1855, when the death of Gauss caused extensive revisions in the mathematical map of Europe.

It had been assumed that Dirichlet was contented at Berlin, then the mathematical capital of the world. But when Gauss died, Dirichlet could not resist the temptation of succeeding the Prince of Mathematicians and his own former master as professor at Göttingen. Even today the glory of being a "successor of Gauss" has an almost irresistible attraction for mathematicians who might easily earn more money in other positions, and until quite recently Göttingen could choose whom it would. The high esteem in which Kummer was held by his fellow mathematicians can be judged by the fact that he was the unanimous choice to succeed Dirichlet at Berlin. Since the age of twenty nine he had been a corresponding member of the Royal Berlin Academy. He now (1855) succeeded Dirichlet in both the University and the Academy, and was also appointed professor at the Berlin War College.

Kummer was one of those rarest of all scientific geniuses who are first class in the most abstract mathematics, the applications of mathematics to practical affairs, including war, which is the most unblushingly practical of all human idiocies, and finally in the ability to do experimental physics of a high degree of excellence. His finest work was in the theory of numbers where his profound originality led him to inventions of the very first order of importance, but in other fields—analysis, geometry, and applied physics—he also did outstanding work. Although Kummer's advance in the higher arithmetic was of the pioneering sort that justifies comparing him with the creators of non-Euclidean geometry, we somehow get the impression on reviewing his life of eighty three years, that splendid as his achievement was, he did not accomplish all that he must have had in him. Possibly his lack of personal ambition (an instance is given presently), his easy-going geniality, and his broad sense of humor prevented him from winding himself in an attempt to beat the record.

The nature of what Kummer did in the theory of numbers has been described in the chapter on Kronecker: he *restored the fundamental theorem of arithmetic to those algebraic number fields which arise in the attempt to prove Fermat's Last Theorem and in the Gaussian theory of cyclotomy, and he effected this restoration by the creation of an entirely new species of numbers, his so-called "ideal numbers."* He also carried on the work of Gauss on the law of biquadratic reciprocity and sought the laws of reciprocity for degrees higher than the fourth.

As has already been mentioned in preceding chapters, Kummer's "ideal numbers" are now largely displaced by Dedekind's "ideals," which will be described when we come to them, so it is not necessary to attempt here the almost impossible feat of explaining in untechnical language what Kummer's "numbers" are. But what he accomplished by means of them can be stated with sufficient accuracy for an account like the present: Kummer *proved* that $x^p + y^p = z^p$, where p is a prime, is impossible in integers x,y,z, all different from zero, for a whole very extensive class of primes p. He did not succeed in proving Fermat's theorem for *all* primes; certain slippery "exceptional primes" eluded Kummer's net—and still do. Nevertheless the step ahead which he took so far surpassed everything that all his predecessors had done that Kummer became famous almost in spite of himself. He was awarded a prize for which he had not competed.

The report in full of the French Academy of Sciences on the competition for its "Grand Prize" in 1857 ran as follows. "Report on the competition for the grand prize in mathematical sciences. Already set in the competition for 1853 and prorogued to 1856. The committee, having found no work which seemed to it worthy of the prize among those submitted to it in competition, proposed to the Academy to award it to M. Kummer, for his beautiful researches on complex numbers composed of roots of unity* and integers. The Academy adopted this proposal."

Kummer's earliest work on Fermat's Last Theorem is dated October, 1835. This was followed by further papers in 1844–47, the last of which was entitled *Proof of Fermat's Theorem on the Impossibility of $x^p + y^p = z^p$ for an Infinite† Number of Primes p.* He continued to add improvements to his theory, including its application to the laws of higher reciprocity, till 1874, when he was sixty four years old.

Although these highly abstract researches were the field of his greatest interest, and although he said of himself, "To describe my personal scientific attitude more exactly, I may conveniently designate it as *theoretical* . . . ; I have particularly striven for that mathematical knowledge which finds its proper sphere in mathematics without reference to applications," Kummer was no narrow specialist. Somewhat like Gauss, he appeared to take equal pleasure in both pure and applied science. Gauss indeed, through his works, was Kummer's real teacher, and the apt pupil proved his mettle by extending his master's work on the hypergeometric series, adding to what Gauss had done substantial developments which today are of great use in the theory of those differential equations which recur most frequently in mathematical physics.

Again, the magnificent work of Hamilton on systems of rays (in optics) inspired Kummer to one of his own most beautiful inventions, that of the surface of the fourth degree which is known by his name and which plays a fundamental part in the geometry of Euclidean

---

*If $x^p + y^p = z^p$, then $x^p = z^p - y^p$, and resolving $z^p - y^p$, into its p factors of the first degree, we get

$$x^p = (z-y)(z-ry)(z-r^2y) \ldots (z-r^{p-1}y),$$

in which r is a "p th root of unity" (other than 1), namely $r^p - 1 = 0$, with r not equal to 1. The algebraic integers in the field of degree p generated by r are those which Kummer introduced into the study of Fermat's equation, and which led him to the invention of his "ideal numbers" to restore unique factorization in the field—an integer in such a field is not uniquely the product of primes in the field for *all* primes p.

†The "infinite" in Kummer's title is still (1936) unjustified; "many" should be put for "infinite."

space when that space is four-dimensional (instead of three-dimensional, as we ordinarily imagine it), as happens when straight lines instead of points are taken as the irreducible elements out of which the space is constructed. This surface (and its generalizations to higher spaces) occupied the center of the stage in a whole department of nineteenth century geometry; it was found (by Cayley) to be representable (parametrically—see the chapter on Gauss) by means of the quadruply periodic functions to which Jacobi and Hermite devoted some of their best efforts.

Quite recently (since 1934) it has been observed by Sir Arthur Eddington that Kummer's surface is a sort of cousin to Dirac's wave equation in quantum mechanics (both have the same finite group; Kummer's surface is the wave surface in space of four dimensions).

To complete the circle, Kummer was led back by his study of systems of rays to physics, and he made important contributions to the theory of atmospheric refraction. In his work at the War College he astonished the scientific world by proving himself a first-rate experimenter in his work on ballistics. With characteristic humor Kummer excused himself for this bad fall from mathematical grace: "When *I* attack a problem experimentally," he told a young friend, "it is a proof that the problem is mathematically impregnable."

Remembering his own struggles to get an education and his mother's sacrifices, Kummer was not only a father to his students but something of a brother to their parents. Thousands of grateful young men who had been helped on their way by Kummer at the University of Berlin or the War College remembered him all their lives as a great teacher and a great friend. Once a needy young mathematician about to come up for his doctor's examination was stricken with smallpox and had to return to his home in Posen near the Russian border. No word came from him, but it was known that he was desperately poor. When Kummer heard that the young man was probably unable to afford proper care, he sought out a friend of the student, gave him the requisite money and sent him off to Posen to see that what was necessary was done. In his teaching Kummer was famous for his homely similes and philosophical asides. Thus, to drive home the importance of a particular factor in a certain expression, he observed that "If you neglect this factor you will be like a man who in eating a plum swallows the pit and spits out the pulp."

The last nine years of Kummer's life were spent in complete retirement. "Nothing will be found in my posthumous papers," he said, thinking of the mass of work which Gauss left to be edited after his death. Surrounded by his family (nine children survived him), Kummer gave up mathematics for good when he retired, and except for occasional trips to the scenes of his boyhood lived in the strictest seclusion. He died after a short attack of influenza on May 14, 1893, aged eighty three.

Kummer's successor in arithmetic was Julius Wilhelm Richard Dedekind (he dropped the first two names when he grew up), one of the greatest mathematicians and one of the most original Germany—or any other country—has produced. Like Kummer, Dedekind had a long life (October 6, 1831–February 12, 1916), and he remained mathematically active to within a short time of his death. When he died in 1916 Dedekind had been a mathematical classic for well over a generation. As Edmund Landau (himself a friend and follower of Dedekind in some of his work) said in his commemorative address to the Royal Society of Göttingen in 1917: "Richard Dedekind was not only a great mathematician, but one of the wholly great in the history of mathematics, now and in the past, the last hero of a great epoch, the last pupil of Gauss, for four decades himself a classic, from whose works not only we, but our teachers and the teachers of our teachers, have drawn."

Richard Dedekind, the youngest of the four children of Julius Levin Ulrich Dedekind, a professor of law, was born in Brunswick, the natal place of Gauss.* From the age of seven to sixteen Richard studied at the Gymnasium in his home town. He gave no early evidence of unmistakable mathematical genius; in fact his first loves were physics and chemistry, and he looked upon mathematics as the handmaiden—or scullery slut—of the sciences. But he did not wander long

---

*No adequate biography of Dedekind has yet appeared. A life was to have been included in the third volume of his collected works (1932), but was not, owing to the death of the editor in chief (Robert Fricke). The account here is based on Landau's commemorative address. Note that, following the good old Teutonic custom of some German biographers, Landau omits all mention of Dedekind's mother. This no doubt is in accordance with the theory of the "three K's" propounded by the late Kaiser of Germany and heartily endorsed by Adolf Hitler: "A woman's whole duty is comprised in the three big K's—Kissing, Kooking [cooking is spelt with a K in German], and Kids." Still, one would like to know at least the maiden name of a great man's mother.

in darkness. By the age of seventeen he had smelt numerous rats in the alleged reasoning of physics and had turned to mathematics for less objectionable logic. In 1848 he entered the Caroline College—the same institution that gave the youthful Gauss an opportunity for self-instruction in mathematics. At the college Dedekind mastered the elements of analytic geometry, "advanced" algebra, the calculus, and "higher" mechanics. Thus he was well prepared to begin serious work when he entered the University of Göttingen in 1850 at the age of nineteen. His principal instructors were Moritz Abraham Stern (1807–1894), who wrote extensively on the theory of numbers, Gauss, and Wilhelm Weber the physicist. From these three men Dedekind got a thorough grounding in the calculus, the elements of the higher arithmetic, least squares, higher geodesy, and experimental physics.

In later life Dedekind regretted that the mathematical instruction available during his student years at Göttingen, while adequate for the rather low requirements for a state teacher's certificate, was inconsiderable as a preparation for a mathematical career. Subjects of living interest were not touched upon, and Dedekind had to spend two years of hard labor after taking his degree to get up by himself elliptic functions, modern geometry, higher algebra, and mathematical physics—all of which at the time were being brilliantly expounded at Berlin by Jacobi, Steiner, and Dirichlet. In 1852 Dedekind got his doctor's degree (at the age of twenty one) from Gauss for a short dissertation on Eulerian integrals. There is no need to explain what this was: the dissertation was a useful, independent piece of work, but it betrayed no such genius as is evident on every page of many of Dedekind's later works. Gauss' verdict on the dissertation will be of interest: "The memoir prepared by Herr Dedekind is concerned with a research in the integral calculus, which is by no means commonplace. The author evinces not only a very good knowledge of the relevant field, but also such an independence as augurs favorably for his future achievement. As a test essay for admission to the examination I find the memoir completely satisfying." Gauss evidently saw more in the dissertation than some later critics have detected; possibly his close contact with the young author enabled him to read between the lines. However, the report, even as it stands, is more or less the usual perfunctory politeness customary in accepting a passable dissertation, and we do not know whether Gauss really foresaw Dedekind's penetrating originality.

In 1854 Dedekind was appointed lecturer (*Privatdozent*) at Göttingen, a position which he held for four years. On the death of Gauss in 1855 Dirichlet moved from Berlin to Göttingen. For the remaining three years of his stay at Göttingen, Dedekind attended Dirichlet's most important lectures. Later he was to edit Dirichlet's famous treatise on the theory of numbers and add to it the epoch-making "Eleventh Supplement" containing an outline of his own theory of algebraic numbers. He also became a friend of the great Riemann, then beginning his career. Dedekind's university lectures were for the most part elementary, but in 1857-8 he gave a course (to two students, Selling and Auwers) on the Galois theory of equations. This was probably the first time that the Galois theory had appeared formally in a university course. Dedekind was one of the first to appreciate the fundamental importance of the concept of a group in algebra and arithmetic. In this early work Dedekind already exhibited two of the leading characteristics of his later thought, abstractness and generality. Instead of regarding a finite group from the standpoint offered by its representation in terms of substitutions (see chapters on Galois and Cauchy), Dedekind defined groups by means of their postulates (substantially as described in Chapter 15) and sought to derive their properties from this distillation of their essence. This is in the modern manner: abstractness and therefore generality. The second characteristic, generality, is, as just implied, a consequence of the first.

At the age of twenty six Dedekind was appointed (in 1857) ordinary professor at the Zurich polytechnic, where he stayed five years, returning in 1862 to Brunswick as professor at the technical high school. There he stuck for half a century. The most important task for Dedekind's official biographer—provided one is unearthed—will be to explain (not explain away) the singular fact that Dedekind occupied a relatively obscure position for fifty years while men who were not fit to lace his shoes filled important and influential university chairs. To say that Dedekind preferred obscurity is one explanation. Those who believe it should leave the stock market severely alone, for as surely as God made little lambs they will be fleeced.

Till his death (1916) in his eighty fifth year Dedekind remained fresh of mind and robust of body. He never married, but lived with his sister Julie, remembered as a novelist, till her death in 1914. His other sister, Mathilde, died in 1860; his brother became a distinguished jurist.

Such are the bare facts of any importance in Dedekind's material career. He lived so long that although some of his work (his theory of irrational numbers, described presently) had been familiar to all students of analysis for a generation before his death, he himself had become almost a legend and many classed him with the shadowy dead. Twelve years before his death, Teubner's *Calendar for Mathematicians* listed Dedekind as having died on September 4, 1899, much to Dedekind's amusement. The day, September 4, might possibly prove to be correct, he wrote to the editor, but the year certainly was wrong. "According to my own memorandum I passed this day in perfect health and enjoyed a very stimulating conversation on 'system and theory' with my luncheon guest and honored friend Georg Cantor of Halle."

Dedekind's mathematical activity impinged almost wholly on the domain of number in its widest sense. We have space for only two of his greatest achievements and we shall describe first his fundamental contribution, that of the "Dedekind cut," to the theory of irrational numbers and hence to the foundations of analysis. This being of the very first importance we may recall briefly the nature of the matter. If a, b are common whole numbers, the fraction a/b is called a rational number; if no whole numbers m, n exist such that a certain "number" N is expressible as m/n, then N is called an irrational number. Thus $\sqrt{2}$, $\sqrt{3}$, $\sqrt{6}$ are irrational numbers. If an irrational number be expressed in the decimal notation the digits following the decimal point exhibit no regularities—there is no "period" which repeats, as in the decimal representations of a rational number, say $13/11$, $= 1.181818$ . . . , where the "18" repeats indefinitely. How then, if the representation is entirely lawless, are decimals equivalent to irrationals to be defined, let alone manipulated? Have we even any clear conception of what an irrational number is? Eudoxus thought he had, and Dedekind's definition of equality between numbers, rational or irrational, is identical with that of Eudoxus (see Chapter 2).

If two rational numbers are equal, it is no doubt obvious that their square roots are equal. Thus 2×3 and 6 are equal; so also then are $\sqrt{2 \times 3}$ and $\sqrt{6}$. But it is *not* obvious that $\sqrt{2} \times \sqrt{3} = \sqrt{2 \times 3}$, and hence that $\sqrt{2} \times \sqrt{3} = \sqrt{6}$. The un-obviousness of this simple assumed equality, $\sqrt{2} \times \sqrt{3} = \sqrt{6}$, taken for granted in school arithmetic, is evident if we visualize what the equality implies: the "lawless" square roots of 2, 3, 6 are to be extracted, the first two of these are then to be multiplied together, and the result is to come out equal to the third. As not one of these three roots can be extracted exactly, no matter to how many decimal places the computation is carried, it is clear that the verification by multiplication as just described will never be complete. The whole human race toiling incessantly through all its existence could never *prove* in this way that $\sqrt{2} \times \sqrt{3} = \sqrt{6}$. Closer and closer approximations to equality would be attained as time went on, but finality would continue to recede. To make these concepts of "approximation" and "equality" precise, or to replace our first crude conceptions of irrationals by sharper descriptions which will obviate the difficulties indicated, was the task Dedekind set himself in the early 1870's—his work on *Continuity and Irrational Numbers* was published in 1872.

The heart of Dedekind's theory of *irrational* numbers is his concept of the "cut" or "section" (*Schnitt*): a cut separates *all* rational numbers into *two* classes, so that each number in the *first* class is *less than* each number in the *second* class; every such cut which does not "correspond" to a rational number "defines" an irrational number. This bald statement needs elaboration, particularly as even an accurate exposition conceals certain subtle difficulties rooted in the theory of the mathematical infinite, which will reappear when we consider the life of Dedekind's friend Cantor.

Assume that some rule has been prescribed which separates *all* rational numbers into *two* classes, say an "upper" class and a "lower" class, such that each number in the *lower* class is *less than* every number in the *upper* class. (Such an assumption would not pass unchallenged today by all schools of mathematical philosophy. However, for the moment, it may be regarded as unobjectionable.) On this assumption one of three mutually exclusive situations is possible.

(A) There may be a number in the *lower* class which is *greater* than every other number in that class.

(B) There may be a number in the *upper* class which is *less* than every other number in that class.

(C) *Neither* of the numbers (*greatest* in [A], *least* in [B]) described in (A), (B) may exist.

The possibility which leads to irrational numbers is (C). For, if (C) holds, the assumed rule "defines" a definite break or "cut" in the set of all rational numbers. The upper and lower classes strive, as it were, to meet. But in order for the classes to meet the cut must be filled with some "number," and, by (C), no such filling is possible.

Here we appeal to intuition. All the distances measured from any fixed point along a given straight line "correspond" to "numbers" which "measure" the distances. If the cut is to be left unfilled, we must picture the straight line, which we may conceive of as having been traced out by the *continuous* motion of a point, as now having an unbridgeable gap in it. This violates our intuitive notions, so we say, by definition, that each cut *does define* a number. The number thus defined is not rational, namely it is irrational. To provide a manageable scheme for operating with the *irrationals* thus *defined by cuts* (of the kind [C]) we now consider the *lower class of rationals* in (C) as being equivalent to the irrational which the cut defines.

One example will suffice. The *irrational* square root of 2 is defined by the cut whose upper class contains *all* the positive rational numbers whose squares are greater than 2, and whose lower class contains *all* other *rational* numbers.

If the somewhat elusive concept of cuts is distasteful two remedies may be suggested: devise a definition of irrationals which is less mystical than Dedekind's and fully as usable; follow Kronecker and, denying that irrational numbers exist, reconstruct mathematics without them. In the present state of mathematics some theory of irrationals is convenient. But, from the very nature of an irrational number, it would seem to be necessary to understand the mathematical infinite thoroughly before an adequate theory of irrationals is possible. The appeal to infinite classes is obvious in Dedekind's definition of a cut. Such classes lead to serious logical difficulties.

It depends upon the individual mathematician's level of sophistication whether he regards these difficulties as relevant or of no consequence for the consistent development of mathematics. The courageous analyst goes boldly ahead, piling one Babel on top of another and trusting that no outraged god of reason will confound him and all his works, while the critical logician, peering cynically at the foundations of his brother's imposing skyscraper, makes a rapid mental calculation predicting the date of collapse. In the meantime all are busy and all seem to be enjoying themselves. But one conclusion appears to be inescapable: without a consistent theory of the mathematical infinite there is no theory of irrationals; without a theory of irrationals there is no mathematical analysis in any form even remotely resembling what we now have; and finally, without analysis the major part of mathematics—including geometry and most of applied mathematics—as it now exists would cease to exist.

The most important task confronting mathematicians would therefore seem to be the construction of a satisfactory theory of the infinite. Cantor attempted this, with what success will be seen later. As for the Dedekind theory of irrationals, its author seems to have had some qualms, for he hesitated over two years before venturing to publish it. If the reader will glance back at Eudoxus' definition of "same ratio" (Chapter 2) he will see that "infinite difficulties" occur there too, specifically in the phrase "any whatever equimultiples." Nevertheless some progress has been made since Eudoxus wrote; we are at least beginning to understand the nature of our difficulties.

The other outstanding contribution which Dedekind made to the concept of "number" was in the direction of algebraic numbers. For the nature of the fundamental problem concerned we must refer to what was said in the chapter on Kronecker concerning algebraic number fields and the resolution of algebraic *integers* into their *prime* factors. The crux of the matter is that in *some* such fields resolution into prime factors is *not unique* as it is in common arithmetic; Dedekind restored this highly desirable uniqueness by the invention of what he called *ideals*. An ideal is not a number, but an infinite class of numbers, so again Dedekind overcame his difficulties by taking refuge in the infinite.

The concept of an ideal is not hard to grasp, although there is one twist—*the more inclusive class divides the less inclusive*, as will be explained in a moment—which shocks common sense. However, common sense was made to be shocked; had we nothing less dentable than shock-proof common sense we should be a race of mongoloid imbeciles. An ideal must do at least two things: it must leave common (rational) arithmetic substantially as it is, and it must force the recalcitrant algebraic integers to obey that fundamental law of arithmetic—*unique* decomposition into primes—which they defy.

The point about a more inclusive class dividing a less inclusive refers to the following phenomenon (and its generalization, as stated presently). Consider the fact that 2 divides 4—*arithmetically*, that is, *without remainder*. Instead of this obvious fact, which leads nowhere if followed into algebraic number fields, we replace 2 by the *class* of *all* its integer multiples, ..., $-8, -6, -4, -2, 0, 2, 4, 6, 8, \ldots$ As a matter of convenience we denote this class by (2). In the same way (4) denotes the class of *all* integer multiples of 4. Some of the numbers in (4) are ..., $-16, -12, -8, -4, 0, 8, 12, 16, \ldots$ It is now obvious that (2) is the more inclusive class; in fact (2) contains *all* the numbers in (4) and in addition (to mention only two) -6 and 6. The fact that (2) contains (4) is symbolized by writing $(2)|(4)$. It can be seen quite easily that if m, n are any common whole numbers then $(m)|(n)$ *when, and only when, m divides n.*

This might suggest that the notion of common arithmetical divisibility be replaced by that of class inclusion as just described. But this replacement would be futile if it failed to preserve the characteristic properties of arithmetical divisibility. That it does so preserve them can be seen in detail, but one instance must suffice. If m divides n, and n divides l, then m divides l—for example, 12 divides 24 and 24 divides 72, and 12 does in fact divide 72. Transferred to classes, as above, this becomes: if $(m)|(n)$ and $(n)|(l)$, then $(m)|(l)$ or, in English, if the class (m) contains the class (n), and if the class (n) contains the class (l), then the class (m) contains the class (l)—which obviously is true. The upshot is that the replacement of numbers by their corresponding classes does what is required when we add the definition of "multiplication": $(m) \times (n)$ is defined to be the class (mn); $(2) \times (6) = (12)$. Notice that the last is a definition; it is not meant to follow from the meanings of (m) and (n).

Dedekind's ideals for algebraic numbers are a generalization of what precedes. Following his usual custom Dedekind gave an *abstract* definition, that is, a definition based upon essential properties rather than one contingent upon some particular mode of representing, or picturing, the thing defined.

Consider the set (or class) of *all* algebraic *integers* in a given algebraic number field. In this all-inclusive set will be subsets. A subset is called an *ideal* if it has the two following properties.

A. The *sum* and *difference* of any two integers in the subset are also in the subset.

B. If any integer in the subset be multiplied by any integer in the all-inclusive set, the resulting integer is in the subset.

An ideal is thus an infinite *class* of integers. It will be seen readily that (m), (n), ..., previously defined, are ideals according to A, B. As before, if one ideal contains another, the first is said to divide the second.

It can be proved that every ideal is a class of integers all of which are of the form

$$x_1a_1 + x_2a_2 + \ldots + x_na_n,$$

where $a_1, a_2, \ldots, a_n$ are *fixed* integers of the field of degree n concerned, and each of $x_1, x_2, \ldots, x_n$ may be any integer whatever in the field. This being so, it is convenient to symbolize an ideal by exhibiting only the fixed integers $a_1, a_2, \ldots, a_n$, and this is done by writing $(a_1, a_2, \ldots, a_n)$ as the symbol of the ideal. The order in which $a_1, a_2, \ldots, a_n$ are written in the symbol is immaterial.

"Multiplication" of ideals must now be defined: the *product* of the two ideals $(a_1, \ldots, a_n)$, $(b_1, \ldots, b_n)$ is the ideal whose symbol is $(a_1b_1, \ldots, a_1b_n, \ldots, a_nb_n)$, in which all possible products a_1b_1, etc., obtained by multiplying an integer in the first symbol by an integer in the second occur. For example, the product of (a_1, a_2) and (b_1, b_2) is $(a_1b_1, a_1b_2, a_2b_1, a_2b_2)$. It is always possible to reduce any such product-symbol (for a field of degree n) to a symbol containing at most n integers.

One final short remark completes the synopsis of the story. An ideal whose symbol contains *but one* integer, such as (a_1), is called a *principal* ideal. Using as before the notation $(a_1)|(b_1)$ to signify that (a_1) contains (b_1), we can see without difficulty that $(a_1)|(b_1)$ *when*, and *only when*, the integer a_1 *divides* the integer b_1. As before, then, the concept of arithmetical divisibility is here—for algebraic integers —completely equivalent to that of class inclusion. A *prime* ideal is one which is not "divisible by"—included in—any ideal except the all-inclusive ideal which consists of *all* the algebraic integers in the given field. Algebraic integers being now replaced by their corresponding principal ideals, it is proved that a given ideal is a product of prime ideals in one way only, precisely as in the "fundamental theorem of arithmetic" a rational integer is the product of primes in one way only. By the above equivalence of arithmetical divisibility for algebraic integers and class inclusion, the fundamental theorem of arithmetic has been restored to integers in algebraic number fields.

Anyone who will ponder a little on the foregoing bare outline of Dedekind's creation will see that what he did demanded penetrating insight and a mind gifted far above the ordinary good mathematical mind in the power of abstraction. Dedekind was a mathematician after Gauss' own heart: "*At nostro quidem judicio hujusmodi veritates ex notionibus potius quam ex notationibus hauriri debeant*" (But in our opinion such truths [arithmetical] should be derived from notions rather than from notations). Dedekind always relied on his head rather than on an ingenious symbolism and expert manipulations of formulas to get him forward. If ever a man put notions into mathematics, Dedekind did, and the wisdom of his preference for creative ideas over sterile symbols is now apparent although it may not have been during his lifetime. The longer mathematics lives the more abstract—and therefore, possibly, also the more practical—it becomes.

CHAPTER TWENTY EIGHT

The Last Universalist

POINCARE

A scientist worthy of the name, above all a mathematician, experiences in his work the same impression as an artist; his pleasure is as great and of the same nature.—HENRI POINCARÉ

IN THE *History of his Life and Times* the astrologer William Lilly (1602–1681) records an amusing—if incredible—account of the meeting between John Napier (1550–1617), of Merchiston, the inventor of logarithms, and Henry Briggs (1561–1631) of Gresham College, London, who computed the first table of common logarithms. One John Marr, "an excellent mathematician and geometrician," had gone "into Scotland before Mr. Briggs, purposely to be there when these two so learned persons should meet. Mr. Briggs appoints a certain day when to meet in Edinburgh; but failing thereof, the lord Napier was doubtful he would not come. It happened one day as John Marr and the lord Napier were speaking of Mr. Briggs: 'Ah John (said Merchiston), Mr. Briggs will not now come.' At the very moment one knocks at the gate; John Marr hastens down, and it proved Mr. Briggs to his great contentment. He brings Mr. Briggs up into my lord's chamber, where almost *one quarter of an hour was spent*, each beholding other with admiration, *before one word was spoke.*"

Recalling this legend Sylvester tells how he himself went after Briggs' world record for flabbergasted admiration when, in 1885, he called on the author of numerous astonishingly mature and marvellously original papers on a new branch of analysis which had been swamping the editors of mathematical journals since the early 1880's.

"I quite entered into Briggs' feelings at his interview with Napier," Sylvester confesses, "when I recently paid a visit to Poincaré [1854–1912] in his airy perch in the Rue Gay-Lussac. . . . In the presence of that mighty reservoir of pent-up intellectual force my tongue at first refused its office, and it was not until I had taken some time (it may be two or three minutes) to peruse and absorb as it were the idea of his external youthful lineaments that I found myself in a condition to speak."

Elsewhere Sylvester records his bewilderment when, after having toiled up the three flights of narrow stairs leading to Poincaré's "airy perch," he paused, mopping his magnificent bald head, in astonishment at beholding a mere boy, "so blond, so young," as the author of the deluge of papers which had heralded the advent of a successor to Cauchy.

A second anecdote may give some idea of the respect in which Poincaré's work is held by those in a position to appreciate its scope. Asked by some patriotic British brass hat in the rabidly nationalistic days of the World War—when it was obligatory on all academic patriots to exalt their esthetic allies and debase their boorish enemies—who was the greatest man France had produced in modern times, Bertrand Russell answered instantly, "Poincaré." "What! *That* man?" his uninformed interlocutor exclaimed, believing Russell meant Raymond Poincaré, President of the French Republic. "Oh," Russell explained when he understood the other's dismay, "I was thinking of Raymond's cousin, *Henri* Poincaré."

Poincaré was the last man to take practically all mathematics, both pure and applied, as his province. It is generally believed that it would be impossible for any human being starting today to understand comprehensively, much less do creative work of high quality in more than two of the four main divisions of mathematics—arithmetic, algebra, geometry, analysis, to say nothing of astronomy and mathematical physics. However, even in the 1880's, when Poincaré's great career opened, it was commonly thought that Gauss was the last of the mathematical universalists, so it may not prove impossible for some future Poincaré once more to cover the entire field.

As mathematics evolves it both expands and contracts, somewhat like one of Lemaître's models of the universe. At present the phase is one of explosive expansion, and it is quite impossible for any man to familiarize himself with the entire inchoate mass of mathematics that has been dumped on the world since the year 1900. But already in certain important sectors a most welcome tendency toward contraction is plainly apparent. This is so, for example, in algebra, where the wholesale introduction of postulational methods is making the subject at once more abstract, more general, and less disconnected. Unexpected similarities—in some instances amounting to disguised identity—are being disclosed by the modern attack, and it is conceivable that the next generation of algebraists will not need to know much that is now considered valuable, as many of these particular, difficult things will have been subsumed under simpler general principles of wider scope. Something of this sort happened in classical mathematical physics when relativity put the complicated mathematics of the ether on the shelf.

Another example of this contraction in the midst of expansion is the rapidly growing use of the tensor calculus in preference to that of numerous special brands of vector analysis. Such generalizations and condensations are often hard for older men to grasp at first and frequently have a severe struggle to survive, but in the end it is usually realized that general methods are essentially simpler and easier to handle than miscellaneous collections of ingenious tricks devised for special problems. When mathematicians assert that such a thing as the tensor calculus is easy—at least in comparison with some of the algorithms that preceded it—they are not trying to appear superior or mysterious but are stating a valuable truth which any student can verify for himself. This quality of inclusive generality was a distinguishing trait of Poincaré's vast output.

If abstractness and generality have obvious advantages of the kind indicated, it is also true that they sometimes have serious drawbacks for those who must be interested in details. Of what immediate use is it to a working physicist to know that a particular differential equation occurring in his work is solvable, because some pure mathematician has proved that it is, when neither he nor the mathematician can perform the Herculean labor demanded by a numerical solution capable of application to specific problems?

To take an example from a field in which Poincaré did some of his most original work, consider a homogeneous, incompressible fluid mass held together by the gravitation of its particles and rotating about an axis. Under what conditions will the motion be stable and what will be the possible shapes of such a stably rotating fluid? MacLaurin, Jacobi, and others proved that certain ellipsoids will be stable; Poincaré, using more intuitive, "less arithmetical" methods than his predecessors, once thought he had determined the criteria for the stability of a pear-shaped body. But he had made a slip. His methods were not adapted to numerical computation and later workers, including G. H. Darwin, son of the famous Charles, undeterred by the horrific jungles of algebra and arithmetic that must be cleared out of the way before a definite conclusion can be reached, undertook a decisive solution.*

The man interested in the evolution of binary stars is more comfortable if the findings of the mathematicians are presented to him in a form to which he can apply a calculating machine. And since Kronecker's fiat of "no construction, no existence," some pure mathematicians themselves have been less enthusiastic than they were in Poincaré's day for existence theorems which are not constructive. Poincaré's scorn for the kind of detail that users of mathematics demand and must have before they can get on with their work was one of the most important contributory causes to his universality. Another was his extraordinarily comprehensive grasp of all the machinery of the theory of functions of a complex variable. In this he had no equal. And it may be noted that Poincaré turned his universality to magnificent use in disclosing hitherto unsuspected connections between distant branches of mathematics, for example between (continuous) groups and linear algebra.

One more characteristic of Poincaré's outlook must be recalled for completeness before we go on to his life: few mathematicians have had the breadth of philosophical vision that Poincaré had, and none is his superior in the gift of clear exposition. Probably he had always been deeply interested in the philosophical implications of science and mathematics, but it was only in 1902, when his greatness as a technical mathematician was established beyond all cavil, that he turned as a side interest to what may be called the popular appeal of mathematics and let himself go in a sincere enthusiasm to share with nonprofessionals the meaning and human importance of his subject. Here his liking for the general in preference to the particular aided him in

---

*This famous question of the "piriform body," of considerable importance in cosmogony, was apparently settled in 1905 by Liapounoff, whose conclusion was confirmed in 1915 by Sir James Jeans: they found that the motion is unstable. Few have had the courage to check the calculations. After 1915 Leon Lichtenstein, a fellow-countryman of Liapounoff, made a general attack on the problem of rotating fluid masses. The problem seems to be unlucky; both L's had violent deaths.

telling intelligent outsiders what is of more than technical importance in mathematics without talking down to his audience. Twenty or thirty years ago workmen and shopgirls could be seen in the parks and cafés of Paris avidly reading one or other of Poincaré's popular masterpieces in its cheap print and shabby paper cover. The same works in a richer format could also be found—well thumbed and evidently read—on the tables of the professedly cultured. These books were translated into English, German, Spanish, Hungarian, Swedish, and Japanese. Poincaré spoke the universal languages of mathematics and science to all in accents which they recognized. His style, peculiarly his own, loses much by translation.

For the literary excellence of his popular writings Poincaré was accorded the highest honor a French writer can get, membership in the literary section of the Institut. It has been somewhat spitefully said by envious novelists that Poincaré achieved this distinction, unique for a man of science, because one of the functions of the (literary) Academy is the constant compilation of a definitive dictionary of the French language, and the universal Poincaré was obviously the man to help out the poets and grammarians in their struggle to tell the world what automorphic functions are. Impartial opinion, based on a study of Poincaré's writings, agrees that the mathematician deserved no less than he got.

Closely allied to his interest in the philosophy of mathematics was Poincaré's preoccupation with the psychology of mathematical creation. How do mathematicians make their discoveries? Poincaré will tell us later his own observations on this mystery in one of the most interesting narratives of personal discovery that was ever written. The upshot seems to be that mathematical discoveries more or less make themselves after a long spell of hard labor on the part of the mathematician. As in literature—according to Dante Gabriel Rossetti—"a certain amount of fundamental brainwork" is necessary before a poem can mature, so in mathematics there is no discovery without preliminary drudgery, but this is by no means the whole story. All "explanations" of creativeness that fail to provide a recipe whereby a gifted human being can create are open to suspicion. Poincaré's excursion into practical psychology, like some others in the same direction, failed to bring back the Golden Fleece, but it did at least suggest that such a thing is not wholly mythical and may some day be found when human beings grow intelligent enough to understand their own bodies.

Poincaré's intellectual heredity on both sides was good. We shall not go farther back than his paternal grandfather. During the Napoleonic campaign of 1814 this grandfather, at the early age of twenty, was attached to the military hospital at Saint-Quentin. On settling in 1817 at Rouen he married and had two sons: Léon Poincaré, born in 1828, who became a first-rate physician and a member of a medical faculty; and Antoine, who rose to the inspector-generalship of the department of roads and bridges. Léon's son Henri, born on April 29, 1854, at Nancy, Lorraine, became the leading mathematician of the early twentieth century; one of Antoine's two sons, Raymond, went in for law and rose to the presidency of the French Republic during the World War; Antoine's other son became director of secondary education. A great-uncle who had followed Napoleon into Russia disappeared and was never heard of after the Moscow fiasco.

From this distinguished list it might be thought that Henri would have exhibited some administrative ability, but he did not, except in his early childhood when he freely invented political games for his sister and young friends to play. In these games he was always fair and scrupulously just, seeing that each of his playmates got his or her full share of officeholding. This perhaps is conclusive evidence that "the child is father to the man" and that Poincaré was constitutionally incapable of understanding the simplest principle of administration, which his cousin Raymond applied intuitively.

Poincaré's biography was written in great detail by his fellow countryman Gaston Darboux (1842–1917), one of the leading geometers of modern times, in 1913 (the year following Poincaré's death). Something may have escaped the present writer, but it seems that Darboux, after having stated that Poincaré's mother "coming from a family in the Meuse district whose [the mother's] parents lived in Arrancy, was a very good person, very active and very intelligent," blandly omits to mention her maiden name. Can it be possible that the French took over the doctrine of "the three big K's"—noted in connection with Dedekind—from their late instructors after the kultural drives of Germany into France in 1870 and 1914? However, it can be deduced from an anecdote told later by Darboux that the family name *may* have been Lannois. We learn that the mother devoted her entire attention to the education of her two young children, Henri and his younger sister (name not mentioned). The sister was to become the wife of Émile Boutroux and the mother of a mathematician (who died young).

Due partly to his mother's constant care, Poincaré's mental development as a child was extremely rapid. He learned to talk very early, but also very badly at first because he thought more rapidly than he could get the words out. From infancy his motor coordination was poor. When he learned to write it was discovered that he was ambidextrous and that he could write or draw as badly with his left hand as with his right. Poincaré never outgrew this physical awkwardness. As an item of some interest in this connection it may be recalled that when Poincaré was acknowledged as the foremost mathematician and leading popularizer of science of his time he submitted to the Binet tests and made such a disgraceful showing that, had he been judged as a child instead of as the famous mathematician he was, he would have been rated—by the tests—as an imbecile.

At the age of five Henri suffered a bad setback from diphtheria which left him for nine months with a paralyzed larynx. This misfortune made him for long delicate and timid, but it also turned him back on his own resources as he was forced to shun the rougher games of children his own age.

His principal diversion was reading, where his unusual talents first showed up. A book once read—at incredible speed—became a permanent possession, and he could always state the page and line where a particular thing occurred. He retained this powerful memory all his life. This rare faculty, which Poincaré shared with Euler who had it in a lesser degree, might be called visual or spatial memory. In temporal memory—the ability to recall with uncanny precision a sequence of events long passed—he was also unusually strong. Yet he unblushingly describes his memory as "bad." His poor eyesight perhaps contributed to a third peculiarity of his memory. The majority of mathematicians appear to remember theorems and formulas mostly by eye; with Poincaré it was almost wholly by ear. Unable to see the board distinctly when he became a student of advanced mathematics, he sat back and listened, following and remembering perfectly without taking notes—an easy feat for him, but one incomprehensible to most mathematicians. Yet he must have had a vivid memory of the "inner eye" as well, for much of his work, like a good deal of Riemann's, was of the kind that goes with facile space-intuition and acute visualization. His inability to use his fingers skilfully of course handicapped him in laboratory exercises, which seems a pity, as some of his own work in mathematical physics might have been closer to reality had he mastered the art of experiment. Had Poincaré been as strong in practical science as he was in theoretical he might have made a fourth with the incomparable three, Archimedes, Newton, and Gauss.

Not many of the great mathematicians have been the absentminded dreamers that popular fancy likes to picture them. Poincaré was one of the exceptions, and then only in comparative trifles, such as carrying off hotel linen in his baggage. But many persons who are anything but absentminded do the same, and some of the most alert mortals living have even been known to slip restaurant silver into their pockets and get away with it.

One phase of Poincaré's absentmindedness resembles something quite different. Thus (Darboux does not tell the story, but it should be told, as it illustrates a certain brusqueness of Poincaré's later years), when a distinguished mathematician had come all the way from Finland to Paris to confer with Poincaré on scientific matters, Poincaré did not leave his study to greet his caller when the maid notified him, but continued to pace back and forth—as was his custom when mathematicizing—for three solid hours. All this time the diffident caller sat quietly in the adjoining room, barred from the master only by flimsy portières. At last the drapes parted and Poincaré's buffalo head was thrust for an instant into the room. *"Vous me dérangez beaucoup"* (You are disturbing me greatly) the head exploded, and disappeared. The caller departed without an interview, which was exactly what the "absentminded" professor wanted.

Poincaré's elementary school career was brilliant, although he did not at first show any marked interest in mathematics. His earliest passion was for natural history, and all his life he remained a great lover of animals. The first time he tried out a rifle he accidentally shot a bird at which he had not aimed. This mishap affected him so deeply that thereafter nothing (except compulsory military drill) could induce him to touch firearms. At the age of nine he showed the first promise of what was to be one of his major successes. The teacher of French composition declared that a short exercise, original in both form and substance, which young Poincaré had handed in, was "a little masterpiece," and kept it as one of his treasures. But he also advised his pupil to be more conventional—stupider—if he wished to make a good impression on the school examiners.

Being out of the more boisterous games of his schoolfellows, Poincaré invented his own. He also became an indefatigable dancer. As all his lessons came to him as easily as breathing he spent most of his time on amusements and helping his mother about the house. Even at this early stage of his career Poincaré exhibited some of the more suspicious features of his mature "absentmindedness": he frequently forgot his meals and almost never remembered whether or not he had breakfasted. Perhaps he did not care to stuff himself as most boys do.

The passion for mathematics seized him at adolescence or shortly before (when he was about fifteen). From the first he exhibited a lifelong peculiarity: his mathematics was done in his head as he paced restlessly about, and was committed to paper only when all had been thought through. Talking or other noise never disturbed him while he was working. In later life he wrote his mathematical memoirs at one dash without looking back to see what he had written and limiting himself to but a very few erasures as he wrote. Cayley also composed in this way, and probably Euler, too. Some of Poincaré's work shows the marks of hasty composition, and he said himself that he never finished a paper without regretting either its form or its substance. More than one man who has written well has felt the same. Poincaré's flair for classical studies, in which he excelled at school, taught him the importance of both form and substance.

The Franco-Prussian war broke over France in 1870 when Poincaré was sixteen. Although he was too young and too frail for active service, Poincaré nevertheless got his full share of the horrors, for Nancy, where he lived, was submerged by the full tide of the invasion, and the young boy accompanied his physician-father on his rounds of the ambulances. Later he went with his mother and sister, under terrible difficulties, to Arrancy to see what had happened to his maternal grandparents, in whose spacious country garden the happiest days of his childhood had been spent during the long school vacations. Arrancy lay near the battlefield of Saint-Privat. To reach the town the three had to pass "in glacial cold" through burned and deserted villages. At last they reached their destination, only to find that the house had been thoroughly pillaged, "not only of things of value but of things of no value," and in addition had been defiled in the bestial manner made familiar to the French by the 1914 sequel to 1870. The grandparents had been left nothing; their evening meal on the day they viewed the great purging was supplied by a poor woman who had refused to abandon the ruins of her cottage and who insisted on sharing her meager supper with them.

Poincaré never forgot this, nor did he ever forget the long occupation of Nancy by the enemy. It was during the war that he mastered German. Unable to get any French news, and eager to learn what the Germans had to say of France and for themselves, Poincaré taught himself the language. What he had seen and what he learned from the official accounts of the invaders themselves made him a flaming patriot for life but, like Hermite, he never confused the mathematics of his country's enemies with their more practical activities. Cousin Raymond, on the other hand, could never say anything about *les Allemands* (the Germans) without an accompanying scream of hate. In the bookkeeping of hell which balances the hate of one patriot against that of another, Poincaré may be checked off against Kummer, Hermite against Gauss, thus producing that perfect zero implied in the scriptural contract "an eye for an eye and a tooth for a tooth."

Following the usual French custom Poincaré took the examinations for his first degrees (bachelor of letters, and of science) before specializing. These he passed in 1871 at the age of seventeen—after almost failing in mathematics! He had arrived late and flustered at the examination and had fallen down on the extremely simple proof of the formula giving the sum of a convergent geometrical progression. But his fame had preceded him. "Any student other than Poincaré would have been plucked," the head examiner declared.

He next prepared for the entrance examinations to the School of Forestry, where he astonished his companions by capturing the first prize in mathematics without having bothered to take any lecture notes. His classmates had previously tested him out, believing him to be a trifler, by delegating a fourth-year student to quiz him on a mathematical difficulty which had seemed particularly tough. Without apparent thought, Poincaré gave the solution immediately and walked off, leaving his crestfallen baiters asking "How does he do it?" Others were to ask the same question all through Poincaré's career. He never seemed to think when a mathematical difficulty was submitted to him by his colleagues: "The reply came like an arrow."

At the end of this year he passed first into the École Polytechnique.

Several legends of his unique examination survive. One tells how a certain examiner, forewarned that young Poincaré was a mathematical genius, suspended the examination for three quarters of an hour in order to devise "a 'nice' question"—a refined torture. But Poincaré got the better of him and the inquisitor "congratulated the examinee warmly, telling him he had won the highest grade." Poincaré's experiences with his tormentors would seem to indicate that French mathematical examiners have learned something since they ruined Galois and came within an ace of doing the like by Hermite.

At the Polytechnique Poincaré was distinguished for his brilliance in mathematics, his superb incompetence in all physical exercises, including gymnastics and military drill, and his utter inability to make drawings that resembled anything in heaven or earth. The last was more than a joke; his score of *zero* in the entrance examination in drawing had almost kept him out of the school. This had greatly embarrassed his examiners: ". . . a zero is eliminatory. In everything else [but drawing] he is absolutely without an equal. If he is admitted, it will be as first; but can he be admitted?" As Poincaré was admitted the good examiners probably put a decimal point before the zero and placed a 1 after it.

In spite of his ineptitude for physical exercises Poincaré was extremely popular with his classmates. At the end of the year they organized a public exhibition of his artistic masterpieces, carefully labelling them in Greek, "this is a horse," and so on—not always accurately. But Poincaré's inability to draw also had its serious side when he came to geometry, and he lost first place, passing out of the school second in rank.

On leaving the Polytechnique in 1875 at the age of twenty one Poincaré entered the School of Mines with the intention of becoming an engineer. His technical studies, although faithfully carried out, left him some leisure to do mathematics, and he showed what was in him by attacking a general problem in differential equations. Three years later he presented a thesis, on the same subject, but concerning a more difficult and yet more general question, to the Faculty of Sciences at Paris for the degree of doctor of mathematical sciences. "At the first glance," says Darboux, who had been asked to examine the work, "it was clear to me that the thesis was out of the ordinary and amply merited acceptance. Certainly it contained results enough to supply material for several good theses. But, I must not be afraid to say, if an accurate idea of the way Poincaré worked is wanted, many points called for corrections or explanations. Poincaré was an intuitionist. Having once arrived at the summit he never retraced his steps. He was satisfied to have crashed through the difficulties and left to others the pains of mapping the royal roads* destined to lead more easily to the end. He willingly enough made the corrections and tidying-up which seemed to me necessary. But he explained to me when I asked him to do it that he had many other ideas in his head; he was already occupied with some of the great problems whose solution he was to give us."

Thus young Poincaré, like Gauss, was overwhelmed by the host of ideas which besieged his mind but, unlike Gauss, his motto was not "Few, but ripe." It is an open question whether a creative scientist who hoards the fruits of his labor so long that some of them go stale does more for the advancement of science than the more impetuous man who scatters broadcast everything he gathers, green or ripe, to fall where it may to ripen or rot as wind and weather take it. Some believe one way, some another. As a decision is beyond the reach of objective criteria everyone is entitled to his own purely subjective opinion.

Poincaré was not destined to become a mining engineer, but during his apprenticeship he showed that he had at least the courage of a real engineer. After a mine explosion and fire which had claimed sixteen victims he went down at once with the rescue crew. But the calling was uncongenial and he welcomed the opportunity to become a professional mathematician which his thesis and other early work opened up to him. His first academic appointment was at Caen on December 1, 1879, as Professor of Mathematical Analysis. Two years later he was promoted (at the age of twenty seven) to the University of Paris where, in 1886, he was again promoted, taking charge of the course in mechanics and experimental physics (the last seems rather strange, in view of Poincaré's exploits as a student in the laboratory). Except for trips to scientific congresses in Europe and a visit to the United States in 1904 as an invited lecturer at the St. Louis Exposition, Poin-

---

*"There is no royal road to Geometry," as Menaechmus is said to have told Alexander the Great when the latter wished to conquer geometry in a hurry.

caré spent the rest of his life in Paris as the ruler of French mathematics.

Poincaré's creative period opened with the thesis of 1878 and closed with his death in 1912—when he was at the apex of his powers. Into this comparatively brief span of thirty four years he crowded a mass of work that is sheerly incredible when we consider the difficulty of most of it. His record is nearly five hundred papers on *new* mathematics, many of them extensive memoirs, and more than thirty books covering practically all branches of mathematical physics, theoretical physics, and theoretical astronomy as they existed in his day. This leaves out of account his classics on the philosophy of science and his popular essays. To give an adequate idea of this immense labor one would have to be a second Poincaré, so we shall presently select two or three of his most celebrated works for brief description, apologizing here once for all for the necessary inadequacy.

Poincaré's first successes were in the theory of differential equations, to which he applied all the resources of the analysis of which he was absolute master. This early choice for a major effort already indicates Poincaré's leaning toward the applications of mathematics, for differential equations have attracted swarms of workers since the time of Newton chiefly because they *are* of great importance in the exploration of the physical universe. "Pure" mathematicians sometimes like to imagine that all their activities are dictated by their own tastes and that the applications of science suggest nothing of interest to them. Nevertheless some of the purest of the pure drudge away their lives over differential equations that first appeared in the translation of physical situations into mathematical symbolism, and it is precisely these practically suggested equations which are the heart of the theory. A particular equation suggested by science may be generalized by the mathematicians and then be turned back to the scientists (frequently without a solution in any form that they can use) to be applied to new physical problems, but first and last the motive is scientific. Fourier summed up this thesis in a famous passage which irritates one type of mathematician, but which Poincaré endorsed and followed in much of his work.

"The profound study of nature," Fourier declared, "is the most fecund source of mathematical discoveries. Not only does this study, by offering a definite goal to research, have the advantage of excluding vague questions and futile calculations, but it is also a sure means of molding analysis itself and discovering those elements in it which it is essential to know and which science ought always to conserve. These fundamental elements are those which recur in all natural phenomena." To which some might retort: No doubt, but what about arithmetic in the sense of Gauss? However, Poincaré followed Fourier's advice whether he believed in it or not—even his researches in the theory of numbers were more or less remotely inspired by others closer to the mathematics of physical science.

The investigations on differential equations led out in 1880, when Poincaré was twenty six, to one of his most brilliant discoveries, a generalization of the elliptic functions (and of some others). The nature of a (uniform) periodic function of a single variable has frequently been described in preceding chapters, but to bring out what Poincaré did, we may repeat the essentials. The trigonometric function $\sin z$ has the period 2π, namely, $\sin(z + 2\pi) = \sin z$; that is, when the variable z is increased by 2π, the sine function of z returns to its initial value. For an elliptic function, say $E(z)$, there are *two* distinct periods, say p_1 and p_2, such that $E(z + p_1) = E(z)$, $E(z + p_2) = E(z)$. Poincaré found that *periodicity* is merely a special case of a more general property: the value of certain functions is restored when the variable is replaced by any one of a *denumerable* infinity of linear fractional transformations of itself, and all these transformations form a group. A few symbols will clarify this statement.

Let z be replaced by $\dfrac{az + b}{cz + d}$. Then, for a *denumerable infinity* of sets of values of a, b, c, d, there are uniform functions of z, say $F(z)$ is one of them, such that

$$F\left(\frac{az + b}{cz + d}\right) = F(z).$$

Further, if a_1, b_1, h_1, d_1 and a_2, b_2, c_2, d_2 are any two of the sets of values of a, b, c, d, and if z be replaced first by $\dfrac{a_1 z + b_1}{c_1 z + d_1}$, and then, in this, z be replaced by $\dfrac{a_2 z + b_2}{c_2 z + d_2}$, giving, say, $\dfrac{Az + B}{Cz + D}$, then not only do we have

$$F\left(\frac{a_1 z + b_1}{c_1 z + d_1}\right) = F(z), \quad F\left(\frac{a_2 z + b_2}{c_2 z + d_2}\right) = F(z),$$

but also

$$F\left(\frac{Az + B}{Cz + D}\right) = F(z).$$

Further the set of all the substitutions

$$z \to \frac{az + b}{cz + d}$$

(the arrow is read "is replaced by") which leave the value of $F(z)$ unchanged as just explained *form a group*: the result of the successive performance of two substitutions in the set,

$$z \to \frac{a_1 z + b_1}{c_1 z + d_1}, \quad z \to \frac{a_2 z + b_2}{c_2 z + d_2},$$

is in the set; there is an "identity substitution" in the set, namely $z \to z$ (here $a = 1, b = 0, c = 0, d = 1$); and finally each substitution has a unique "inverse"—that is, for each substitution in the set there is a single other one which, if applied to the first, will produce the identity substitution. In summary, using the terminology of previous chapters, we see that $F(z)$ is *a function which is invariant under an infinite group of linear fractional transformations*. Note that the infinity of substitutions is a *denumerable* infinity, as first stated: the substitutions can be counted off 1,2,3, . . . , and are *not* as numerous as the points on a line. Poincaré actually constructed such functions and developed their most important properties in a series of papers in the 1880's. Such functions are called *automorphic*.

Only two remarks need be made here to indicate what Poincaré achieved by this wonderful creation. First, his theory includes that of the elliptic functions as a detail. Second, as the distinguished French mathematician Georges Humbert said, Poincaré found two memorable propositions which "gave him the keys of the algebraic cosmos":

Two automorphic functions*invariant under the same group are connected by an algebraic equation;

Conversely, the coordinates of a point on any algebraic curve can be expressed in terms of automorphic functions, and hence by uniform functions of a single parameter (variable).

An algebraic curve is one whose equation is of the type $P(x,y) = 0$, where $P(x,y)$ is a polynomial in x and y. As a simple example, the equation of the circle whose center is at the origin—(0,0)—and whose

---

*Poincaré called some of his functions "Fuchsian," after the German mathematician Lazarus Fuchs (1833–1902) one of the creators of the modern theory of differential equations, for reasons that need not be gone into here. Others he called "Kleinian" after Felix Klein—in ironic acknowledgment of disputed priority.

radius is a, is $x^2 + y^2 = a^2$. According to the second of Poincaré's "keys," it must be possible to express x,y as automorphic functions of a single parameter, say t. It is; for if $x = a \cos t$ and $y = a \sin t$, then, squaring and adding, we get rid of t (since $\cos^2 t + \sin^2 t = 1$), and find $x^2 + y^2 = a^2$. But the trigonometric functions $\cos t$, $\sin t$ are special cases of elliptic functions, which in turn are special cases of automorphic functions.

The creation of this vast theory of automorphic functions was but one of many astonishing things in analysis which Poincaré did before he was thirty. Nor was all his time devoted to analysis; the theory of numbers, parts of algebra, and mathematical astronomy also shared his attention. In the first he recast the Gaussian theory of binary quadratic forms (see chapter on Gauss) in a geometrical shape which appeals particularly to those who, like Poincaré, prefer the intuitive approach. This of course was not all that he did in the higher arithmetic, but limitations of space forbid further details.

Work of this caliber did not pass unappreciated. At the unusually early age of thirty two (in 1887) Poincaré was elected to the Academy. His proposer said some pretty strong things, but most mathematicians will subscribe to their truth: "[Poincaré's] work is above ordinary praise and reminds us inevitably of what Jacobi wrote of Abel—that he had settled questions which, before him, were unimagined. It must indeed be recognized that we are witnessing a revolution in Mathematics comparable in every way to that which manifested itself, half a century ago, by the accession of elliptic functions."

To leave Poincaré's work in pure mathematics here is like rising from a banquet table after having just sat down, but we must turn to another side of his universality.

Since the time of Newton and his immediate successors astronomy has generously supplied mathematicians with more problems than they can solve. Until the late nineteenth century the weapons used by mathematicians in their attack on astronomy were practically all immediate improvements of those invented by Newton himself, Euler, Lagrange, and Laplace. But all through the nineteenth century, particularly since Cauchy's development of the theory of functions of a complex variable and the investigations of himself and others on the convergence of infinite series, a huge arsenal of untried weapons had been accumulating from the labors of pure mathematicians. To Poincaré, to whom analysis came as naturally as thinking, this vast pile of unused mathematics seemed the most natural thing in the world to use in a new offensive on the outstanding problems of celestial mechanics and planetary evolution. He picked and chose what he liked out of the heap, improved it, invented new weapons of his own, and assaulted theoretical astronomy in a grand fashion it had not been assaulted in for a century. He *modernized* the attack; indeed his campaign was so extremely modern to the majority of experts in celestial mechanics that even today, forty years or more after Poincaré opened his offensive, few have mastered his weapons and some, unable to bend his bow, insinuate that it is worthless in a practical attack. Nevertheless Poincaré is not without forceful champions whose conquests would have been impossible to the men of the pre-Poincaré era.

Poincaré's first (1889) great success in mathematical astronomy grew out of an unsuccessful attack on "the problem of n bodies." For $n = 2$ the problem was completely solved by Newton; the famous "problem of three bodies" ($n = 3$) will be noticed later; when n exceeds 3 some of the reductions applicable to the case $n = 3$ can be carried over.

According to the Newtonian law of gravitation two particles of masses m, M at a distance D apart attract one another with a force proportional to $\frac{m \times M}{D^2}$. Imagine n material particles distributed in any manner in space; the masses, initial motions, and the mutual distances of all the particles are assumed known at a given instant. If they attract one another according to the Newtonian law, *what will be their positions and motions (velocities) after any stated lapse of time?* For the purposes of mathematical astronomy the stars in a cluster, or in a galaxy, or in a cluster of galaxies, may be thought of as material particles attracting one another according to the Newtonian law. The "problem of n bodies" thus amounts—in one of its applications—to asking what will be the aspect of the heavens a year from now, or a billion years hence, it being assumed that we have sufficient observational data to describe the general configuration *now*. The problem of course is tremendously complicated by radiation—the masses of the stars do not remain constant for millions of years; but a complete, calculable solution of the problem of n bodies in its Newtonian form would probably give results of an accuracy sufficient for all human purposes—the human race will likely be extinct long before radiation can introduce observable inaccuracies.

This was substantially the problem proposed for the prize offered by King Oscar II of Sweden in 1887. Poincaré did not solve the problem, but in 1889 he was awarded the prize anyhow by a jury consisting of Weierstrass, Hermite, and Mittag-Leffler for his general discussion of the differential equations of dynamics and an attack on the problem of three bodies. The last is usually considered the most important case of the n-body problem, as the Earth, Moon, and Sun furnish an instance of the case $n = 3$. In his report to Mittag-Leffler, Weierstrass wrote, "You may tell your Sovereign that this work cannot indeed be considered as furnishing the complete solution of the question proposed, but that it is nevertheless of such importance that *its publication will inaugurate a new era in the history of Celestial Mechanics*. The end which His Majesty had in view in opening the competition may therefore be considered as having been attained." Not to be outdone by the King of Sweden, the French Government followed up the prize by making Poincaré a Knight of the Legion of Honor—a much less expensive acknowledgment of the young mathematician's genius than the King's 2500 crowns and gold medal.

As we have mentioned the problem of three bodies we may now report one item from its fairly recent history; since the time of Euler it has been considered one of the most difficult problems in the whole range of mathematics. Stated mathematically, the problem boils down to solving a system of nine simultaneous differential equations (all linear, each of the second order). Lagrange succeeded in reducing this system to a simpler. As in the majority of physical problems, the solution is not to be expected in *finite* terms; *if a solution exists at all* it will be given by *infinite series*. The solution will "exist" if these series satisfy the equations (formally) and moreover *converge* for certain values of the variables. The central difficulty is to prove the convergence. Up till 1905 various special solutions had been found, but the existence of anything that could be called general had not been proved.

In 1906 and 1909 a considerable advance came from a rather unexpected quarter—Finland, a country which sophisticated Europeans even today consider barely civilized, especially for its queer custom of paying its debts, and which few Americans thought advanced beyond the Stone Age till Paavo Nurmi ran the legs off the United

States. Excepting only the rare case when all three bodies collide simultaneously, Karl Frithiof Sundman of Helsingfors, utilizing analytical methods due to the Italian Levi-Civita and the French Painlevé, and making an ingenious transformation of his own, *proved* the existence of a solution in the sense described above. Sundman's solution is not adapted to numerical computation, nor does it give much information regarding the actual motion, but that is not the point of interest here: a problem which had not been known to be solvable was proved to be so. Many had struggled desperately to prove this much; when the proof was forthcoming, some, humanly enough, hastened to point out that Sundman had done nothing much because he had not solved some problem other than the one he had. This kind of criticism is as common in mathematics as it is in literature and art, showing once more that mathematicians are as human as anybody.

Poincaré's most original work in mathematical astronomy was summed up in his great treatise *Les méthodes nouvelles de la mécanique céleste* (New methods of celestial mechanics; three volumes, 1892, 1893, 1899). This was followed by another three-volume work in 1905–1910 of a more immediately practical nature, *Leçons de mécanique céleste*, and a little later by the publication of his course of lectures *Sur les figures d'équilibre d'une masse fluide* (On the figures of equilibrium of a fluid mass), and a historical-critical book *Sur les hypothèses cosmogoniques* (On cosmological hypotheses).

Of the first of these works Darboux (seconded by many others) declares that it did indeed start a new era in celestial mechanics and that it is comparable to the *Mécanique céleste* of Laplace and the earlier work of D'Alembert on the precession of the equinoxes. "Following the road in analytical mechanics opened up by Lagrange," Darboux says, " . . . Jacobi had established a theory which appeared to be one of the most complete in dynamics. For fifty years we lived on the theorems of the illustrious German mathematician, applying them and studying them from all angles, but without adding anything essential. It was Poincaré who first shattered these rigid frames in which the theory seemed to be encased and contrived for it vistas and new windows on the external world. He introduced or used, in the study of dynamical problems, different notions: the first, which had been given before and which, moreover, is applicable not solely to mechanics, is that of *variational equations*, namely, linear differential equations that determine solutions of a problem infinitely near to a given solution; the second, that of *integral invariants*, which belong entirely to him and play a capital part in these researches. Further fundamental notions were added to these, notably those concerning so-called 'periodic' solutions, for which the bodies whose motion is studied return after a certain time to their initial positions and original relative velocities."

The last started a whole department of mathematics, the investigation of *periodic orbits:* given a system of planets, or of stars, say, with a complete specification of the initial positions and relative velocities of all members of the system at a stated epoch, it is required to determine under what conditions the system will return to its initial state at some later epoch, and hence continue to repeat the cycle of its motions indefinitely. For example, is the solar system of this recurrent type, or if not, would it be were it isolated and not subject to perturbations by external bodies? Needless to say the general problem has not yet been solved completely.

Much of Poincaré's work in his astronomical researches was qualitative rather than quantitative, as befitted an intuitionist, and this characteristic led him, as it had Riemann, to the study of analysis situs. On this he published six famous memoirs which revolutionized the subject as it existed in his day. The work on analysis situs in its turn was freely applied to the mathematics of astronomy.

We have already alluded to Poincaré's work on the problem of rotating fluid bodies—of obvious importance in cosmogony, one brand of which assumes that the planets were once sufficiently like such bodies to be treated as if they actually were without patent absurdity. Whether they were or not is of no importance for the mathematics of the situation, which is of interest in itself. A few extracts from Poincaré's own summary will indicate more clearly than any paraphrase the nature of what he mathematicized about in this difficult subject.

"Let us imagine a [rotating] fluid body contracting by cooling, but slowly enough to remain homogeneous and for the rotation to be the same in all its parts.

"At first, very approximately a sphere, the figure of this mass will become an ellipsoid of revolution which will flatten more and more, then, at a certain moment, it will be transformed into an ellipsoid with three unequal axes. Later, the figure will cease to be an ellipsoid and will become pear-shaped until at last the mass, hollowing out more and more at its 'waist,' will separate into two distinct and unequal bodies.

"The preceding hypothesis certainly can not be applied to the solar system. Some astronomers have thought that it might be true for certain double stars and that double stars of the type of Beta Lyrae might present transitional forms analogous to those we have spoken of."

He then goes on to suggest an application to Saturn's rings, and he claims to have proved that the rings can be stable only if their density exceeds 1/16 that of Saturn. It may be remarked that these questions were not considered as fully settled as late as 1935. In particular a more drastic mathematical attack on poor old Saturn seemed to show that he had not been completely vanquished by the great mathematicians, including Clerk Maxwell, who have been firing away at him off and on for the past seventy years.

Once more we must leave the banquet having barely tasted anything and pass on to Poincaré's voluminous work in mathematical physics. Here his luck was not so good. To have cashed in on his magnificent talents he should have been born thirty years later or have lived twenty years longer. He had the misfortune to be in his prime just when physics had reached one of its recurrent periods of senility, and he was so thoroughly saturated with nineteenth century theories when physics began to recover its youth—after Planck, in 1900, and Einstein, in 1905, had performed the difficult and delicate operation of endowing the decrepit roué with its first pair of new glands—that he had barely time to digest the miracle before his death in 1912. All his mature life Poincaré seemed to absorb knowledge through his pores without a conscious effort. Like Cayley, he was not only a prolific creator but also a profoundly erudite scholar. His range was probably wider than ever Cayley's, for Cayley never professed to be able to understand everything that was going on in applied mathematics. This unique erudition may have been a disadvantage when it came to a question of living science as opposed to classical.

Everything that boiled up in the melting pots of physics was grasped instantly as it appeared by Poincaré and made the topic of several purely mathematical investigations. When wireless telegraphy was invented he seized on the new thing and worked out its mathematics.

While others were either ignoring Einstein's early work on the (special) theory of relativity or passing it by as a mere curiosity, Poincaré was already busy with its mathematics, and he was the first scientific man of high standing to tell the world what had arrived and urge it to watch Einstein as probably the most significant phenomenon of the new era which he foresaw but could not himself usher in. It was the same with Planck's early form of the quantum theory. Opinions differ, of course; but at this distance it is beginning to look as if mathematical physics did for Poincaré what Ceres did for Gauss; and although Poincaré accomplished enough in mathematical physics to make half a dozen great reputations, it was not the trade to which he had been born and science would have got more out of him if he had stuck to pure mathematics—his astronomical work was nothing else. But science got enough, and a man of Poincaré's genius is entitled to his hobbies.

We pass on now to the last phase of Poincaré's universality for which we have space: his interest in the rationale of mathematical creation. In 1902 and 1904 the Swiss mathematical periodical *L'Enseignement Mathématique* undertook an enquiry into the working habits of mathematicians. Questionnaires were issued to a number of mathematicians, of whom over a hundred replied. The answers to the questions and an analysis of general trends were published in final form in 1912.* Anyone wishing to look into the "psychology" of mathematicians will find much of interest in this unique work and many confirmations of the views at which Poincaré had arrived independently before he saw the results of the questionnaire. A few points of general interest may be noted before we quote from Poincaré.

The early interest in mathematics of those who were to become great mathematicians has been frequently exemplified in preceding chapters. To the question "At what period . . . and under what circumstances did mathematics seize you?" 93 replies to the first part were received: 35 said before the age of ten; 43 said eleven to fifteen; 11 said sixteen to eighteen; 3 said nineteen to twenty; and the lone laggard said twenty six.

Again, anyone with mathematical friends will have noticed that

---

*Enquête de "L'Enseignement Mathématique" sur la méthode de travail des mathématiciens. Available either in the periodical or in book form (8 + 137 pp.) by Gauthier-Villars, Paris.

some of them like to work early in the morning (I know one very distinguished mathematician who begins his day's work at the inhuman hour of five a.m.), while others do nothing till after dark. The replies on this point indicated a curious trend—possibly significant, although there are numerous exceptions: mathematicians of the northern races prefer to work at night, while the Latins favor the morning. Among night-workers prolonged concentration often brings on insomnia as they grow older and they change—reluctantly—to the morning. Felix Klein, who worked day and night as a young man, once indicated a possible way out of this difficulty. One of his American students complained that he could not sleep for thinking of his mathematics. "Can't sleep, eh?" Klein snorted. "What's chloral for?" However, this remedy is not to be recommended indiscriminately; it probably had something to do with Klein's own tragic breakdown.

Probably the most significant of the replies were those received on the topic of inspiration versus drudgery as the source of mathematical discoveries. The conclusion is that "Mathematical discoveries, small or great . . . are never born of spontaneous generation. They always presuppose a soil seeded with preliminary knowledge and well prepared by labor, both conscious and subconscious."

Those who, like Thomas Alva Edison, have declared that genius is ninety nine per cent perspiration and only one per cent inspiration, are not contradicted by those who would reverse the figures. Both are right; one man remembers the drudgery while another forgets it all in the thrill of apparently sudden discovery but both, when they analyze their impressions, admit that without drudgery and a flash of "inspiration" discoveries are not made. If drudgery alone sufficed, how is it that many gluttons for hard work who seem to know everything about some branch of science, while excellent critics and commentators, never themselves make even a small discovery? On the other hand, those who believe in "inspiration" as the sole factor in discovery or invention—scientific or literary—may find it instructive to look at an early draft of any of Shelley's "completely spontaneous" poems (so far as these have been preserved and reproduced), or the successive versions of any of the greater novels that Balzac inflicted on his maddened printer.

Poincaré stated his views on mathematical discovery in an essay first published in 1908 and reproduced in his *Science et Méthode*. The genesis of mathematical discovery, he says, is a problem which should interest psychologists intensely, for it is the activity in which the human mind seems to borrow least from the external world, and by understanding the process of mathematical thinking we may hope to reach what is most essential in the human mind.

How does it happen, Poincaré asks, that there are persons who do not understand mathematics? "This should surprise us, or rather it would surprise us if we were not so accustomed to it." If mathematics is based only on the rules of logic, such as all normal minds accept, and which only a lunatic would deny (according to Poincaré), how is it that so many are mathematically impermeable? To which it may be answered that no exhaustive set of experiments substantiating mathematical incompetence as the normal human mode has yet been published. "And further," he asks, "how is error possible in mathematics?" Ask Alexander Pope: "To err is human," which is as unsatisfactory a solution as any other. The chemistry of the digestive system may have something to do with it, but Poincaré prefers a more subtle explanation—one which could not be tested by feeding the "vile body" hasheesh and alcohol.

"The answer seems to me evident," he declares. Logic has very little to do with discovery or invention, and memory plays tricks. Memory however is not so important as it might be. His own memory, he says without a blush, is bad: "Why then does it not desert me in a difficult piece of mathematical reasoning where most chess players [whose "memories" he assumes to be excellent] would be lost? Evidently because it is guided by the general course of the reasoning. A mathematical proof is not a mere juxtaposition of syllogisms; it is syllogisms *arranged in a certain order*, and the order is more important than the elements themselves." If he has the "intuition" of this order, memory is at a discount, for each syllogism will take its place automatically in the sequence.

Mathematical creation however does not consist merely in making new combinations of things already known; "anyone could do that, but the combinations thus made would be infinite in number and most of them entirely devoid of interest. To create consists precisely in avoiding useless combinations and in making those which are useful and which constitute only a small minority. Invention is discernment, selection." But has not all this been said thousands of times before? What artist does not know that selection—an intangible—is one of the secrets of success? We are exactly where we were before the investigation began.

To conclude this part of Poincaré's observations it may be pointed out that much of what he says is based on an assumption which may indeed be true but for which there is not a particle of scientific evidence. To put it bluntly he assumes that the majority of human beings are mathematical imbeciles. Granting him this, we need not even then accept his purely romantic theories. They belong to inspirational literature and not to science. Passing to something less controversial we shall now quote the famous passage in which Poincaré describes how one of his own greatest "inspirations" came to him. It is meant to substantiate his theory of mathematical creation. Whether it does or not may be left to the reader.

He first points out that the technical terms need not be understood in order to follow his narrative: "What is of interest to the psychologist is not the theorem but the circumstances.

"For fifteen days I struggled to prove that no functions analogous to those I have since called *Fuchsian functions* could exist; I was then very ignorant. Every day I sat down at my work table where I spent an hour or two; I tried a great number of combinations and arrived at no result. One evening, contrary to my custom, I took black coffee; I could not go to sleep; ideas swarmed up in clouds; I sensed them clashing until, to put it so, a pair would hook together to form a stable combination. By morning I had established the existence of a class of Fuchsian functions, those derived from the hypergeometric series. I had only to write up the results, which took me a few hours.

"Next I wished to represent these functions by the quotient of two series; this idea was perfectly conscious and thought out; analogy with elliptic functions guided me. I asked myself what must be the properties of these series if they existed, and without difficulty I constructed the series which I called thetafuchsian.

"I then left Caen, where I was living at the time, to participate in a geological trip sponsored by the School of Mines. The exigencies of travel made me forget my mathematical labors; reaching Coutances we took a bus for some excursion or another. The instant I put my foot on the step the idea came to me, apparently with nothing whatever in my previous thoughts having prepared me for it, that the transformations which I had used to define Fuchsian functions were identical with those of non-Euclidean geometry. I did not make the verification; I should not have had the time, because once in the bus I resumed an interrupted conversation; but I felt an instant and complete certainty. On returning to Caen I verified the result at my leisure to satisfy my conscience.

"I then undertook the study of certain arithmetical questions without much apparent success and without suspecting that such matters could have the slightest connection with my previous studies. Disgusted at my lack of success, I went to spend a few days at the seaside and thought of something else. One day, while walking along the cliffs, the idea came to me, again with the same characteristics of brevity, suddenness, and immediate certainty, that the transformations of indefinite ternary quadratic forms were identical with those of non-Euclidean geometry.

"On returning to Caen, I reflected on this result and deduced its consequences; the example of quadratic forms showed me that there were Fuchsian groups other than those corresponding to the hypergeometric series; I saw that I could apply to them the theory of thetafuchsian functions, and hence that there existed thetafuchsian functions other than those derived from the hypergeometric series, the only ones I had known up till then. Naturally I set myself the task of constructing all these functions. I conducted a systematic siege and, one after another, carried all the outworks; there was however one which still held out and whose fall would bring about that of the whole position. But all my efforts served only to make me better acquainted with the difficulty, which in itself was something. All this work was perfectly conscious.

"At this point I left for Mont-Valérien, where I was to discharge my military service. I had therefore very different preoccupations. One day, while crossing the boulevard, the solution of the difficulty which had stopped me appeared to me all of a sudden. I did not seek to go into it immediately, and it was only after my service that I resumed the question. I had all the elements, and had only to assemble and order them. So I wrote out my definitive memoir at one stroke and with no difficulty."

Many other examples of this sort of thing could be given from his own work, he says, and from that of other mathematicians as reported in *L'Enseignement Mathématique*. From his experiences he believes that this semblance of "sudden illumination [is] a manifest sign of previous long subconscious work," and he proceeds to elaborate his theory of the subconscious mind and its part in mathematical creation. Conscious work is necessary as a sort of trigger to fire off the accumulated dynamite which the subconscious has been excreting—he does not put it so, but what he says amounts to the same. But what is gained in the way of rational explanation if, following Poincaré, we foist off on the "subconscious mind," or the "subliminal self," the very activities which it is our object to understand? Instead of endowing this mysterious agent with a hypothetical tact enabling it to discriminate between the "exceedingly numerous" possible combinations presented (how, Poincaré does not say) for its inspection, and calmly saying that the "subconscious" rejects all but the "useful" combinations because it has a feeling for symmetry and beauty, sounds suspiciously like solving the initial problem by giving it a more impressive name. Perhaps this is exactly what Poincaré intended, for he once defined mathematics as the art of giving the same name to different things; so here he may be rounding out the symmetry of his view by giving different names to the same thing. It seems strange that a man who could have been satisfied with such a "psychology" of mathematical invention was the complete skeptic in religious matters that Poincaré was. After Poincaré's brilliant lapse into psychology skeptics may well despair of ever disbelieving anything.

During the first decade of the twentieth century Poincaré's fame increased rapidly and he came to be looked upon, especially in France, as an oracle on all things mathematical. His pronouncements on all manner of questions, from politics to ethics, were usually direct and brief, and were accepted as final by the majority. As almost invariably happens after a great man's extinction, Poincaré's dazzling reputation during his lifetime passed through a period of partial eclipse in the decade following his death. But his intuition for what was likely to be of interest to a later generation is already justifying itself. To take but one instance of many, Poincaré was a vigorous opponent of the theory that all mathematics can be rewritten in terms of the most elementary notions of classical logic; something more than logic, he believed, makes mathematics what it is. Although he did not go quite so far as the current intuitionist school he seems to have believed, as that school does, that at least some mathematical notions precede logic, and if one is to be derived from the other it is logic which must come out of mathematics, not the other way about. Whether this is to be the ultimate creed remains to be seen, but at present it appears as if the theory which Poincaré assailed with all the irony at his command is not the final one, whatever may be its merits.

Except for a distressing illness during his last four years Poincaré's busy life was tranquil and happy. Honors were showered upon him by all the leading learned societies of the world, and in 1906, at the age of fifty two, he achieved the highest distinction possible to a French scientist, the Presidency of the Academy of Sciences. None of all this inflated his ego, for Poincaré was truly humble and unaffectedly simple. He knew of course that he was without a close rival in the days of his maturity, but he could also say without a trace of affectation that he knew nothing compared to what is to be known. He was happily married and had a son and three daughters in whom he took much pleasure, especially when they were children. His wife was a great-granddaughter of Étienne Geoffroy Saint-Hilaire, remembered as the antagonist of that pugnacious comparative anatomist Cuvier. One of Poincaré's passions was symphonic music.

At the International Mathematical Congress of 1908, held at Rome, Poincaré was prevented by illness from reading his stimulating (if premature) address on *The Future of Mathematical Physics*. His trouble was hypertrophy of the prostate, which was relieved by the Italian surgeons, and it was thought that he was permanently cured. On his return to Paris he resumed his work as energetically as ever. But in 1911 he began to have presentiments that he might not live long, and on December 9 wrote asking the editor of a mathematical journal whether he would accept an unfinished memoir—contrary to the usual custom—on a problem which Poincaré considered of the highest importance: " . . . at my age, I may not be able to solve it, and the results obtained, susceptible of putting researchers on a new and unexpected path, seem to me too full of promise, in spite of the deceptions they have caused me, that I should resign myself to sacrificing them . . . " He had spent the better part of two fruitless years trying to overcome his difficulties.

A proof of the theorem which he conjectured would have enabled him to make a striking advance in the problem of three bodies; in particular it would have permitted him to prove the existence of an infinity of periodic solutions in cases more general than those hitherto considered. The desired proof was given shortly after the publication of Poincaré's "unfinished symphony" by a young American mathematician, George David Birkhoff (1884–   ).

In the spring of 1912 Poincaré fell ill again and underwent a second operation on July 9. The operation was successful, but on July 17 he died very suddenly from an embolism while dressing. He was in the fifty ninth year of his age and at the height of his powers—"the living brain of the rational sciences," in the words of Painlevé.

CHAPTER TWENTY NINE

Paradise Lost?

CANTOR

Mathematics, like all other subjects, has now to take its turn under the microscope and reveal to the world any weaknesses there may be in its foundations.
—F. W. WESTAWAY

THE CONTROVERSIAL TOPIC of *Mengenlehre* (theory of sets, or classes, particularly of infinite sets) created in 1874–1895 by Georg Cantor (1845–1918) may well be taken, out of its chronological order, as the conclusion of the whole story. This topic typifies for mathematics the general collapse of those principles which the prescient seers of the nineteenth century, foreseeing everything but the grand débâcle, believed to be fundamentally sound in all things from physical science to democratic government.

If "collapse" is perhaps too strong to describe the transition the world is doing its best to enjoy, it is nevertheless true that the evolution of scientific ideas is now proceeding so vertiginously that evolution is barely distinguishable from revolution.

Without the errors of the past as a deep-seated focus of disturbance the present upheaval in physical science would perhaps not have happened; but to credit our predecessors with all the inspiration for what our own generation is doing, is to give them more than their due. This point is worth a moment's consideration, as some may be tempted to say that the corresponding "revolution" in mathematical thinking, whose beginnings are now plainly apparent, is merely an echo of Zeno and other doubters of ancient Greece.

The difficulties of Pythagoras over the square root of 2 and the paradoxes of Zeno on continuity (or "infinite divisibility") are—so far as we know—the origins of our present mathematical schism. Mathematicians today who pay any attention to the philosophy (or foundations) of their subject are split into at least two factions, apparently beyond present hope of reconciliation, over the validity of the reasoning used in mathematical analysis, and this disagreement can be traced back through the centuries to the Middle Ages and thence to ancient Greece. All sides have had their representatives in all ages of mathematical thought, whether that thought was disguised in provocative paradoxes, as with Zeno, or in logical subtleties, as with some of the most exasperating logicians of the Middle Ages. The root of these differences is commonly accepted by mathematicians as being a matter of temperament: any attempt to convert an analyst like Weierstrass to the skepticism of a doubter like Kronecker is bound to be as futile as trying to convert a Christian fundamentalist to rabid atheism.

A few dated quotations from leaders in the dispute may serve as a stimulant—or sedative, according to taste—for our enthusiasm over the singular intellectual career of Georg Cantor, whose "positive theory of the infinite" precipitated, in our own generation, the fiercest frog-mouse battle (as Einstein once called it) in history over the validity of traditional mathematical reasoning.

In 1831 Gauss expressed his "horror of the actual infinite" as follows. "I protest against the use of infinite magnitude as something completed, which is never permissible in mathematics. Infinity is merely a way of speaking, the true meaning being a limit which certain ratios approach indefinitely close, while others are permitted to increase without restriction."

Thus, if x denotes a real number, the fraction $1/x$ diminishes as x increases, and we can find a value of x such that $1/x$ differs from zero by any preassigned amount (other than zero) which may be as small as we please, and as x continues to increase, the difference *remains* less than this preassigned amount; the *limit* of $1/x$, "as x tends to infinity," is zero. The symbol of infinity is ∞; the assertion $1/\infty = 0$ is nonsensical for two reasons: "division by infinity" is an operation which is *undefined*, and hence has no meaning; the second reason was stated by Gauss. Similarly $1/0 = \infty$ is meaningless.

Cantor agrees and disagrees with Gauss. Writing in 1886 on the problem of the actual (what Gauss called completed) infinite, Cantor says that "in spite of the essential difference between the concepts of the *potential* and the *actual* 'infinite,' the former meaning a *variable* finite magnitude increasing beyond all finite limits (like x in $1/x$ above), while the latter is a *fixed, constant* magnitude lying beyond all finite magnitudes, it happens only too often that they are confused."

Cantor goes on to state that misuse of the infinite in mathematics had justly inspired a horror of the infinite among careful mathematicians of his day, precisely as it did in Gauss. Nevertheless he maintains that the resulting "uncritical rejection of the legitimate actual infinite is no lesser a violation of the nature of things [whatever that may be—it does not appear to have been revealed to mankind as a whole], which must be taken as they are"—however that may be. Cantor thus definitely aligns himself with the great theologians of the Middle Ages, of whom he was a deep student and an ardent admirer.

Absolute certainties and complete solutions of age-old problems always go down better if well salted before swallowing. Here is what Bertrand Russell had to say in 1901 about Cantor's Promethean attack on the infinite.

"Zeno was concerned with three problems. . . . These are the problem of the infinitesimal, the infinite, and continuity. . . . From his day to our own, the finest intellects of each generation in turn attacked these problems, but achieved, broadly speaking, nothing. . . . Weierstrass, Dedekind, and Cantor . . . have completely solved them. Their solutions . . . are so clear as to leave no longer the slightest doubt of difficulty. This achievement is probably the greatest of which the age can boast. . . . The problem of the infinitesimal was solved by Weierstrass, the solution of the other two was begun by Dedekind and definitely accomplished by Cantor."*

The enthusiasm of this passage warms us even today, although we know that Russell in the second edition (1924) of his and A. N. Whitehead's *Principia Mathematica* admitted that all was not well with the Dedekind "cut" (see Chapter 27), which is the spinal cord of analysis. Nor is it well today. More is done for or against a particular creed in science or mathematics in a decade than was accomplished in a century of antiquity, the Middle Ages, or the late renaissance. More good minds attack an outstanding scientific or mathematical problem today than ever before, and finality has become the private property of fundamentalists. Not one of the finalities in Russell's remarks of 1901 has survived. A quarter of a century ago those who were unable to see the great light which the prophets assured them was blazing overhead like the noonday sun in a midnight sky were called merely stupid. Today for every competent expert on the side of the prophets there is an equally competent and opposite expert against them. If

---

*Quoted from R. E. Moritz' *Memorabilia Mathematica*, 1914. The original source is not accessible to me.

there is stupidity anywhere it is so evenly distributed that it has ceased to be a mark of distinction. We are entering a new era, one of doubt and decent humility.

On the doubtful side about the same time (1905) we find Poincaré. "I have spoken . . . of our need to return continually to the first principles of our science, and of the advantages of this for the study of the human mind. This need has inspired two enterprises which have assumed a very prominent place in the most recent development of mathematics. The first is Cantorism. . . . Cantor introduced into science a new way of considering the mathematical infinite . . . but it has come about that we have encountered certain paradoxes, certain apparent contradictions that would have delighted Zeno the Eleatic and the school of Megara. So each must seek the remedy. I for my part—and I am not alone—think that the important thing is never to introduce entities not completely definable in a finite number of words. Whatever be the cure adopted, we may promise ourselves the joy of the physician called in to treat a beautiful pathologic case."

A few years later Poincaré's interest in pathology for its own sake had abated somewhat. At the International Mathematical Congress of 1908 at Rome, the satiated physician delivered himself of this prognosis: "Later generations will regard *Mengenlehre* as a disease from which one has recovered."

It was Cantor's greatest merit to have discovered in spite of himself and against his own wishes in the matter that the "body mathematic" is profoundly diseased and that the sickness with which Zeno infected it has not yet been alleviated. His disturbing discovery is a curious echo of his own intellectual life. We shall first glance at the facts of his material existence, not of much interest in themselves, perhaps, but singularly illuminative in their later aspects of his theory.

Of pure Jewish descent on both sides, Georg Ferdinand Ludwig Philipp Cantor was the first child of the prosperous merchant Georg Waldemar Cantor and his artistic wife Maria Bohm. The father was born in Copenhagen, Denmark, but migrated as a young man to St. Petersburg, Russia, where the mathematician Georg Cantor was born on March 3, 1845. Pulmonary disease caused the father to move in 1856 to Frankfurt, Germany, where he lived in comfortable retirement till his death in 1863. From this curious medley of nationalities it is possible for several fatherlands to claim Cantor as their son.

Cantor himself favored Germany, but it cannot be said that Germany favored him very cordially.

Georg had a brother Constantin, who became a German army officer (very few Jews ever did), and a sister, Sophie Nobiling. The brother was a fine pianist; the sister an accomplished designer. Georg's pent-up artistic nature found its turbulent outlet in mathematics and philosophy, both classical and scholastic. The marked artistic temperaments of the children were inherited from their mother, whose grandfather was a musical conductor, one of whose brothers, living in Vienna, taught the celebrated violinist Joachim. A brother of Maria Cantor was a musician, and one of her nieces a painter. If it is true, as claimed by the psychological proponents of drab mediocrity, that normality and phlegmatic stability are equivalent, all this artistic brilliance in his family may have been the root of Cantor's instability.

The family were Christians, the father having been converted to Protestantism; the mother was born a Roman Catholic. Like his archenemy Kronecker, Cantor favored the Protestant side and acquired a singular taste for the endless hairsplitting of medieval theology. Had he not become a mathematician it is quite possible that he would have left his mark on philosophy or theology. As an item of interest that may be noted in this connection, Cantor's theory of the infinite was eagerly pounced on by the Jesuits, whose keen logical minds detected in the mathematical imagery beyond their theological comprehension indubitable proofs of the existence of God and the self-consistency of the Holy Trinity with its three-in-one, one-in-three, co-equal and co-eternal. Mathematics has strutted to some pretty queer tunes in the past 2500 years, but this takes the cake. It is only fair to say that Cantor, who had a sharp wit and a sharper tongue when he was angered, ridiculed the pretentious absurdity of such "proofs," devout Christian and expert theologian though he himself was.

Cantor's school career was like that of most highly gifted mathematicians—an early recognition (before the age of fifteen) of his greatest talent and an absorbing interest in mathematical studies. His first instruction was under a private tutor, followed by a course in an elementary school in St. Petersburg. When the family moved to Germany, Cantor first attended private schools at Frankfurt and the Darmstadt nonclassical school, entering the Wiesbaden Gymnasium in 1860 at the age of fifteen.

Georg was determined to become a mathematician, but his practical father, recognizing the boy's mathematical ability, obstinately tried to force him into engineering as a more promising bread-and-butter profession. On the occasion of Cantor's confirmation in 1860 his father wrote to him expressing the high hopes he and all Georg's numerous aunts, uncles, and cousins in Germany, Denmark, and Russia had placed on the gifted boy: "They expect from you nothing less than that you become a Theodor Schaeffer and later, perhaps, if God so wills, a shining star in the engineering firmament." When will parents recognize the presumptuous stupidity of trying to make a cart horse out of a born racer?

The pious appeal to God which was intended to blackjack the sensitive, religious boy of fifteen into submission in 1860 would today (thank God!) rebound like a tennis ball from the harder heads of our own younger generation. But it hit Cantor pretty hard. In fact it knocked him out cold. Loving his father devotedly and being of a deeply religious nature, young Cantor could not see that the old man was merely rationalizing his own absurd ambition. Thus began the first warping of Georg Cantor's acutely sensitive mind. Instead of rebelling, as a gifted boy today might do with some hope of success, Georg submitted till it became apparent even to the obstinate father that he was wrecking his son's disposition. But in the process of trying to please his father against the promptings of his own instincts Georg Cantor sowed the seeds of the self-distrust which was to make him an easy victim for Kronecker's vicious attack in later life and cause him to doubt the value of his work. Had Cantor been brought up as an independent human being he would never have acquired the timid deference to men of established reputation which made his life wretched.

The father gave in when the mischief was already done. On Georg's completion of his school course with distinction at the age of seventeen, he was permitted by "dear papa" to seek a university career in mathematics. "My dear papa!" Georg writes in his boyish gratitude: "You can realize for yourself how greatly your letter delighted me. The letter fixes my future. . . . Now I am happy when I see that it will not displease you if I follow my feelings in the choice. I hope you will live to find joy in me, dear father; since my soul, my whole being, lives in my vocation; what a man desires to do, and that to which an inner compulsion drives him, that will he accomplish!" Papa no doubt deserves a vote of thanks, even if Georg's gratitude is a shade too servile for a modern taste.

Cantor began his university studies at Zurich in 1862, but migrated to the University of Berlin the following year, on the death of his father. At Berlin he specialized in mathematics, philosophy, and physics. The first two divided his interests about equally; for physics he never had any sure feeling. In mathematics his instructors were Kummer, Weierstrass, and his future enemy Kronecker. Following the usual German custom, Cantor spent a short time at another university, and was in residence for one semester of 1866 at Göttingen.

With Kummer and Kronecker at Berlin the mathematical atmosphere was highly charged with arithmetic. Cantor made a profound study of the *Disquisitiones Arithmeticae* of Gauss and wrote his dissertation, accepted for the Ph.D. degree in 1867, on a difficult point which Gauss had left aside concerning the solution in integers x, y, z of the indeterminate equation

$$ax^2 + by^2 + cz^2 = 0,$$

where a, b, c are any given integers. This was a fine piece of work, but it is safe to say that no mathematician who read it anticipated that the conservative author of twenty two was to become one of the most radical originators in the history of mathematics. Talent no doubt is plain enough in this first attempt, but genius—no. There is not a single hint of the great originator in this severely classical dissertation.

The like may be said for all of Cantor's earliest work published before he was twenty nine. It was excellent, but might have been done by any brilliant man who had thoroughly absorbed, as Cantor had, the doctrine of rigorous proof from Gauss and Weierstrass. Cantor's first love was the Gaussian theory of numbers, to which he was attracted by the hard, sharp, clear perfection of the proofs. From this, under the influence of the Weierstrassians, he presently branched off into rigorous analysis, particularly in the theory of trigonometric series (Fourier series).

The subtle difficulties of this theory (where questions of convergence of infinite series are less easily approachable than in the theory of power series) seem to have inspired Cantor to go deeper for the foundations of analysis than any of his contemporaries had cared to look, and he was led to his grand attack on the mathematics and philosophy of the infinite itself, which is at the bottom of all questions concerning continuity, limits, and convergence. Just before he was thirty, Cantor published his first revolutionary paper (in Crelle's *Journal*) on the theory of infinite sets. This will be described presently. The unexpected and paradoxical result concerning the set of *all* algebraic numbers which Cantor established in this paper and the complete novelty of the methods employed immediately marked the young author as a creative mathematician of extraordinary originality. Whether all agreed that the new methods were sound or not is beside the point; it was universally admitted that a man had arrived with something fundamentally new in mathematics. He should have been given an influential position at once.

Cantor's material career was that of any of the less eminent German professors of mathematics. He never achieved his ambition of a professorship at Berlin, possibly the highest German distinction during the period of Cantor's greatest and most original productivity (1874–1884, age twenty nine to thirty nine). All his active professional career was spent at the University of Halle, a distinctly third-rate institution, where he was appointed *Privatdozent* (a lecturer who lives by what fees he can collect from his students) in 1869 at the age of twenty four. In 1872 he was made assistant professor and in 1879—before the criticism of his work had begun to assume the complexion of a malicious personal attack on himself—he was appointed full professor. His earliest teaching experience was in a girls' school in Berlin. For this curiously inappropriate task he had qualified himself by listening to dreary lectures on pedagogy by an uninspired mathematical mediocrity before securing his state license to teach children. More social waste.

Rightly or wrongly, Cantor blamed Kronecker for his failure to obtain the coveted position at Berlin. When two academic specialists disagree violently on purely scientific matters, they have a choice, if discretion seems the better part of valor, of laughing their hatreds off and not making a fuss about them, or of acting in any of the number of belligerent ways that other people resort to when confronted with situations of antagonism. One way is to go at the other in an efficient, underhand manner, which often enables one to gain his spiteful end under the guise of sincere friendship. Nothing of the sort here! When

Cantor and Kronecker fell out, they disagreed all over, threw reserve to the dogs, and did everything but slit the other's throat. Perhaps after all this is a more decent way of fighting—if men must fight— than the sanctimonious hypocrisy of the other. The object of any war is to destroy the enemy, and being sentimental or chivalrous about the unpleasant business is the mark of an incompetent fighter. Kronecker was one of the most competent warriors in the history of scientific controversy; Cantor, one of the least competent. Kronecker won. But, as will appear later, Kronecker's bitter animosity toward Cantor was not wholly personal but at least partly scientific and disinterested.

The year 1874 which saw the appearance of Cantor's first revolutionary paper on the theory of sets was also that of his marriage, at the age of twenty nine, to Vally Guttmann. Two sons and four daughters were born of this marriage. None of the children inherited their father's mathematical ability.

On their honeymoon at Interlaken the young couple saw a lot of Dedekind, perhaps the one first-rate mathematician of the time who made a serious and sympathetic attempt to understand Cantor's subversive doctrine.

Himself somewhat of a *persona non grata* to the leading German overlords of mathematics in the last quarter of the nineteenth century, the profoundly original Dedekind was in a position to sympathize with the scientifically disreputable Cantor. It is sometimes imagined by outsiders that originality is always assured of a cordial welcome in science. The history of mathematics contradicts this happy fantasy: the way of the transgressor in a well established science is likely to be as hard as it is in any other field of human conservatism, even when the transgressor is admitted to have found something valuable by overstepping the narrow bounds of bigoted orthodoxy.

Both Dedekind and Cantor got what they might have expected had they paused to consider before striking out in new directions. Dedekind spent his entire working life in mediocre positions; the claim— now that Dedekind's work is recognized as one of the most important contributions to mathematics that Germany has ever made—that Dedekind *preferred* to stay in obscure holes while men who were in no sense his intellectual superiors shone like tin plates in the glory of public and academic esteem, strikes observers who are themselves "Aryans" but not Germans as highly diluted eyewash.

The ideal of German scholarship in the nineteenth century was the lofty one of a thoroughly coordinated "safety first," and perhaps rightly it showed an extreme Gaussian caution toward radical originality—the new thing might conceivably be not quite right. After all an honestly edited encyclopaedia is in general a more reliable source of information about the soaring habits of skylarks than a poem, say Shelley's, on the same topic.

In such an atmosphere of cloying alleged fact, Cantor's theory of the infinite—one of the most disturbingly original contributions to mathematics in the past 2500 years—felt about as much freedom as a skylark trying to soar up through an atmosphere of cold glue. Even if the theory was totally wrong—and there are some who believe it cannot be salvaged in any shape resembling the thing Cantor thought he had launched—it deserved something better than the brickbats which were hurled at it chiefly because it was new and unbaptized in the holy name of orthodox mathematics.

The pathbreaking paper of 1874 undertook to establish a totally unexpected and highly paradoxical property of the set of *all* algebraic numbers. Although such numbers have been frequently described in preceding chapters, we shall state once more what they are, in order to bring out clearly the nature of the astounding fact which Cantor proved—in saying "proved" we deliberately ignore for the present all doubts as to the soundness of the reasoning used by Cantor.

If r satisfies an algebraic equation of degree n with rational integer (common whole number) coefficients, and if r satisfies no such equation of degree less than n, then r is an algebraic number of degree n.

This can be generalized. For it is easy to prove that any root of an equation of the type

$$c_0 x^n + c_1 x^{n-1} + \ldots + c_{n-1} x + c_n = 0,$$

in which the c's are any given *algebraic* numbers (as defined above), is itself an algebraic number. For example, according to this theorem, all roots of

$$(1 - 3\sqrt{-1})x^2 - (2 + 5\sqrt{17})x + \sqrt[3]{90} = 0$$

are algebraic numbers, since the coefficients are. (The first coefficient satisfies $x^2 - 2x + 10 = 0$, the second, $x^2 - 4x - 421 = 0$, the third, $x^3 - 90 = 0$, of the respective degrees 2, 2, 3.)

Imagine (if you can) the set of *all* algebraic numbers. *Among* these will be *all* the positive rational integers 1, 2, 3, . . . , since any one of them, say n, satisfies an algebraic equation, $x - n = 0$, in which the coefficients (1, and $-n$) are rational integers. But *in addition to these* the set of *all* algebraic numbers will include *all* roots of *all* quadratic equations with rational integer coefficients, and *all* roots of *all* cubic equations with rational integer coefficients, and so on, indefinitely. Is it not *intuitively evident* that the set of *all* algebraic numbers will contain *infinitely more* members than its *sub*-set of the rational integers 1, 2, 3, . . . ? It might indeed be so, but it happens to be false.

Cantor proved that the set of all rational integers 1, 2, 3, . . . contains precisely as many members as the "infinitely more inclusive" set of *all* algebraic numbers.

A proof of this paradoxical statement cannot be given here, but the kind of device—that of "one-to-one correspondence"—upon which the proof is based can easily be made intelligible. This should induce in the philosophical mind an understanding of what a *cardinal number* is. Before describing this simple but somewhat elusive concept it will be helpful to glance at an expression of opinion on this and other definitions of Cantor's theory which emphasizes a distinction between the attitudes of some mathematicians and many philosophers toward all questions regarding "number" or "magnitude."

"A mathematician never defines magnitudes in themselves, as a philosopher would be tempted to do; he defines their equality, their sum and their product, and these definitions determine, or rather constitute, all the mathematical properties of magnitudes. In a yet more abstract and more formal manner he *lays down* symbols and at the same time *prescribes* the rules according to which they must be combined; these rules suffice to characterize these symbols and to give them a mathematical value. Briefly, he creates mathematical entities by means of arbitrary conventions, in the same way that the several chessmen are defined by the conventions which govern their moves and the relations between them."* Not all schools of mathe-

---

*L. Couturat, *De l'infini mathématique*, Paris, 1896, p.49. With the caution that much of this work is now hopelessly out of date, it can be recommended for its clarity to the general reader. An account of the elements of Cantorism by a leading Polish expert which is within the comprehension of anyone with a grade-school education and a taste for abstract reasoning is the *Leçons sur*

matical thought would subscribe to these opinions, but they suggest at least one "philosophy" responsible for the following *definition* of cardinal numbers.

Note that the initial stage in the definition is the description of "same cardinal number," in the spirit of Couturat's opening remarks; "cardinal number" then arises phoenix-like from the ashes of its "sameness." It is all a matter of *relations* between concepts not explicitly defined.

Two sets are said to have *the same cardinal number* when all the things in the sets can be *paired off* one-to-one. After the pairing there are to be no unpaired things in either set.

Some examples will clarify this esoteric definition. It is one of those trivially obvious and fecund nothings which are so profound that they are overlooked for thousands of years. The sets (x, y, z), (a, b, c) have *the same cardinal number* (we shall not commit the blunder of saying "Of course! Each contains *three letters*") *because* we can *pair off* the things x, y, z in the first set with those, a, b, c in the second as follows, x with a, y with b, z with c, and having done so, find that none remain unpaired in either set. Obviously there are other ways for effecting the pairing. Again, in a Christian community practising technical monogamy, if twenty married couples sit down together to dinner, the set of husbands will have the same cardinal number as the set of wives.

As another instance of this "obvious" sameness, we recall Galileo's example of the set of all squares of positive integers and the set of all positive integers:

$$1^2, 2^2, 3^2, 4^2, \ldots, n^2, \ldots$$
$$1, 2, 3, 4, \ldots, n, \ldots$$

The "paradoxical" distinction between this and the preceding examples is apparent. If all the wives retire to the drawing room, leaving their spouses to sip port and tell stories, there will be precisely twenty human beings sitting at the table, just half as many as there were before. But if all the squares desert the natural numbers, there are just as many left as there were before. Dislike it or not as we may

---

les nombres transfinis, by Waclaw Sierpinski, Paris, 1928. The preface by Borel supplies the necessary danger signal. The above extract from Couturat is of some historical interest in connection with Hilbert's program. It anticipates by thirty years Hilbert's statement of his formalist creed.

(we should not, if we are rational animals), the crude miracle stares us in the face that *a part of a set may have the same cardinal number as the entire set*. If anyone dislikes the "pairing" definition of "same cardinal number," he may be challenged to produce a comelier. Intuition (male, female, or mathematical) has been greatly overrated. Intuition is the root of all superstition.

Notice at this stage that a difficulty of the first magnitude has been glossed. *What is a set, or a class?* "That," in the words of Hamlet, is "the question." We shall return to it, but we shall not answer it. Whoever succeeds in answering that innocent question to the entire satisfaction of Cantor's critics will quite likely dispose of the more serious objections against his ingenious theory of the infinite and at the same time establish mathematical analysis on a non-emotional basis. To see that the difficulty is not trivial, try to imagine the set of *all* positive rational integers 1, 2, 3, . . . , and ask yourself whether, with Cantor, you can hold this totality—which is a "class"—in your mind as a definite object of thought, as easily apprehended as the class x, y, z of three letters. Cantor requires us to do just this thing in order to reach the *transfinite* numbers which he created.

Proceeding now to the definition of "cardinal number," we introduce a convenient technical term: two sets or classes whose members can be paired off one-to-one (as in the examples given previously) are said to be *similar*. *How many* things are there in the set (or class) x, y, z? Obviously three. But what is "three"? An answer is contained in the following definition: "The *number* of things in a given class is the *class* of all classes that are similar to the given class."

This definition gains nothing from attempted explanation; it must be grasped as it is. It was proposed in 1879 by Gottlob Frege, and again (independently) by Bertrand Russell in 1901. One advantage which it has over other definitions of "cardinal number of a class" is its applicability to both finite and infinite classes. Those who believe the definition too mystical for mathematics can avoid it by following Couturat's advice and not attempting to *define* "cardinal number." However, that way also leads to difficulties.

Cantor's spectacular result that the class of all algebraic numbers is similar (in the technical sense defined above) to its sub-class of all the positive rational integers was but the first of many wholly unexpected properties of infinite classes. Granting for the moment that his reasoning in reaching these properties is sound, or, if not unobjectionable in the form in which Cantor left it, that it can be made rigorous, we must admit its power.

Consider for example the "existence" of transcendental numbers. In an earlier chapter we saw what a tremendous effort it cost Hermite to prove the transcendence of *a particular* number of this kind. Even today there is no general method known whereby the transcendence of any number which we suspect is transcendental can be proved; each new type requires the invention of special and ingenious methods. It is suspected, for example, that the number (it is a constant, although it looks as if it might be a variable from its definition) which is defined as the limit of

$$1 + \frac{1}{2} + \frac{1}{3} + \ldots + \frac{1}{n} - \log n$$

as *n* tends to infinity, is transcendental, but we cannot prove that it is. What is required is to show that this constant is not a root of *any* algebraic equation with rational integer coefficients.

All this suggests the question "How many transcendental numbers are there?" Are they *more* numerous than the integers, or the rationals, or the algebraic numbers as a whole, or are they *less* numerous? Since (by Cantor's theorem) the integers, the rationals, and *all* algebraic numbers are equally numerous, the question amounts to this: can the transcendental numbers be counted off 1, 2, 3, . . . ? Is the class of all transcendental numbers *similar* to the class of all positive rational integers? The answer is no; the transcendentals are *infinitely more numerous than the integers*.

Here we begin to get into the controversial aspects of the theory of sets. The conclusion just stated was like a challenge to a man of Kronecker's temperament. Discussing Lindemann's proof that π is transcendental (see Chapter 24), Kronecker asked, "Of what use is your beautiful investigation regarding π? Why study such problems, since irrational [and hence transcendental] numbers do not exist?" We can imagine the effect on such a skepticism of Cantor's proof that the transcendentals are infinitely more numerous than the integers 1, 2, 3, . . . which, according to Kronecker, are the noblest work of God and the *only* numbers that *do* "exist."

Even a summary of Cantor's proof is out of the question here, but something of the kind of reasoning he used can be seen from the following simple considerations. If a class is similar (in the above technical sense) to the class of all positive rational integers, the class is said to be *denumerable*. The things in a denumerable class can be counted off 1, 2, 3, . . . ; the things in a non-denumerable class can *not* be counted off 1, 2, 3, . . . : there will be more things in a non-denumerable class than in a denumerable class. Do non-denumerable classes exist? Cantor proved that they do. In fact the class of all points on any line-segment, no matter how small the segment is (provided it is more than a single point), is non-denumerable.

From this we see a hint of why the transcendentals are non-denumerable. In the chapter on Gauss we saw that any root of any algebraic equation is representable by a point on the plane of Cartesian geometry. All these roots constitute the set of all algebraic numbers, which Cantor proved to be denumerable. But if the points on a mere line-segment are non-denumerable, it follows that *all* the points on the Cartesian plane are likewise non-denumerable. The algebraic numbers are spotted over the plane like stars against a black sky; the dense blackness is the firmament of the transcendentals.

The most remarkable thing about Cantor's proof is that it provides no means whereby a single one of the transcendentals can be constructed. To Kronecker any such proof was sheer nonsense. Much milder instances of "existence proofs" roused his wrath. One of these in particular is of interest as it prophesied Brouwer's objection to the full use of classical (Aristotelian) logic in reasoning about infinite sets.

A polynomial $ax^n + bx^{n-1} + \ldots + l$, in which the coefficients a, b, . . . l are rational numbers is said to be *irreducible* if it cannot be factored into a product of two polynomials both of which have rational number coefficients. Now, it is a meaningful statement to most human beings to assert, as Aristotle would, that a given polynomial either *is* irreducible or *is not* irreducible.

Not so for Kronecker. Until some definite process, capable of being carried out in a *finite* number of nontentative steps, is provided whereby we can settle the reducibility of any given polynomial, we have no logical right, according to Kronecker, to use the concept of irreducibility in our mathematical proofs. To do otherwise, according to him, is to court inconsistencies in our conclusions and, at best, the use of "irreducibility" without the process described, can give us only a Scotch verdict of "not proven." All such *non-constructive* reasoning is—according to Kronecker—illegitimate.

As Cantor's reasoning in his theory of infinite classes is largely non-constructive, Kronecker regarded it as a dangerous type of mathematical insanity. Seeing mathematics headed for the madhouse under Cantor's leadership, and being passionately devoted to what he considered the truth of mathematics, Kronecker attacked "the positive theory of infinity" and its hypersensitive author vigorously and viciously with every weapon that came to his hand, and the tragic outcome was that not the theory of sets went to the asylum, but Cantor. Kronecker's attack broke the creator of the theory.

In the spring of 1884, in his fortieth year, Cantor experienced the first of those complete breakdowns which were to recur with varying intensity throughout the rest of his long life and drive him from society to the shelter of a mental clinic. His explosive temper aggravated his difficulty. Profound fits of depression humbled himself in his own eyes and he came to doubt the soundness of his work. During one lucid interval he begged the authorities at Halle to transfer him from his professorship of mathematics to a chair of philosophy. Some of his best work on the positive theory of the infinite was done in the intervals between one attack and the next. On recovering from a seizure he noticed that his mind became extraordinarily clear.

Kronecker perhaps has been blamed too severely for Cantor's tragedy; his attack was but one of many contributing causes. Lack of recognition embittered the man who believed he had taken the first —and last—steps toward a rational theory of the infinite and he brooded himself into melancholia and irrationality. Kronecker however does appear to have been largely responsible for Cantor's failure to obtain the position he craved in Berlin. It is usually considered not quite sporting for one scientist to deliver a savage attack on the work of a contemporary to his students. The disagreement can be handled objectively in scientific papers. Kronecker laid himself out in 1891 to criticize Cantor's work to his students at Berlin, and it became obvious that there was no room for both under one roof. As Kronecker was already in possession, Cantor resigned himself to staying out in the cold.

However, he was not without some comfort. The sympathetic Mittag-Leffler not only published some of Cantor's work in his journal

Paradise Lost? 571

(*Acta Mathematica*) but comforted Cantor in his fight against Kronecker. In one year alone Mittag-Leffler received no less than fifty two letters from the suffering Cantor. Of those who believed in Cantor's theories, the genial Hermite was one of the most enthusiastic. His cordial acceptance of the new doctrine warmed Cantor's modest heart: "The praises which Hermite pours out to me in this letter . . . on the subject of the theory of sets are so high in my eyes, so unmerited, that I should not care to publish them lest I incur the reproach of being dazzled by them."

With the opening of the new century Cantor's work gradually came to be accepted as a fundamental contribution to all mathematics and particularly to the foundations of analysis. But unfortunately for the theory itself the paradoxes and antinomies which still infect it began to appear simultaneously. These may in the end be the greatest contribution which Cantor's theory is destined to make to mathematics, for their unsuspected existence in the very rudiments of logical and mathematical reasoning about the infinite was the direct inspiration of the present critical movement in all deductive reasoning. Out of this we hope to derive a mathematics which is both richer and "truer"—freer from inconsistency—than the mathematics of the pre-Cantor era.

Cantor's most striking results were obtained in the theory of *nondenumerable* sets, the simplest example of which is the set of all points on a line-segment. Only one of the simplest of his conclusions can be stated here. Contrary to what intuition would predict, two unequal line-segments contain the *same number* of points. Remembering that two sets contain the same number of things if, and only if, the things in them can be paired off one-to-one, we easily see the reasonableness of Cantor's conclusion. Place the unequal segments AB, CD as in the figure. The line OPQ cuts CD in the point P, and AB in Q; P and Q are thus paired off. As OPQ rotates about O, the point P traverses CD, while Q simultaneously traverses AB, and each point of CD has one, and only one, "paired" point of AB.

An even more unexpected result can be proved. Any line-segment, no matter how small, contains as many points as an infinite straight line. Further, the segment contains as many points as there are in an entire plane, or in the whole of three-dimensional space, or in the whole of space of n dimensions (where n is *any* integer greater than

(*Acta Mathematica*) but comforted Cantor in his fight against Kronecker. In one year alone Mittag-Leffler received no less than fifty two letters from the suffering Cantor. Of those who believed in Cantor's theories, the genial Hermite was one of the most enthusiastic. His cordial acceptance of the new doctrine warmed Cantor's modest heart: "The praises which Hermite pours out to me in this letter . . . on the subject of the theory of sets are so high in my eyes, so unmerited, that I should not care to publish them lest I incur the reproach of being dazzled by them."

With the opening of the new century Cantor's work gradually came to be accepted as a fundamental contribution to all mathematics and particularly to the foundations of analysis. But unfortunately for the theory itself the paradoxes and antinomies which still infect it began to appear simultaneously. These may in the end be the greatest contribution which Cantor's theory is destined to make to mathematics, for their unsuspected existence in the very rudiments of logical and mathematical reasoning about the infinite was the direct inspiration of the present critical movement in all deductive reasoning. Out of this we hope to derive a mathematics which is both richer and "truer"—freer from inconsistency—than the mathematics of the pre-Cantor era.

Cantor's most striking results were obtained in the theory of *nondenumerable* sets, the simplest example of which is the set of all points on a line-segment. Only one of the simplest of his conclusions can be stated here. Contrary to what intuition would predict, two unequal line-segments contain the *same number* of points. Remembering that two sets contain the same number of things if, and only if, the things in them can be paired off one-to-one, we easily see the reasonableness of Cantor's conclusion. Place the unequal segments AB, CD as in the figure. The line OPQ cuts CD in the point P, and AB in Q; P and Q are thus paired off. As OPQ rotates about O, the point P traverses CD, while Q simultaneously traverses AB, and each point of CD has one, and only one, "paired" point of AB.

An even more unexpected result can be proved. Any line-segment, no matter how small, contains as many points as an infinite straight line. Further, the segment contains as many points as there are in an entire plane, or in the whole of three-dimensional space, or in the whole of space of n dimensions (where n is *any* integer greater than zero) or, finally, in a space of a denumerably infinite number of dimensions.

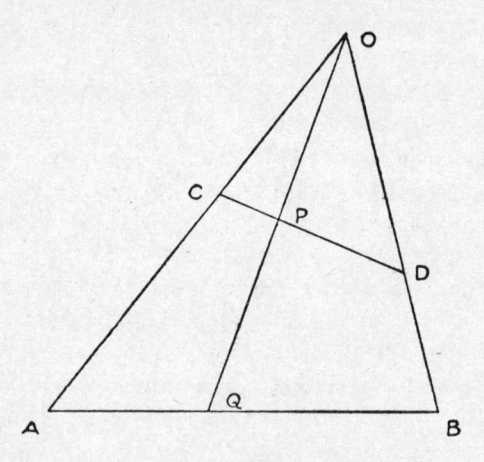

In all this we have not yet attempted to define a *class* or a *set*. Possibly (as Russell held in 1912) it is not necessary to do so in order to have a clear conception of Cantor's theory or for that theory to be consistent with itself—which is enough to demand of any mathematical theory. Nevertheless present disputes seem to require that some clear, self-consistent definition be given. The following used to be thought satisfactory.

A set is characterized by three qualities: it contains all things to which a certain definite property (say redness, or volume, or taste) belongs; no thing not having this property belongs to the set; each thing in the set is recognizable as the same thing and as different from all other things in the set—briefly, each thing in the set has a permanently recognizable individuality. The set itself is to be grasped as a whole. This definition may be too drastic for use. Consider, for example, what happens to Cantor's set of all transcendental numbers under the third demand.

At this point we may glance back over the whole history of mathematics—or as much of it as is revealed by the treatises of the master mathematicians in their purely technical works—and note two modes of expression which recur constantly in nearly all mathematical exposition. The reader perhaps has been irritated by the repetitious use of phrases such as "we can find a whole number greater than 2," or "we can choose a number less than n and greater than $n - 2$." The choice of such phraseology is not merely stereotyped pedantry. There is a reason for its use, and careful writers mean exactly what they say when they assert that "we can find, etc." They mean that *they can do what they say*.

In sharp distinction to this is the other phrase which is reiterated over and over again in mathematical writing: "There exists." For example, some would say "there exists a whole number greater than 2," or "there exists a number less than n and greater than $n - 2$." The use of such phraseology definitely commits its user to the creed which Kronecker held to be untenable, *unless*, of course, the "existence" is proved by a *construction*. The existence is not proved for the sets (as defined above) which appear in Cantor's theory.

These two ways of speaking divide mathematicians into two types: the "we can" men believe (possibly subconsciously) that mathematics is a purely human invention; the "there exists" men believe that mathematics has an extra-human "existence" of its own, and that "we" merely come upon the "eternal truths" of mathematics in our journey through life, in much the same way that a man taking a walk in a city comes across a number of streets with whose planning he had nothing whatever to do.

Theologians are "exist" men; cautious skeptics for the most part "we" men. "There exist an infinity of even numbers, or of primes," say the advocates of extra-human "existence"; "produce them," say Kronecker and the "we" men.

That the distinction is not trivial can be seen from a famous instance of it in the New Testament. Christ asserted that the Father "exists"; Philip demanded "Show us the Father and it sufficeth us." Cantor's theory is almost wholly on the "existence" side. Is it possible that Cantor's passion for theology determined his allegiance? If so, we shall have to explain why Kronecker, also a connoisseur of Christian theology, was the rabid "we" man that he was. As in all such questions ammunition for either side can be filched from any pocket.

A striking and important instance of the "existence" way of looking at the theory of sets is afforded by what is known as Zermelo's postulate (stated in 1904). "For every set M whose elements are sets P (that is, M is a set of *sets*, or a class of *classes*), the sets P being non-empty and non-overlapping (no two contain things in common), there exists at least one set N which contains precisely one element from each of the sets P which constitute M." Comparison of this with the previously stated definition of a set (or class) will show that the "we" men would not consider the postulate self-evident if the set M consisted, say, of an infinity of non-overlapping line segments. Yet the postulate seems reasonable enough. Attempts to prove it have failed. It is of considerable importance in all questions relating to continuity.

A word as to how this postulate came to be introduced into mathematics will suggest another of the unsolved problems of Cantor's theory. A set of distinct, *countable* things, like all the bricks in a certain wall, can easily be *ordered*; we need only count them off 1, 2, 3, . . . in any of dozens of different ways that will suggest themselves. But how would we go about *ordering* all the points on a straight line? They cannot be counted off 1, 2, 3, ... The task appears hopeless when we consider that between *any* two points of the line "we can find," or "there exists" *another* point of the line. If every time we counted two adjacent bricks another sprang into being between them in the wall our counting would become slightly confused. Nevertheless the points on a straight line do appear to have some sort of order; we can say whether one point is to the right or the left of another, and so on. Attempts to order the points of a line have not succeeded. Zermelo proposed his postulate as a means for making the attempt easier, but it itself is not universally accepted as a reasonable assumption or as one which it is safe to use.

Cantor's theory contains a great deal more about the actual infinite and the "arithmetic" of transfinite (infinite) numbers than what has been indicated here. But as the theory is still in the controversial stage, we may leave it with the statement of a last riddle. Does there "exist," or can we "construct," an infinite set which is not similar (technical sense of one-to-one matching) either to the set of all the positive rational integers or to the set of all points of a line? The answer is unknown.

Cantor died in a mental hospital in Halle on January 6, 1918, at the age of seventy three. Honors and recognition were his at the last, and even the old bitterness against Kronecker was forgotten. It was no doubt a satisfaction to Cantor to recall that he and Kronecker had become at least superficially reconciled some years before Kronecker's death in 1891. Could Cantor have lived till today he might have taken a just pride in the movement toward more rigorous thinking in *all* mathematics for which his own efforts to found analysis (and the infinite) on a sound basis were largely responsible.

Looking back over the long struggle to make the concepts of *real number*, *continuity*, *limit*, and *infinity* precise and consistently usable in mathematics, we see that Zeno and Eudoxus were not so far in time from Weierstrass, Dedekind, and Cantor as the twenty four or twenty five centuries which separate modern Germany from ancient Greece might seem to imply. There is no doubt that we have a clearer conception of the nature of the difficulties involved than our predecessors had, because we see the same unsolved problems cropping up in new guises and in fields the ancients never dreamed of, but to say that we have disposed of those hoary old difficulties is a gross mis-statement of fact. Nevertheless the net score records a greater gain than any which our predecessors could rightfully claim. We are going deeper than they ever imagined necessary, and we are discovering that some of the "laws"—for instance those of Aristotelian logic—which they accepted in their reasoning are better replaced by others —pure conventions—in our attempts to correlate our experiences. As has already been said, Cantor's revolutionary work gave our present activity its initial impulse. But it was soon discovered— twenty one years before Cantor's death—that his revolution was either too revolutionary or not revolutionary enough. The latter now appears to be the case.

The first shot in the counter-revolution was fired in 1897 by the Italian mathematician Burali-Forti who produced a flagrant contradiction by reasoning of the type used by Cantor in his theory of infinite sets. This particular paradox was only the first of several, and as it would require lengthy explanations to make it intelligible, we shall state instead Russell's of 1908.

We have already mentioned Frege, who gave the "class of all classes similar to a given class" definition of the cardinal number of the given class. Frege had spent years trying to put the mathematics of numbers on a sound logical basis. His life work is his *Grundgesetze der Arithmetik* (The Fundamental Laws of Arithmetic), of which the first volume was published in 1893, the second in 1903. In this work the concept of sets is used. There is also a considerable use of more or less sarcastic invective against previous writers on the foundations of arithmetic for their manifest blunders and manifold stupidities. The second volume closes with the following acknowledgment.

"A scientist can hardly encounter anything more undesirable than to have the foundation collapse just as the work is finished. I was put in this position by a letter from Mr. Bertrand Russell when the work was almost through the press."

Russell had sent Frege his ingenious paradox of "the set of all sets which are not members of themselves." Is this set a member of itself? Either answer can be puzzled out with a little thought to be wrong. Yet Frege had freely used "sets of all sets."

Many ways were proposed for evading or eliminating the contradictions which began exploding like a barrage in and over the Frege-Dedekind-Cantor theory of the real numbers, continuity, and the infinite. Frege, Cantor, and Dedekind quit the field, beaten and disheartened. Russell proposed his "vicious circle principle" as a remedy: "Whatever involves all of a collection must not be one of the collection"; later he put forth his "axiom of reducibility," which, as it is now practically abandoned, need not be described. For a time these restoratives were brilliantly effective (except in the opinion of the German mathematicians, who never swallowed them). Gradually, as the critical examination of all mathematical reasoning gained headway, physic was thrown to the dogs and a concerted effort was begun to find out what really ailed the patient in his irrational and real number system before administering further nostrums.

The present effort to understand our difficulties originated in the work of David Hilbert (1862–) of Göttingen in 1899 and in that of L. E. J. Brouwer (1881–) of Amsterdam in 1912. Both of these men and their numerous followers have the common purpose of putting mathematical reasoning on a sound basis, although in several respects their methods and philosophies are violently opposed. It seems unlikely that both can be as wholly right as each appears to believe he is.

Hilbert returned to Greece for the beginning of his philosophy of mathematics. Resuming the Pythagorean program of a rigidly and fully stated set of postulates from which a mathematical argument must proceed by strict deductive reasoning, Hilbert made the program of the *postulational* development of mathematics more precise than it had been with the Greeks, and in 1899 issued the first edition of his classic on the foundations of geometry. One demand which Hilbert thing must have a certain property or must not have that property, as for example in the assertion that a number is prime or is not prime) is legitimately usable only when applied to *finite* sets. Aristotle devised his logic as a body of working rules for finite sets, basing his method on human experience of *finite* sets, and there is no reason whatever for supposing that a logic which is adequate for the *finite* will continue to produce consistent (not contradictory) results when applied to the *infinite*. This seems reasonable enough when we recall that the very definition of an infinite set emphasizes that a *part* of an *infinite* set may contain precisely *as many* things as the *whole* set (as we have illustrated many times), a situation which *never* happens for a finite set when "part" means *some, but not all* (as it does in the definition of an infinite set).

Here we have what some consider the root of the trouble in Cantor's theory of the actual infinite. For the *definition* of a set (as stated some time back), by which *all* things having a certain quality are "united" to form a "set" (or "class"), is not suitable as a basis for the theory of sets, in that the definition either is *not constructive* (in Kronecker's sense) or *assumes* a constructibility which no mortal can produce. Brouwer claims that the use of the law of excluded middle in such a situation is at best merely a heuristic guide to propositions which *may be* true, but which are not necessarily so, even when they have been deduced by a rigid application of Aristotelian logic, and he says that numerous false theories (including Cantor's) have been erected on this rotten foundation during the past half century.

Such a revolution in the rudiments of mathematical thinking does not go unchallenged. Brouwer's radical move to the left is speeded by an outraged roar from the reactionary right. "What Weyl and Brouwer are doing [Brouwer is the leader, Weyl being his companion in revolt] is mainly following in the steps of Kronecker," according to Hilbert, the champion of the status quo. "They are trying to establish mathematics by jettisoning everything which does not suit them and setting up an embargo. The effect is to dismember our science and to run the risk of losing a large part of our most valuable possessions. Weyl and Brouwer condemn the general notions of irrational numbers, of functions—even of such functions as occur in the theory of numbers—Cantor's transfinite numbers, etc., the theorem that an infinite set of positive integers has a least, and even the 'law of excluded middle,' as for example the assertion: Either there is only a made, and which the Greeks do not seem to have thought of, was that the proposed postulates for geometry shall be *proved* to be self-consistent (free of internal, concealed contradictions). To produce such a proof for geometry it is shown that any contradiction in the geometry developed from the postulates would imply a contradiction in arithmetic. The problem is thus shoved back to proving the consistency of arithmetic, and there it remains today.

Thus we are back once more asking the sphinx to tell us what a number is. Both Dedekind and Frege fled to the infinite—Dedekind with his infinite classes defining irrationals, Frege with his class of all classes similar to a given class defining a cardinal number—to interpret the numbers that puzzled Pythagoras. Hilbert, too, would seek the answer in the infinite which, he believes, is necessary for an understanding of the finite. He is quite emphatic in his belief that Cantorism will ultimately be redeemed from the purgatory in which it now tosses. "This [Cantor's theory] seems to me the most admirable fruit of the mathematical mind and indeed one of the highest achievements of man's intellectual processes." But he admits that the paradoxes of Burali-Forti, Russell, and others are not resolved. However, his faith surmounts all doubts: "No one shall expel us from the paradise which Cantor has created for us."

But at this moment of exaltation Brouwer appears with something that looks suspiciously like a flaming sword in his strong right hand. The chase is on: Dedekind, in the role of Adam, and Cantor disguised as Eve at his side, are already eyeing the gate apprehensively under the stern regard of the uncompromising Dutchman. The postulational method for securing freedom from contradiction proposed by Hilbert will, says Brouwer, accomplish its end—produce no contradictions, but "nothing of mathematical value will be attained in this manner; a false theory which is not stopped by a contradiction is none the less false, just as a criminal policy unchecked by a reprimanding court is none the less criminal."

The root of Brouwer's objection to the "criminal policy" of his opponents is something new—at least in mathematics. He objects to an unrestricted use of Aristotelian logic, particularly in dealing with *infinite* sets, and he maintains that such logic is bound to produce contradictions when applied to sets which cannot be definitely *constructed* in Kronecker's sense (a rule of procedure must be given whereby the things in the set can be produced). The law of "excluded middle" (a finite number of primes or there are infinitely many. These are examples of [to them] forbidden theorems and modes of reasoning. I believe that impotent as Kronecker was to abolish irrational numbers (Weyl and Brouwer do permit us to retain a torso), no less impotent will their efforts prove today. No! Brouwer's program is not a revolution, but merely the repetition of a futile *coup de main* with old methods, but which was then undertaken with greater verve, yet failed utterly. Today the State [mathematics] is thoroughly armed and strengthened through the labors of Frege, Dedekind, and Cantor. The efforts of Brouwer and Weyl are foredoomed to futility."

To which the other side replies by a shrug of the shoulders and goes ahead with its great and fundamentally new task of reestablishing mathematics (particularly the foundations of analysis) on a firmer basis than any laid down by the men of the past 2500 years from Pythagoras to Weierstrass.

What will mathematics be like a generation hence when—we hope —these difficulties will have been cleared up? Only a prophet or the seventh son of a prophet sticks his head into the noose of prediction. But if there is any continuity at all in the evolution of mathematics—and the majority of dispassionate observers believe that there is—we shall find that the mathematics which is to come will be broader, firmer, and richer in content than that which we or our predecessors have known.

Already the controversies of the past third of a century have added new fields—including totally new logics—to the vast domain of mathematics, and the new is being rapidly consolidated and coordinated with the old. If we may rashly venture a prediction, what is to come will be fresher, younger in every respect, and closer to human thought and human needs—freer of appeal for its justification to extra-human "existences"—than what is now being vigorously refashioned. The spirit of mathematics is eternal youth. As Cantor said, "The essence of mathematics resides in its freedom"; the present "revolution" is but another assertion of that freedom.

* * *

Baffled and beaten back she works on still,
  Weary and sick of soul she works the more,
Sustained by her indomitable will:
  The hands shall fashion and the brain shall pore
And all her sorrow shall be turned to labour,
Till death the friend-foe piercing with his sabre
  That mighty heart of hearts ends bitter war.
                              —JAMES THOMSON.

Index

Abel, Niels Henrik, 3, 164, 167, 223, 229-30, 260, 270-2, 295, chap. 17, 328-9, 335-7, 362, 364, 366, 368, 377, 381, 407-8, 412, 415, 418, 437, 448, 452-3, 457, 460, 473, 476-8, 541
Abelian integral, 338, 419
Adams, John Couch, 350
Airy, G. B., 198, 345, 351, 353
Alexander, J. W., 268
Alexander the Great, 537
algebra, 13, 21, 40, 52-4, 61-2, 64-5, 97, 123, 140, 149, 163-6, 168, 199, 208, 211, 213, 222-3, 225, 227, 231-6, 238, 253-4, 260-1, 272, 282-4, 287, 309-14, 317, 321-2, 324, 328, 330, 334-5, 337-8, 347-9, 354-60, 364, 368, 372, 376, 386, 388-91, 394-5, 398-404, 408, 437-9, 442-5, 448, 450-1, 453, 456-7, 459-61, 463-4, 470, 472-9, 482, 489, 493, 511, 517-8, 527-29, 540-1, 564-5, 568-9
algebraic forms, 394, 458
algebraic integers, 470-2, 514, 518, 522, 524
algebraic numbers, 462, 464, 469-71, 474, 477-8, 482, 523-4, 562, 564-5, 567-9
algebraic number field, 470-4, 513, 522-4
algorithm, 140
Ampère, A. M., 318

analysis, 13, 16, 22-3, 54, 64, 70, 87, 117-8, 139-40, 144, 150-2, 154-5, 161-3, 166, 168-9, 171, 174, 176, 183, 187, 198, 201-2, 207, 213, 222-4, 236-7, 248-52, 268, 270-2, 275, 286, 311-2, 316, 319, 322, 327, 334-5, 338-9, 346, 364, 377, 400, 406-8, 413-4, 416-7, 419, 421-3, 429, 446, 451, 456-7, 460-1, 469, 474-5, 477-81, 488-90, 494, 510, 513, 519, 521-2, 526-7, 537-8, 541-2, 544, 547, 556-7, 561, 567, 571, 575, 579
analysis situs, 263, 267, 492, 545
Antoinette, Marie, 166, 170
Apollonius, 7, 27, 78, 317, 400
Appell, Paul, 454
Arago, F. J. D., 139, 150, 190, 193, 205
Archimedes, 7, 19-20, 28-34, 59, 102, 114, 120, 147, 153, 162, 218, 220, 230, 237, 240-1, 255-6, 400, 459, 533
Archytas, 25
Aristotle, 20, 25, 78, 210, 278, 569, 575, 577-8
arithmetic, 21, 64, 118, 152, 161-2, 168, 185, 208, 221, 223, 225-6, 227-9, 231-2, 234-8, 240-1, 248, 252-4, 256, 260-3, 267, 271-2, 276, 284, 356, 368, 408, 419, 438, 446, 451, 455-8, 470-4, 477-8, 481, 485, 488, 500, 510, 513, 517-8, 522-5, 527-9, 541, 551, 561, 574-5, 577

arithmetical theory of forms, 356
Arnauld, A., 83, 128-9
associative, associativity, 278-9, 280, 356
Ausonius, 39
Austen, Jane, 380
axioms, 20-1, 305-6, 333, 419, 503, 576
Ayscough, Rev. Wm., 91-2

Babbage, Charles, 438
Bachet de Méziriac, 71
Baillet, A., 39
Balzac, Honoré de, 548
Barrow, I., 96, 106-7, 118
Bartels, Johann Martin, 222-4
Bauer, Heinrich, 328
Beethoven, L. v., 405
Berkeley, Bishop, 90, 345
Bernoullis, 115, 126, 132, chap. 8, 143-5, 155-6
Berthollet, Count Claude-Louis, 183-4, 189-90, 193-4, 196, 273
Bertrand, J. L. F., 453
Bessel, Friedrich Wilhelm, 245, 248, 250-1, 331
Biot, J. B., 181
Birkhoff, George David, 553
Bismarck, O. E. L., Prince von, 467
Blake, William, 10
Bliss, G. A., 133-4
Boeckh, P. A., 328-9
Bohr, N., 19
Bolyai, Johann, 231
Bolyai, Wolfgang, 220, 231, 243, 474, 505
Boole, George, 118, 121, 124, 213, 354, 389, 390, 406, chap. 23, 448, 486
Boole, Mary, 446-7
Borchardt, C. W., 332, 421, 426, 464, 501
Borel, Émile, 454, 566
boundary values, 339
Boutroux, Émile, 532
Brahe, Tycho, 110

branches, branch points, 493-5
Brewster, Sir David, 112, 269, 500
Brianchon, C. J., 216
Briggs, Henry, 426
Brinkley, John, 343-5
Brochard, Jeanne, 36
Brooke, Rupert, 398
Brouwer, L. E. J., 19, 24, 278, 569, 576-9
Bruno, Giordano, 46
Bunsen, R. W., 424-5
Burali-Forti, C., 575, 577
Burnet, John, 25
Butler, Samuel, 340
Byron, George Gordon, Lord, 257, 380

calculus, 5-7, 13, 23, 32-3, 56-7, 59, 61, 64, 95, 97-107, 109, 113-4, 117-8, 121, 125-6, 132-7, 139-40, 147, 149, 151, 154-6, 162, 169, 186, 208, 251, 270, 284, 286, 322, 324, 328, 330, 334, 348, 361, 364, 407, 442, 448, 488-9, 507, 517
calculus of variations, 437
calculus, tensor, 213, 256, 448, 528
Campanella, Tammaso, 46
Cantor, Georg F. L. P., 354, 406-8, 446, 482, 519-20, 522, chap. 29
Cantor, Marie Bohm, 558
Cantor, Moritz, 17, 19, 25
Cantor, Vally Guttman, 563
Cantor, Waldemar Georg, 558
Cardan, H., 323
Carnot, Lazare-Nicolas-Marguerite, 207, 285
Catherine the Great, 141, 145-7, 149-50
Cauchy, Augustin-Louis, 152, 164-5, 167, 169, 223, 250-1, 260, chap. 15, 317-8, 320, 335, 351, 368-9, 376-8, 407, 416, 457, 473, 482, 488-90, 518, 527, 541
causality, 306
Cavalieri, B., 118

Index

Cayley, Arthur, 3–4, 213–4, 271, 282, 359, 368, chap. 21, 437, 448, 459, 474, 515, 534, 546
Chanute (French ambassador), 50–1
characteristic, 349
Charles I, 92
Charlet, Father, 36–7, 50
Chasles, Michel, 139
Chevalier, Auguste, 374, 377
Christ, Jesus, 573
Christine, Queen of Sweden, 49, 51, 82
Christoffel, E. B., 256, 391, 480
Cicero, 55, 363
class, 567–9, 570, 572–5, 577–8
Clifford, Wm. K., 294, 490, 503–4
Colburn, Zerah, 66, 342–3
Coleridge, Samuel Taylor, 344–5, 484
Columbus, Christopher, 338
combination, 280
commutative, 356, 360, 402
complex number, variable, 233–4, 248–50, 253, 260, 263, 267, 284, 293, 333–5, 356–60, 407, 457–9 487–93, 495, 510, 514, 529, 541
complex units, 474
Condorcet, N. C. de, 152, 180, 187–8
congruence, 225–7, 235–6, 252–3, 293
conjugates, 459
Conon, 30
continuity, continuous, 13, 22–3, 27, 50, 117, 140, 162, 200, 208–9, 211, 216, 234, 237, 268, 283, 286, 407, 430, 456, 488, 491, 504–5, 510, 520–1, 555, 557, 562, 569, 574–6, 579
convergence, 151, 222–3, 274, 286–7, 430–2, 487, 535, 541, 543, 561, 562
Copernicus, Nicolas, 46, 294, 306
Corneille, Pierre, 76
corpus, 355
correlation, 289

Cotes, R., 503
Couturat, L., 121, 565–7
covariant, 395
Crelle, August Leopold, 314, 316–7, 319–20, 326, 413, 416, 418–9, 421, 432, 562
Cromwell, Oliver, 93
curvature, 264–5, 500, 504–5, 507–9
cuts, 494–5, 519–21, 557
cyclotomy, 513

D'Alembert, Jean le Rond, 148–9, 155–62, 173, 187, 489, 544
Darboux, Gaston, 454, 531, 533, 536, 544
Darwin, Charles, 16, 137
Darwin, G. H., 529
Dedekind, Julie, 518
Dedekind, Richard, 19, 25, 223, 238, 253, 406–8, 468, 470, 472, 474, 489, 500, 502, chap. 27, 557, 563, 575–7, 579
Delambre, J. B. J., 154
De Long, Claire, 57
De Long, Louise, 58
De Morgan, A., 147, 354, 384, 387, 434, 438, 440–1
denumerable infinity, 539–40, 569, 572
Desargues, G., 74, 77, 79, 183, 208, 213, 285, 398
Descartes, René, 7, 14–5, 20, 32, chap. 3, 56–9, 62–4, 73–4, 80, 82, 93, 111, 121–2, 129, 140, 147, 211–2, 233, 245, 265–6, 347, 349, 376, 418, 457–8, 481, 505, 512
determinants, 338, 388
Dickens, Charles, 380, 436
Dickson, L. E., 261, 355
Diderot, Denis, 146–7
Diophantus, 70–1, 152
Dirac, P. A. M., 19, 515
Dirichlet, P. G. Lejeune, 236–7, 318, 332, 405, 458, 468–9, 473, 489, 496, 499, 500, 512, 517, 518

discrete, 13-4, 21, 24, 117, 140, 162, 283, 505
discriminant, 389-90, 401
distance, 505-9
distributive law, 356
divergent, 287
Dositheus, 30
duality, 214, 216-7
Dumas, Alexandre, 373
Dürer, Albert, 17

Eddington, Sir Arthur S., 484, 515
Edgeworth, Maria, 344
Edison, Thomas A., 548
Einstein, Albert, 5, 7, 15, 19, 138, 154, 213, 255-6, 305, 320, 350, 391, 448, 490-1, 546-7, 556
Eisenstein, F. M. G., 68, 237, 253, 390, 455-6, 468, 489
Elijah, 69
Elisabeth, Princess, 41, 49, 51, 129
elliptic functions, see functions, elliptic
elliptic integrals, 317, 322, 324, 336
equations, 52, 65, 68, 70, 136, 150, 155, 157, 162-5, 179, 186, 192, 198-9, 201, 211, 225, 232-5, 252, 265-6, 270-2, 282-3, 309-14, 317-8, 322, 328, 334-6, 338, 347, 354-6, 364, 368, 372, 376, 389-90, 404, 408, 416, 430, 438, 445, 450-3, 456-7, 459-64, 470, 472-4, 476-8, 482, 489-90, 492-3, 499, 514-5, 518, 528, 536, 538-40, 543-4, 561, 564-5, 568-9
Eratosthenes, 30-1
Essenbeck, Nees von, 240
Euclid, 7, 14, 19-20, 27, 75, 127, 153, 165, 176, 215-6, 223, 266, 299-303, 305-6, 314, 351, 358, 379, 399-400, 443, 454, 474, 514
Eudoxus, 19, 25-7, 407, 480, 483, 519, 522, 575
Euler, Albert, 149
Euler, Catharina Gsell, 145

Euler, Léonard, 69, 115, 132-3, 135-7, chap. 9, 155-6, 158-60, 175, 223-6, 242, 245, 263, 271, 277, 284, 308-9, 328-9, 354, 368, 378, 443, 450, 488-9, 492, 517, 532, 534, 541, 543
Euler, Marguerite Brucker, 143
Euler, Paul, 143
existence, 482
extrema, 304

factorials, 417, 462
factorization, 334, 569
factorization, unique, 514, **522**
factors, prime, 522
Ferdinand II, 39
Ferdinand, Duke of Brunswick, **224**, 231-2, 241, 243-6, 248
Fermat, Clément-Samuel, 58
Fermat, Dominique, 57
Fermat, Pierre, 7, 35, chap. 4, 73-4, 86, 102, 118-9, 133-4, 140, 152, 161-2, 228, 236-8, 253, 261, 284, 311, 337, 465, 472-3, 488, 510, 513-4
field, 355, 357, 469-70, 473, 514, 522
Fleming, Admiral, 50
Foncenex, D. le, 154
formalism, 287
Forsyth, A. R., 402
Fourier, Jean-Baptiste-Joseph, 105, 176, 178, chap. 12, 318, 338-9, 496, 538-9, 561
fractions, continued, 368
Franklin, Fabian, 397
Frederick the Great, 141, 148, 153, 159-60, 165, 244
Frege, Gottlob, 567, 575-7, 579
Fresnel, A. J., 351
Fricke, Robert, 516
Fuchs, Lazarus, 540
functions, 98-9, 101-4, 140, 169, 179, 201, 223, 232, 248-50, 252, 260, 263, 265, 275-7, 284, 318-20, 322-4, 329, 333-5, 338, 348-9, 364, 377, 407, 413, 430, 455,

457, 460-2, 480, 487-95, 508, 515, 529, 540-1, 550-1, 578
functions, Abelian, 338, 398, 408, 416, 419-20, 451, 455, 492, 495, 499-500
functions, automorphic, 530, 540-1
functions, elliptic, 127, 202, 228-9, 247, 253, 259-60, 282, 320, 322, 324, 329, 331, 333, 335-8, 372, 382, 386, 408, 413-4, 430, 453, 455, 461, 469, 477-9, 517, 539-41, 550
functions, multiple periodic, 408, 452-3, 455, 515, 539

Galileo, 16, 20, 26, 36, 41, 45-6, 80, 84, 93, 121, 129, 291, 566
Galois, Adélaïde-Marie Demante, 362
Galois, Évariste, 3, 164-5, 167, 270, 312, 362, chap. 20, 381, 408, 437, 449, 450, 476-7, 518, 536
Galois, Nicolas-Gabriel, 362, 369-70
Galton, Francis, 137, 202, 323
Gauss, Carl Friedrich, 3-4, 20, 28, 64, 67, 69, 72, 105, 108, 120, 146, 151, 161-2, 167, 186, chap. 14, 271, 284, 286, 297, 308, 313-4, 328, 331, 333, 335, 337, 354, 356, 359, 377, 380, 389, 405-7, 443, 448, 450, 455-8, 472-4, 478, 486, 488-89, 491-3, 495-8, 501, 507-8, 510, 512-8, 525, 527, 533, 535, 537, 539, 541, 547, 550-7, 561, 569
Gauss, Dorothea Benz, 219-20, 224
Gauss, Friederich, 219
Gauss, Johanne Osthof, 243
Gauss, Minna Waldeck, 243
Gelfond, Alexis, 463-4
Gelon, 29
geodesic, 302, 305-6, 507
geometry, 13-4, 20-2, 27, 31, 33, 40, 52-5, 59-62, 66-7, 74-8, 84-5, 96, 109, 123, 127, 140, 147, 154, 183-4, 186-7, 207-17, 223, 227-8, 233, 236, 242, 254, 256, 260, 263-8, 272, 277, 282, 284, 286, 294, 299-306, 317, 322, 324, 346-9, 355, 358, 364, 367-9, 381-2, 398-400, 403, 443, 448, 451, 458, 464, 469, 474, 478-81, 485, 489-91, 493, 498, 500, 503-4, 506, 508, 510, 513-7, 522, 527, 531, 535, 537, 541, 569, 577
geometry, analytic, 5-7, 32, 40-1, 52-7, 59-64, 93, 123, 133, 139, 147, 154, 199-200, 211-2, 233, 248, 399, 517
geometry, descriptive, 183, 185-6, 207, 240, 452
geometry, foundations, 497-8, 502-3
geometry, n-dimensional, 399-400
geometry, non-Euclidean, 6, 108, 127, 186, 214, 223, 230-1, 240, 256, 260, 264, 266, 294, 297, 300-6, 358-60, 379, 398-9, 439, 473-4, 493-5, 503-5, 507-9, 513, 550-1
geometry, projective, 78, 206-10, 212-7, 398-9
Germain, Sophie, 253, 261-3
Gibbon, Edward, 257
Gilbert, Wm., 36
Gilman, Arthur, 394
Goethe, J. W. von, 257
Goldbach, C., 405
Gounod, Charles François, 386
Grassmann, Hermann, 123
Gregory, D. F., 438
Gregory, James, 125
Grote, Geo., 381
groups, 6, 68, 165, 268, 271-2, 278-81, 283, 451, 495, 515, 518, 540, 551
groups, abstract, 282, 375-6
groups, continuous, 268, 529
Gudermann, Christof, 412-3, 424

Hachette, J. N. P., 318, 320
Hadamard, Jacques, 54, 212
Halley, Edmund, 105, 108-10, 153

Index

Halphen, Georges, 403
Hamilton, Eliza, 345
Hamilton, James, 340-1
Hamilton, Sarah Hutton, 340
Hamilton, Sir Wm., 438-9, 440-1
Hamilton, William Rowan, 16, 29, 66, 123, 133-4, 155, 260-1, 270, 331, 334, 338, chap. 19, 514
Hansteen, Christoph, 319
Hardy, G. H., 488
Hegel, G. W. F., 239-41, 439, 469
Heiberg, J. L., 31
Heisenberg, W., 19, 402
Helmholtz, H. von, 424
Henry, C., 86
Herbart, J. F., 491
Hermite, Charles, 4, 164, 183, 273, 307, 320, 368, 422, chap. 24, 476-7, 501, 515, 535-6, 543, 568, 571
Hermitian forms, 458-9
Herschel, Sir William, 239, 438
Hertz, Heinrich, 16
Hieron II, 29, 33
Hilbert, David, 63, 239, 443, 463, 566, 576-8
Hipparchus, 109, 361
Hitler, Adolf, 516
Holmboë, Bernt Michael, 308-9, 312, 315, 318, 325, 329
Hooke, Robert, 107-8
Horace, 307, 320, 397
Humbert, Georges, 540
Humboldt, Alexander von, 204, 242, 245, 259, 333
Hume, David, 146
Huntington, E. V., 444
Huygens, Christian, 84, 107, 123-5, 129, 134, 350-1

ideals, 408, 510, 513, 522-4
ideal, prime, 524
identical operation, identity, 279-80
imaginaries, 356, 372, 399, 493
infinite, 6, 22-5, 27, 69-71, 75, 89, 123, 151, 168-9, 199, 201, 211-2, 216, 222-4, 236-7, 250, 274, 283, 286-7, 309, 351, 355, 394, 400, 407-8, 446, 448, 462-3, 480, 487-8, 492, 514, 520-3, 539-41, 543, 549, 553, 555-8, 561-2, 564-71, 573-9
invariance, invariants, 6, 78, 165, 230, 283, 378-9, 382-3, 386, 388, 390-2, 395, 398, 401, 403, 437, 448, 459, 509, 540, 545
irreducibles, 472, 515, 569

Jacobi, Carl Gustav Jacob, 16, 21, 52, 140, 229-30, 259-60, 270, 320-4, chap. 18, 346-77, 408, 418, 420, 448, 452-6, 463, 468-9, 489, 512, 515, 517, 528, 541, 544
Jacobi, M. H., 327
James II, 111
Jansen, Cornelius, 79
Jeans, Sir James H., 17, 21, 177, 529
Jeffreys, George, 111
Jerrard, G. B., 460
Joachim, Joseph, 559
Jourdain, P. E. B., 185

Kant, I., 177, 240, 345, 358-9
Kelvin, Lord (William Thomson), 16, 183, 293, 452
Kepler, J., 28, 93, 105, 109, 121
Kingsley, Chas., 385
Kipling, Rudyard, 324
Kirchoff, G. R., 424
Klein, Felix, 206, 214, 282, 379, 398, 419, 540, 548
Kneser, Adolf, 467
Königsberger, L., 424-5
Kowalewski, Sonja, 261, chap. 22
Kronecker, Leopold, 19, 164, 218, 234, 238, 293, 406-8, 419, 422, 431, chap. 25, 501, 512-3, 521, 529, 556, 559-63, 568-71, 573, 577-9
Kummer, Ernst Eduard, 238, 253, 467-70, 473-4, 476, 478-9, 486, 501, chap. 27, 510, 535, 561

Lacroix, S. F., 318
Lagrange, Joseph-Louis, 5, 9, 62, 115, 133, 142, 148–9, chap. 10, 174, 177–8, 180–1, 185–6, 197, 223–6, 237, 247, 262–3, 270–1, 274–7, 284, 308, 311, 328–9, 338, 346, 364, 381, 389, 400, 437, 450, 541, 543–4
Lagrange, Marie-Thérèse Gros, 153
Lamarck, J. B. A. F., 181
Lamb, Horace, 14
Landau, Edmund, 516
Laplace, Pierre-Simon, Marquis, 104–5, 111, 156, 163, 169–70, chap. 11, 184, 197, 209, 223, 225, 238, 240–2, 246–8, 270, 273–5, 286, 318, 329, 338, 343, 381, 412, 437, 450, 541, 544
Lavoisier, A. L., 166–7, 169, 184
least squares, theory of, 259, 260, 263
Lefschetz, S., 268
Legendre, Adrien-Marie, 174, 197, 225–7, 236–7, 259–60, 270, 277, 284, 312, 317–8, 320–4, 335–7, 364, 452, 487, 488
Leibniz, Gottfried Wilhelm, 7, 13, 15–6, 18, 32, 56, 59, 68, 76, 78, 87, 98, 103, 113–6, chap. 7, 133–5, 140–1, 169, 223, 231, 434, 443
Lemaître, Father, 527
Lemonnier, P. C., 169
Leverrier, U. J. J., 290, 350, 367
Levi-Civita, T., 256, 391, 544
Liapounoff, A., 529
Libri, G., 321
Lichtenstein, Leon, 529
Lie, S., 391
Lilly, Wm., 526
limits, 430
Lincoln, Abraham, 434
Lindemann, F., 314, 464–5, 568
Linus (of Liége), 107
Liouville, Joseph, 335–6, 451, 463, 476
Listing, J. B., 491

Littrow, J. J. von, 295, 333
Lloyd, Humphrey, 351
Lobatchewsky, Nikolas Ivanovitch, chap. 16, 359, 377, 379, 474, 503
Lobatchewsky, Praskovia Ivanovna, 294
Locke, John, 111–2
logic, symbolic, 6, 13, 121, 124, 439, 440–6
Lotze, R. H., 439
Lucas (of Liége), 107 8
MacLaurin, C., 528
MacMahon, P. A., 378
Macaulay, Thos. B., 257, 381
magnitude, 565
Malus, Étienne-Louis, 277–8, 348, 351
manifold, 265, 504–7
mapping, conformal, 266–7
Marcellus, 28, 33–4
Marie, Abbé, 158, 163
Marr, John, 526
matrices, 379, 398, 400–2
Maurice, Prince of Orange, 38
Maxwell, James Clerk, 267, 350, 433, 500, 546
Menaechmus, 537
Mendelssohn, Felix, 468
Mercator, N., 107, 125
Méré, Gombaud Antoine, Chevalier de, 86–7, 89
Mersenne, P., 37, 45, 62, 76, 80
Mill, John Stuart, 35
Milton, John, 36
Minkowski, H., 7
Mittag-Leffler, G. M., 316, 324, 419, 422, 428, 543, 570, 571
modulus, 225–7, 235–6, 252, 293
Monge, Gaspard, 171, chap. 12, 207, 263, 285, 366
Monge, Jacques, 184, 287
monogenic, 249, 250–1
Montagu, Charles, 112
More, L. T., 64
Morgan, John Pierpont, 466

Moritz, R. E., 557
Mozart, W. A. C., 405
multiformity, 492

Napier, John, 526
Napoleon Bonaparte, 124, 153, 168, 170, 179–82, 184, 187, 190–6, 202–7, 209, 224, 243, 246–7, 251, 258, 271, 274–5, 277, 285, 348, 362, 422, 511, 531
Newton, Hannah Ayscough, 90
Newton, Isaac, 5, 7, 13, 18–20, 26, 28–9, 32, 36, 56, 58–60, 64, 73, chap. 6, 118–21, 123, 125–6, 129–30, 133, 135, 138–40, 142–3, 147, 153–5, 157–8, 166–75, 179–80, 198, 201, 213, 218, 220, 223, 225, 230, 237–8, 240–1, 248, 254–6, 263, 267, 270, 289, 308–9, 338, 343–4, 349–50, 354, 361, 400, 404, 416, 433, 443, 478, 490–1, 500, 503–4, 533, 538, 541–2
Nightingale, Florence, 387–8
Noether, Emmy, 261
non-denumerable, 569, 571
normal, 264–5
number, 6, 13–6, 21–7, 87–8, 125, 140, 169, 208, 222, 225–9, 231–7, 250, 252–4, 256, 265–6, 271, 284, 318, 334, 337–8, 347, 354, 356–7, 364, 376, 386, 405, 419, 431, 438, 446, 450, 456, 458–9, 462–4, 470–2, 474, 478, 481–2, 505–6, 510, 513–4, 519–21, 523–4, 556, 564–9, 573, 575–6, 578
number, cardinal, 565–7, 575, 577
numbers, ideal, 473–4, 513–4
numbers, irrational, 22, 27, 407, 431–2, 456, 460, 463–4, 480, 482, 519–22, 568, 576–9
numbers, negative, 356–7, 482
numbers, prime, 471–2, 487–8, 510, 513–4, 573, 578–9
numbers, theory of, 13, 57–8, 64–72, 98, 140, 161–2, 176, 225–9, 231–8, 241, 248, 252–4, 260–3, 267, 271–2, 284, 299, 318, 328, 331, 334, 337–8, 372, 386, 405–6, 408, 438, 455–9, 469–74, 476, 478–9, 482, 487–9, 510–1, 513–4, 517–24, 539, 541, 561, 578
numbers, transcendental, 459, 462–4, 567–9, 572
numbers, transfinite, 566–7, 574, 578
Nurmi, Paavo, 543

Olbers, H. W. M., 238, 246, 259, 262
Oldenburg, H., 108
order, 281, 574
Oscar II, King of Sweden, 543

Painlevé, Paul, 454, 544, 554
parameter, 266, 540, 541
parametric representation, 265, 515
partitions, theory of, 386, 405
Pascal, Antoinette Bégone, 73
Pascal, Blaise, 3, 9, 36, 57, chap. 5, 115, 119, 125, 128, 183, 208, 212, 216, 285, 398
Pascal, Étienne, 73
Pascal, Gilberte (Madame Périer), 57, 73, 75, 80–2
Pascal, Jacqueline, 73–5, 80–3
Pastoret, M. de, 180
Paul, Jean, *see* Richter, J. P. F.
Peacock, G., 438
Peel, Sir Robert, 354
Peirce, Benjamin, 354
Pepys, Samuel, 112
periodicity, 199–202, 229, 335–6, 461, 519, 539, 545, 553
permanence of form, 355
permutation, 279, 282
Peter the Great, 129, 141, 145
Pfaff, Johann Friedrich, 231, 242
Pheidias, 29
Phidias, 322
Philip, Apostle, 573
Philippe, Louis, 372–3
Piazzi, Giuseppe, 239

Index

Picard, Émile, 454
Picard, Jean, 335
Pindar, 329
Planck, M., 546–7
Plato, 4, 16, 20–1, 24–7, 31–2, 240
Plücker, J., 399, 412, 474
Plutarch, 28, 33
Poincaré, Henri, 9, 17, 158, 270, 378, 428, 447–8, 454, 461, 483, chap. 28, 558
Poincaré, Raymond, 527, 531, 535
Poinsot, L., 277
Poisson, S. D., 318
Poncelet, Jean-Victor, 191, chap. 13, 285, 398
Pope, Alexander, 549
postulate, 20–1, 23, 279–80, 282, 299, 301, 303, 305, 354–7, 438, 443–5, 474, 477, 518, 574, 576–7
power series, 429–30, 561
probability, mathematical theory of, 6, 73, 83, 86–9, 133–4, 136, 155, 172, 176–7, 318, 441–2, 451
problem of n bodies, 542–3
progression, 355, 359, 535
pseudo-sphere, 304–6
Ptolemy, 109, 176, 361
Pythagoras, 16–7, 20–22, 266, 398, 407, 457, 474, 480, 506–7, 555, 576–7, 579

quadratic forms, 335, 388, 456, 478, 502, 541, 551
quantics, 394–5, 398
quantum theory, 63, 88, 107, 350, 459, 500, 515, 547
quaternions, 260, 352, 354–5, 357, 359–61

radicals, 372, 376–7, 453, 460, 477
Ramanujan, Srinivasa, 328
ratio, anharmonic or cross, 213–4
ratios, 27, 348, 522
rays, systems of, 346–51
reciprocants, 403
reciprocity, 226, 235, 252–3, 513–4

relativity, 6–7, 54, 63, 154, 186, 263, 305–6, 350, 360, 378, 391, 399, 437, 448, 480, 500, 502, 506–7, 509, 528, 547
Ricci, G., 256, 391
Richard, Louis-Paul-Émile, 367–8, 450
Richelieu, Cardinal, 42, 52, 76
Richelot, F. J., 421
Richter, J. P. F. (Jean Paul), 257
Riemann, Georg Friedrich Bernhard, 4, 186, 256, 264, 266, 300, 306, 324, 379, 391–2, 406, 408, 474, chap. 26, 518, 532, 545
roots of unity, 514
Rosenhain, J. G., 324
Rossetti, Dante Gabriel, 528
Rostand, Edmond, 59
Rousseau, J. J., 258
Russell, Bertrand A. W., 14, 17, 118–9, 121, 239, 433–4, 446, 527, 557, 567, 572, 575–7

Savoy, Duke of, 42
Schaeffer, Theodor, 560
Schelling, F. W. J. von, 240
Scherk, H. F., 511
Schiller, J. C. F. von, 257, 362
Schumacher, H. C., 239, 260
Scott, Sir Walter, 257, 380
Seneca, 363
sequence, 13, 431–2, 480
series, 149, 151, 201, 221–3, 237, 251–2, 274, 286–7, 413–4, 430, 487, 496, 499, 541, 543, 550, 561, 568
series, hypergeometric, 514, 550–1
Serret, J. A., 368
sets, 555, 562–7, 569–78
Shakespeare, Wm., 36, 230, 257, 381
Shelley, Percy Bysshe, 548, 564
Sierpinski, Waclaw, 566
Smith, Rev. Barnabas, 91
Smith, H. J. S., 68, 403
Snellius, W., 351

Index

Sophie of Brandenburg, 129
Southey, Robert, 344
space, 15, 54, 63, 99, 150, 154, 185, 208, 217, 240, 264, 347, 358–60, 379, 391, 398–400, 491, 500, 503–4, 506–9, 515, 533, 571
Spinoza, Benedict de, 126, 486
squares, least, 517
Steiner, J., 27, 317, 468–9, 489, 517
Stern, M. A., 489, 491, 517
Storey, Miss, 92
Sturm, C., 453
subgroups, 281
substitutions, 278, 280, 282–3, 495, 518, 540
Sundman, Karl Frithiof, 544
Swinburne, Algernon Charles, 353
Sylvester, James Joseph, 76, 213, chap. 21, 437, 448, 459, 490, 526–7
symbolic reasoning, 6

Tait, P. G., 183
Tannery, P., 86
Tchebycheff, P., 426
tensor analysis, 360
Terquem, O., 367
Thackeray, Wm. M., 380
Thomson, James, 17, 580
Thomson, William, *see* Kelvin, Lord
time, 15, 54, 63, 99–100, 264, 347, 349, 359, 391, 399, 500
topology, 491–2, 495
Torricelli, E., 80
tractrix, 304
transposition, 283
trigonometry, 140, 201, 252, 322–4, 430, 453, 455, 461, 541, 561
Tuscany, Duke of, 46

uniformity, 249, 250–1

unique operation, 279

Vanini, L., 46
Veblen, O., 268
vector, 358–60
vector analysis, 360, 528
Vere, Aubrey de, 344
Vernier, P., 366
Virgil, 149
Voltaire, 129, 148
Volterra, V., 406, 428

Wakeford, E. K., 398
Wallis, J., 118
Waltershausen, Sartorius von, 245, 258
Weber, Heinrich, 467
Weber, Wilhelm, 255, 267, 490–1, 496–8, 517
Weierstrass, Karl W. T., 19, 25, 30, 223, 250, 270, 273, 316, 324, 332, chap. 22, 474–5, 477–83, 500–1, 512, 543, 556–7, 561, 575, 579
Westaway, F. W., 555
Weyl, H., 578–9
Whitehead, A. N., 34, 118–9, 239, 446, 557
Wilson, John, 162
Wilson, Woodrow, 485
Wolfskehl, Paul, 72
Wordsworth, William, 344–5, 402
world-lines, 391
Wren, Sir Christopher, 84, 105

Young, John Wesley, 215

Zeno, 19, 24–5, 49, 119, 407, 555–8, 575
Zermelo, E., 573–4
Zolotareff, G., 162

ABOUT THE AUTHOR

ERIC TEMPLE BELL *was born in 1883 in Aberdeen, Scotland. His early education was obtained in England. Coming to the United States in 1902, he entered Stanford University and took his A.B. degree in 1904. In 1908 he was teaching fellow at the University of Washington, where he took his A.M. degree in 1909. In 1911 he entered Columbia University, where he took his Ph.D. degree in 1912. He returned to the University of Washington as instructor in mathematics and became full professor in 1921. During the summers of 1924-28 he taught at the University of Chicago, and in 1926 (first half) at Harvard University, when he was appointed Professor of Mathematics at the California Institute of Technology.*

Dr. Bell was a former President of the Mathematical Association of America, a former Vice President of the American Mathematical Society and of the American Association for the Advancement of Science. He was on the editorial staffs of the Transactions of the American Mathematical Society, *the* American Journal of Mathematics, *and the* Journal of the Philosophy of Science. *He belonged to The American Mathematical Society, the Mathematical Association of America, the Circolo Matematico di Palermo, the Calcutta Mathematical Society, Sigma Xi, and Phi Beta Kappa, and was a member of the National Academy of Sciences of the United States. He won the Bôcher Prize of the American Mathematical Society for his research work. His twelve published books include* The Purple Sapphire *(1924),* Algebraic Arithmetic *(1927),* Debunking Science, *and* Queen of the Sciences *(1931),* Numerology *(1933), and* The Search for Truth *(1934).*

Dr. Bell died in December 1960, just before the publication of his latest book, The Last Problem.